Biology of the
Turbellaria

BIOLOGY OF THE TURBELLARIA, ed. by Nathan W. Riser and
M. Patricia Morse. McGraw-Hill, 1973. 530p il tab bibl
(McGraw-Hill series in the invertebrates) 73-13695. 25.00.
ISBN 0-07-052947-7. C.I.P.
A collection of 21 papers presented at the Libbie H. Hyman Memorial
Symposium held in 1970. The different aspects of the biology of
turbellarians covered in this volume (ecology, physiology, phylogenetic
relationships, ultrastructure, and biogeography) reflect the advances
made in turbellariology in the 1960s all over the world. Well-labelled
illustrations, schemes of phylogenetic relationships, maps of geographic
distributions, and micrographs make the book more useful. The
bibliography at the end of each chapter is an added asset. The book
updates the information available from *Platyhelminthes and rhynchocoela,* the second volume of L. H. Hyman's classic work *The invertebrates* (3v., 1940–51). The present work is of limited use for under-
graduate students, except for those deeply involved in the study of
invertebrates. Any library used by graduate students should find
place for this volume by the side of the six volumes on the invertebrates
written by Hyman.

McGRAW-HILL SERIES IN THE INVERTEBRATES

Libbie H. Hyman Memorial Volume

Biology of the Turbellaria

Edited by
Nathan W. Riser
M. Patricia Morse
Marine Science Institute
Northeastern University

McGraw-Hill Book Company
New York   St. Louis   San Francisco   Düsseldorf   Johannesburg
Kuala Lumpur   London   Mexico   Montreal   New Delhi
Panama   Rio de Janeiro   Singapore   Sydney   Toronto

Biology of the Turbellaria

1234567890MAMM79876543

This book was set in Times Roman by Black Dot, Inc. The editors were James R. Young, Jr., and Carol First; the cover was designed by Pencils Portfolio, Inc.; and the production supervisor was Thomas J. LoPinto. New drawings were done by John Cordes, J & R Technical Services, Inc.
The printer and binder was The Maple Press Company.

Library of Congress Cataloging in Publication Data
Main entry under title:

Biology of the Turbellaria.

 (McGraw-Hill series in the invertebrates)
 At head of title: Libbie H. Hyman memorial volume.
 Papers prepared for a Libbie H. Hyman memorial symposium and presented at meetings of the American Association for the Advancement of Science held in Chicago, Dec. 1970, sponsored by the Division of Invertebrate Zoology of the American Society of Zoologists and co-sponsored by the Society of Systematic Zoology and the American Microscopical Society.
 "Bibliography of Libbie H. Hyman": p.
 Includes bibliographies.

 1. Turbellaria—Congresses. I. Hyman, Libbie Henrietta, 1888–1969. II. Riser, Nathan W., ed.
III. Morse, M. Patricia, ed. IV. American Association for the Advancement of Science. V. American Society of Zoologists. Division of Invertebrate Zoology.
VI. Society of Systematic Zoology. VII. American Microscopical Society.
QL391.P7B55 595'.123 73–13695
ISBN 0–07–052947–7

Contents

Preface

A Libbie H. Hyman Memorial Symposium arranged by M. Patricia Morse, Nathan W. Riser, and Horace W. Stunkard, sponsored by the Division of Invertebrate Zoology of the American Society of Zoologists and cosponsored by the Society of Systematic Zoology and the American Microscopical Society, was presented at the meetings of the American Association for the Advancement of Science in December 1970 in Chicago, Illinois.

With the aid of grant GB-22862 from the National Science Foundation, participants were brought from eleven countries. The interaction of these specialists with the American participants and specialists was immediately fruitful not only through the exchange of ideas, but more significantly through personal cooperation and collaboration, including aid to one another's students and associates.

While the information presented at the Symposium was primarily new, some was in preliminary form and some was already committed for publication. Thus, not all the papers were submitted for inclusion in this volume.

Nathan W. Riser
M. Patricia Morse

In Memoriam

Libbie Henrietta Hyman, 1888–1969

In a memorial note for Professor Charles Manning Child, with whom she was associated for more than 20 years, Libbie Hyman wrote (1955; *Science,* **121:** 717–718), "He ever adhered to the highest standards of scientific integrity and thereby won the respect of all who knew him." The same eulogistic appraisal may appropriately be applied to her own life and work. As a person and as a scholar, her career bears witness of devotion to the highest ideals.

She was born in Des Moines, Iowa, December 6, 1888, the third of four children and only daughter of Mr. and Mrs. Joseph Hyman, Jewish immigrants from eastern Europe. The family name was "invented," according to Libbie, to replace a Russian one. From her father, she learned Russian as a child. At the age of 14, he had escaped from Russian Poland and made his way to London, where he worked as a tailor, the family trade. Eventually he migrated to the United States and with David Goldman operated a clothing and tailoring establishment in Des Moines. Later, with a brother, he opened a store in Sioux Falls, South Dakota, but it was not a success; he returned to Iowa and settled in Fort Dodge, where he lived until his death in 1907. The store there was never profitable and was a source of constant worry. The fourth child was born in South Dakota.

Libbie grew up in Fort Dodge, where she attended the public schools and graduated from the high school in 1905 as valedictorian of her class. The next year she returned to the high school to study advanced German, and she took a job in a factory, pasting labels on boxes. Mary Crawford, a Radcliffe graduate who taught English and German in the high school, took an interest in the girl and secured a scholarship for her at the University of Chicago. Libbie entered the University in 1906, and after basic courses in physical and biological sciences, she majored in zoology. She graduated in 1910 with the B.S. degree.

In her senior year she had taken a course with Professor Child and, at his suggestion, entered the graduate school of the University. At the time, she received an appointment as a graduate assistant, to direct laboratory work in elementary zoology and in comparative vertebrate anatomy. The training and experience as a laboratory assistant paid off handsomely later.

Her research program was directed by Professor Child, and she was awarded the Ph.D. degree in 1915, with a dissertation "Analysis of the Process of Regeneration in Certain Microdrilous Oligochaetes." She had a high regard for Professor Child and accepted an appointment as his research assistant, a position she held until 1931, when he was nearing retirement. He was an advocate of the organismal theory, maintaining that a plant or animal is an integrated individual, not merely the sum of its constituent parts, and that gradients of physiological activity, as measured by O_2 consumption, CO_2 production, and differences in electric potential, determine and maintain organic integrity. As his research assistant, Libbie conducted various physiological experiments on planarians and other lower organisms, designed to produce evidence that could support the ideas of Professor Child. But Libbie did not regard the work as of outstanding importance and decided that she was not the research type. She was entertaining an idea, a dream, a hope that she could accomplish a great project. She resigned the assistantship and spent the next 15 months touring western Europe.

While a graduate assistant at the University of Chicago, she felt the need of her students in elementary zoology for a better laboratory manual. Accordingly, she wrote *A Laboratory Manual for Elementary Zoology,* which was published by the University of Chicago Press, 1919. The first printing was rapidly exhausted and the book had a wide and enthusiastic reception. Also, to fill the need for a better laboratory manual for courses in comparative vertebrate anatomy, she wrote *A Laboratory Manual for Comparative Vertebrate Anatomy,* published by the University of Chicago Press, 1922. These books were adopted so extensively that the income from sales was substantial. In 1929 she published a revised edition of the manual for elementary zoology and in 1942 a completely new *Comparative Vertebrate Anatomy,* "thoroughly revised and considerably expanded with the intention that the book shall now serve as a text as well as a laboratory manual." These books are so excellent that they still are used by many colleges and universities. Her personal tastes were modest, and the income from the manuals made Libbie financially independent. Accordingly, she resigned the assistantship at Chicago and never thereafter had a salary.

Although she had written an excellent and successful manual for comparative anatomy, Libbie cared little for the vertebrates and in the preface of *Comparative Vertebrate Anatomy* admitted that the task was arduous, "For I confess, apologetically, that I am not a student of vertebrate anatomy."

Instead, her real interest was with the invertebrates and especially the little ones, the protozoans, sponges, coelenterates, and turbellarians. She admired the exquisite colors, the delicacy, symmetry, and manifold variety of these forms. In addition to the four books published by the University of Chicago Press and the six volumes of *The Invertebrates,* published by the McGraw-Hill Book Company, she published 136 papers, most of them on the morphology, physiology, development, systematics, and bionomics of the lower invertebrates. While a research assistant with Professor Child, Libbie realized that in English there was no major monographic series comparable to Bronn's *Klassen und Ordnungen des Tierreichs,* the beautifully illustrated *Traité de Zoologie Concrète* of Delage and Hérouard, or the Kükenthal-Krumbach *Handbuch der Zoologie.* The success of her earlier writing encouraged her to try to do a monographic study in English. Accordingly, when she returned from Europe, she took an apartment near The American Museum of Natural History in order to use its superb library. At first she worked at home, but in 1937 she was appointed a research associate at the Museum and provided with office and laboratory space. Endowed with a photographic memory, a facility for translation, and familiarity with European languages, she read and made reference card-abstracts of world literature. She could process a prodigious amount of information and keep it organized in her head. When the data were thoroughly integrated, she sat at her typewriter and wrote the text. The expression was just as it emerged from her mind. She did not make carbon copies of anything. She never had a secretary, an assistant, or a technician. Her histological preparations, her illustrations, and text were entirely her own.

The first volume of *The Invertebrates, Protozoa through Ctenophora,* appeared in 1940. The preface clearly stated the ideas, purposes, plans, and procedures of the author. The coverage is comprehensive, the treatment is authoritative, and the many illustrations were designed for clarity and simplicity. To execute the drawings from living or prepared material, Libbie spent several summers at the Marine Biological Laboratory in Woods Hole, Massachusetts, and other marine stations. Volumes II and III were published in 1951, volume IV in 1955, volume V in 1959, and volume VI in 1967. Progress on the final volume was retarded by Libbie's failing health, and much of the work was done when she was unable to walk across the room without assistance. The preface of volume I contains the statement, "It is obviously impossible for any one person to have a comprehensive firsthand knowledge of the entire range of invertebrates, and consequently a work of this kind is essentially a compilation from the literature." But *The Invertebrates* is more than a compilation; it involves incisive analysis, judicious evaluation, and masterly integration of information. On January 14, 1967, the McGraw-Hill Book Company, publishers of *The Invertebrates,* held a reception in the Portrait Room of the Museum in honor of Libbie Hyman, at which tributes were expressed by officers of the publishers and of the Museum, and by colleagues. She received a special leather-bound library edition of *The Invertebrates*.

Libbie's home life was unhappy. Her mother, born Sabina Neumann in Stettin, Germany, came to Des Moines, where she had a brother. Here she worked for the Posner family. Mrs. Posner was a sister of Joseph Hyman, a bachelor in his late forties. Libbie's parents were married in 1884, the mother 20 years younger than the father. The mother was domineering and demanding, the father distressed by financial difficulties, and the home atmosphere depressing. When Libbie went to Chicago, she stayed with an aunt and uncle, but after the death of her father in 1907, her mother moved the family to Chicago and Libbie lived at home until the death of her mother in 1929. During this time she was subjected to the ill temper and disapproval of her mother and bachelor brothers. It was to escape from this situation that she went to Europe in 1931.

As a child she roamed the fields and woods of Iowa, collecting and identifying flowers, butterflies, and moths. All her life she loved music, flowers, and birds, even pigeons, which found a friend, food, and water at the window of her laboratory. She wanted to live in the country, and in 1941 she purchased a house in Millwood, Westchester County, some 20 miles north of New York City. Here she found full play for her passion for flowers and gardening. Also, she maintained subscriptions to the Metropolitan Opera and the New York Philharmonic, whose performances she regularly attended. But the time consumed in commuting and gardening did not advance the scientific program, and in 1952 she sold the house in the country and returned to a hotel apartment near the Museum, where she lived the rest of her life. She died there on Sunday, August 3, 1969.

Libbie Hyman was a most unusual person. Utterly honest, independent, uninhibited, outspoken; to many she presented a formidable exterior. She could puncture pomposity, deflate windbags, and expose stuffed shirts with alacrity. Proud, but above vanity, she scorned superficiality, ostentation, injustice, and fraud, and her poignant epithets were keen and to the point. Her close friends were those she respected, and they knew her gentle, warm, and generous nature. She made large contributions to charitable and philanthropic organizations and encouraged young investigators. It was once rumored that certain students received degrees from the University of Chicago because Libbie virtually wrote their theses. Questioned, she retorted that she never wrote a thesis for anyone, but admitted that she did "help some."

Respected, admired, she lived to receive many honors. She was a member of Phi Beta Kappa, Sigma Xi, the American Microscopical Society, American Society of Naturalists, Marine Biological Laboratory of Woods Hole, Massachusetts, American Society of Zoologists (vice-president, 1953), Society of Systematic Zoology (president, 1959), American Society of Limnology and Oceanography, Society of Protozoologists, and the National Academy of Science. A member of the editorial board of several publications, she was editor of *Systematic Zoology* (1959–1963). She received the Sc.D. degree from the University of Chicago, 1941; Goucher College, 1958; Coe College, 1959; and the LL.D. degree from Upsala College, 1963. She was awarded the Daniel Giraud Elliot Medal of the National Academy, 1951, the Gold Medal of the Linnean Society of London, 1960, and on April 9, 1969, at the Centennial Celebration of The American Museum of Natural History, she was awarded its Gold Medal for Distinguished Achievement in Science. This was her last official visit to the Museum.

With the publication of volume VI of *The Invertebrates,* Libbie's monumental contribution has ended. No one person can carry on her project, but the publishers intend to continue the series with each major group handled by a different author. The preface of her last volume concludes with the statement, "I now retire from the field, satisfied that I have accomplished my original purpose—to stimulate the study of the invertebrates."

Horace W. Stunkard
American Museum of Natural History

Bibliography of
Libbie H. Hyman

1916

1. An Analysis of the Process of Regeneration in Certain Microdrilous Oligochaetes. *J. Exptl. Zool.*, **20**:99–163.
2. On the Action of Certain Substances on Oxygen Consumption. I. The Action of Potassium Cyanide. *Am. J. Physiol.*, **40**:238–248.

1917

3. Metabolic Gradients in Amoeba and Their Relation to the Mechanism of Amoeboid Motion. *J. Exptl. Zool.*, **24**:55–99.

1918

4. Suggestions Regarding the Causes of Bioelectric Phenomena. *Science,* **48**:518–524.

1919

5. Physiological Studies on Planaria. I. Oxygen Consumption in Relation to Feeding and Starvation. *Am. J. Physiol.*, **49**:377–402.
6. Physiological Studies on Planaria. II. Oxygen Consumption in Relation to Regeneration. *Am. J. Physiol.*, **50**:67–81.
7. Physiological Studies on Planaria. III. Oxygen Consumption in Relation to Age (Size) Differences. *Biol. Bull.*, **37**:388–403.
8. On the Action of Certain Substances on Oxygen Consumption, II. Action of Potassium Cyanide on Planaria. *Am. J. Physiol.*, **48**:340–371.
9. With C. M. Child. The Axial Gradients in Hydrozoa. I. Hydra. *Biol. Bull.*, **36**:183–223.

10. On the Action of Certain Substances on Oxygen Consumption. III. Action of Potassium Cyanide on Some Coelenterates and Annelids. *Biol. Bull.*, **37**:404–415.

1920

11. Physiological Studies on Planaria. IV. A Further Study of Oxygen Consumption during Starvation. *Am. J. Physiol.*, **53**:399–420.
12. The Axial Gradients in Hydrozoa. III. Experiments on the Gradient of *Tubularia*. *Biol. Bull.*, **38**:353–403.
13. On the Action of Certain Substances on Oxygen Consumption. IV. Further Experiments on the Action of Potassium Cyanide on Invertebrates. *Publ. Puget Sound Biol. Sta.*, **2**:387–393.

1921

14. The Metabolic Gradients of Vertebrate Embryos. I. Teleost Embryos. *Biol. Bull.*, **40**:32–73.
15. With A. E. Galigher. Direct Demonstration of the Existence of a Metabolic Gradient in Annelids. *J. Exptl. Zool.*, **34**:1–16.

1922

16. With A. W. Bellamy. Studies on the Correlation between Metabolic Gradients, Electrical Gradients, and Galvanotaxis. I. *Biol. Bull.*, **43**:313–347.

1923

17. Physiological Studies on Planaria. V. Oxygen Consumption of Pieces with Respect to Length, Level, and Time after Section. *J. Exptl. Zool.*, **37**:47–68.
18. Some Notes on the Fertilization Reaction in Echinoderm Eggs. *Biol. Bull.*, **45**:254–278.

1924

19. Some Technical Methods for *Hydra* and *Planaria*. *Trans. Am. Microscop. Soc.*, **43**:68–71.
20. With B. H. Willier and S. A. Rifenburgh. Physiological Studies of Planaria VI. A Respiratory and Histochemical Study of the Source of the Increased Metabolism after Feeding. *J. Exptl. Zool.*, **40**:473–494.

1925

21. With B. H. Willier and S. A. Rifenburgh. A Histochemical Study of Intracellular Digestion in Triclad Flatworms. *J. Morphol.*, **40**:299–340.

22. On the Action of Certain Substances on Oxygen Consumption. VI. The Action of Acids. *Biol. Bull.*, **49**:288–322.
23. Respiratory Differences along the Axis of the Sponge *Grantia*. *Biol. Bull.*, **48**: 379–388.
24. The Reproductive System and Other Characters of *Planaria dorotocephala* Woodworth. *Trans. Am. Microscop. Soc.*, **44**:51–89.
25. Methods of Securing and Cultivating Protozoa. I. General Statements and Methods. *Trans. Am. Microscop. Soc.*, **44**:216–221.

1926

26. Note on the Destruction of *Hydra* by a Chydorid Cladoceran, *Anchistropus minor* Birge. *Trans. Am. Microscop. Soc.*, **45**:298–301.
27. With C. M. Child. Studies on the Axial Gradients of *Corymorpha palma*. I. Respiratory, Electric, and Reconstitutional Gradients. *Biol. Generalis*, **2**:355–374.
28. The Metabolic Gradients of Vertebrate Embryos. II. The Brook Lamprey. *J. Morphol.*, **42**:111–141.

1927

29. The Metabolic Gradients of Vertebrate Embryos. III. The Chick. *Biol. Bull.*, **52**:1–38.
30. The Metabolic Gradients of Vertebrate Embryos. IV. The Heart. *Biol. Bull.*, **52**:39–50.

1928

31. Miscellaneous Observations on Hydra with Special Reference to Reproduction. *Biol. Bull.*, **54**:65–108.
32. Studies on the Morphology, Taxonomy, and Distribution of North American Triclad Turbellaria. I. *Procotyla fluviatilis*, Commonly But Erroneously Known as *Dendrocoelum lacteum*. *Trans. Am. Microscop. Soc.*, **47**:222–255.

1929

33. Taxonomic Studies on the Hydras of North America. I. *Trans. Am. Microscop. Soc.*, **48**:242–255.
34. The Effect of Oxygen Tension on Oxygen Consumption in *Planaria* and Some Echinoderms. *Physiol. Zool.*, **2**:505–534.
35. Studies on the Morphology, Taxonomy, and Distribution of North American Triclad Turbellaria. II. On the Distinctions between *Planaria agilis* and *Planaria dorotocephala* with Notes on the Distribution of *agilis* in the Western United States. *Trans. Am. Microscop. Soc.*, **48**:406–415.

1930

36. Taxonomic Studies on the Hydras of North America. II. The Characters of *Pelmatohydra oligactis* (Pallas), *Trans. Am. Microscop. Soc.*, **49**:322–333.

1931

37. Taxonomic Studies on the Hydras of North America. III. Rediscovery of *Hydra carnea* L. Agassiz (1850), with a Description of Its Characters. *Trans. Am. Microscop. Soc.*, **50**:20–29.
38. Methods of Securing and Cultivating Protozoa. II. Paramecium and Other Ciliates. *Trans. Am. Microscop. Soc.*, **50**:50–57.
39. Studies on the Morphology, Taxonomy, and Distribution of North American Triclad Turbellaria. III. On *Polycelis coronata* (Girard). *Trans. Am. Microscop. Soc.*, **50**:124–135.
40. Taxonomic Studies on the Hydras of North America. IV. Description of Three New Species with a Key to the Known Species. *Trans. Am. Microscop. Soc.*, **50**:302–315.
41. Studies on the Morphology, Taxonomy, and Distribution of North American Triclad Turbellaria. IV. Recent European Revisions of the Triclads and Their Application to the American Forms, with a Key to the Latter and New Notes on Distribution. *Trans. Am. Microscop. Soc.*, **50**:316–335.
42. Studies on the Morphology, Taxonomy, and Distribution of North American Triclad Turbellaria. V. Descriptions of Two New Species. *Trans. Am. Microscop. Soc.*, **50**:336–343.

1932

43. Studies on the Correlation between Metabolic Gradients, Electrical Gradients, and Galvanotaxis. II. Galvanotaxis of the Brown Hydra and Some Non-fissioning Planarians. *Physiol. Zool.*, **5**:185–190.
44. Relation of Oxygen Tension to Oxygen Consumption in *Nereis virens*. *J. Exptl. Zool.*, **61**:209–221.
45. The Axial Respiratory Gradient: Experimental and Critical. *Physiol. Zool.*, **5**:566–592.

1934

46. Report on Triclad Turbellaria from Indian Tibet. *Mem. Connecticut Acad.*, 10, art. 2:5–12.
47. With W. A. Castle. Observations on *Fonticola velata* (Stringer), Including a Description of the Reproductive System. *Trans. Am. Microscop. Soc.*, **53**:154–171.

1935

48. Studies on the Morphology, Taxonomy, and Distribution of North American Triclad Turbellaria. VI. A New Dendrocoelid from Montana, *Dendrocoelopsis vaginatus* n. sp. *Trans. Am. Microscop. Soc.,* **54:**338–344.
49. Fragmentation in the Naid *Nais paraguayensis* Michaelson, 1905. *Anat. Record,* **64,** Suppl. 1: 79. Abstract.

1936

50. Studies on the Rhabdocoela of North America. I. On *Macrostomum tubum* (von Graff), 1882. *Trans. Am. Microscop. Soc.,* **55:**14–20.
51. Observations on Protozoa. I. The Impermanence of the Contractile Vacuole in *Amoeba vespertilio.* II. Structure and Mode of Food Ingestion of Peranema. *Quart. J. Microscop. Sci.,* **70:**43–57.

1937

52. Peranema and "Grantia," *Science,* **85:**454.
53. Reproductive System and Copulation in *Amphiscolops langerhansi* (Turbellaria acoela). *Biol. Bull.,* **72:**319–326.
54. Studies on the Morphology, Taxonomy, and Distribution of North American Triclad Turbellaria. VII. The Two Species Confused under the Name *Phagocata gracilis,* the Validity of the Generic Name *Phagocata* Leidy 1847, and Its Priority over *Fonticola* Komarek 1926. *Trans. Am. Microscop. Soc.,* **56:**298–310.
55. Studies on the Morphology, Taxonomy, and Distribution of North American Triclad Turbellaria. VIII. Some Cave Planarians of the United States. *Trans. Am. Microscop. Soc.,* **56:**457–477.

1938

56. The Vacuolar System of the *Euglenida. Beih. Bot. Centralblatt,* **58**(A): 379–382.
57. The Water Content of Medusae. *Bull. Mount Desert Island Biol. Lab.,* 1937: 23.
58. The Water Content of Medusae, *Science,* **87:**166–167.
59. With C. D. Van Cleave. Annotated Bibliography of the Scientific Publications of Professor Charles Manning Child. *Physiol. Zool.,* **11:**105–125.
60. The Fragmentation of *Nais paraguayensis. Physiol. Zool.,* **11:**126–143.
61. Taxonomic Studies on the Hydras of North America. V. Description of *Hydra cauliculata* n. sp., with Some Notes on Other Species, Especially *Hydra littoralis. Am. Museum Novitates,* no. 1003, 9 pp.
62. North American Rhabdocoela and Alloeocoela. II. Rediscovery of *Hydrolimax grisea* Haldeman. *Am. Museum Novitates,* no. 1004, 19 pp.
63. Faunal Notes. *Bull. Mount Desert Island Biol. Lab.,* 1937: 24–25.
64. North American Rhabdocoela and Alloeocoela. III. *Mesostoma arctica,* n. sp., from Northern Canada. *Am. Museum Novitates,* no. 1005, 8 pp.
65. "Land Planarians from Yucatan." In A. S. Pearse (ed.), *Fauna of the Caves of Yucatan.* Carnegie Institute of Washington, Publication no. 491: 23–32.

1939

66. New Species of Flatworms from North, Central, and South America. *Proc. U.S. Natl. Museum,* **86:**419–439.
67. Land Planarians from the Hawaiian Islands. *Arch. Zool. Exptl. Gen.,* **80,** Notes et Revue, no. 3: 116–124.
68. Some Polyclads of the New England Coast, Especially of the Woods Hole Region. *Biol. Bull.,* **76:**127–152.
69. A New Polyclad Genus of the Family Discocoelidae. *Vestnik Ceskoslovenske Zoologiche Spoelcnosti Praze, Sbornik Praci 90 Narozeninam Professor Franz Vejdovsky:* 237–246.
70. Acoel and Polyclad Turbellaria from Bermuda and the Sargassum. *Bull. Bingham Oceanog. Collection,* **7:**1–36.
71. Polyclad Worms Collected on the Presidential Cruise of 1938. *Smithsonian Inst. Misc. Collections,* **98**(17), 13 pp.
72. North American Rhabdocoela and Alloeocoela. IV. *Mesostoma macropenis,* new species, from Douglas Lake, Michigan. *Am. Midland Naturalist,* **21:**646–650.
73. North American Rhabdocoela and Alloeocoela. V. Two New Mesostomine Rhabdocoels. *Am. Midland Naturalist,* **22:**629–636.
74. North American Triclad Turbellaria. IX. The Priority of *Dugesia* Girard 1850 over *Euplanaria* Hesse 1897 with Notes on American Species of *Dugesia. Trans. Am. Microscop. Soc.,* **58:**264–275.
75. North American Triclad Turbellaria. X. Additional Species of Cave Planarians. *Trans. Am. Microscop. Soc.,* **58:**276–284.

1940

76. Land Planarians from the Palau and Caroline Islands, Micronesia. *Ann. Mag. Nat. Hist.,* (11) **5:**345–362.
77. Aspects of Regeneration in Annelids. *Am. Naturalist,* **74:**513–527.
78. Observations and Experiments on the Physiology of Medusae. *Biol. Bull.,* **99:**282–296.
79. The Polyclad Flatworms of the Atlantic Coast of the United States and Canada. *Proc. U.S. Natl. Museum,* **89:**449–495.
80. Revision of the Work of Pearse and Walker on Littoral Polyclads of New England and Adjacent Parts of Canada. *Bull. Mount Desert Island Biol. Lab.,* 14–20.
81. Native and Introduced Land Planarians in the United States. *Science,* **92:**105–106.

1941

82. Environmental Control of Sexual Reproduction in a Planarian. Abstract. *Anat. Record,* **81:** Suppl., p. 108.

83. Terrestrial Flatworms from the Canal Zone, Panama. *Am. Museum Novitates,* no. 1105, 11 pp.
84. Lettuce as a Medium for the Continous Culture of a Variety of Small Laboratory Animals. *Trans. Am. Microscop. Soc.,* **60:**365–370.
85. Small Animal Cultures Maintained on Lettuce Leaves. *Educational Focus* (Bausch and Lomb Optical Company), 12, no. 3: 14–19.

1942

86. Mealworms, Their Importance and Culture. *The Aquarium, Philadelphia,* **11:** 51–52.
87. The Transition from the Unicellular to the Multicellular Individual. *Biological Symposia,* **8:**27–42.

1943

88. Water Content of Medusae; Sexuality in a Planarian. *Nature,* **151:**140.
89. On a Species of *Macrostomum* (Turbellaria: Rhabdocoela) Found in Tanks of Exotic Fishes. *Am. Midland Naturalist,* **30:**322–335.
90. Endemic and Exotic Land Planarians in the United States with a Discussion of Necessary Changes of Names in the Rhynchodemidae. *Am. Museum Novitates,* no. 1241, 21 pp.

1944

91. Marine Turbellaria from the Atlantic Coast of North America. *Am. Museum Novitates,* no. 1266, 15 pp.

92. A New Hawaiian Polyclad Flatworm Associated with *Teredo. Occasional Papers Bernice P. Bishop Museum,* **18:**73–75.

1945

93. North American Triclad Turbellaria. XI. New, Chiefly Cavernicolous Planarians. *Am. Midland Naturalist,* **34:**475–484.

1946

94. The Nature of the Eosinophilous Spheres in the Intestimal Epithelium of Planarians: a Correction. *Trans. Am. Microscop. Soc.,* **65:**276–277.

1947

95. Two New Hydromedusae from the California Coast. *Trans. Am. Microscop. Soc.,* **66:**262–268.

1950

96. A New Hawaiian Polyclad, *Stylochoplana inquilina,* with Commensal Habits. *Occasional Papers Bernice P. Bishop Museum,* **20:**55–58.

1951

97. North American Triclad Turbellaria. XII. Synopsis of the Known Species of Planarians of North America. *Trans. Am. Microscop. Soc.,* **70:**154–167.

1952

98. Further Notes on the Turbellarian Fauna of the Atlantic Coast of the United States. *Biol. Bull.,* **103:**195–200.

1953

99. The Polyclad Flatworms of the Pacific Coast of North America. *Bull. Am. Museum Nat. Hist.,* **100:**265–392.
100. North American Triclad Turbellaria. 14. A New, Probably Exotic, Dendrocoelid. *Am. Museum Novitates,* no. 1629, 6 pp.
101. Some Polyclad Flatworms from Galapagos Islands. *Allan Hancock Pacific Expeditions,* **15:**183–210.
102. Posterior Growth in Annelids. *Am. Naturalist,* **87:**395–396.

1954

103. The Polyclad Genus *Pseudoceros* with Special Reference to the Indo-Pacific Region. *Pacific Sci.,* **8:**219–225.
104. Some Land Planarians of the United States and Europe with Remarks on Nomenclature. *Am. Museum Novitates,* no. 1667, 21 pp.
105. North American Triclad Turbellaria. XIII. Three New Cave Planarians. *Proc. U.S. Natl. Museum,* **103:**563–573.
106. A New Marine Triclad from the Coast of California. *Am. Museum Novitates,* no. 1679, 5 pp.
107. Some Polyclad Flatworms from the Hawaiian Islands. *Pacific Sci.,* **8:**331–336.
108. Free-living Flatworms (Turbellaria) of the Gulf of Mexico. In P. S. Galtsoff (ed.), *Gulf of Mexico, Its Origin, Waters, and Marine Life.* U.S. Fish and Wildlife Service, Fishery Bulletin, **89:**301–302.

1955

109. The Polyclad Flatworms of the Pacific Coast of North America: Additions and Corrections. *Am. Museum Novitates,* no. 1704, 11 pp.
110. Descriptions and Records of Fresh-water Turbellaria from the United States. *Am. Museum Novitates,* no. 1714, 36 pp.

111. Some Polyclad Flatworms from the West Indies and Florida. *Proc. U.S. Natl. Museum*, **104:**115–150.
112. Miscellaneous Marine and Terrestrial Flatworms from South America. *Am. Museum Novitates*, no. 1742, 33 pp.
113. How Many Species? *Systematic Zool.*, **4:**142–143.
114. Some Polyclad Flatworms from Polynesia and Micronesia. *Proc. U.S. Natl. Museum*, **105:**65–82.
115. A Further Study of the Polyclad Flatworms of the West Indian Region. *Bull. Marine Sci. Gulf Caribbean*, **5:**259–268.
116. *Charles Manning Child, 1865–1954.* Biographical Memoirs, National Academy of Science, Vol. XXX, pp. 73–103.

1956

117. Textbook Planarians and the Reality. *Am. Biol. Teacher*, **18:**124–127.
118. North American Triclad Turbellaria. XV. Three New Species. *Am. Museum Novitates*, no. 1808, 14 pp.

1957

119. North American Rhabdocoela and Alloeocoela. IV. A Further Study of *Mesostoma. Am. Museum Novitates*, no. 1829, 15 pp.
120. A Few Turbellarians from Trinidad and the Canal Zone with Corrective Remarks. *Am. Museum Novitates*, no. 1862, 8 pp.

1958

121. Notes on the Biology of the Five-lunuled Sand Dollar. *Biol. Bull.*, **114:**54–56.
122. The Occurrence of Chitin in the Lophophorate Phyla. *Biol. Bull.*, **114:**106–112.
123. Turbellaria. *British, Australian, and New Zealand Antarctic Expedition 1929–1931.* Reports, series B (Zoology and Botany), 6, part 12: 279–290.

1959

124. A Further Study of Micronesian Polyclad Flatworms. *Proc. U.S. Natl. Museum*, **108:**543–597.
125. On Two Freshwater Planarians from Chile. *Am. Museum Novitates*, no. 1932, 11 pp.
126. Some Australian Polyclads (Turbellaria). *Records Australian Museum*, **25:**1–17.
127. Some Turbellaria from the Coast of California. *Am. Museum Novitates*, no. 1943, 17 pp.

1960

128. New and Known Umagillid Rhabdocoels from Echinoderms. *Am. Museum Novitates*, no. 1984, 14 pp.

129. Second Report on Hawaiian Polyclads. *Pacific Sci.,* **14**:308–309.
130. Cave Planarians in the United States. *Am. Midland Naturalist,* **64**:10–11.

1962

131. Some Land Planarians from Caribbean Countries. *Am. Museum Novitates,* no, 2110, 25 pp.

1963

132. North American Triclad Turbellaria. XVI. Fresh-water Planarians from the Vicinity of Portland, Oregon. *Am. Museum Novitates,* no. 2123, 5 pp.
133. Notes on a Didymozoid Trematode from the Bahama Islands. *Bull. Marine Sci. Gulf Caribbean,* **13**:193–196.

1964

134. North American Rhabdocoela and Alloeocoela. VII. A New Seriate Alloeocoel, with Corrective Remarks on Alloeocoels. *Trans. Am. Microscop. Soc.,* **83**: 248–251.

1966

135. Further Notes on the Occurrence of Chitin in Invertebrates. *Biol. Bull.,* **130** (1): 94–95.

Books by
Libbie H. Hyman

A Laboratory Manual for Elementary Zoology. Chicago: University of Chicago Press, 1919.
A Laboratory Manual for Comparative Vertebrate Anatomy. Chicago: University of Chicago Press, 1922.
A Laboratory Manual for Elementary Zoology, 2d ed., Chicago: University of Chicago Press, 1929.
The Invertebrates. Volume I. *Protozoa through Ctenophora.* New York: McGraw-Hill Book Company, 1940.
Comparative Vertebrate Anatomy. Chicago: University of Chicago Press, 1942.
The Invertebrates. Volume II. *Platyhelminthes and Rhynchocoela.* New York: McGraw-Hill Book Company, 1951.
The Invertebrates. Volume III. *Acanthocephala, Aschelminthes, and Entoprocta.* New York: McGraw-Hill Book Company, 1951.

The Invertebrates. Volume IV. *Echinodermata*. New York: McGraw-Hill Book Company, 1955.

The Invertebrates. Volume V. *Smaller Coelomate Groups*. New York: McGraw-Hill Book Company, 1959.

The Invertebrates. Volume VI. *Mollusca I*. New York: McGraw-Hill Book Company, 1967.

compiled by
William K. Emerson
Department of Living Invertebrates
The American Museum of Natural History

Biology of the
Turbellaria

Chapter 1

On the Anatomy and Affinities of the Turbellarian Orders

Tor G. Karling
Swedish Museum of Natural History, Section for
Invertebrate Zoology, Stockholm, Sweden

The phylogeny of the turbellarian orders will be discussed on the basis of a diagram influenced by the sister-group principles of Hennig (1966).[1] A revision of the turbellarian system is outside my actual approach and entanglement in theoretical discussions on the affinities of the Turbellaria to other Metazoa is avoided.

Today, as earlier, comparative anatomy must be the basis for the turbellarian system; information of systematic bearing from other fields of

[1] I thank my friend Dr. Lars Brundin for valuable discussions on systematic principles.

turbellarian research is sparse. Another question is to what degree this basis is reliable. The turbellarian orders are kept together by similarities based on characters of highly different evolutionary age, and opinions diverge as to what is primitive and what is derived. Moreover, it is not easy to state what is homology and what is analogy. These difficulties are pronounced in animal taxa with a low degree of differentiation in general, such as the turbellarians.

The principal idea in Hennig's approach is the absolute demand for monophyly of every group (cf. also Brundin, 1967). A monophyletic group is constituted by the species—all the species—with an ancestral species in common, which is not at the same time the ancestral species of species outside the group. Two groups originating from the same ancestral species are *sister groups*. One of two sister groups is more conservative, *plesiomorph*; the other is more derived, *apomorph*. The attributes plesiomorph and apomorph are also used to indicate primitive and derived conditions respectively. A special homology characterizing a group (or species) is one of its *autoapomorphies*. For the taxa of a monophyletic group it can be at the same time an exclusive *synapomorphy*. These taxa have further nonexclusive homologies in common with species of other groups, *symplesiomorphies*. The systematic value of the common homologies thus diminishes with the further hierarchic ramifications: a synapomorphy for a larger taxon, e.g., an order, is a plesiomorphy for every one of its suborders. Thus, the application of these ideas depends upon recognition of what is apomorph and what is plesiomorph in a series of transformation.

THE TURBELLARIAN ARCHETYPE

By a comparative analysis of all the turbellarian orders, an analysis impossible to review here as a whole, I have arrived at an archetype which I place at the base of the hierarchic diagram of the Turbellaria (Fig. 1). The attributes of this archetype, listed below, are at the same time the basic plesiomorphies for every order. Progress in the field of turbellarian research during the last decades has forced me to revise the pictures of the turbellarian archetype and the turbellarian phylogeny which I outlined in the year 1940. A third archetype was presented by Peter Ax in his excellent review on turbellarian anatomy and phylogeny at the Asilomar Symposium in 1960 (published 1963).

1 Epidermis one-layered, entirely ciliated, cellular, with rhabdites and intracellular nuclei. Basement membrane lacking or weakly developed. Subepidermal musculature with an outer layer of circular fibers and an inner layer of longitudinal fibers. This organization of the body wall is the original in all orders except the Nemertodermatida. EM studies on rhabdites can con-

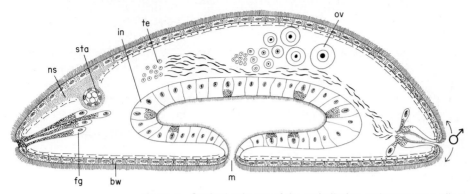

Figure 1 Sagittal scheme of the turbellarian archetype according to the author. (*Original.*) For abbreviations see Fig. 2.

tribute to our understanding of the phylogeny in the Turbellaria; today only scattered observations are available (Reisinger, 1969, with references).

2   Nervous system epidermal-subepidermal. This type of nervous system has been found in many species of different acoel genera (Dörjes, 1968) and in the two nemertodermatid genera (*Nemertoderma*, Westblad, 1937; *Meara*, Westblad, 1949), i.e., in the taxa with the most primitive conditions. Most authors regard the epidermal location—even in the Turbellaria—as a primitive condition (Hyman, 1951), but Ax regards it as secondary, "a parallel persistence of juvenile characters" (1963, pp. 201–202). The acceptance of an epidermal nervous system as primitive in species of different genera in the Acoela requires that these species represent "direct descendants of a common stem type with an epidermal system and that they all belong together systematically (phylogenetically)" (loc. cit.). However, if we regard the epidermal condition as a basic plesiomorphy, the common stem type is the turbellarian archetype, the internal nervous system arising from trends in several evolutionary lines.

3   Eyes lacking or epidermal. The typical internal turbellarian pigment-cup eyes evolve from epidermal "pigment spots" evidently separately in different lines. Epidermal eyes occur in the primitive taxa Acoela and Microstomidae as well as in the genus *Gnosonesima* (basic plesiomorphic feature, Karling, 1968, with references). The eyes are secondarily lost in many taxa.

4   Weakly differentiated statocyst containing a variable number of statoliths. A statocyst belongs to the fundamental qualities in the Acoela, the Nemertodermatida, a part of the Catenulida, and in the Proseriata. It is sometimes secondarily lost in these taxa. I regard as primitive a statocyst with epithelial wall and a variable number of statoliths (cf. below, the taxa with statocyst).

5   Simple mouth pore or pharynx simplex. These conditions can be realized secondarily in some parasites. From a phylogenetic point of view

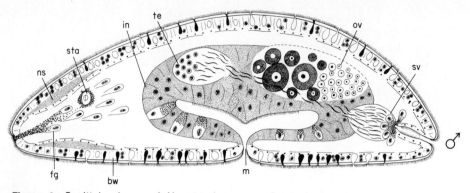

Figure 2 Sagittal scheme of *Nemertoderma* sp. (*Original after several figures in Westblad, 1937, and Karling, 1967.*)
Abbreviations (applicable also to Figs. 4 and 5): ac, male accessory organ; br, brain; bw, body wall; fg, frontal glands; gi, genitointestinal communication; in, intestine; m, mouth; ns, nervous system; ov, ovary; ph, pharynx; pi, pigment; pr, prostatic vesicle; sb, sensory bristle; sta, statocyst; sty, stylet; sv, seminal vesicle; te, testis.

there is no reason to keep the simple mouth pore and the pharynx simplex apart, these conditions being variations on the same theme; i.e., the pharynx simplex can arise from trends in several lines.

6 Sac-shaped intestine with ciliated epithelial wall, no anal pore. Some authors regard the lack of an anal pore in the Turbellaria as a secondary condition, an opinion lacking support in the turbellarian anatomy (Reisinger, 1961; Karling, 1965). Several species in different groups have a weakly differentiated, mostly temporary anal pore (Karling, 1966a). There is thus anatomical evidence that the turbellarians can give rise to taxa with a true digestive canal.

7 Two alternatives must be considered: (*a*) no excretory organs; (*b*) some kind of primitive (diffuse) protonephridia. No methods have hitherto revealed any kinds of excretory organs in the Acoela, not even in the limnic species *Oligochoerus limnophilus* Ax and Dörjes, 1966. Opinions differ as to the primitiveness of this phenomenon (Westblad, 1948, p. 65: primitive; Ax, 1961, pp. 24–31, and Reisinger, 1968, p. 20: secondary). I accept preliminarily the assertion that an excretory system is lacking also in *Nemertoderma* and *Meara* (cf. above), but no EM studies have hitherto been carried out on these subjects. EM studies have revealed a complicated structure of the turbellarian flame bulbs, and further studies in this field will certainly contribute to our understanding of the affinities between the turbellarian taxa (Reisinger, 1970, with references).

8 Female gonads homocellular (ovaries), sometimes with nutritive cells, eggs entolecithal. This type of female gonad characterizes the grade (formerly order; cf. below, Nemertodermatida, etc.) Archoophora. Reduction

of the vitelline parts in heterocellular gonads can in some cases give rise to homocellular gonads (Karling, 1967).

9 With male genital pore in a variable position (primitive copulatory organ, internal fertilization), no female pore (common oviduct). We do not know any turbellarian without male genital pore and with external fertilization. A female pore is lacking in the Acoela, the Nemertodermatida, and the Catenulida, the primitive mode of copulation being injection (impregnation) and the primitive mode of egg laying being rupture of body wall or gut wall. The oviduct is secondarily lost in some higher taxa, e.g., *Bresslauilla* (Ax, 1963, p. 205).

The archetype presented by Ax (1963) principally differs from that presented here in the following respects:

1 Pharynx simplex. As shown above (point 5), this difference can be disregarded.

2 "Internal nervous system with a brain, several pairs of longitudinal cords and transverse commissures (orthogon)," cf. above, point 2.

3 "Single statocyst with one statolith." Ax believes that "in the few instances with two or three statoliths . . . a secondary multiplication has taken place" (op. cit., p. 202). I think the evolution starts with a low grade of stability and leads to more fixed conditions, such as in the Acoela and the Proseriata.

4 A pair of protonephridia. My alternative *b* presupposes protonephridia in the archetype, but I find it easier to derive the unpaired system of the Catenulida (as well as the paired condition in most other taxa) from a diffuse system of emunctories (Ax, op. cit., pp. 203–204).

5 Separate male and female gonopores, oviducts, a simple bursa, and a vagina. This high level of organization is based on Ax's assumption that the turbellarians are highly reduced Spiralia (op. cit., pp. 211–215). Like most other specialists, he holds to the theory, first presented by Meixner (1926), that the first female pore was a vagina and that "the atrium femininum and the vagina of the Macrostomida are respectively homologous with the bursa and the vagina of the Acoela" (op. cit., p. 204). Regarding the Acoela as extremely reduced, he reverses the evolutionary direction Acoela → Macrostomida accepted by most authors. However, the real basis for this vagina-oviduct theory is rather weak (Karling, 1940, p. 210). In most turbellarians the female apparatus opens behind the male apparatus, separately or into a common atrium. In the Prolecithophora, evidently the most primitive Neoophora, the common oviduct always opens from behind into a common atrium (also in *Protomonotresis centrophora* Reisinger, 1923, originally described as lacking female gonoducts; Reisinger, *in litt.*) and that seems to be the primary condition also in the Proseriata, Tricladida, and Rhabdocoela, and perhaps also in the Polycladida, though in this order the female pore is mostly removed from the male one. Also in the Lecithoepitheliata the female apparatus is situated as a whole behind the male organs. The bursal organs arise independently of the common duct (Prolecithophora, many Proseriata and Rhabdocoela) or as a derivative of that duct. A bursal organ separated from the common duct seems to have a positive selective value, and a trend toward a separation of these structures appears in many lines. Also in the two families of the Macrostomida (Macrostomidae, Microstomidae) this trend is realized, a separate vagina being differentiated in some genera and species (Papi, 1950; Westblad, 1953). Further, we can regard the female ducts from another point of view: are they all homologous? If the common oviduct arises from a vagina in the Macrostomida, this mode of evolution is not in any way necessary in other lines (cf. below).

ON THE ANATOMY AND AFFINITIES OF THE TURBELLARIAN ORDERS

As seen in the diagram (Fig. 3), I have here abandoned the division of the Turbellaria into two groups, the Archoophora and the Neoophora. The main common attribute of the Archoophora, the homocellular female gonads (the ovaries), is a primitive attribute of the Turbellaria, a *symplesiomorphy*, the group Archoophora being thus a paraphyletic grade (*Stadiengruppe*), not a monophyletic taxon. The turbellarian orders with heterocellular female gonads, i.e., with germaria and yolk glands, constitute the group Neoophora, a group appearing monophyletic from the diagram. I shall later return to this question. I thus regard the Nemertodermatida, Acoela, Catenulida, Macrostomida and Haplopharyngida, earlier brought together in the Archoophora, as orders. The Polycladida too have sometimes been regarded as Archoophora.

In the first alternative here presented I regard the lack of excretory organs as primitive, the Acoela and Nemertodermatida thus constituting the plesiomorph sister group, all the other orders the apomorph group with excretory organs as synapomorphy. The systematic relationship of the orders Nemertodermatida and Acoela, and thus the position of these orders in the phylogenetic diagram, cannot be definitely fixed today owing to our deficient knowledge on some points and also owing to divergent opinions about the primitiveness of some structures. The two genera of the Nemertodermatida, *Nemertoderma* and *Meara*, evidently stand closest to the turbellarian archetype, but their organizations diverge at some important points and I am not convinced that they belong together.

The statocyst of the Acoela consists of a thin wall with two matrix nuclei in dorsal position. The single statolith is about plane-convex with one single matrix nucleus on its dorsal side (*Convoluta* type, Luther, 1912; Westblad, 1940; Ax and Dörjes, 1966). The nuclei are difficult to see on squeezed material owing to their dorsal position. Two statoliths are sometimes seen in

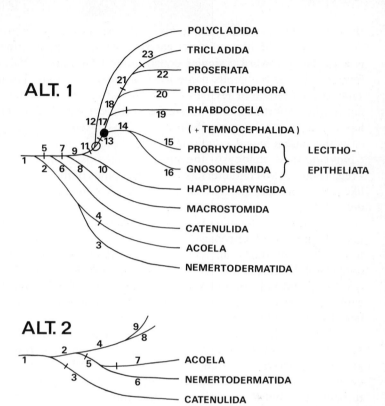

Figure 3 Phylogenetic diagram of the turbellarian orders (and the suborders Prorhynchida and Gnosonesimida). A short bar indicates the comparatively apomorph sister group; a ring indicates the common ancestral species of the taxa with complex pharynx; a black dot indicates the common ancestral species of the taxa with heterocellular ovaries. Other explanations in the text.

species of the Hofsteniidae (Steinböck, 1966, pp. 75–76). The Nemertodermatida mostly have two statoliths, but Riedl has seen three statoliths in *Nemertoderma* (1960). The wall of the statocyst is epithelial with a variable number of cellular nuclei, concentrated to a vertical girdle between the statoliths in *Meara* (Westblad, 1937, 1949; my own observations).

The parenchymal condition of the digestive tract in the Acoela is one of the most-discussed features in turbellarian anatomy. *Nemertoderma* and *Meara* have a true intestine, which in *Nemertoderma* is in part ciliated (Fig. 2) (Karling, 1967, pp. 5–8). In the Acoela a central lumen is often found in the digestive tract. According to my own observations *Archiproporus minimus* An der Lan, 1936 has a distinct gastroderm, and in many other species there is a tendency toward a similar construction (Antonius, 1968; Dörjes has informed me that the genus name *Archiproporus* is valid). These observations support the opinion of Hyman (1959) that absence of a lumen in the acoels is not of any importance. The existence or nonexistence of an epithelial gut wall has a greater phylogenetic bearing but has often been exaggerated. In any case the parenchymal condition of the digestive tract remains—besides the *Convoluta*-type statocyst—the principal diagnostic character of the Acoela. I regard these two characters as synapomorphies of the Acoela. The nemertine-like epidermis and the two statoliths are synapomorphies of the Nemerto-dermatida.

The order Catenulida is rather specialized, most of its taxa being limnic, mainly with a sexual propagation. The statocyst is often difficult to see, owing to its extremely thin wall. The number of statoliths varies. I have seen two to four statoliths in *Rhynchoscolex* species (cf. also Rixen, 1961). Sterrer reports one to six statoliths in a marine catenulid (1966). The internal structure of the catenulid statocyst is not sufficiently known. Marcus has found a variable number of cells in its matrix (1945).

I derive the unpaired excretory system of the Catenulida from a diffuse system of emunctories, others deriving it from the paired system of most other groups (cf. above, the archetype of Ax). I regard the dorsally situated male copulatory apparatus of the Catenulida as homologous with that of other turbellarians, but this opinion can be criticized and I myself have earlier regarded this apparatus as independently evolved (1940). Recent studies on a *Stenostomum* species (Borkott, 1970) have revealed a highly aberrant structure of the male apparatus.

Synapomorphies of the Catenulida are the unpaired excretory system and the dorsorostral position of the male copulatory organ. Group 7 (Fig. 3, alternative 1) is the apomorph sister group compared with the Catenulida with the synapomorphies common oviduct and paired excretory system.

The orders hitherto discussed can be arranged in another way, here presented as a second alternative, if we provide the archetype with some kind of excretory system, perhaps irregularly arranged emunctories. The synapomorphy of group 5 (Fig. 3, alternative 2), the Acoela and the Nemertoderma-tida, would then be the loss of the excretory organs. It is difficult to find out true synapomorphies for group 2, i.e., all turbellarians except the Catenulida. The synapomorphies of the Catenulida are the same as in the first alternative.

I have earlier regarded it necessary from a practical point of view to include the family Haplopharyngidae in the order Macrostomida (1965), but this measure is incompatible with the ideas of plesiomorphy and apomorphy, the Macrostomida s. lat. lacking true synapomorphies. The Haplopharyngida have the caudal position of the female apparatus (common oviduct, female genital canal, indeed the existence of this canal is not granted) in common with most higher taxa. However, I would not like to say whether this condition or the opposite position of the outleading genital apertures is the primitive one, and consequently none of the actual lines has been indicated as apomorph compared with its sister group (cf. above, Ax's archetype, point 5). The Macrostomida s. str. have the synapomorphy of common oviduct in front of the male apparatus, but this character is sometimes indistinct in taxa with common atrium. The synapomorphy of group 9 (Fig. 3) is the position of the common oviduct behind the male atrial apparatus, but this condition can also here be indistinct or secondarily altered. The differentiation of a proboscis is the most pronounced autoapomorphy of the Haplopharyngida. The synapomorphy of the apomorph sister group 11 (Fig. 3) is the complex pharynx, characteristic of all the higher turbellarian orders. The first stage in the transformation of this pharynx type is the plicate pharynx. However, the homology of all kinds of complex pharynges is not warranted.

The order Lecithoepitheliata contains the isolated suborders Prorhynchida *mihi* and Gnosonesimida *mihi* with the lecithoepitheliate type of female gonads as synapomorphy, i.e., with a common layer for ovocytes and vitellocytes and the growing ovocytes enclosed by vitellocytes (Figs. 4 and 5). Prorhynchida is only a new name for Typhlocoela Steinböck (1923) [abandoned in favor of Lecithoepitheliata (Reisinger, 1924; Steinböck, 1925)] with the families Prorhynchidae and Hofsteniidae. The family Gnosonesimidae was later (Reisinger, 1926) included in the Lecithoepitheliata. Steinböck (1966) accepted the opinion that the Hofsteniidae belonged to the Acoela.

According to Steinböck (1966) and Reisinger (1968) the prorhynchid bulbous pharynx has arisen from the simple pharynx of *Hofstenia* type without the intermediate plicate step. Such an evolution is perhaps not impossible, as we can see in the acoel species *Diopisthoporus brachypharyngeus* (Dörjes, 1968), but the construction of the pharynx in this species is far from that of all other known bulbous pharynx types; I would call it a bulbiform simple

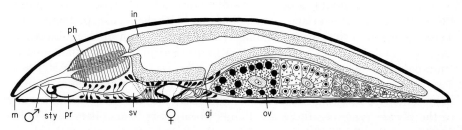

Figure 4 Sagittal scheme of *Geocentrophora sphyrocephala*. (*From Steinböck, 1927.*) For abbreviations see Fig. 2.

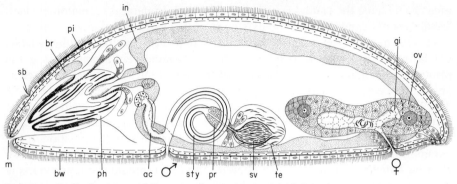

Figure 5 Sagittal scheme of *Gnosonesima brattstroemi*. (*Original after Karling, 1968.*) For abbreviations see Fig. 2.

pharynx. Anatomically the prorhynchid pharynx is nothing but a strongly muscular variable pharynx (Karling, 1940). The coniform pharynx, recently described by Karling (1968) in the genus *Gnosonesima* is a type of its own, evidently directly evolved from a plicate pharynx.

Reisinger and Steinböck base their derivation of the Lecithoepitheliata directly from acoel turbellarians mainly on the construction of the pharynx and the female gonads. Here I regard the complex pharynx, including the pharynges of Prorhynchida and Gnosonesimida, throughout as a homologous structure and moreover the female gonads of all the "Lecithophora" also as homologous, but I admit that a definite position in these questions is difficult to take. However, how can the highly differentiated excretory organs of the Prorhynchida be explained if we accept the evolution of this group directly from the Acoela? A more acceptable alternative could be to let the Lecitho-epitheliata (or its two suborders separately, cf. Karling, 1968) branch off from the archoophoran stem in the vicinity of Haplopharyngida.

To indicate one of the two lecithoepitheliate suborders as the apo-morph sister group would give a false impression of close phylogenetic affinity between the two taxa, a measure thus omitted here.

There is no doubt about the rather close affinity between the orders Prolecithophora, Proseriata, Tricladida, and Rhabdocoela (including the Temnocephalida), their heterocellular female gonads with different layers for ovocytes and yolk cells—the latter also producing shell substance—con-stituting a true synapomorphy for group 17 (Fig. 3) (Karling, 1940; Ax, 1963). The structure of the pharynx in different taxa of the first three orders plays an important part in fixing the origin of the Rhabdocoela. It is fundamentally of the plicate type, but there are trends toward a bulbous pharynx in many families of the Prolecithophora, and in the Proseriata there is at least one genus (*Ciliopharyngiella* Ax, 1952) with a bulbous pharynx. Here I have resolved the controversy about the origin of the Rhabdocoela (Ax, 1963) by letting the rhabdocoelan line branch off from the common stem of the other three orders.

The synapomorphies of the comparatively apomorph rhabdocoelan sister group are the specific construction of the rosulate and doliiform pharynx types (diaphragm strongly developed, circular muscle layer next to the pharynx lumen, Karling, 1940, p. 172) and the compact gonads mostly provided with a tunica propria. In some species these gonads can be divided secondarily into follicles.

The order Prolecithophora can be characterized as neoophoran turbellarians with fusiform or cylindrical shape, plicate to variable pharynx, sac-shaped intestine, and primary follicular to compact gonads, mostly without a tunica propria. For the specialist it appears as a very homogeneous group, but the characters here listed do not constitute good synapomorphies. This fact indicates that this group has been ranked on too high a level in the hierarchic diagram, but a more satisfactory position for the group in question could not yet be found. This situation also complicates the establishment of synapomorphies for the whole plesiomorph group 18 (compared with the apomorph sister group Rhabdocoela): testes and vitellaria diffuse—follicular, plicate pharynx with trends toward a bulbous type, penis fundamentally of the papilla type. Group 21 with the orders Proseriata and Tricladida is rather homogeneous, sometimes regarded as an order (Seriata, Metamerata), and the same is true for its two branches. Its synapomorphies are the fundamentally serial arrangement (i.e., in longitudinal rows) of the testes and yolk follicles, the backwards-directed tubiform pharynx and the elongate body shape. These features can be secondarily altered (one pair of testes, ruffled pharynx with vertical axis).

The synapomorphy of the Proseriata is the statocyst of *Monocelis* type (sometimes secondarily lost). This statocyst differs in some aspects from the acoelan statocyst (*Convoluta* type). In its frontal wall a pair of nuclei are easily seen in live material, and often also another pair in a more caudal position. The single statolith within the statocyst is globular or frontally somewhat conical with two frontal groups of matrix cells, containing two to four nuclei each (Ax, 1956; Giesa, 1966; my own observations on representatives of all proseriate families). The isolated occurrence of a statocyst speaks in favor of an independent evolution of this structure in the Proseriata, just as in the rhabdocoel *Lurus evelinae* Marcus, 1950. Otherwise we must accept an independent loss of the statocyst in most turbellarian orders, which, however, is a theoretical possibility (Karling, 1966b).

The Tricladida constitute the apomorph sister group compared with the Proseriata. They have the synapomorphies: tricladoid intestine, germaria anteriorly situated in the female gonads, and mostly a flattened body shape. This organization is sometimes secondarily changed.

From a purely anatomical viewpoint the order Polycladida belongs to the grade Archoophora owing to their homocellular female gonads, but they are more highly differentiated than the other orders of this grade. I have repeatedly maintained the idea of a rather close relationship between the Polycladida and the complex around the Prolecithophora, but the diagram is made without attention to the hypothesis of a total reduction of the yolk glands in the Polycladida. A parallel evolution of the order Polycladida with the Prolecithophora complex has been indicated by the direction of the poly-cladid branch 12. The Polycladida and the Prolecithophora complex are both relatively conservative branches of the same stem, the two lines Rhabdocoela and Lecithoepitheliata being more advanced. Transferring the Lecithoepithe-liata to a lower level would bring the Polycladida into a position closer to the Prolecithophora complex. The synapomorphies of the Polycladida are the polycladoid intestine and the (secondarily) follicular ovaries.

DISCUSSION

Perhaps the diagram here presented, and consequently also its author, appear highly conservative. I have given all the species with complex pharynx a common ancestral species (here indicated with a ring) and in the same way all the species with heterocellular female gonads a common ancestral species (indicated with a dot), though there are facts indicating that the complex pharynx as well as the gonads in question can have arisen more than once. Thus, the monophyly of the Neoophora is in no way warranted. There are many other subjects for criticism. The theoretical basis of the diagram, the search for sister groups, makes a fixed connection of all branches in the hier-archic tree necessary, a measure often avoided before. I have presented one partial alternative to the diagram and several more alternatives may have been desirable, but I have contented myself with pointing out the problems. The diagram may be regarded not as a demonstration of a fixed opinion but as an exponent of an unsatisfactory situation. Then also the question arises as to the correctness of the methodics, sharply criticized during the last years (Mayr, 1969). According to the definition of the concept "monophyly," the class Turbellaria in its whole is not monophyletic, its common ancestral species being at the same time the ancestral species for the classes Trematoda and Cestoda (derivatives of the Turbellaria Rhabdocoela). Here we find one of the points where ideas strongly diverge. Mayr rejects the "prospective postulate" of Hennig (that a monophyletic group must include all the species with an ancestral species in common). He says: "If a descendant group . . . evolves more rapidly than the other collateral lines, it not only can but must be ranked in a higher category than its sister groups. This does not violate the principle of monophyly, retrospectively defined" (loc. cit.). Thus, the question is how to define monophyly. According to Hennig the process of phylogenesis begins with the splitting up of an ancestral species into daughter species, these species giving rise to sister groups with true synapomorphies. Breaking out one of the groups of the turbellarian diagram, e.g., the Macrostomida, to a separate class leaves a heterogeneous rest—an artificial conglomerate—without the criteria of an exclusive organization (synapomorphies), its only features in common being nonexclusive (symplesiomorphies), inherited from the archetype, thus also in common with the Macrostomida. Replacing in this example the Macrostomida with the Trematoda leaves in the same way an artificial conglomerate, a heterogeneous turbellarian class, if the Trematoda and other derivatives of the Turbellaria are not degraded to turbellarian taxa. The phylum Platyhelminthes is evidently a monophyletic entity based on the turbellarian archetype. Its groups Turbellaria, Trematoda, and Cestoda are not sister groups or groups on the same hierarchic level and cannot consequently (according to Hennig) be of the same systematic rank, i.e., classes. Here we meet the same state of affairs as within the Turbellaria: taxa on different phylogenetic levels holding today the same absolute rank (orders, cf. introduction). The "cladistic" principles express without doubt the processes of phylogenesis in a clear and logical way, but their strict application leads to a breakdown of the classification of today. My intention is only to point out the controversy of the different systematic points of views in regard to the Turbellaria; I have no solution of the controversy at hand.

Focusing on the class Turbellaria only, we find that the search for sister groups throws a sharp light on our insufficient knowledge of the phylogenetic connections between the turbellarian taxa. In the future increasing information from EM studies, biochemistry, sperm morphology, etc., will elucidate many of these obscure points.

SUMMARY

The structure and phylogeny of the turbellarian orders are discussed on the basis of a hierarchic diagram according to the sister-group schema of Hennig (1966).

A theoretical archetype is reconstructed at the root of the turbellarian stem with the following anatomical plesiomorphies (primitive characters) of all the Turbellaria: one-layered, cellular, entirely ciliated epidermis with rhabdites and intracellular nuclei; basement membrane lacking or thin; two-layered subepidermal musculature; epidermal-subepidermal nervous system; eyes lacking or epidermal; weakly differentiated statocyst with a variable number of statoliths; simple mouth pore; sac-shaped intestine with epithelial ciliated wall, no anal pore; no excretory organs (alternatively some kind of primitive protonephridia); homocellular female gonads, entolecithal eggs, no female pore; male genital pore, internal fertilization.

The turbellarian orders here considered are: Nemertodermatida, Acoela, Catenulida, Macrostomida *mihi*, Haplopharyngida *mihi*, Lecithoepitheliata (with the suborders Prorhynchida *mihi* and Gnosonesimida *mihi*), Rhabdocoela (with the Temnocephalida), Prolecithophora, Proseriata, Tricladida, Polycladida.

The sister-group principle demanding fixed connections of all the branches in the hierarchic diagram necessitates the controversial phylogenetic solutions that the complex pharynx and the heterocellular female gonads have arisen only once (ring and dot respectively in the diagram).

Also several other solutions are more or less controversial: the Nemertodermatida stand closest to the archetype; the unpaired excretory system of the Catenulida has arisen from a diffuse system of emunctories, not from a paired system; the origin of the oviduct from a vagina and thus the close relationship Macrostomidae-Convolutidae are called in question; the establishment of the complex Prolecithophora-Proseriata-Tricladida as sister group to the Rhabdocoela.

The direct connection of the Lecithoepitheliata with the Acoela is not accepted, although some conditions speak in favor of it. The close affinity between the two lecithoepitheliate suborders here nominated, Prorhynchida and Gnosonesimida, is questionable.

Consequent realization of Hennig's principle that sister groups must have the same absolute systematic rank is impossible without a thorough revision of the whole turbellarian system. The main obstacle for such a revision is our deficient knowledge of the series of transformation, i.e., of the true homologies in the Turbellaria.

BIBLIOGRAPHY

An der Lan, H. 1936. Acoela 1. Ergebniss einer . . . Reise in Grönland 1926. 7. *Vid. Medd. Dansk Naturh. Foren.*, **99**:289–330.

Antonius, A. 1968. Faunistische Studien am Roten Meer im Winter 1961/62. IV. Neue Convolutidae und eine Bearbeitung des Verwandtschaftskreises *Convoluta* (Turbellaria Acoela). *Zool. Jahrb. (Syst.)*, **95**:297–394.

Ax, P. 1952. *Ciliopharyngiella intermedia* nov. gen. nov. spec., Repräsentant einer neuen Turbellarien-Familie des marinen Mesopsammon. *Zool. Jahrb. (Syst.)*, **81**:286–312.

———. 1956. Monographie der Otoplanidae (Turbellaria). *Akad. Wiss. Lit. Mainz Abhandl. Math. Nat. Kl.*, 1955, **13**:499–796.

———. 1961. Verwandtschaftsbeziehungen und Phylogenie der Turbellarien. *Ergeb. Biol.*, **24**:1–68.

———. 1963. "Relationships and Phylogeny of the Turbellaria." In E. C. Dougherty (ed.), *The Lower Metazoa*, pp. 191–224. Berkeley and Los Angeles: University of California Press.

———— and J. Dörjes. 1966. *Oligochoerus limnophilus* nov. spec., ein kaspisches Faunenelement als erster Süsswasservertreter der Turbellaria Acoela in Flüssen Mitteleuropas. *Intern. Rev. Ges. Hydrobiol.*, **51**(1):15–44.

Borkott, H. 1970. Geschlechtliche Organisation, Fortpflanzungsverhalten und Ursachen der sexuellen Vermehrung von *Stenostomum sthenum* nov. spec. (Turbellaria, Catenulida). *Z. Morphol. Tiere*, **67**:183–262.

Brundin, L. 1967. "Application of Phylogenetic Principles in Systematics and Evolutionary Theory." *In* T. Ørvig (ed.), *Current Problems of Lower Vertebrate Phylogeny*, pp. 473–495. New York: Interscience.

Dörjes, J. 1968. Die Acoela (Turbellaria) der deutschen Nordseeküste und ein neues System der Ordnung. *Z. Zool. Syst. Evolut.-forsch.*, **6**:56–452.

Giesa, S. 1966. Die Embryonalentwicklung von *Monocelis fusca* Oersted (Turbellaria, Proseriata). *Z. Morphol. Ökol.*, **57**:137–230.

Hennig, W. 1966. *Phylogenetic Systematics*. Urbana: University of Illinois Press. 263 pp.

Hyman, L. H. 1951. *The Invertebrates*. II. *Platyhelminthes and Rhynchocoela. The Acoelomate Bilateria*. New York: McGraw-Hill. 550 pp.

————. 1959. *The Invertebrates*. V. *Smaller Coelomate Groups*. New York: McGraw-Hill. 783 pp.

Karling, T. G. 1940. Zur Morphologie und Systematik der Alloeocoela Cumulata und Rhabdocoela Lecithophora (Turbellaria). *Acta Zool. Fenn.*, **26**:1–260.

————. 1965. *Haplopharynx rostratus* Meixner (Turbellaria) mit den Nemertinen verglichen. *Z. Zool. Syst. Evolut.-forsch.*, **3**(1–2):1–18.

————. 1966a. On the Defecation Apparatus in the Genus *Archimonocelis* (Turbellaria, Monocelididae). *Sarsia*, **24**:37–44.

————. 1966b. Marine Turbellaria from the Pacific Coast of North America. IV. Coelogynoporidae and Monocelididae. *Arkiv. Zool.*, **18**(11):493–528.

————. 1967. Zur Frage von dem systematischen Wert der Kategorien Archoophora und Neoophora (Turbellaria). *Comment. Biol. Soc. Sci. Fenn.*, **30**(3):1–11.

————. 1968. On the Genus *Gnosonesima* Reisinger (Turbellaria). *Sarsia*, **33**:81–108.

Luther, A. 1912. Studien über acöle Turbellarien aus dem finnischen Meerbusen. *Acta Soc. F. Fl. Fenn.*, **36**:1–60.

Marcus, E. 1945. Sôbre Catenulida Brasileiros. *Bol. Fac. Filosof. Ciências Letras Univ. São Paulo, Zool.*, **10**:3–133.

————. 1950. Turbellaria Brasileiros (8). *Bol. Fac. Filosof. Ciências Letras Univ. São Paulo, Zool.*, **15**:5–192.

Mayr, E. 1969. *Principles of Systematic Zoology*. New York: McGraw-Hill. 428 pp.

Meixner, J. 1926. Beitrag zur Morphologie und zum System der Turbellaria-Rhabdocoela. II. Über *Typhlorhynchus nanus* Laidlaw und die parasitischen Rhabdocoelen nebst Nachträgen zu den Calyptorhynchia. *Z. Morphol. Ökol.*, **5**:577–624.

Papi, F. 1950. Sulle affinità morfologiche nella fam. Macrostomidae (Turbellaria). *Boll. Zool.*, **17**:461–468.

Reisinger, E. 1923. *Protomonotresis centrophora* n. gen. n. sp. eine neue Süsswasser-Alloeocoele aus Steiermark. *Zool. Anz.*, **58**:1–12.

————. 1924. Zur Anatomie von *Hypotrichina* (=*Genostoma*) *tergestina* Cal. nebst einem Beitrag zur Systematik der Alloeocoelen. *Zool. Anz.*, **60**:137–149.

————. 1926. Zur Turbellarienfauna der Antarktis. *Deut. Südpolar-Exp.* 1901–1903. 18 Zool.: 415–462.

————. 1961. Allgemeine Morphologie der Metazoen. Morphologie der Coelenteraten, acoelomaten und pseudocoelomaten Würmer. *Fortschr. Zool.*, **13**:1–82.

————. 1968. *Xenoprorhynchus* ein Modellfall für progressiven Funktionswechsel. *Z. Zool. Syst. Evolut.-forsch.*, **6**:1–55.

————. 1969. Ultrastrukturforschung und Evolution. *Phys.-Med. Ges. Würzburg*, **77**:1–43.

————. 1970. Zur Problematik der Evolution der Coelomaten. *Z. Zool. Syst. Evolut.-forsch.*, **8**(2):81–109.

Riedl, R. 1960, Über einige nordatlantische und mediterrane *Nemertoderma*-Funde. *Zool. Anz.*, **165**:232–248.

Rixen, J.-Uve. 1961. Kleinturbellarien aus dem Litoral der Binnengewässer Schleswig-Holsteins. *Arch. Hydrobiol.*, **57**:464–538.

Steinböck, O. 1923. Eine neue Gruppe allöocöler Turbellarien: Alloeocoela typhlocoela (Familie Prorhynchidae). *Zool. Anz.*, **58**:233–242.

————. 1925. Zur Systematik der Turbellaria Metamerata, zugleich ein Beitrag zur Morphologie des Tricladen-Nervensystems. *Zool. Anz.*, **64**:165–192.

————. 1966. Die Hofsteniiden (Turbellaria Acoela). *Z. Zool. Syst. Evolut.-forsch.*, **4**(1–2):58–195.

Sterrer, Wolfgang. 1966. New Polylithophorous Marine Turbellaria. *Nature*, **210** (5034):436.

Westblad, E. 1937. Die Turbellarien-Gattung *Nemertoderma* Steinböck. *Acta Soc. F. Fl. Fenn.*, **60**:45–89.

————. 1940. Studien über Skandinavische Turbellaria Acoela 1. *Arkiv Zool.*, **32A**(20): 1–28.

————. 1948. Studien über Skandinavische Turbellaria Acoela 5. *Arkiv Zool.*, **41A**(7): 1–82.

————. 1949. On *Meara stichopi* (Bock) Westblad, a New Representative of Turbellaria Archoophora. *Arkiv Zool.*, **1**(5):43–57.

————. 1953. Marine Macrostomida (Turbellaria) from Scandinavia and England. *Arkiv Zool.*, **4**(23):391–408.

History of the Study of Turbellaria in North America

Roman Kenk
Department of Invertebrate Zoology, National
Museum of Natural History, Smithsonian Institution,
Washington, D.C.

The first turbellarians of North America were described and named by Samuel S. Haldeman, a naturalist of manifold interests in geology, chemistry, entomology, conchology, archaeology, and even philology. These species were a freshwater triclad, *Planaria gracilis*, described in 1840, and an alloeocoel, *Hydrolimax grisea*, published in 1843, both observed in eastern Pennsylvania. These findings were followed by studies of two of Haldeman's colleagues and friends, both associated with the important center of early faunistic study, the Academy of Natural Sciences of Philadelphia: Joseph Leidy, a very diligent collector and student of invertebrates who in his lifetime published over 800 papers, and Charles Girard. In the decade from 1847 to 1857 these two zoologists described a number of turbellarian species, representatives of all major systematic groups, Microturbellaria (i.e., Acoela and the orders formerly designated as Rhabdocoela and Alloeocoela), freshwater, marine, and terrestrial triclads, and polyclads. Their descriptions were based chiefly on external characters as was the general practice of the time.

After a lapse of about 20 years, Addison Emory Verrill, connected with Harvard College, began an extensive investigation of marine invertebrates, particularly turbellarians, along the Atlantic coast under the auspices of the United States Commission of Fish and Fisheries. He published a series of papers describing polyclads and microturbellarians between 1874 and 1902. In 1885 Wyllis A. Silliman reported on freshwater forms of Monroe County, New York, both microturbellarians and triclads, among them several new species.

The first modern study of American Turbellaria, applying the sectioning technique, was carried out by the German zoologist J. Kennel, who in 1888 published a paper on the Turbellaria of Trinidad in the West Indies. Soon after this appeared an important study by William McMichael Woodworth of Harvard, describing the anatomy of what he considered to be *Phagocata gracilis* (actually *P. woodworthi*) from Massachusetts. Within the next years Woodworth extended his work to other triclads, polyclads, and some microturbellarians. A description of the acoel *Polychoerus caudatus* from New England was published in 1892 by Edward L. Mark. In the same year, Harvey N. Ott of the University of Michigan reported on the anatomy of *Stenostoma leucops*. In 1893 appeared the first revision of the Turbellaria of North America, by Girard.

From here on we may trace the investigations of North American turbellarians by the major taxonomic groups. A neorhabdocoel, *Graffilla gemellipara*, from the Woods Hole region, Massachusetts, was described by Edwin Linton in 1910. An important milestone in the exploration of the Microturbellaria, both freshwater and marine, was a paper by the eminent Austrian turbellarian specialist Ludwig von Graff (1911), who had spent 3 months in New England and New York State and presented a critical revision of the species then known from the United States as well as descriptions of 27 new species. The freshwater species were again reviewed by Caroline E. Stringer (1918) in Ward and Whipple's *Fresh-Water Biology*. Additional information on the morphology and ecology of the Turbellaria of the Mississippi basin was published by Ruth Higley.

The next contributions to our knowledge of the Microturbellaria originated from the University of Virginia, where William Allison Kepner embarked on an intensive investigation principally of the local fauna. He himself published over 30 papers on Turbellaria. Kepner was a dynamic teacher whose enthusiasm and sincerity influenced the lives and careers of many of his students. Some of the zoologists who received their training under Kepner are today still active in turbellarian studies. I mention here only a few of Kepner's students who have published in this field: E. Ruffin Jones, author of the chapters on Microturbellaria in the second edition of Ward and Whipple's *Fresh-Water Biology*; John William Nuttycombe; Chaucey McLean Gilbert; Margaret A. Stirewalt; Trenton K. Ruebush (prepared a key to the North American microturbellarian genera); and Frederick F. Ferguson (published a revision of the Macrostomida).

Libbie H. Hyman began her studies of the Microturbellaria in the 1930s and in the course of 30 years published a series of papers on both marine and freshwater species, but never attempted a revision of these still very incompletely known groups. Additional information on Microturbellaria was furnished by Helen M. Costello and Donald P. Costello (*Polychoerus carmelensis* from the coast of California), Wayland J. Hayes (from Wisconsin), Eugene N. Kozloff (marine forms of the Pacific Coast), and Julian T. Darlington (from Georgia). In recent years our knowledge of the marine species of the Pacific was greatly enriched by the studies of Tor G. Karling of the Stockholm Museum and Peter Ax of the University of Göttingen, who did their own collecting on the coast of California.

Investigations of the order Tricladida proceeded at a rather slow pace. Concerning the freshwater planarians or Paludicola, I may mention here only a few contributions which were made after Woodworth's classic study: Winterton C. Curtis on *Planaria simplissima*, Nettie M. Stevens on *Planaria morgani*, Caroline E. Stringer on *Planaria velata*, Leon D. Peaslee on the anatomy of *Phagocata gracilis*, Charles Manning Child on the life history of *Planaria velata*, and William A. Castle on *P. velata* and *Hymanella*. Libbie H. Hyman in 1931 presented a critical review of the paludicolous triclads, revising their systematic arrangement according to modern views. During the next 30 years she added to our knowledge of this group by describing or redescribing individual species, reporting on their geographic distribution, and providing keys and revisions from time to time (1951, 1953a, 1955b, 1959). Some of the new species described by Hyman were sent to her by various collectors and were not all in the best state of preservation, which in the study of this group is of utmost importance. Hence, some of these species will need further anatomical analysis.

During this time appeared contributions by other authors: James William Buchanan on *Sphalloplana percoeca*; a paper of mine on the triclads of Virginia, followed by similar reports on Michigan and Alaska; Julian T. Darlington on planarians from Georgia; Paul S. Stokely on Jefferson County, Ohio; and recently contributions by Robert W. Mitchell on planarians from Texas caves, the Japanese zoologist Masaharu Kawakatsu on species from

Lake Tahoe, California and Nevada, Ian R. Ball on planarians from eastern Canada, Charlotte Holmquist of the Stockholm Museum on species from Alaska and northern Canada, and Jerry H. Carpenter on a planarian from Utah.

The marine triclads or Maricola of the Atlantic coast were studied primarily by Julius F. Wilhelmi (1909), a German zoologist who worked for a limited time on the coast of New England. Hyman (1954) described the first Pacific coast representative of the Maricola, *Procerodes pacifica*, from California. Additional species were studied by Diva Diniz Corréa in Florida and John T. Holleman in California and Oregon.

Although the first North American terricolous triclad had been reported by Leidy in 1851, there was no significant study of this group until Hyman (1943, 1954) reviewed all pertinent scattered data. More recently Robert E. Ogren published a paper on the sexual reproduction of *Rhynchodemus sylvaticus*.

Concerning the last order of Turbellaria, the polyclads, after Girard's and Verrill's investigations, individual species of the Atlantic coast were studied by William Morton Wheeler, by the Italian zoologist Arturo Palombi, and more intensively by Arthur S. Pearse of Duke University. The first critical review of the accumulated findings was done by Hyman, who published several papers on the polyclads of the Atlantic coast (1904), the Gulf of Mexico, and the West Indies (1955c). Polyclads of the Pacific were described by Marianne Plehn, Harold Heath, Eleanor S. Boone, and Daniel Freeman. Again it was Hyman (1953b, 1955a) who consolidated our knowledge of this fauna and added about 49 new species to the faunal list.

To sum up the present situation, we may say that today we have a fairly good, though by no means complete, picture of the North American triclads and polyclads, while there are still many gaps in our knowledge of the microturbellarians. Much of the credit for the orderly arrangement of what we know is due to the perseverance and encyclopedic knowledge of Libbie H. Hyman, who herself described over 100 species of North American Turbellaria. The geographic coverage of the faunal investigations is somewhat uneven, with rather scant studies in the western half of the continent. Certain ecological habitats have hardly been touched, such as the subterranean interstitial fauna or psammon on which we have only fragmentary information by Louise F. Bush on the beaches of Miami, Florida, and by Peter Ax on the sand fauna of the Pacific shores.

So far we have discussed only studies of the morphology, ecology, geographic distribution, and taxonomy of the North American turbellarians. At about the turn of this century, during the great upswing of physiological researches, Turbellaria were also widely used in investigations of general biological principles. It is impossible, within the frame of this report, to enumerate even the more important contributions in these fields. Among them are the classic studies of animal regeneration by Thomas Hunt Morgan and many other workers, the exploration of physiological gradients by Charles Manning Child and his school, the studies of various taxes (formerly called tropisms) by Raymond Pearl, numerous investigations on the biochemistry and metabolic pathways in turbellarians, and many other inquiries into general and specific biological phenomena. May I mention only a few of the workers who at present are using turbellarians as experimental objects: Krystyna D. Ansevin of Columbia University (regeneration and tissue culture); Jay Boyd Best, Colorado State University (biochemical and behavioral studies, ultrastructure); Frank A. Brown, Jr., Northwestern University (behavior in magnetic fields, biological rhythms); Stuart J. Coward, University of Georgia (regeneration, biochemistry); Reed Adams Flickinger, State University of New York (various aspects of morphogenesis); Allan L. Jacobson, University of California, Los Angeles (behavioral studies, particularly classical conditioning); Marie M. Jenkins, Madison College, Harrisonburg, Virginia (physiology, life history, and reproduction of *Dugesia*); Thomas L. Lentz, Yale University (ultrastructure); James V. McConnell, psychologist at the University of Michigan (biochemical correlates of learning and memory); Edith Krugelis MacRae, University of Illinois College of Medicine (biochemistry and ultrastructure); Gordon Marsh, University of Iowa (various aspects of regeneration); James L. Oschman, Western Reserve (fine structure of *Convoluta*); C. Vowinckel, McGill University, Montreal (reproductive physiology); and John H. Welsh, Harvard University (biochemistry). Thus we may confidently look forward to further progress in our knowledge of the North American Turbellaria.

BIBLIOGRAPHY

Only papers containing faunal reviews are listed here. They cite references to most publications mentioned in the preceding article.

Ball, I. R. 1969. An Annotated Checklist of the Freshwater Tricladida of the Nearctic and Neotropical Regions. *Can. J. Zool.*, **47**:59–64.

Ferguson, F. F. 1954. Monograph of the Macrostomine Worms of Turbellaria. *Trans. Am. Microscop. Soc.*, **73**:137–164.

Girard, C. 1893. Recherches sur les Planariés et les Némertiens de l'Amérique du Nord. *Ann. Sci. nat. Zool.*, (7)**15**:145–310.

Graff, L. von. 1911. Acoela, Rhabdocoela und Alloeocoela des Ostens der Vereinigten Staaten von Amerika. *Z. Wiss. Zool.*, **99**:1–108, pls. 1–6.

Hyman, L. H. 1940. The Polyclad Flatworms of the Atlantic Coast of the United States and Canada. *Proc. U.S. Natl. Museum*, **89**:449–495.

———. 1943. Endemic and Exotic Land Planarians in the United States with a Discussion of Necessary Changes of Names in the Rhynchodemidae. *Am. Museum Novitates*, no. 1241, 21 pp.

——. 1951. North American Triclad Turbellaria. XII. Synopsis of the Known Species of Fresh-water Planarians of North America. *Trans. Am. Microscop. Soc.*, **70:**154–167.

——. 1953a. "Turbellaria (Flatworms)." In Robert W. Pennak (ed.), *Freshwater Invertebrates of the United States*, pp. 114–141. New York: Ronald Press.

——. 1953b. The Polyclad Flatworms of the Pacific Coast of North America. *Bull. Am. Museum Nat. Hist.*, **100:**267–392.

——. 1954a. Some Land Planarians of the United States and Europe, with Remarks on Nomenclature. *Am. Museum Novitates*, no. 1667, 21 pp.

——. 1954b. A New Marine Triclad from the Coast of California. *Am. Museum Novitates*, no. 1679, 5 pp.

——. 1955a. The Polyclad Flatworms of the Pacific Coast of North America: Additions and Corrections. *Am. Museum Novitates,* no. 1704, 11 pp.

——. 1955b. Descriptions and Records of Freshwater Turbellaria from the United States. *Am. Museum Novitates*, no. 1714, 36 pp.

——. 1955c. Some Polyclad Flatworms from the West Indies and Florida. *Proc. U.S. Natl. Museum*, **104:**115–150.

——. 1959. "Order Tricladida." In H. B. Ward and G. C. Whipple (ed. by W. T. Edmondson), *Fresh-water Biology*, 2d ed., pp. 326–334. New York: John Wiley & Sons.

Jones, E. R. 1959. "Catenulida, Macrostomida, Neorhabdocoela, Alloeocoela." In H. B. Ward and G. C. Whipple (ed. by W. T. Edmondson), *Fresh-water Biology*, 2d ed., pp. 334–365. New York: John Wiley & Sons.

Nuttycombe, J. W., and A. J. Waters. 1938. The American Species of the Genus *Stenostomum. Proc. Am. Phil. Soc.*, **79:**213–301.

Ruebush, T. K. 1941. A Key to the American Freshwater Turbellarian Genera, Exclusive of the Tricladida. *Trans. Am. Microscop. Soc.*, **60:**29–40.

Stringer, C. E. 1918. "The Free-living Flatworms (Turbellaria)." In H. B. Ward and G. C. Whipple (ed. by W. T. Edmondson), *Fresh-Water Biology*, pp. 323–364. New York: John Wiley & Sons.

Wilhelmi, J. 1909. "Tricladen." *Fauna und Flora des Golfes von Neapel*. Monog. 32, xii + 405 pp., 16 pls.

Chapter 3

A New Group of Turbellaria-Typhloplanoida with a Proboscis and Its Relationship to Kalyptorhynchia

Reinhard M. Rieger
University of North Carolina, Department of Zoology,
Chapel Hill, and Institute of Marine Sciences,
Morehead City, North Carolina

During faunistic investigations on marine interstitial Turbellaria since 1969 on the North Carolina coast, I found several new species of Turbellaria-Typhloplanoida which caught my attention because of their having a small terminal proboscis. Similar, or perhaps homologous, proboscis structures might be those of some members of the freshwater group Typhloplanidae-Protoplanellinae.

In 1924 Meixner briefly mentioned a new type of proboscis in Typhloplanoida. He defined a new genus—*Haplorhynchella*—and named the species, according to its type location in a swampy meadow near Graz,

Austria, *paludicola*. Unfortunately no drawings of this species exist. The type material was damaged in World War II, and, as I have found out from Professor Reisinger, there is now a nice house on the type location.

About 10 years later a new monotypic genus *Microcalyptorhynchus* (Kepner and Ruebush, 1935a; see also Kepner and Ruebush, 1935b), was placed in the kalyptorhynchid family Koinocystidae. However, Reisinger (1937) pointed out that this is not a kalyptorhynchid, but rather is closely related to *Haplorhynchella* Meixner, 1924.

Finally, Ruebush (1939) described the monotypic protoplanellid genus *Prorhynchella minuta* from several freshwater ponds of Connecticut, which is characterized by a small proboscis at the anterior end of the body and was considered by the author also to have a systematic position near *Haplorhynchella*.

Unfortunately the fine structure of the proboscis, pharynx, and genital organs is insufficiently known, in *Microcalyptorhynchus* as well as in *Prorhynchella*, thus making a comparison with the newly discovered marine forms very difficult.

The eight new species described in this paper allow a preliminary classification and provide new facts for the phylogenetic discussion of the evolution of the eukalyptorhynchid "conorhynch."

I would like to thank Professor T. G. Karling for advising me on early literature and for very valuable discussion about phylogenetic problems in Turbellaria. He also gave me unpublished material from new species in this group, which we hope to describe soon in a joint paper. I am also grateful to my wife, Gunde R. Rieger, who prepared all the serial sections and carefully read the manuscript.

Financial support was provided by the National Science Foundation grant XA120−09 (R. J. Riedl, principal investigator), by the North Carolina Board of Science and Technology (RA012−09), by the Hochschuljubilaeumsstiftung der Stadt Wien, Austria, and by the National Science Foundation grant for *R/V Eastward*, Duke University, at the Duke University Marine Laboratory, Beaufort, North Carolina.

METHODS, MATERIAL, AND DISTRIBUTION

To separate the animals from the sand, an approved magnesium chloride method was used (Sterrer, 1968). For microanatomical studies the animals were fixed in Bouin. The serial sections were obtained by a combined paraffin-celloidin embedding method, described by Antonius (1965). For staining Azocarmin-Pasini, after Kohashi, was used.

About 50 specimens have been studied live or in serial sections:

Species	Number studied live	Number studied in serial sections
Genus *Kytorhynchella* gen. n.:		
K. *meixneri* sp. n.	25	10
K. *karlingi* sp. n.	3	
K. *riedli* sp. n.	1	1
Genus *Kytorhynchus* (*Kytorhynchus*) gen. n. subgen. n.:		
K. *oculatus* sp. n.	3	2
K. sp. I.	2	1
K. sp. II	1	1
Genus *Kytorhynchus* (*Kytorhynchoides*) gen. n. subgen. n.:		
K. *microstylus* sp. n.	11	6
K. *macrostylus* sp. n.	1	1

Current investigations of deeper sandy bottoms (20 to 200 m in depth) show that we still have to expect further new forms in this group. This is also stressed by the discovery of new species in other parts of the world.

Unfortunately no material of the freshwater forms (*Haplorhynchella, Microcalyptorhynchus, Prorhynchella*) was available.

For the description of body proportions and positions of organs, a relative scale will be used: 100 units (U) being equivalent to the total length of a specimen. Positions or proportions will then be defined starting at the anterior tip of the animal (Rieger and Sterrer, 1968).

So far we can already state that the new group has a worldwide distribution. Marine forms were found not only on the Atlantic coast of North Carolina, but also in Bermuda (*Kytorhynchus microstylus*), Pacific Grove, California (*Kytorhynchus* sp.), and the Skagerrak Sea (another new species of the genus *Kytorhynchus*). For the last two records I must thank Professor T. G. Karling, Stockholm. Furthermore, after this paper was written, I discovered several new species of the genus *Kytorhynchus* on the Florida Keys.

Two of the eight species from North Carolina (*Kytorhynchella meixneri* and *Kytorhynchella karlingi*) occur in the eulittoral fine sand flats between low- and mid-tide level, and are most abundant at a depth of 0.5 to 2 cm, which in this case is slightly above the redox discontinuity layer. Both species were also found in a somewhat different form (see page 55) in subtidal sands offshore (10 to 60 m). The other species were found in subtidal sediments off Beaufort, North Carolina. Some specimens of *Kytorhynchus microstylus* were obtained from upper subtidal sediments (1 to 2 m) in Bermuda.

In the shallow locations, it is obvious that these forms are closely related in their distribution to lenitic situations with sand very rich in organic matter. This occurrence in proximity to strongly reducing sediments could, according to the recent investigations of Fenchel and Riedl (1970), also be assumed for the species found in deeper subtidal porous sediments.

GENERAL ANATOMY

Habitus, Body Wall, Protonephridia, and Nervous System

Habitus: All species studied are more or less cylindrical in shape (Figs. 1*b*–*g*, 7*e*, *f*). In the petri dish they swim upwards into the water rather than crawl or creep between the sand grains. A very characteristic shape is shown in Figs. 1*e* and 7*e*; it is very reminiscent of that of many little eukalyptorhynchids.

The animals are generally fairly small, about 0.6 mm, only *Kytorhynchus macrostylus* being larger (1.2 mm long).

Colors were not observed, although, in the genus *Kytorhynchus*, the numerous small granules in the epidermis give the animals a slight brownish appearance.

The epidermis (ep) (see list of Abbreviations at end of chapter), the well-developed basement membrane (bm), and subepidermal muscle sheath with circular (cm), diagonal (dm), and longitudinal (lm) fibers show normal features (Fig. 13*a*, *c*). The nuclei are more or less lobed and scarce. Only in squeeze preparations were polygonal cell borders in the epidermis of *Kytorhynchella karlingi* seen. Besides the fine granules in the epidermis, and the dermal rhabdites, I found adenal rhabdites in the caudal part of the body in the genus *Kytorhynchella* as well (Fig. 1*a*: rh). The dermal rhabdites are either single or in clusters of two to three (Fig. 12).

For the excretory system I can only state that *Kytorhynchella meixneri* has two lateral stems of protonephridia. Their openings remain unknown.

The brain (br), situated medioventrally behind the proboscis, consists of a central fibrous cord surrounded by the nerve cell bodies (Figs. 1, 3, 5, 8, 10, 11). The latter are concentrated on the sides of the brain extending into anterior and posterior gangliarlobes (Fig. 9*a*), which cause a pillow shape of the brain in squeeze preparation. The central mass of nerve fibers is crescent-shaped and extends ventrocaudally into the main body nerves. Numerous nerves originating from the anterior part of the brain lead to the proboscis.

The eyes (e) in the genus *Kytorhynchella* and in *Kytorhynchus oculatus* and *Kytorhynchus* sp. I are considerably anterior to the brain and are connected with the latter by thin optical nerves (Figs. 8, 10: on).

Proboscis

The proboscis (pr) is a terminal invagination of the epidermis, the basement membrane, and the subepidermal muscle sheath (Fig. 12). The layers of the

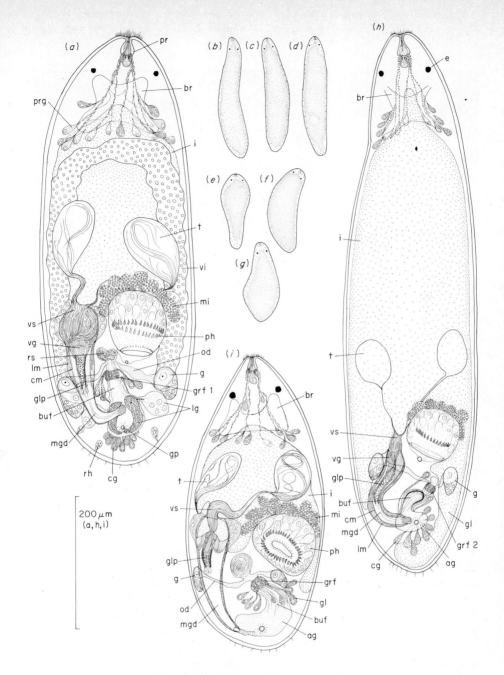

Figure 1 Schemes of the organization of the three species of *Kytorhynchella*, from living animals. (a) *Kytorhynchella meixneri*, dorsal view; (b–g) free-swimming animals of *Kytorynchella meixneri;* (h) *Kytorhynchella riedli*, dorsal view; (i) *Kytorhynchella karlingi*, ventral view.

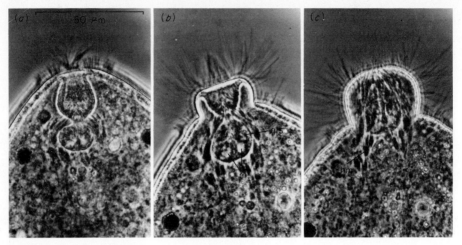

Figure 2 Different stages in the protruding of the proboscis of
Kytorhynchella meixneri (by squeezing).

subepidermal muscle sheath follow the invaginated epithelium without in-
version (inner circular muscle, outer longitudinal muscles; see Riedl, 1954).
Some of the longitudinal muscle fibers of the body wall are modified into retractor muscles (rm) of the organ. They insert at the proximal part of the
invagination and extend from there in a circle to the body wall (see also
Fig. 9d–f).

The shape of the proboscis is either that of an urn with a little sac extending caudally at its proximal end (*Kytorhynchella*, Fig. 1), or deltoid (*Kytorhynchus* s. str., Fig. 3), or triangular (*Kytorhynchus* [*Kytorhynchoides*]
Fig. 5). The little sac of the proboscis in *Kytorhynchella* is nothing other
than the proximal part of the invagination which is separated from the distal
part by a sphincter and obviously has a glandular function (Figs. 2, 4a, 12a).

In *Kytorhynchella* the epithelium of the proboscis bears long sensory
hairs in the whole urn-shaped distal part (Fig. 12a), whereas the proximal
glandular sac lacks any cilia or sensory hairs. In *Kytorhynchus oculatus* the
whole proboscis is provided with very long but scarce sensory hairs (Figs.
4b, 12b). In the other species, the sensory hairs are limited to the most distal
part of the proboscis near the opening (Figs. 3b, c, 5a, b).

The nuclei of the proboscis epithelium are either sunken in the glandular
sac (*Kytorhynchella*) or lie in a circle in the middle (*Kytorhynchus oculatus*)
or basal region (*Kytorhynchus* sp. II and *microstylus*) of the proboscis (Figs.
9b–d, 12). They are mostly four in number. Some specimens of *Kytorhynchella meixneri* had only three nuclei, whereas one specimen of *Kytorhynchus
oculatus* and two of *Kytorhynchus microstylus* showed six nuclei. In *Kyto-*

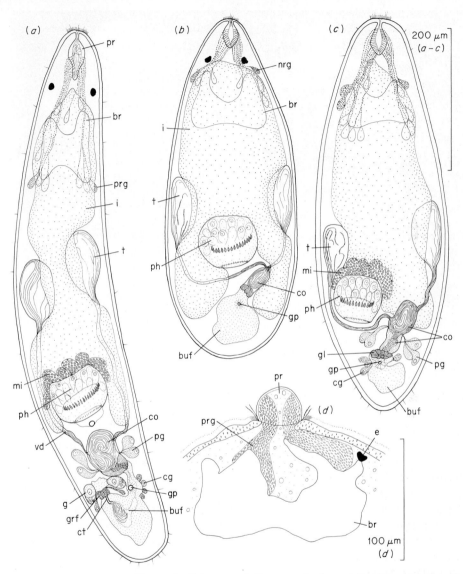

Figure 3 Schemes of the organization of the three species of *Kytorhynchus* (s. str.) dorsal view from living animals. (*a*) *Kytorhynchus* (s. str.) *oculatus;* (*b*) *Kytorhynchus* sp. I; (*c*) *Kytorhynchus* sp. II; (*d*) protruded (by squeezing) proboscis of *Kytorhynchus* sp. I.

rhynchus macrostylus no nuclei at all could be found; the cause for this might be the high density of secretion granules in the whole proboscis epithelium (pre) of this particular species (Fig. 9*e*).

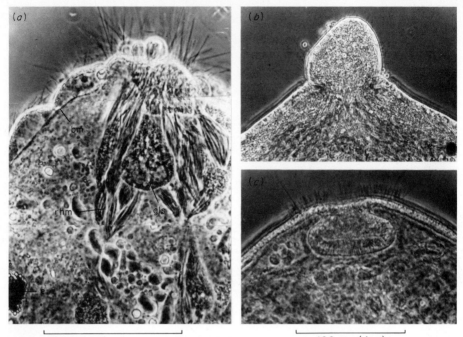

Figure 4 (a) Proboscis of *Kytorhynchella karlingi* (littoral form, see page 55) in the process of being squeezed out; (b) protruded proboscis of *Kytorhynchus* (s. str.) *oculatus*; (c) proboscis of *Kytorhynchus* (*Kytorhynchoides*) *microstylus*.

Proceeding from *Kytrorhynchella* to *Kytorhynchus macrostylus*, one finds a strong increase in the number of retractor muscles. Parallel to this is an increase in the glandular function of the organ and a reduction of the sensory hairs (Fig. 12). The long, slender secretion granules of the proboscis glands of *Kytorhynchella* very much resemble true rhammites (rhm); in the genus *Kytorhynchus* they are round to oval in shape (Fig. 4). Owing to the action of the circular muscles in the proximal part of the proboscis, one can observe in living specimens of *Kytorhynchus microstylus* that the basal part of the invagination can be folded into a small cone thus enclosed by its lateral walls (Fig. 5c, d).

I never saw the animals protrude this organ completely. The existence of strong retractor muscles and the fact that one can easily protrude the proboscis by squeezing (Figs. 2, 4b) strongly suggest that this can be done. The great number of proboscis glands also leads to the assumption that this organ is used not only as a sensory pit—which might have been the primary function—but also as a glandular muscular proboscis for prey capture.

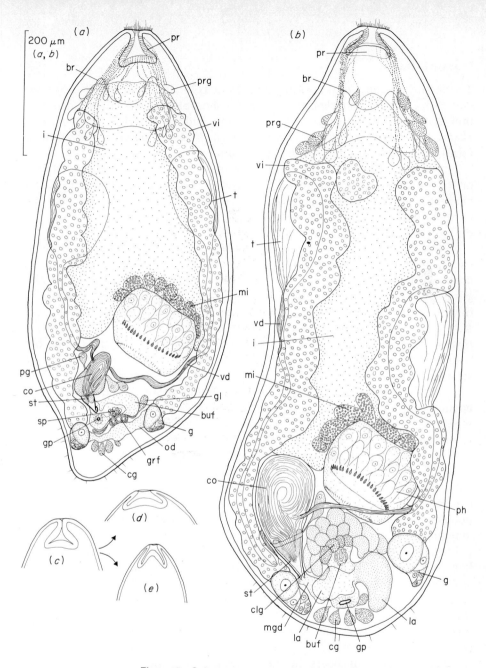

Figure 5 Schemes of the organization of the two species of *Kyto-rhynchus* (*Kytorhynchoides*), dorsal view from living animals. (a) *Kytorhynchus* (*Kytorhynchoides*) *microstylus*; (b) *Kytorhynchus* (*Kytorhynchoides*) *macrostylus*; (c–e) movements of the proboscis observed in *Kytorhynchus* (*Kytorhynchoides*) *microstylus*.

Digestive Tract

The mouth opening (mo) is situated at about U75, and is surrounded by several strong circular muscles. It leads into a short pharyngeal cavity, lined by a thin epithelium and surrounded by fine longitudinal muscles. In *Kytorhynchella riedli* and in one specimen of *Kytorhynchus microstyla* (Bermuda form), I found some nuclei in the epithelium.

Pharynx: A well-developed ciliation occurs on the distal protruded part of the pharynx in *Kytorhynchella riedli* (Fig. 14a). In the other forms, the pharynx epithelium in this position is covered with a very thin striated layer of small granules which might resemble certain stages of reduction of cilia or, more probably, a layer of microvilli (Fig. 14b–f).

In all species studied the pharynx lumen is divided by a characteristic circular groove in about mid-pharynx. The inner pharynx epithelium is, except for the circular groove, a homogeneous, intensely stained layer. In the groove itself it is higher (Fig. 14), containing all nuclei of the epithelium (*Kytorhynchella*, six to eight nuclei; *Kytorhynchus microstylus*, two to six; *K. macrostylus*, four). The specimens of *Kytorhynchus oculatus* and *Kytorhynchus* sp. II did not show any nuclei in the inner epithelium. Some big granules in the epithelium of the circular groove in these species suggest strongly that one has to consider such structures to be degenerated nuclei.

No extrapharyngeal glands have been found in any of the species. The openings of the intrapharyngeal glands form two very distinct and relatively separated circles (Fig. 14). The proximal one contains more gland openings. In one section of *Kytorhynchus oculatus* some of the glands appeared to open also into the proximal part of the pharynx.

The muscle system of the pharynx is characterized by relatively thin but numerous fibers in all layers. Most interestingly *Kytorhynchella meixneri* and *K. riedli* clearly show, in the proximal part of the pharynx, three layers of inner muscles (Figs. 13b, 14a, b): underneath the inner epithelium first longitudinal fibers (ilm 1), then circular ones (icm), and finally a second layer of longitudinal muscles (ilm 2). The longitudinal fibers next to the inner epithelium join, at the edge of the circular groove, the longitudinal fibers outside the circular ones, and then branch into the distal part of the pharynx. There the inner circular muscles are next to the epithelium, as is the case the length of the pharynx in *Kytorhynchus* (*Kytorhynchoides*) (Figs. 14d, f, 11b). In *Kytorhynchus oculatus* and *Kytorhynchus* sp. II the arrangement of the inner pharynx muscles seems to be the same as in *Kytorhynchella* (Fig. 14c, e).

The number of fibers in the innermost longitudinal muscle layer in *Kytorhynchella meixneri* and *K. riedli* ranges between 60 and 70; in the second layer of inner longitudinal muscles, between 40 and 60. (The last numbers refer only to *Kytorhynchella meixneri*.) For the remaining forms investigated in this respect, one can state with certainty that the number of inner longitudinal fibers considerably exceeds 30.

Proceeding from the pharynx of *Kytorhynchella* to that of the forms in *Kytorhynchus* (*Kytorhynchoides*), one sees a trend to concentrate and to singularize the fibers of the radial and inner pharynx muscle layers (Fig. 14).

Around the distal part of the pharynx, slightly above the proximal edge of the pharyngeal cavity, many muscle fibers insert, connecting the pharynx with several parts of the body wall. Actually these muscles (em) break through the pharynx membrane and branch, thus crossing the inner longitudinal fibers, into the distal part of the pharynx (Fig. 14).

The esophagus (es) is distinct, surrounded by circular and longitudinal muscles. Directly in front of it one finds a circle of large granular clubs (Figs. 8, 10, 11: mi).

The gut wall (i) is more or less syncytial and surrounds a central cavity. In one of the sectioned specimens of *Kytorhynchella meixneri* I found the cuticle of a digested nematode, which suggests that the animals are carnivorous.

Genital Organs

The genital pore (gp) is situated at about U90 and is surrounded by large cement glands. It opens into a common genital atrium (ag), lined by an epithelium and surrounded by inner circular and outer longitudinal muscles (Figs. 8, 10, 11). In *Kytorhynchus macrostylus* the circular muscles, or here rather spiral muscles, form a very strong sphincter above the genital opening (Fig. 16c). In some parts of the atrium the epithelium bears long cilia (see *Kytorhynchella meixneri*, Fig. 16a, b). This latter species shows two masses of secretory granules in the lateral walls of the atrium (Figs. 1a, 13c, 16b: lg). Apparently these granules originate from huge glandular sacs which open from both sides into the genital atrium.

The genital atrium in the species of the genus *Kytorhynchus* is horizontally depressed (especially the middle part); the nuclei are partially sunken into the parenchyma (Figs. 15b, 16c).

Male organs

All species are clearly protandric. The paired, compact testes (t) are either dorsal (*Kytorhynchella*) or ventral (*Kytorhynchus*) to the vitellaria (Figs. 1, 3, 5, 8, 10, 11). The sperms are long and provided with two flagella (results from *Kytorhynchella meixneri* and *Kytorhynchus microstylus* only, Fig. 7a).

The vasa deferentia (vd) lead, dorsolateral to the pharynx, to the male copulatory organ (co). The proximal part of the latter contains an unpaired

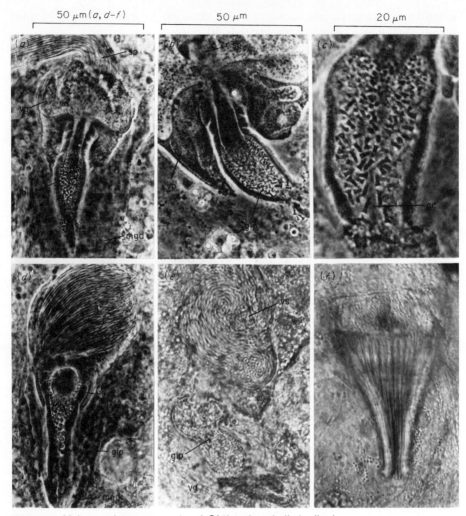

Figure 6 Male copulatory organ. (a–c) Of *Kytorhynchella karlingi*
(a and b) of the littoral form, see page 55); (d) of *Kytorhynchella
meixneri*; (e) of *Kytorhynchus* sp. II; (f) stylet of *Kytorhynchus
(Kytorhynchoides) macrostylus*.

seminal vesicle (vs), which is large, especially in the genus *Kytorhynchus*
(Figs. 1, 3, 5, 8, 10, 11). Distal to the seminal vesicle, the copulatory organ
is filled with secretion of the prostatic glands which lie outside of the bulbus.

In *Kytorhynchella* this prostatic vesicle (vg) seems to be only a specially
differentiated part of the original male genital canal which forms distally a
glandular penis papilla (Fig. 1: glp). *Kytorhynchella riedli* seems to be the most

primitive in this respect (Figs. 1*h*, 8*b*). The glandular penis papilla of *Kytorhynchella meixneri*—and most probably also *Kytorhynchella karlingi*—must be considered to be a fold in the epithelium of the male genital canal (Fig. 16*a*). The rounded or nail-shaped granules in the middle of the penis papilla (Fig. 6*a–d*) were first considered to be cuticularized parts of the cirrus, but later proved to be only secretion granules. The male copulatory organ of this genus is connected to the genital atrium by a fairly long male genital canal.

The male copulatory organ in the genus *Kytorhynchus* differs from *Kytorhynchella* mainly in the much less developed prostatic vesicle and in the much shorter male genital canal (Figs. 10, 11, 16*c*). The species of *Kytorhynchus* (*Kytorhynchoides*) have a cuticular stylet which is funnel-shaped and has marked longitudinal folds (Figs. 6*f*, 7*b–d*). This stylet is closely attached to the male genital canal—one could almost say it lines this canal—so that each fold has a corresponding groove in the wall of the male genital canal (Fig. 15*a*, stylet artificially shrunken).

In all species the male genital canal and the copulatory organ are enclosed by the basement membrane and two muscle layers which also cover the geni-

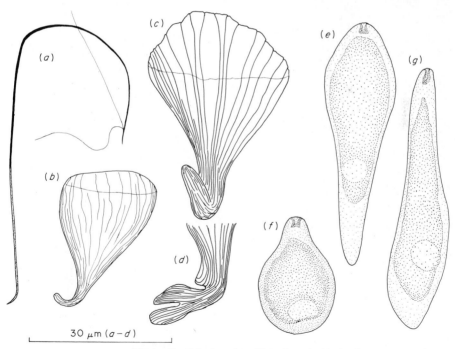

30 μm (*a–d*)

Figure 7 *Kytorhynchus (Kytorhynchoides) microstylus.* (a) Sperm; (b–d) stylet; (e–g) swimming animals. (Parts c and d refer to the Bermuda form.)

tal atrium. In *Kytorhynchella* these latter are an inner circular and an outer longitudinal layer; in *Kytorhynchus* the layers are opposing spirals (Fig. 16*a*, *c*). The male canal opens rostrally (*Kytorhynchus*), dorsally (*Kytorhynchella meixneri*), or laterally (*Kytorhynchella riedli*) into the atrium.

Female gonad

The germinative zone of the ventrolaterally (*Kytorhynchella*) or dorsolaterally (*Kytorhynchus*) situated germovitellaria lies in the caudal part of

Figure 8 Scheme of the organization obtained from serial sections, in lateral view. (*a*) Of *Kytorhynchella meixneri*; (*b*) of *Kytorhynchella riedli*.

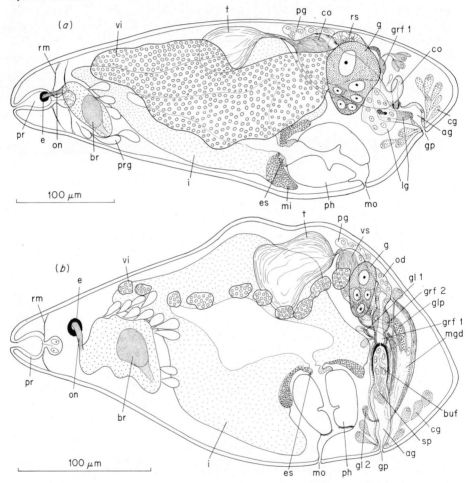

the body in a vertical position. Vitellarium (vi) and germarium (g) are con-nected to each other throughout a broad area (Figs. 1*a*, 5*a*, *b*, 8*a*, 11*b*, 15*d*, *e*). In younger specimens one finds the vitellaria in the form of two lateral rows of cell groups which are not always connected to each other (Fig. 8*b*).

In all species a protrusion of the genital atrium (which must be considered the primary female genital canal) extends between the two germaria. The dif-ferences in this organ between the species are considerable; nevertheless— owing to some special features—this female genital canal must be considered homologous in all species.

In *Kytorhynchella* this canal enters the genital atrium rostrally (Figs. 1, 8, 16*b*). Two regions can be distinguished: a distal one (buf) which his-tologically is similar to the genital atrium (this is especially true for *Kytorhyn-chella meixneri* [Fig. 16*b*]; in *Kytorhynchella riedli* the basement membrane of this distal part is thicker, and leaves only a very narrow canal open at the transition to the proximal region) and a proximal region (grf)—separated from the distal one by a constriction—consisting of one (*K. karlingi*) or two spheri-cal parts filled with secretion granules (Figs. 8, 16*b*). One of these parts in particular is seen in squeeze preparation as a glandular girdle (Fig. 1). In *Kytorhynchella meixneri* and *K. karlingi* the secretion granules of the more distal girdle (grf 1) are produced in huge glands surrounding the female genital duct (Fig. 16*b*). In *Kytorhynchella riedli* the main girdle is the more proximal one (grf 2), whereas the distal one is not filled with secretory granules (Fig. 8*b*). The gland openings here have long, slender secretory bodies (gl 1) con-sisting of fine granules, very reminiscent of the second type of shell glands found by Luther (1943) in *Promesostoma marmoratum* (Schultze) (see also Luther, 1904, p. 124). *Kytorhynchella meixneri* and *Kytorhynchella karlingi* bear a small seminal receptacle (rs) at the end of the female duct; in *Kytorhyn-chella riedli* the sperms seem to be stored in the distal part of the female duct itself (Figs. 1, 8, 16*a*, *b*). The germaria of all three species are connected to the genital duct by more or less parenchymatic oviducts (Figs. 1, 15*f*: od).

In *Kytorhynchus* (*Kytorhynchoides*) this schema of the female system is essentially the same. Nevertheless considerable differences do exist in the various parts of the system. In both species the distal part (buf) of the female genital canal is no longer a simple protrusion of the epithelial wall of the genital atrium; it is much larger and is lined with a syncytium which must be derived from the atrium wall (Figs. 5*a*, *b*, 11, 16*b*). In *Kytorhynchus macrostylus* this part is further divided into a central part and two lateral lobes (Figs. 5*b*, 16*b*: la). A constriction still separates the distal region (buf) from a proximal, glandular one which—as in *Kytorhynchus microstylus*—still has the same or similar shape and structure as in the genus *Kytorhynchella* (Figs. 5*a*, 11*a*: grf). *Kytorhynchus macrostylus* shows a strongly developed con-strictor muscle, and the proximal glandular region is separated into many

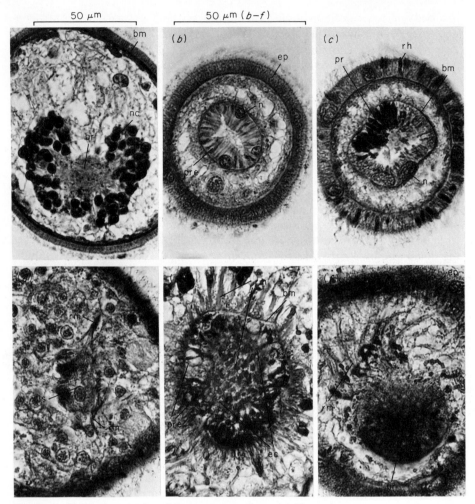

Figure 9 Photomicrographs of transverse sections. (a) Through
the brain of *Kytorhynchus* (s. str.) *oculatus*; (b) through the proboscis of the same species; (c) through the proboscis of *Kytorhynchus*
(*Kytorhynchoides*) *microstylus*; (d) through the proximal glandular
sac of the proboscis of *Kytorhynchella meixneri*; (e) through the
proximal part—with end cone—of the proboscis of *Kytorhynchus*
(*Kytorhynchoides*) *macrostylus*; (f) through the proximal region
of the proboscis of *Kytorhynchus* (*Kytorhynchoides*) *microstylus*.

huge, oval gland cells (cgl) to which the germaria are directly attached, thus
lacking an oviduct (Figs. 5*b*, 11*b*, 16*c*). The typical slender shell glands (gl 1)
described for *Kytorhynchella riedli* are also present here and enter the genital

canal near the constriction between the distal and proximal regions (Figs. 11*a*, 16*b*). The sperms are stored in the distal part of the canal.

Kytorhynchus s. str.—studied only in *Kytorhynchus oculatus*—agrees with *Kytorhynchus* (*Kytorhynchoides*) *microstylus* in nearly all details (Figs. 3*a*, 10). But it possesses a cuticular tube (ct) in the transition of the distal and proximal region. This tube should be considered a ductus spermaticus, since it is clearly connected to the sperm mass present in the distal part of the female genital canal. Furthermore the germaria are completely surrounded by a syncytial tissue, which connects the germaria to the glandular girdle of the female genital canal (grf) and has an extension (x) connecting each germarium with the atrium genitale itself (Figs. 10, 15*c*). (I must mention here that this latter connection was very clear on the right side and less visible on the left side in both of the specimens studied.) The same slender shell glands (gl 1) as those found in almost all the other species are present in *Kytorhynchus oculatus* as well. Unfortunately one could not determine whether these glands open into the glandular part of the primary female duct or somewhere into the syncytium around the germaria.

SYSTEMATIC POSITION

Although the pharynx shows a very special primitive, bulbous type, there is no doubt that the group has to be placed in the Neorhabdocoela-Typhloplanoida.

In discussing the position of the new forms within the Typhloplanoida, one should be aware that present higher systematic grouping (the family and subfamily levels) in this suborder rests on a weak base and needs a careful revision. This was very cautiously stated by Karling (1957, p. 32), who thus

Figure 10 Scheme of the organization of *Kytorhynchus* (s. str.) *oculatus*, obtained from serial sections, laterial view.

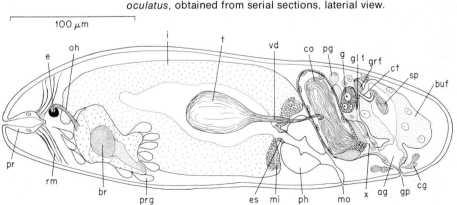

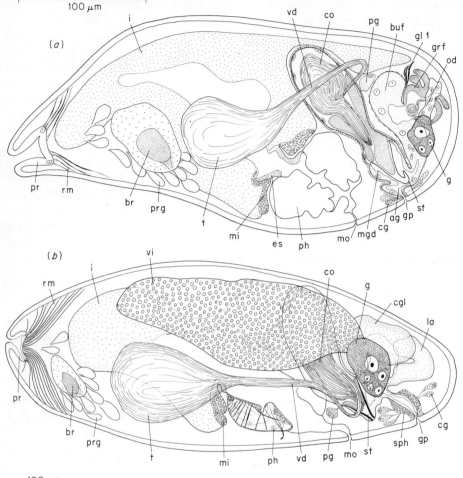

Figure 11   Schemes of the organization in a lateral view, obtained from serial sections. (*a*) *Kytorhynchus* (*Kytorhynchoides*) *microstylus* (Bermuda form); (*b*) *Kytorhynchus* (*Kytorhynchoides*) *macrostylus*.

follows the thinking of Luther (1943, 1946, 1948, 1950). Luther (1948, p. 117) mentioned that in spite of the achieved progress in the systematics of the Typhloplanoida, one still has to consider the present system a preliminary one. In particular, the numerous findings of new marine genera (summarized in Ax, 1960) in the family Typhloplanidae demonstrated the difficulties of

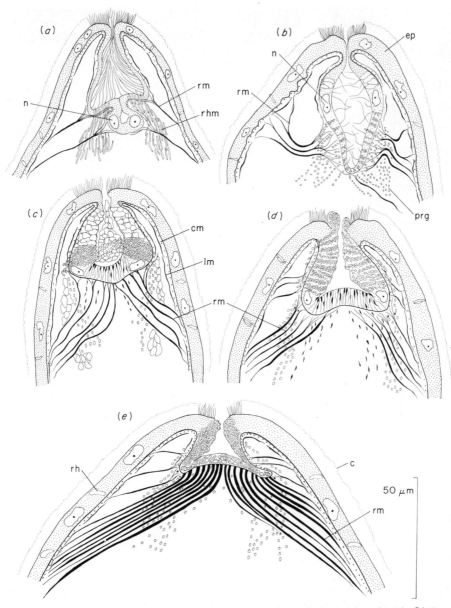

Figure 12 Sagittal sections through the proboscis. (a) Of *Kyto-rhynchella meixneri;* (b) *Kytorhynchus* (s. str.) *oculatus;* (c) *Kyto-rhynchus* sp. II; (d) *Kytorhynchus* (*Kytorhynchoides*) *microstylus;* (e) *Kytorhynchus* (*Kytorhynchoides*) *macrostylus.*

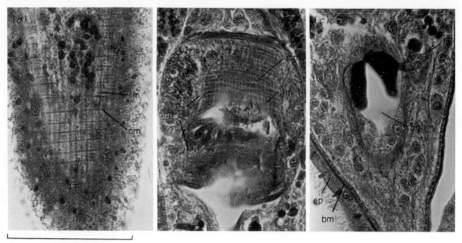

50 μm

Figure 13 Photomicrographs of frontal sections of *Kytorhynchella meixneri*. (a) Tangential through the skin; (b) Longitudinal through the pharynx; (c) longitudinal through the genital atrium.

the present system (see, for example, Luther, 1946, p. 34, about the systematic position of *Thalassoplanella* Luther, 1946).

Den Hartog (1964) split the family Trigonostomidae Luther, 1948 into the two families Promesostomidae and Trigonostomidae. Counting this recent change, we now have six families in the suborder Typhloplanoida (diagnoses, see Luther, 1962, 1963; Den Hartog, 1964); the ectoparasitic family Typhlorhynchidae Meixner, 1925b, is not considered here; during the preparation of this manuscript, two papers—Ax and Heller, 1970, and Ax, 1971—have been published on the systematics of the families Trigonostomidae and Promesostomidae describing two new subfamilies Mariplanellinae and Adenorhynchinae (Ax and Heller, 1970).

Byrsophlebidae
Trigonostomidae   (Subfam. Trigonostominae, subfam. Paramesostominae, subfam. Mariplanellinae)
Promesostomidae   (Subfam. Brinkmanniellinae, subfam. Promesostominae, subfam. Adenorhynchinae)
Typhloplanidae   (Subfam. Protoplanellinae, subfam. Typhloplaninae, subfam. Rhynchomesostominae, subfam. Ascophorinae, subfam. Olisthanellinae, subfam. Mesostominae, subfam. Phaenochorinae, subfam. Opisthominae)
Carcharodopharyngidae
Solenopharyngidae

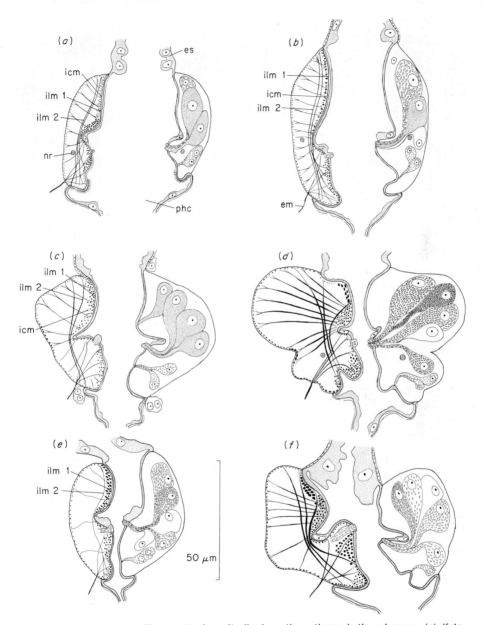

Figure 14 Longitudinal sections through the pharynx. (a) *Kytorhynchella riedli;* (b) *Kytorhynchella meixneri;* (c) *Kytorhynchus (s. str.) oculatus;* (d) *Kytorhynchus (Kytorhynchoides) microstylus* (Bermuda form); (e) *Kytorhynchus* sp. II; (f) *Kytorhynchus (Kytorhynchoides) microstylus.*

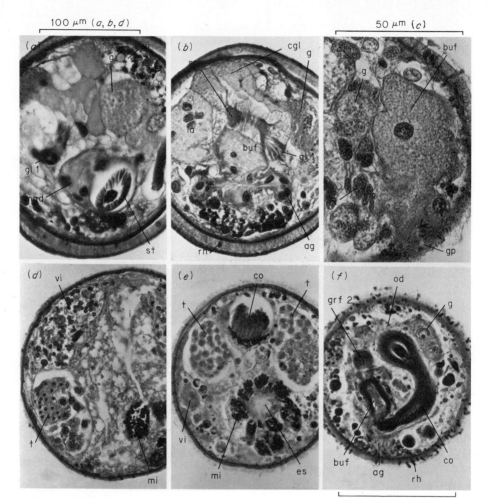

Figure 15 Photomicrographs of sections. (a, b) Transverse section in the genital region of *Kytorhynchus* (*Kytorhynchoides*) *macrostylus*; (c) sagittal section through the posterior body end of *Kytorhynchus* (s. str.) *oculatus*; (d) transverse section in front of the pharynx of *Kytorhynchus* (*Kytorhynchoides*) *microstylus*; (e) transverse section in front of the pharynx of *Kytorhynchella meixneri*; (f) transverse section in the genital region of *Kytorhynchella riedli*.

I am not sure how much progress was made by simply splitting the Trigonostomidae into two families. Especially the family Promesostomidae (*sensu* Den Hartog, 1964) still seems to be fairly heterogeneous, because

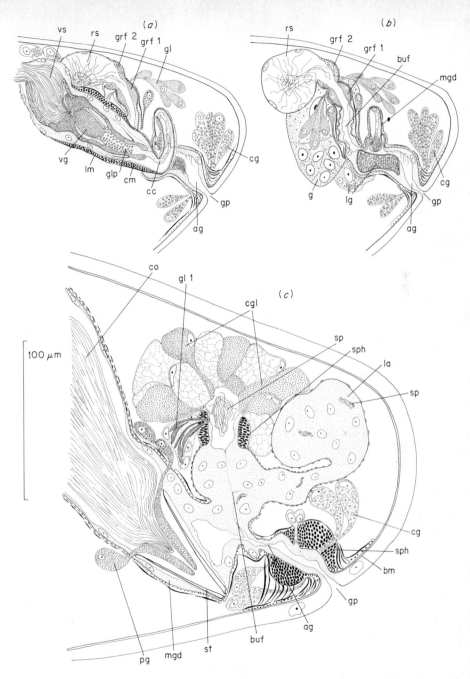

Figure 16 Scheme of the organization of the genital system, obtained from transverse sections. (*a*, *b*) Of *Kytorhynchella meixneri* (in *b* without the male copulatory organ); (*c*) of *Kytorhynchus* (*Kytorhynchoides*) *macrostylus*.

we still do not know more about the anatomy of the Brinkmanniellinae (see Luther, 1962, p. 40). I, therefore, agree very much with Papi (1959, p. 47), who suggested for the family Typhloplanidae that for the groups of the clearly related genera, new systematic groups be defined, and that the uncertain rest be left in the existing system until more is known about these forms.

Because of the position and special structure of the pharynx (page 32), the presence of a terminal proboscis in the form of a differentiated terminal invagination, the presence of paired germovitellaria, and also some very specific features of the female genital canal (proximal glandular girdle, special type of shell glands [page 37]), we have to consider the described new species to be members of one systematic entity. Within this group one can define two clearly separated subgroups: the forms of the genus *Kytorhynchella* with a specialized proboscis, a glandular penis, and testes located dorsal to the vitellaria, and the forms of the genus *Kytorhynchus* with the more primitive proboscis and testes ventral to the vitellaria. The latter subgroup is more heterogeneous. It might be necessary in the future to place the forms around *Kytorhynchus oculatus* and around *Kytorhynchus microstylus* in different genera. Meanwhile I distinguish only two subgenera in the genus *Kytorhynchus*. These are based mainly on the lack or presence of a cuticularized stylet.

At the present time one cannot find a *clear* relationship of the new group to one of the families of the Typhloplanoida. Because of this and because the present arrangement of the families in this suborder should be reestablished, as mentioned previously, I suggest that a separate family be established for the new group under the name Kytorhynchidae.

In the definition (page 53) the new family shows a closer relationship to the genera of the Promesostomidae-Brinkmanniellinae (*sensu* Den Hartog, 1964) with the exception of *Brinkmanniella* Luther, 1943 itself. The main diagnostic characters of the Brinkmanniellinae are paired germovitellaria and the presence of only one connection between the female gonad and the atrium genitale. Between the genera the differences in the fine structure of the female genital canal as well as in the male copulatory organ (especially in the bulbus) are still considerable, and make the subfamily heterogeneous. For example, the female genital canal of *Westbladiella* Luther, 1943 (see also Riedl, 1954), *Einarella* Luther, 1948, *Memyla* Marcus, 1952, *Wydula* Marcus, 1954, and probably also that of *Tvaerminnea* Luther, 1943 (see also Ax, 1956) resembles much more the general situation in the family Typhloplanidae (see, for example, *Notomonoophorum* Luther, 1948, or *Haloplanella* Luther, 1946) than it does the very simple situation in *Brinkmanniella* (see also Luther, 1948; Marcus, 1951; Westblad, 1952); the undifferentiated female genital canal in *Brinkmanniella* seems to be more similar to the situation in the Prolecithophora (see Karling, 1940, 1962a, b).

The main characters used in systematics will be discussed in more detail in the following paragraphs.

1   Of the special structure of the pharynx in the Kytorhynchidae, the internal longitudinal muscles are of interest. Within the Typhloplanoida, longitudinal muscles next to the internal pharynx epithelium are only reported, to my knowledge, for the pharynx of *Notomonoophorum coecum* Luther, 1948, *Proxenetes cochlear* Luther, 1948, and probably *Brinkmanniella falclandica* Westblad, 1952 (see also *Mesostoma ehrenbergii* in Luther, 1904, p. 46). According to the discussion of the bulbous pharynx in Karling (1940), one could consider this feature a primitive one; this might be of interest for the comparison of the pharynx variabilis and pharynx rosulatus. The taxonomic value of this situation in the muscles, however, is questionable because there is already in the pharynx plicatus a high variability in this respect (see Ax, 1961, fig. 5).

The arrangement of the inner longitudinal muscles and gland openings in the Kytorhynchidae (see Fig. 14) suggests further that the whole inner pharynx region distal to the circular groove in about mid-pharynx is homologous to the grasping border (*Greifwulst*) of the pharynx rosulatus.

2   As for the different position of the testes in the genera *Kytorhynchella* and *Kytorhynchus*, in the family Typhloplanidae this feature, together with the opening of the protonephridia, is very important for the definition of the subfamilies (see Luther, 1948, 1963). Therefore one might, after having found more species, use this character for distinguishing subgroups within the new family *Kytorhynchidae*. At the present time we do not have enough material to solve this problem.

3   Discussing the systematic value of the copulatory organ, Luther wrote at the end of his paper in 1948, p. 118: "Ich stelle mir vor, dass bei künftigen Verbesserungen des Systems der Typhloplanoida dem Typus des Kopulationsorganes, speziell dem Vorhandensein oder Fehlen eines kutikularen Stiletts, mehr Aufmerksamkeit wird geschenkt werden müssen als bis jetzt." Our new group shows that the first of Luther's statements will be very valuable indeed for further classification. The second statement is somewhat questionable; whereas "the type of the copulatory organ" is the same in all members of the new group (see page 35), we find "the presence and absence of a cuticular stylet" within one genus, or in other words, within a closely related group of species. Therefore the fine structure of the whole copulatory organ, rather than the presence or absence of a stylet, should be used in the future for classification.

4   The female genital canal (with the typical proximal glandular girdle) of the new group (especially in the genus *Kytorhynchella*) is so similar to the female duct of some members of the family Typhloplanidae, that one could assume a common origin (see *Pratoplana salsa* Ax, 1960; *Notomonoophorum coecum* Luther, 1948, further Luther, 1904, p. 123). On the other hand the new group shows a tendency to stress the bursal function of the female genital canal which led, in *Kytorhynchus oculatus*, probably to a complete loss of the efferent function of the female genital canal and to the development of secondary efferent ducts of the female gonad (pages 37–39). This arrangement and the cuticular mouthpiece of *Kytorhynchus oculatus* are very reminiscent of the situation in the family Trigonostomidae (*sensu* Den Hartog, 1964).

The "bursa with mouthpiece" in *Kytorhynchus oculatus*, however, according to the present interpretation, is homologous with the efferent female genital canal of the family Trigonostomidae and must be considered an analogous structure to the bursa-receptacle-mouthpiece complex of the latter group.

5   Finally one should touch on the proboscis problem within the Typhloplanoida. The tendency to modify the anterior part of the body into a proboscis is widespread in this suborder (see Meixner, 1938, p. 25; Luther, 1904, p. 31). Some of these proboscis types are similar to the construction in Kytorhynchidae. Therefore, it should be pointed out why the new species are nevertheless taxonomically separate.

Forms with a differentiated anterior tip, which can be invaginated as a whole by the action of strong longitudinal muscles are *Astrotorhynchus bifidus* (see Luther, 1950) and several species of Mesostominae (see Luther, 1904). The little proboscis of *Rhynchomesostoma* (see Luther, 1904, 1963) is even more differentiated and sometimes separated by a muscle septum. Extensively developed *Stäbchenstrassen* forming a very simple proboscis are especially evident in the genus *Adenorhynchus* Meixner (1938). But none of these forms ever have the anterior end permanently invaginated as do the Kytorhynchidae.

Of interest in this discussion is the proboscis of the *Trigonostomum* group, where the proboscis is situated ventrally in front of the mouth. In an anatomical study Meixner (1924)—see also Marcus (1948) and Riedl (1954) —pointed out that this proboscis originated from a ventral invagination of the body wall. One could assume, however, that it is an originally terminal invagination which, because of better function in prey capture, was displaced ventrally toward the mouth opening. This is probably hard to decide in the adult stages (arrangement of muscles) but might be solved by studying the postembryonic development of the animals. But even if this were true, one would still have to be very careful in homologizing the proboscis of the *Trigonostomum* group with that of forms such as *Kytorhynchella* and *Kytorhynchus* because of the striking differences in the characteristics of the families (male copulatory organ and the female genital apparatus). These two organ systems of *Trigonostomum* are without doubt closely related to the situation in the *Proxenetes* group, where the proboscis is lacking (Meixner, 1924; Luther, 1948).

Proboscis structures very similar to those of the Kytorhynchidae—a terminal invagination of the body wall—are found in three freshwater Typhloplanidae-Protoplanellinae, *Haplorhynchella paludicola* Meixner, 1924, *Microcalyptorhynchus virginianus* (Kepner and Ruebush, 1935a), and *Prorhynchella minuta* Ruebush, 1939 (see page 24). *Haplorhynchella* seems to have a proboscis structure very similar, probably homologous, to that of the Kytorhynchidae, as I found out from Professor Reisinger, Austria (pers. comm.). Even less is known about the proboscis of the two American forms, although several sketches are available (Kepner and Ruebush, 1935a, p. 259; Kepner,

Ferguson, and Stirewalt, 1941, p. 245; Ruebush, 1939, p. 205; Ruebush, 1941, p. 35). As far as one can tell from these sketches and from the brief description of this organ given in the papers, it could be the same structure as in the Kytorhynchidae (also in *Microcalyptorhynchus* it is slightly subterminal).

The three freshwater species differ from the new group in the unpaired germarium and in the position of the pharynx—not reported in *Haplorhynchella*—which lies in the anterior body half. *Prorhynchella minuta* shows the most difference; its genital opening is in the anterior body half, and furthermore it shows two lateral, ciliated pits. (This form might be more closely related to the Typhloplanidae-Mesostominae; see Luther, 1904, chapter on *Wimpergrübchen*).

I suggest that these three poorly described freshwater forms with a terminal proboscis should not be put into any existing family until more is known about their general anatomy and fine structure.

THE PROBOSCIS OF THE NEW GROUP AND THE EVOLUTION OF THE EUKALYPTORHYNCHID CONORHYNCH

Graff (1905 and 1904–1908, p. 2081), in reviewing the most important ideas about the eukalyptorhynchid conorhynch, suggested that it might have developed from stages like the proboscis of *Astrotorhynchus* and *Rhynchomesostoma*, where the anterior tip gets invaginated as a whole.

Meixner (1925a, p. 284) mentioned that Graff's hypotheses could be true but that with the same probability one could consider the invagination of the body wall in the *Trigonostomum* group as a stage previous to it. In this case the conorhynch would have arisen from the base of the invagination.

Later Meixner (1938) preferred his second version, formulated in 1925a, "Als eine Weiterbildung dieses Typus [namely, that of the *Trigonostomum* proboscis] ist der Scheidenrüssel der Kalyptorhynchia aufzufassen."

According to Karling (1963, p. 226 and fig. 15-3a–d), the origin of the conorhynch lies in a simple glanduloepidermal organ. From this evolve adhesive pits (*Trigonostomum*, see Meixner, 1924; *Acmostomum*, see Marcus, 1947), followed by extensible structures which are later separated from the surrounding parenchyma by a septum (*Astrotorhynchus* and *Rhynchomesostoma*, see Graff, 1904–1908). In looking at fig. 15-3c in Karling (1963), one gets the impression that the author thought that the conorhynch developed at the base of the adhesive pit. Karling (1968, p. 100), however, states clearly that "the conorhynch is a modified sucker arising from the anterior body end by folding of the body wall. . . ." (See also Karling, 1968, figs. 46–48). This author therefore seems to follow the general idea of Graff (1904–1908).

Further, I would like to mention that Riedl (1954) stated specifically that the kalyptorhynchid conorhynch must be—owing to the radially symmetrical arrangement around the longitudinal body axis—derived from a terminal and not a ventral differentiation (Riedl, 1954, p. 229). In the literature it is quite evident that this must be the general opinion (actually Graff's theory [1904–1908] is already based on this assumption), but I could find it specifically stated only by Karling (1968) as well.

Besides the poorly known freshwater forms *Haplorhynchella*, *Microcalyptorhynchus*, and *Prorhynchella*, we now have, with the description of the new group, a whole systematic entity characterized by having a proboscis in the form of a terminal invagination.

In this group one can see a very interesting trend in the differentiation of the proboscis toward more and more modification of the longitudinal muscles of the subepidermal muscle sheath into retractor muscles of the proboscis (see page 32 and Fig. 12). The proximal circular muscles, in the living *Kytorhynchus microstylus*, temporarily cause a cone shaped projection of the proximal part of the proboscis (Fig. 5e). The sensory function, which might have been the primary one, is changed more and more to that of a glandular-muscular organ, most probably used in prey capture. In *Kytorhynchus microstylus* (and maybe also in *K. macrostylus*) the nuclei of the proboscis epithelium are sunken into the corners of the triangular proboscis (Fig. 12d).

This trend in the kytorhynchid proboscis can be a model for understanding the evolution of the eukalyptorhynchid conorhynch derived from an adhesive pit, as suggested by Meixner (1925a, 1938).

In Fig. 17 the possible pathways of evolution of this organ are summarized.

Following the ideas of Graff (1904–1908) and later Karling (1968), one has to consider the differentiated anterior tip, with strong retractor muscles (stage IA_1), to be the origin for the conorhynch (e.g., in some Mesostominae, see Luther, 1904). The subsequent step (IA_2) would be the development of a circular groove around the anterior body tip, thus permitting retraction of the latter as a whole into the body (e.g., *Astrotorhynchus*, *Rhynchomesostoma*, see Graff, 1904–1908).

On the other hand (as suggested by Meixner, 1925), one can start with a simple, cup-shaped terminal invagination (stage IB_1). Together with a strong increase in retractor muscles and the tendency to protrude the basal part of the invagination, the stages IB_2 and IB_3 are reached (e.g., Kytorhynchidae, see above).

The next step in the evolution of the conorhynch is the same whether one follows $IA_1–IA_2$ or $IB_1–IB_3$; we only have to assume that the temporarily protruded base of the invagination $(IB_2, IB_3$, see Fig. 5e) becomes stationary (stage II). Owing to the action of the circular muscles at the basis of the proboscis, the basement membrane then extends into a fold (or delamination?), thus enclosing the distal part of the retractor muscles and leading to the eukalyptorhynchid conorhynch (stages III and IV).

Which of the two starting points (IA_1 or IB_1 in Fig. 17) one should choose

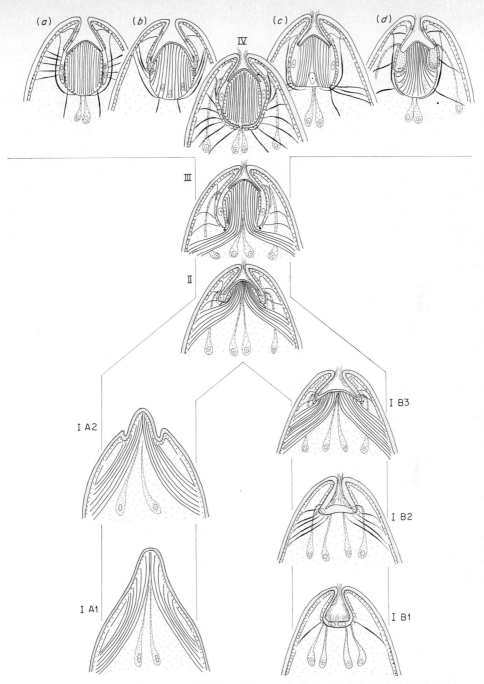

Figure 17 Diagrams demonstrating possible evolutionary pathways of the eukalyptorhynchid conorhynch. Stages $1A_1$–$1A_2$ and $1B_1$–$1B_2$ represented by living forms (see text), stages II–IV hypothetical. (a–d) Schemes of conorhynchs in some families of Eukalyptorhynchia: (a) Polycystididae, (b) Koinocystidae, (c) Cytocystididae, (d) Cicerinidae.

in preference to the other is hard to say. The similarity of the eukalyptorhyn-chid conorhynch to the coniform pharynx of *Gnosonesima* Reisinger, 1926, which one understands as an invagination of the entire anterior body tip (Karling, 1968, figs. 46-48), could be explained by assuming the same evolution of the two different structures. This seems to stress version IA_1-IA_2. However, with the discovery of a clear morphological sequence from a simple terminal invagination with mostly sensory function to a glandular-muscular adhesive pit with a protrusible basis, one must also consider the second version as another possibility. Moreover, I think that the result (stage II in Fig. 17) would not be too different, whether it originates from the first (IA_1-IA_2) or the second (IB_1-IB_2) evolutionary line, because the matrix—the anterior body tip—is the same in both versions.

The origin of the septum around the proboscis bulb is, according to Meixner (1938, p. 25), the basement membrane: "Bei den *Eukalyptorhynchia* handelt es sich um einen einheitlichen Muskelzapfen ('Bulbus') der durch ein Grenzmembran-Ringmuskel-Septum, eine Abspaltung der Basalmembran und der Hautmuskulatur, vom uebrigen Parenchym abgeschlossen und von Binnenlaengsmuskeln erfuellt ist."

My suggestion (see above) that the basement membrane did enclose the posterior part of the proboscis by *folding* is very hypothetical. It depends on whether one thinks that the basement membrane is produced by the basal part of the epidermis—the present general opinion—or by the subepidermal muscle sheath (a suggestion made by Karling in a personal discussion). It might be possible to solve this problem by electron-microscopical study of the postembryonic development of the eukalyptorhynchid conorhynch in primitive (according to Karling, 1953, 1964) families such as the Cytocystididae or Cicerinidae.

The "folding theory" could, at the moment, explain some of the interesting differences of this organ within some eukalyptorhynchid families. (Besides this, one should mention that Karling, 1968, describes the septum of the coniform pharynx of *Gnosonesima*—which according to Karling looks very much like a eukalyptorhynchid conorhynch—as a *double* structure.)

The kalyptorhynchid family Cytocystididae Karling (1964)—primitive in many features (see Karling, 1953)—has four mantle cells in the conorhynch which are connected to the epithelium of the proboscis (Fig. 17*h*). These might be the remaining cells of the ancestral invagination which were retained because of a glandular function. From the fold of the basement membrane, only the external side would have remained, and with it the external muscle sheath (compare III and IV with *h* in Fig. 17).

In the other main groups of Eukalyptorhynchia (the proboscis diagram of three families is given in Fig. 17), the internal part of the basement membrane fold and the inner circular muscle seem to have built the final septum around the conorhynch.

One could also trace back the glandular girdles of the Cicerinidae Meixner (1928), the "juncter" cells in the Koinocystidae Meixner (1925a), and the external coat of cells of many Polycystidae *sensu* Karling (1955) to the cells of the ancestral invagination (see also Karling, 1952a, b, 1953, 1954, 1955).

The Kalyptorhynchia are today thought to be an early side branch of the Typhloplanoida (Karling, 1940, p. 218; Ax, 1963, p. 209).

Besides differences in the interpretation of the origin of the Typhloplanoida (see Karling, 1940; Ax, 1961), which of course influence the discussion of the stem form of the Kalyptorhynchia, one can state that the new family Kytorhynchidae shows many features which put it near the base of the Typhloplanoida. I mention only the pharynx with the numerous small fibers in all muscle layers, with inner longitudinal muscles next to the inner epithelium (see Karling, 1940, p. 166; Ax, 1963, fig. 14-4a), no distinct *Greifwulst*, the paired germovitellaria, the presence primarily of only one connection between the female gonad and the atrium, and a relatively simple male copulatory organ which can be derived directly as a differentiation of the male genital canal.

Moreover, I agree with Karling (1940) that the common gonopore in the Typhloplanoida (and Kalyptorhynchia, see Karling, 1955, 1956) must be considered primitive in comparison with the separated pores and that the latter in the family Byrsophlebidae are to be considered secondary.

Therefore, if the evolution of the eukalyptorhynchid conorhynch followed the pathway IB–IV (sketched in Fig. 17), one could consider the Kytorhynchidae to be a side branch of the ancestral kalyptorhynchs. On the other hand the differences in the genital system between the Kytorhynchidae and Kalyptorhynchia today (internal seminal vesicle, uterus, etc.; see Meixner, 1925a; Karling, 1956, 1964) make comparison between the two groups difficult. Further material of the Kytorhynchidae and ontogenetic investigations of the eukalyptorhynchid conorhynch should be a tool to overcome these problems.

DEFINITION OF THE NEW TAXA

The type material is deposited in The American Museum of Natural History, Department of Living Invertebrates, New York.

Kytorhynchidae fam. n.

Typhloplanoida with a proboscis derived from a terminal invagination of the body wall. With numerous erythrophilous proboscis glands, (modified Rhabdites), with dermal and adenal rhabdites.

Pharynx in the posterior part of the body, oblique and pointing caudoventrally. Pharyngeal cavity poorly developed; without a typical *Greifwulst*, instead divided into a proximal and distal region by a circular groove at about mid-pharynx. Nuclei of the inner pharynx epithelium restricted to the circular groove; pharynx glands open in two clearly separated girdles in the distal region. With a distinct esophagus and a circle of granular clubs around the *Darmund*.

With a single gonopore and a genital atrium in the posterior part of the body. Testis paired, a single seminal vesicle not separated from the prostatic vesicle forming a copulatory bulbus with the latter; sometimes with a penis papilla or stylet in the form of a cuticularized lining of the male genital canal.

Female gonad paired germovitellaria, with the germaria situated at the caudal end. Generally only one connection between the female gonad and the genital atrium functioning as efferent duct of the gonad as well as a bursal organ. This female genital canal divided by a constriction into a distal region—a simple invagination of the atrium wall—functioning as bursal organ, and a proximal, glandular region.

Opening of the excretory system unknown.

Marine.

NAME: *Kytos* (Greek) = urn, referring to the general shape of the proboscis.

Kytorhynchella gen. n.

Proboscis divided into a distal, urn-shaped part with many sensory hairs and a proximal, glandular sac containing the nuclei of the proboscis epithelium; latter mostly four in number. Only a few retractor muscles of the proboscis. Proboscis glands open mainly into the base of the urn-shaped portion. With eyes. Pharynx with two layers of inner longitudinal muscles in the proximal region. Testes mediodorsal to the vitellaria. Male copulatory organ with a glandular penis papilla; male genital canal long, with a strong muscle sheath; two well-developed oviducts connect the female gonad with the primary female genital canal. The proximal glandular region of the latter forming one distinct glandular girdle (sometimes two); distal part lined with an epithelium.

Type of the genus *K. meixneri*.

Kytorhynchella meixneri sp. n. (Figs. 1*a*–*g*, 2*a*–*c*, 6*d*, 8*a*, 9*d*, 12*a*, 13*a*–*c*, 14*b*, 15*e*, 16*a*, *b*)

500 to 700 μm in length, colorless. External pharynx epithelium without ciliation. With two hook-shaped granular masses in the lateral wall of the genital atrium. With a seminal receptacle at the proximal end of the female genital canal. Penis papilla short, filled with round granules.

TYPE LOCALITY: Wrightsville Beach north of Wilmington, North Carolina. U.S.A.; sand flat 300 ft north of the drawbridge on Highway 76; mid-tide level, at low tide, 0.5 to 2 cm depth. (See Riedl, 1970, p. 242, for further details.)

OTHER OCCURRENCE: Wrightsville Beach, sound side; beach in front of the water tower; medium sand rich in organic matter. One specimen was obtained from a sample taken offshore (Beaufort, North Carolina) at a depth of 20 m. This specimen "sublittoral form" differs slightly in measurements and fine structure of the pharynx (more inner circular muscles).

HOLOTYPE: One transverse section, AMNH #746.

PARATYPES: Two sagittal, one frontal and five transverse serial sections, AMNH #747.

Kytorhynchella karlingi sp. n. (Figs. 1*i*, 4*a*, 6*a*–*c*)

350 to 450 μm long; colorless. Female genital canal with a proximal seminal receptacle. Penis papilla short, with nail-shaped granules.

TYPE LOCALITY: Off Beaufort, North Carolina; in front of Shackleford Banks, 11 m deep, fine sand with detritus.

OTHER OCCURRENCE: Wrightsville Beach north of Wilmington, sound side beach in front of the water tower; medium sand, rich in detritus, surface layers. This specimen (littoral form) differs slightly in the shape of the proboscis and the measurement of the granules in the penis papilla (see Figs. 4*a* and 6*a* and *b*).

HOLOTYPE: Whole mount, AMNH #748.

Kytorhynchella riedli sp. n. (Figs. 1*h*, 8*b*, 14*a*, 15*f*)

800 to 900 μm long, colorless. External pharynx epithelium with ciliation. Distal part of the female genital canal with strongly thickened basal membrane, without a seminal receptacle at the proximal end. Penis papilla long, with round granules; male genital canal with especially strong outer longitudinal muscles.

TYPE LOCALITY: Off Beaufort, North Carolina; *R/V Eastward* cruise E–4–70, station 14207, at 34°16′N and 76°24′W; April 1970; 32 m depth; heterogeneous medium sand rich in organic matter. (Courtesy Dr. C. Jenner.)

HOLOTYPE: One transverse serial section, AMNH #749.

Kytorhynchus gen. n.

Proboscis undivided, proboscis glands open into the whole proboscis epithelium, sensory hairs more or less reduced and sometimes limited to the distal region; four to six nuclei in the proboscis epithelium, retractor muscles well-developed. Eyes present or absent. Testes ventral compared to the vitellaria; male genital canal short, sometimes with a cuticularized stylet. Distal part of the primary female duct enlarged and syncytial.

Kytorhynchus (*Kytorhynchus*) subgen. n.

Proboscis more or less rhomboid in shape. Male copulatory organ without cuticularized stylet. With paired secondary efferent ducts of the female gonad. The original female genital canal functioning only as a bursal organ and with a distinct proximal glandular girdle and a cuticularized tube-shaped mouthpiece; female gonads connected to the original female genital canal by a syncytium.

Kytorhynchus (*Kytorhynchus*) *oculatus* sp. n. (Figs. 3*a*, 4*b*, 9*a*, *b*, 10, 12*b*, 14*c*, 15*c*)

700 to 800 μm long, with eyes, slightly brownish in color, proboscis 100 μm long, elongate, with long sensory hairs throughout; pharynx without nuclei in the circular groove in the adult stage; male copulatory organ 100 μm long, 50 μm wide, with two kinds of secretory granules in the prostatic vesicle. Cuticular tube in the female duct 10 to 15 μm long and bent at the proximal end.

TYPE LOCALITY: Off Beaufort, North Carolina; *R/V Eastward* cruise E-4-70, station 14207 at 34°12′N and 76°14′W, 28 m depth; April 1970; and *R/V Eastward* cruise E-10-70, station 14457 at 34°18′N and 16°13.6′ W, 30 m depth; May 1970; sand rich in organic matter. (Courtesy Dr. C. Jenner.)

HOLOTYPE: One sagittal serial section. AMNH #750.

PARATYPE: One transverse serial section, AMNH #751.

From the following two species, *Kytorhynchus* sp. I and *Kytorhynchus* sp. II, only specimens have been studied which did not have a fully developed female system. Therefore it is uncertain whether they belong to the above-defined subgenus *Kytorhynchus* s. str. However, the two forms agree with *Kytorhynchus* s. str. in the shape of the proboscis and in the lack of a cuticularized stylet, both characters which also separate the subgenus *Kytorhynchus* s. str. from *K.* (*Kytorhynchoides*). Therefore they should be mentioned here as an appendage of the subgenus *K.* (*Kytorhynchus*).

Kytorhynchus sp. I (Fig. 3*b*, *d*)

400 to 600 μm long, with eyes, slightly brownish in color, proboscis short (50 μm) and with sensory hairs only in the distal part; male copulatory organ 50 to 90 μm long, 30 μm wide, with only one kind of prostatic gland.

TYPE LOCALITY: Off Beaufort, North Carolina; *R/V Eastward* cruise

E-10-70, station 14460 at 34°25.5′N and 76°20′W, 20 m depth; May 1970; fairly clean, heterogeneous coarse sand. (Courtesy Dr. B. Coull.)

Kytorhynchus sp. II (Figs. 3*c*, 6*e*, 12*c*, 14*d*

600 to 700 μm long, without eyes, slightly brownish in color, proboscis especially rich in proboscis glands, pharynx without nuclei in the circular groove; male copulatory organ (100 μm long, 40 μm wide) with an elongate distal glandular part (prostatic vesicle), in the latter two kinds of secretion granules.

TYPE LOCALITY: Off Beaufort, North Carolina; *R/V Eastward* cruise E-4-70 station 14208 at 34°12.5′N and 76°14′W, 28 m depth; May 1970; heterogeneous coarse sand with detritus. (Courtesy Dr. C. Jenner.)

Kytorhynchus (*Kytorhynchoides*) subgen. n.

Proboscis triangular to cup-shaped; without eyes; male copulatory organ with a cuticularized stylet, the latter funnel-shaped, with longitudinal folds; only one connection between the female gonad and the atrium; the proximal, glandular part of the female genital canal enlarged.

Kytorhynchus (*Kytorhynchoides*) *microstylus* sp. n. (Figs. 4*c*, 5*a*, *c–e*, 7*a–g*, 9*c*, *f*, 11*a*, 12*d*, 14*d*, *f*, 15*d*)

700 to 800 μm long, brownish in color; male copulatory organ 80 to 100 μm long (except stylet), stylet 30 to 35 μm long, with a hook at the distal end, only slightly cuticularized; proximal part of the female genital canal an enlarged glandular girdle.

TYPE LOCALITY: Off Beaufort, North Carolina; *R/V Eastward* cruise E-10-70, station 14457 at 34°18′N and 16°73′W, 30 m depth; and station 14455 at 34°16′N and 76°8′W, 60 m depth; May 1970; heterogeneous coarse sand rich in detritus. (Courtesy Dr. B. Coull.)

OTHER OCCURRENCE: Tobacco Bay, Bermuda; 1.5 m depth, calcareous medium sand rich in organic matter. The specimens obtained there showed several differences in measurement and probably represent a different geographic form (see Figs. 7*c*, *d*, 11*a*, 12*d*, 14*d*).

HOLOTYPE: One transverse serial section, AMNH #753.

PARATYPE: One sagittal serial section (Bermuda form), AMNH #754. One transverse serial section (Bermuda form), AMNH #754.

Kytorhynchus (*Kytorhynchoides*) *macrostylus* sp. n. (Figs. 5*b*, 6*f*, 9*e*, 11*b*, 12*e*, 15*a*, *b*, 16*c*)

1100 to 1200 μm long, dark brownish in color; male copulatory organ

180μm long (except stylet), stylet 80 μm long, without distal hook and without longitudinal folds in the proximal part; proximal part of the female genital canal transformed to a spherical glandular mass, the germaria directly attached to this latter part.

TYPE LOCALITY: Off Beaufort, North Carolina; *R/V Eastward* cruise E-10-70, station 14460 at 34°25.5′N and 76°20′W; 20 m depth; May 1970; heterogeneous coarse sand, fairly clean. (Courtesy Dr. B. Coull.)

HOLOTYPE: One transverse serial section, AMNH #755.

LIST OF ABBREVATIONS

ag	genital atrium
bm	basement membrane
br	brain
buf	bursal region of the female genital duct
c	cilia
cc	central canal
cg	cement glands
cgl	central glands of the female duct
cm	circular muscle
co	copulatory organ
ct	cuticularized tube
dm	diagonal muscle
ec	end cone of the proboscis
em	extrinsic muscle of the pharynx
ep	epidermis
es	esophagus
g	germarium
gl	gland cells
gl 1	slender shell glands
gl 2	gland cells of the genital atrium
glp	glandular penis papilla
gls	glandular sac of proboscis
gp	common genital pore
grf 1, 2	granular girdle of the female genital duct
i	intestine
icm	inner circular muscles of the pharynx
ilm 1, 2	inner longitudinal muscles of the pharynx
la	lateral lobes of the genital atrium
lg	lateral glands of the genital atrium
lm	longitudinal muscle
mgd	male genital duct

mi	Mignot's glands
mo	mouth opening
ms	muscle sheath
n	nucleus
nc	ganglion cells
nr	nerve ring
od	oviduct
on	optical nerve
pg	prostatic glands
ph	pharynx
phc	pharyngeal cavity
pr	proboscis
pre	proboscis epithelium
prg	proboscis glands
rh	rhabdites
rhm	rhammites
rm	retractor muscles of the proboscis
rs	seminal receptacle
sp	sperm
sph	sphincter muscle
st	stylet
t	testis
vd	vas deferens
vg	vesicula granulorum
vi	vitellarium
vs	vesicula seminalis
x	"secondary" female genital duct

BIBLIOGRAPHY

Antonius, A. 1965. Methodischer Beitrag zur mikroskopischen Anatomie und graphischen Rekonstruktion sehr kleiner zoologischer Objekte. *Mikroskopie*, **20**(5/6): 145–153.

Ax, P. 1956. Les Turbellariés des Etangs côtiers du littoral méditerranéen de la France méridionale. *Vie et Milieu, Suppl.* **5**:1–215.

———. 1960. Turbellarien aus salzdurchtränkten Wiesenböden der deutschen Meeresküsten. *Z. wiss. Zool.*, **163**(1/2):210–235.

———. 1961. Verwandtschaftbeziehungen und Phylogenie der Turbellarien. *Ergeb. Biol.*, **24**:1–68.

———. 1963. "Relationship and Phylogeny of the Turbellaria." In E. C. Dougherty (ed.), *The Lower Metazoa*, pp. 191–224. Berkeley and Los Angeles: University of California Press.

————. 1971. Zur Systematik und Phylogenie der Trigonostominae (Turbellaria, Neorhabdocoela). *Mikrofauna des Meeresbodens*, **4**:1–80.

———— and R. Heller. 1970. Neue Neorhabdocoela (Turbellaria) vom Sandstrand der Nordsee-Insel Sylt. *Mikrofauna des Merresbodens*, **2**:1–46.

Den Hartog, C. 1964. A Preliminary Revision of the Proxenetes-group (Trigono-stomidae, Turbellaria). I. *Koninkl Ned. Akad. Wetenschap. Proc*, **C67**: 371–407.

Fenchel, T., and R. Riedl. 1970. The Sulfide System; a New Biotic Community underneath the Oxidized Layer of Marine Sandy Bottoms. *Marine Biol.*, **7**:255–269.

Graff, L. von. 1905. Marine Turbellarien Orotovas und der Küsten Europas. I. Einleitung und Acoela. *Z. Wiss. Zool.*, **78**:190–244.

————. 1904–1908. "Acoela und Rhabdocoelida." In H. G. Bronn (ed.), *Klassen und Ordnung des Tierreichs*, 4: Vermes Ic: Turbellaria I: 1733–2599.

Karling, T. G. 1940. Zur Morphologie und Systematik der Alloeocoela cumulata und Rhabdocoela lecithophora (Turbellaria). *Acta Zool. Fenn.*, **26**:1–260.

————. 1952a. Studien ueber Kalyptorhynchien (Turbellaria). Einige Eukalyptorhynchia. *Acta Zool. Fenn.*, **2**: 1–49.

————. 1952b. Kalyptorhynchia (Turbellaria). *Further Zool. Results Swed. Antarctic Expedition, 1901–1903*, 4/9: 1–50.

————. 1953. *Cytocystis clitellatus* n. gen., n. spec., ein neuer Eukalyptorhynchientypus (Turbellaria). *Arkiv Zool.*, (2)**4**:493–504.

————. 1954. Einige marine Vertreter der Kalyptorhynchien-Familie Koinocystidae. *Arkiv Zool.*, (2)**7**:165–183.

————. 1955. Studien ueber Kalyptorhynchien (Turbellaria). V. Der Verwandtschaftskreis von *Gyratrix* Ehrenberg. *Acta Zool. Fenn.*, **88**:1–39.

————. 1956. Morphologisch-histologische Untersuchung an den maennlichen Atrialorganen der Kalyptorhynchia (Turbellaria). *Arkiv Zool.*, (2)**9**:187–278.

————. 1957. Drei neue Turbellaria Neorhabdocoela aus dem Grundwasser der schwedischen Ostseekueste. *Kgl. Fysiograf. Sallskap. Lund Forh.*, **27**:25–33.

————. 1962a. Marine Turbellaria from the Pacific Coast of North America I. Plagiostomidae. *Arkiv Zool.*, (2)**15**:113–141.

————. 1962b. Marine Turbellaria from the Pacific Coast of North America II. Pseudostomidae and Cylindrostomidae. *Arkiv Zool.*, (2)**15**:181–209.

————. 1963. "Some Evolutionary Trends in Turbellarian Morphology." In E. C. Dougherty (ed.), *The Lower Metazoa*, pp. 225–233. Berkeley and Los Angeles: University of California Press.

————. 1964. Ueber einige neue und ungenuegend bekannte Turbellaria Eukalyptorhynchia. *Zool. Anz.*, **172**(3):159–183.

————. 1968. On the Genus *Gnosonesimia* Reisinger (Turbellaria). *Sarsia*, **33**:81–108.

Kepner, W. A., and T. K. Ruebush. 1935a. *Microrhynchus virginianus* n. gen., n. spec. *Zool. Anz.*, **111**:257–261.

———— and ————. 1935b. Berichtigung. *Zool. Anz.*, **112**:272.

————, F. F. Ferguson, and Margaret A. Stirewalt. 1941. A New Turbellarian (Rhabdocoele) from Beaufort, North Carolina, *Trigonostomum prytcherchi*, n. sp. *J. Elisha Mitchell Sci. Soc.*, **57**:243–251.

Luther, A. 1904. Die Eumesostominen. *Z. Wiss. Zool.*, **77**:1–273.

———. 1943. Untersuchungen an rhabdocoelen Turbellarien IV. Ueber einige Repraesentanten der Familie Proxenetidae. *Acta Zool. Fenn.*, **38**:1–95.

———. 1946. Untersuchungen an rhabdocoelen Turbellarien V. Ueber einige Typhloplaniden. *Acta Zool. Fenn.*, **46**:1–56.

———. 1948. Untersuchungen an rhabdocoelen Turbellarien VII. Ueber einige marine Dalyellioida, VIII. Beitraege zur Kenntnis der Typhloplanoida. *Acta Zool. Fenn.*, **55**:1–131.

———. 1950. Untersuchungen an rhabdocoelen Turbellarien IX. Zur Kenntnis einiger Typhloplaniden, X. Ueber *Astrotorhynchus bifidus* (M'I.). *Acta Zool. Fenn.*, **60**:1–41.

———. 1962. Die Turbellarien Ostfennoskandiens III. Neorhabdocoela 1. Dalyellioida, Typhloplanoida: Byrsophlebidae und Trigonostomidae. *Soc. F. Fl. Fenn. F. F.*, **12**:1–71.

———. 1963. Die Turbellarien Ostfennoskandiens IV. Neorhabdocoela 2. Typhloplanoida: Typhloplanidae, Solenopharyngidae, Carcharodopharyngidae. *Soc. F. Fl. Fenn., F. F.*, **16**:1–163.

Marcus, E. 1947. Turbellarios marinhos do Brasil. *Bol. Fac. Filosof. Ciências Letras Univ. São Paulo, Zool.*, **12**:99–216.

———. 1948. Turbellaria do Brasil. *Bol. Fac. Filosof. Ciências Letras Univ. São Paulo, Zool.*, **13**:111–243.

———. 1951. Turbellaria Brasileiros (9). *Bol. Fac. Filosof. Ciências Letras Univ. São Paulo, Zool.*, **16**:5–216.

———. 1952. Turbellaria Brasileiros (10). *Bol. Fac. Filosof. Ciências Letras Univ. São Paulo, Zool.*, **17**:5–188.

———. 1954. Turbellaria. Reports of the Lund University Chile Expedition 1948–49, 11. *Lunds Univ. Årsskr.*, N. F. Avd. 2, **49**:1–115.

Meixner, J. 1924. Studien zu einer Monographie der Kalyptorhynchia und zum System der Turbellaria Rhabdocoela. *Zool. Anz.*, **60**:89–105, 112–125.

———. 1925a. Beitrag zur Morphologie und zum System der Turbellaria-Rhabdocoela I. Die Kalyptorhynchia. *Z. Morphol. Ökol.*, **3**:255–343.

———. 1925b. Beitrag zur Morphologie und zum System der Turbellaria-Rhabdocoela II. Ueber *Typhlorhynchus nanus* Laidlaw und die parasitischen Rhabdocoelen nebst Nachtraegen zu den Calyptorhynchia. *Z. Morphol. Ökol.*, **5**:577–624.

———. 1928. Aberrante Kalyptorhynchia (Turbellaria Rhabdocoela) aus dem Sande der Kieler Bucht I. *Zool. Anz.*, **77**:229–253.

———. 1938. "Turbellaria (Strudelwuermer) I." In G. Grimpe and E. Wagler (eds.), *Die Tierwelt der Nord- und Ost- see* IVb, 146 pp.

Papi, F. 1959. Ricerche su alcuni generi affini della fam. Typhloplanidae (Turbellaria Neorhabdocoela). *Arch. Zool. Ital.*, **44**:1–51.

Reisinger, E. 1937. Morphologie und Entwicklungsgeschichte der Wirbellosen (excl. Arthropoda). *Fortschr. Zool.*, N. F. **1**:93–110.

Riedl, R. 1954. Neue Turbellarien aus dem mediterranen Felslitoral. Ergebnisse der "Unterwasser-Expedition Austria 1948–49." *Zool. Jahrb. Syst.*, **82**(3–4): 157–244.

———. 1970. On *Labidognathia longicollis* nov. gen., nov. spec., from the West Atlantic coast. *Intern. Rev. Ges. Hydrobiol.*, **55**:227–244.

Rieger, R., and W. Sterrer. 1968. *Megamorion brevicauda* gen. nov., spec. nov., ein Vertreter der Turbellarienordnung Macrostomida aus dem Tiefenschlamm eines norwegischen Fjords. *Sarsia*, **31**:75–100.

Ruebush, T. K. 1939. A New North American Rhabdocoel Turbellarian, *Prorhynchella minuta* n. gen., n. sp.. *Zool. Anz.*, **127**:204–209.

——. 1941. A Key to the American Freshwater Turbellarian Genera, Exclusive of the Tricladida. *Trans. Am. Microscop. Soc.*, **60**:29–40.

Sterrer, W. 1968. Beiträge zur Kenntnis der Gnathostomulida. I. Anatomie und Morphologie des Genus *Pterognathia* Sterrer. *Arkiv Zool.*, **22**:1–125.

Westblad, E. 1952. Turbellaria (excl. Kalyptorhynchia) of the Swedish South Polar Expedition 1901–1903. *Further Zool. Results Swed. Anarctic Expedition, 1901–1903*, **4**:1–55.

Chapter 4

Retronectidae—a New Cosmopolitan Marine Family of Catenulida (Turbellaria)

Wolfgang Sterrer
Bermuda Biological Station for Research, Bermuda
Reinhard Rieger
Department of Zoology, University of North Carolina,
Chapel Hill, and Institute of Marine Sciences,
Morehead City, North Carolina

The turbellarian order Catenulida has long been regarded as restricted to freshwater. In 1959, however, Riedl described *Tyrrheniella sigillata*, a peculiar turbellarian from coastal caves, which he considered the first marine representative of the order.

Since the early 1960s we have found marine catenulids on most European coasts (Sterrer, 1966) and more recently on the American east coast. Additional data have been contributed by Riedl (pers. comm.) for the Red Sea, Swedmark (pers. comm.) for the Baltic, and Bush (1968, fig. 2) for the Woods Hole area. At the present time we know of about 15 species of marine Catenu-

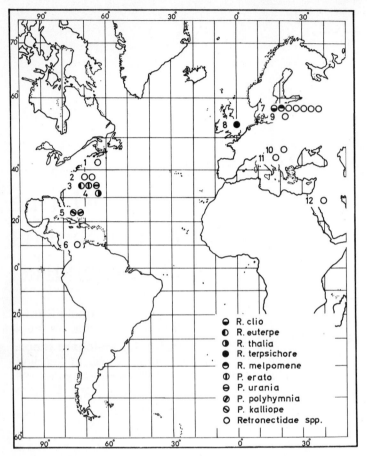

Figure 1 Geographic distribution of Retronectidae. Localities:
1, Woods Hole, Mass.; 2, Morehead City, N.C.; 3, Wilmington, N.C.;
4, off the North Carolina coast; 5, Lower Keys, Florida; 6, Atlantic
coast of Panama; 7, Kristineberg, Swedish west coast; 8, Portaferry,
Northern Ireland; 9, Simrishamn, Swedish Baltic coast; 10, Rovinj,
northern Adriatic; 11, Marina di Carrara, Ligurian coast; 12, Al
Ghardaqa, Red Sea.

lida, which represent almost 20 percent of all known species of this order
(Fig. 1).

We agree with Reisinger (1924), Marcus (1945a, b), and Luther (1960)
that the study of Catenulida is one of the most difficult in the systematics of
Turbellaria. The animals are extremely fragile and are rarely encountered
sexually mature. Hard structures are lacking, and most characters (especially
shape, color, and even the number of statoliths) show a high variability. In

addition, the most refined methods of sample extraction that routinely produce large numbers of other turbellaria, gastrotrichs, and gnathostomulids yield no more than a few specimens of catenulids. The fact that marine catenulids have been recorded by several authors and seem to have worldwide distribution encouraged us to present the following data.

MATERIAL AND METHODS

Qualitative samples of about 10 l volume were taken with a spade in the intertidal, with a hand net or bucket in shallow depths (by skin diving), and with a dredge in deep water. Samples were then brought into the laboratory and treated according to the method used for Gnathostomulida (Sterrer, 1968; Hulings and Gray, 1971). This involves deterioration of sample climate, anesthesia with an isotonic solution of magnesium chloride, and subsequent active migration of specimens through a 63-μm filter.

Most animals were observed alive in squeeze preparation, using a phase-contrast microscope with drawing tube and photomicrographic attachment. Measurements were taken from both photographs and scale drawings.

For microanatomy, animals were fixed and preserved in hot Bouin, or glutaraldehyde, and treated according to Antonius (1965). The staining technique used was Azocarmin-Pasini.

For electron microscopy, specimens were anesthetized in $MgCl_2$ solution, fixed in 6.5 percent glutaraldehyde in phosphate buffer, postfixed in OsO_4, embedded in Epon-Araldite. For staining, uranyl acetate/lead citrate was used. Pictures were taken with a Zeiss EM 9A electron microscope.

For the description of body proportions and positions of organs a relative scale will be used, 100 units (U) being equivalent to the total length of a specimen. Positions (e.g., ovary from U13 to U15) or proportions (e.g., width of body equal to 10U) are then defined starting at the anterior tip (U0) of the animal (Rieger and Sterrer, 1968).

Type specimens, where available, have been deposited at the Department of Living Invertebrates, The American Museum of Natural History, New York.

BIOTOPE AND DISTRIBUTION

As a rule, Retronectidae occur in a rather sheltered environment represented by sand of varying grain size containing a high amount of organic detritus. This sediment can be found from the upper tidal (and even brackish groundwater, see *R. euterpe*) to the intertidal and immediate subtidal (where most species have been found so far), to a depth of 30 m. The only species (*Retronectes* sp. IV) found in a high-energy beach situation occurred between mid and high water level in about 30 cm sediment depth, again demonstrating the group's preference for sheltered conditions.

Quantitative analysis of vertical zonation in the sediment showed that the majority of specimens and species occur in the vicinity of the "black" or redox-discontinuity layer. The biotope of Retronectidae, especially the genus *Paracatenula* gen. n., therefore is identical with the typical biotope of Gnathostomulida (Sterrer, 1971a, b). As an example, of 63 samples of sediment from various geographic locations, and yielding gnathostomulids, 19 also produced Retronectidae; whereas no gnathostomulids were found in only one sample containing Retronectidae.

This biotope, recently described by Fenchel and Riedl (1970) and termed the "sulfide system," is an anaerobic layer underlying the oxidized surface of marine sediments, and is thus of worldwide distribution. So far, Retronectidae have been found in whatever geographical area they (or gnathostomulids) have been looked for (Fig. 1). Most of our findings are from European coasts, and the American east coast; however, the ecological requirements which they share with gnathostomulids let us expect that they also share the worldwide distribution of the latter group.

Since it is our experience that the European and the American coasts of the Atlantic yield different (though often closely related) species, and the whole of the Pacific is still unknown in this respect, we can anticipate a considerable number of new species of Retronectidae to be discovered, some of them possibly of high systematic interest.

GENERAL MORPHOLOGY

The nine species to be described here belong to two genera, *Retronectes* and *Paracatenula*. Species differences, as we see them now, lie in the general shape and length of the body and rostrum, color, statocyst, inclusions and pseudorhabdites, and the sperm (Fig. 2a–ff). The main generic differences are the presence or absence of mouth, pharynx, and gut lumen.

External Features

All species are more or less cylindrical or slightly depressed, sometimes with a rostrum, or a tail; the size ranges from 500 μm to 7 mm. They are colorless, or of dark appearance due to the brownish or greenish granules in the skin, or to the "granular strands" (see "Parenchyma" below). Movement is a slow ciliary gliding, with occasional spiralization (Fig. 2g), and reversal of the ciliary beat.

Epidermis

The epidermis is a one-layered epithelium with very sparse ciliation (Fig. 13h), and microvilli that are occasionally forked (Fig. 6). False, basophil

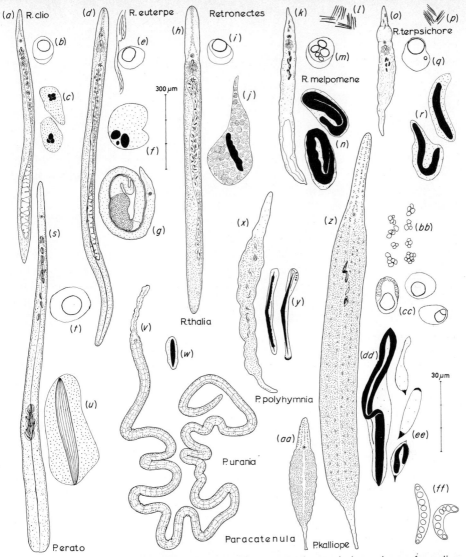

Figure 2 Retronectidae, general morphology drawn from live specimens. All organizations are in one scale, all details in the other. (a–c) *Retronectes clio* sp. n.: (a) organization; (b) statocyst; (c) sperm. (d–g) *Retronectes euterpe* sp. n.: (d) organization; (e) statocyst and part of protonephridia; (f) sperm; (g) "female" specimen. (h–j) *Retronectes thalia* sp. n.: (h) organization; (i) statocyst; (j) sperm. (k–n) *Retronectes melpomene* sp. n.: (k) organization; (l) rhabdoids from rostrum; (m) statocyst; (n) sperm. (o–r) *Retronectes terpsichore* sp. n.: (o) organization; (p) rhabdoids from rostrum; (q) statocyst; (r) sperm. (s–u) *Paracatenula erato* sp. n.: (s) organization; (t) statocyst; (u) sperm. (v, w) *Paracatenula urania* sp. n.: (v) organization; (w) sperm. (x, y) *Paracatenula polyhymnia* sp. n.: (x) organization; (y) sperm. (z, aa–ff) *Paracatenula kalliope* sp. n.: (z) organization; (aa) juvenile specimen; (bb) inclusions from rostrum; (cc) statocysts of three specimens; (dd) sperm; (ee) spermatids; (ff) inclusions from parenchyma.

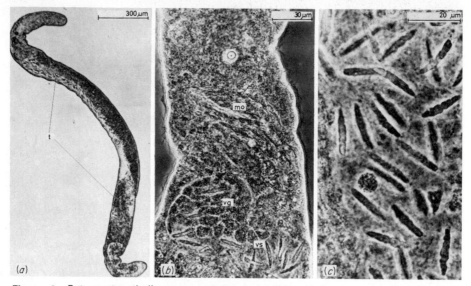

Figure 3 *Retronectes thalia* sp. n., photomicrographs of live specimens. (a) Organization of slightly squeezed animal; (b) statocyst, pharynx, and male copulatory organ; (c) sperm.

adenal rhabdites may occur. There is no basement membrane (see Chap. 6 for another catenulid, but also see Pullen, 1957, p. 585). In electron micrographs the microvilli are embedded in an outer layer (Fig. 7: o), covering the whole surface like a "cuticle." This layer might correspond to the outer layer observed in light microscopy of histological preparations. A lamellar structure encountered in an electron micrograph (Fig. 6: x) agrees very well with what Jennings (Chap. 9) calls indigestible material, and may be an indication of food intake through the epidermis.

Large nuclei, partly or fully sunken underneath the epithelium, were found in the whole body (Fig. 10e: n). They might belong to glandular cells as described by Reisinger (1924) for *Rhynchoscolex*. The body musculature, in *Retronectes* (Fig. 10e), consists of delicate circular fibers (cm) immediately under the skin, and a few fairly strong longitudinal fibers (lm). In *Paracatenula* there is no muscle sheath, but in *P. erato* (Fig. 9: f), circular and longitudinal fibers, possibly muscles, occur in the epithelium.

Nervous System

The fairly complicated brain always shows a central mass of nerve fibers (Fig. 5), with a pair of caudal gangliar lobes, and a varying number and arrangement of frontal lobes (Fig. 11d) which seem to correspond to the "sensory plates" of Stenostomidae, with the nerves toward the center as in *Rhynchoscolex*.

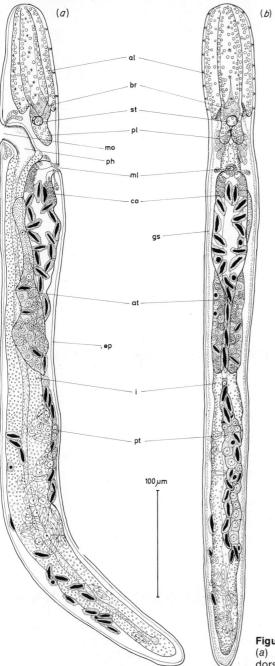

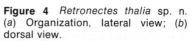

Figure 4 *Retronectes thalia* sp. n. (*a*) Organization, lateral view; (*b*) dorsal view.

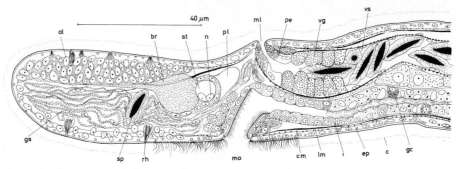

Figure 5 *Retronectes thalia* sp. n. Organization of anterior body region, lateral view.

The number of posterior body nerves is still uncertain. There are no refractile bodies.

A statocyst (st), situated between the two caudal lobes of the brain (Fig. 4*b*), is present in all but two of the species. It is a small vesicle with usually four nuclei in its ventral wall (Figs. 5, 9). Its anterior wall is attached to the central mass of nerve fibers, and it contains one statolith in most species. In some, however, the number of statoliths may vary from one to two (Figs. 2*q*, 12*a*), and up to six (Fig. 15*k*). They are never connected with each other.

Parenchyma and Protonephridia

What may be called a parenchyma is present in the form of two to four granular strands (Figs. 4*b*, 5, 7, 9: gs) which originate in the proboscis, under the brain, and extend throughout the whole body.

Protonephridia were found in only two *Retronectes* species (Figs. 2*e*, 15*r*); they conform to the unpaired dorsal type usually found in catenulids. In *Paracatenula*, a "dorsal cord" (Figs. 9, 10*j*, 11*e*: dc) of fibrous appearance is present—to be interpreted either as a modified excretory canal now in connection with the male organ or, more probably, as a contractile element. It extends throughout the whole body and originates ventrally between the two posterior lobes of the brain. In *P. erato* this dorsal cord is embedded in tissue which seems to be connected with the epidermis. This tissue contains the sperms (Fig. 10*j*).

Digestive Tract

In *Paracatenula*, the mouth opening is lacking, as is the pharynx (Fig. 9). The gut consists of a rodlike arrangement of very large turgescent cells (Figs. 9, 11*e*: i). The cells are filled with little granular bodies, which make the gut look like a storage organ with no digestive function.

In *Rectronectes*, the mouth opening (Fig. 5: mo) is oval, the pharynx is simple and tubular, with basophil glands in a circular arrangement at its distal part. Both pharynx and gut are ciliated. In the pharynx, however, the cilia are especially dense and have very long rootlets (Fig. 7: rl). The cells of the gut contain many vacuoles, and the cell borders are not clearly seen. Between these cells basophil glandular cells can be found.

Genital Organs

Probably the most puzzling feature of the whole group is the reproductive system. In none of the species have we found an indication of asexual reproduction; on the other hand, only one specimen was ever found to contain a mature egg. All the other specimens have male organs that basically conform to the catenulid type: a male pore situated dorsally above or behind the

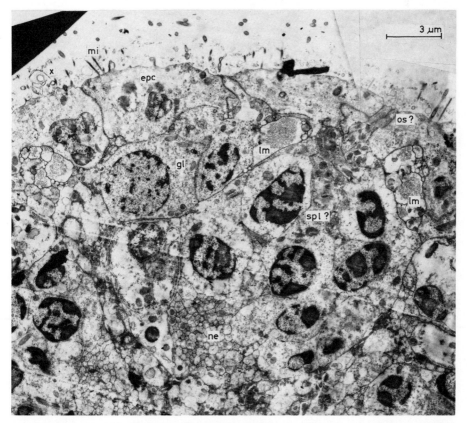

Figure 6 *Retronectes thalia* sp. n. Ultramicrograph of a transverse section in front of the brain.

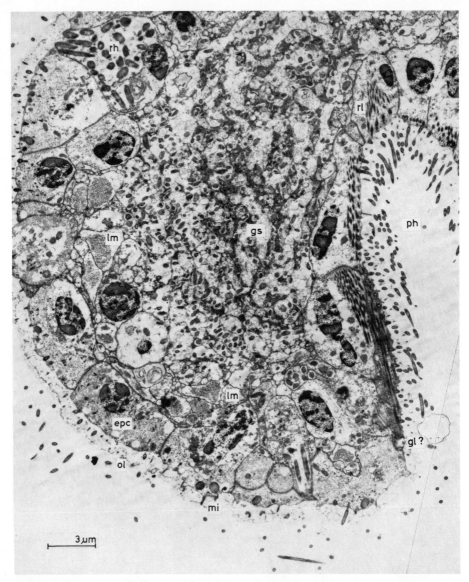

Figure 7 *Retronectes thalia* sp. n. Ultramicrograph of a transverse
section through the pharynx region.

pharynx, and a dorsal male organ and testis (Fig. 4*a*, *b*). The latter, however,
is not compact, but extends far into the posterior body region. It is typically
hourglass-shaped, with an anterior and posterior vesicle connected by a solid,

sometimes interrupted rod of spermatogonia. The anterior part of the organ (Fig. 5), somewhat difficult to subdivide into penis (pe), vesicula granulorum (vg), and vesicula seminalis (vs), is surrounded by a membrane and circular muscles. In *R. thalia* there is an additional pair of lobes, joining ventrolaterally, that must also be regarded as a germinative zone of the testis (Fig. 4*a, b*: at).

What we have described as a "testis" should probably be considered as a gonad with mixed production of eggs and sperms. Although spermatogenesis stages (see below) can be found in the posterior region of the hourglass-shaped organ, the latter generally shows a few very large germinative cells (Fig. 11*c*). The latter probably develop into oocytes during certain stages of the animal's life. This is strongly supported by the fact that in one specimen of *Rectronectes thalia* we found a few germinative cells in the posterior region of the gonad, which, owing to their size and the presence of numerous small erythrophilous granules, must be regarded as oocytes (Fig. 8: oc).

We never found any flagellated sperm; however, the male organ usually contains structures of distinct shape that—without doubt—represent final stages of spermatogenesis. Basically, this sperm consists of a more or less elongated nucleus in a homogeneous matrix, and it can have the appearance

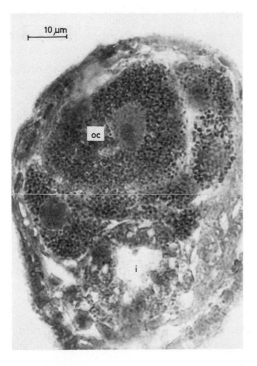

Figure 8 *Retronectes thalia* sp. n. Photomicrograph of a transverse section through the posterior body region (gonad!).

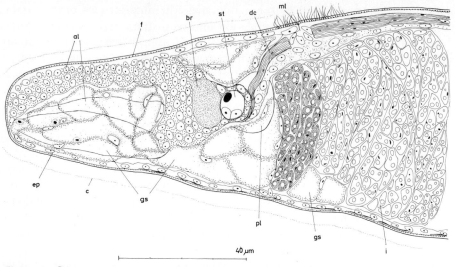

Figure 9 *Paracatenula erato* sp. n. Organization of anterior body region, lateral view.

of conglomerates (Fig. 2*c*), rods (Fig. 2*w*), ribbons (Fig. 2*n, dd*), spicules (Fig. 2*y*), or bananas (Fig. 2*u*).

Spermatogenesis, as fragmentarily recorded from *P. erato*, follows this general pattern (Fig. 10*a–d*):

A large rounded stage contains a round nucleus, and a plasma with a loose arrangement of pearl-string-like spirals of small granules (Fig. 10*a*). In a later stage, both the cell and the nucleus are more elongated, and the pearl strings condensed (Fig. 10*b*). Further elongation takes place, and the granules now appear in agglomerations (Figs. 10*c*, 14*b*). Finally, what we consider to be the sperm shows an elongated nucleus with longitudinal ridges on its surface, whereas the granules are now evenly distributed throughout the plasma (Figs. 10*d*, 14*c*). Whether these "stages" represent spermatogonia, spermatocytes, or spermatids we cannot say, as we never observed meiosis.

Although the structures described above can be termed sperm, we do not know whether they ever act as such or have any direct function in reproduction at all. However, we did find sperm outside the male organ (especially in the posterior part of the gut), even in the rostrum (Fig. 5), that may have originated from a partner.

Discussion

A brief consideration of the main characters shows there is no doubt about the group belonging to Catenulida. The new marine forms are actually much closer to the organization of the freshwater catenulids than the first-described marine catenulid, *Tyrrheniella sigillata*. External features and behavior, especially spiralization and swimming backwards, have been reported for several other representatives, especially *Rhynchoscolex*, by Reisinger (1924).

The brain, with its lobes and sensory plates, conforms to the nervous system of Stenostomidae. A statocyst is widely represented in Catenulida, and a varying number of statoliths within the same species has been reported for *Rhynchoscolex remanei* by Rixen (1961). Our own material shows a similar variability in *Nemertoderma* (Sterrer, 1966) and in the curious dalyelliid-like *Lurus evelinae* Marcus, 1950.

The fact that parenchyma is rather poorly represented in Catenulida has been stated previously. Borkott, in his 1970 paper on *Stenostomum*, reports it completely substituted for by a body fluid that fills the large cavity between skin and gut. The organization of what may be called a parenchyma into "granular strands," however, is new for the order.

The lack of protonephridia, in most species, can be related to their marine habitat. We are quite sure that the "dorsal cord" in *Paracatenula* is a homologous structure; because of its fibrous appearance and the lack of a subepidermal muscle sheath, we suspect it may have to do with the unusual ability to spiralize, as observed in this genus.

The lack of a mouth opening and of a pharynx in *Paracatenula* needs some explanation. Such a situation, known from certain parasites (as, for instance, *Kronborgia*), is rare in free-living animals, and the food intake in such cases is then effected by a specialized epidermis [e.g., in Pogonophora (Southward and Southward, 1971)]. We cannot, of course, completely exclude the possibility that in all four species the lack of mouth and pharynx represents just a temporary stage in the life cycle of the animal; the habitat of our species, however, is a medium with considerable concentrations of dissolved organic matter and high bacterial activity, and, therefore, we are inclined to assume the possibility of food intake through the skin. Furthermore, one of our species was isolated from tubes of onuphid polychaetes, together with an also mouthless nematode.

As far as *Retronectes* is concerned, the term "pharynx simplex" implies that we are talking of a plesiomorph character, poor in details, that may very well be a heterogeneous assemblage of superficially similar structures of different origin. In both cases, pharynx or nonpharynx, we expect to learn a great deal more from further ultrastructural analysis.

Reproduction in Catenulida is most peculiar. Whereas in the rest of the order there is an often-reported tendency to reduce the reproductive system, and especially the male organs (Borkott, 1970), we have in the Retronectidae a suppression of the female organs, without asexual reproduction substituting for it. Whether we are confronted with a completely new mode of reproduction

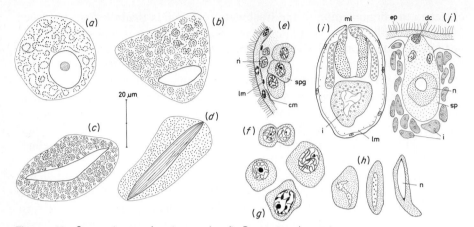

Figure 10 Spermatogenesis stages. (a–d) *Paracatenula erato* sp. n., from a live specimen. (e–h) *Retronectes thalia* sp. n.: (e) epidermis and young spermatogonia, detail from a transverse section through the mid-body region; (f) spermatogonia from the mid-body region; (g) large germinative cells from the posterior part of the "testis"; (h) spermatogenesis stages. (i) *Retronectes clio* sp. n., transverse section through male opening. (j) *Paracatenula erato* sp. n., dorsal cord with sperm, in posterior body region. All figures except part *i* are in the same scale.

—for instance, parthenogenesis from sperm—or with just a fragment of a complex cycle, we do not know. In connection with the possibility of the presence of a mixed gonad (see p. 73) in the marine forms, the recent report of Borkott (1970) is very interesting. According to this author spermatogonia of the proximal part of the testis in *Stenostomum sthenum*, under certain environmental conditions, can develop into fertile eggs. This author considers such an ovary a secondary one, whereas the primary ovary generally develops from the indifferent cells of the gut. If, however, the presence of a mixed gonad with an anterior male and a posterior female part (as in the acoel family Antigonariidae Dörjes, 1968) should prove to be typical for marine Catenulida, then such a gonad situation would have to be considered the more primitive one within the order. It is further suspected that copulation, in Catenulida, does not lead to an exchange of sperm. On the other hand, there is a possibility that the well-known cannibalism represents the true copulation in this group. Riser (pers. comm.) suggests a similar situation in *Childia groenlandica*, where cocoons are routinely laid within 18 hours after cannibalism.

In order to complete the picture, we should also mention the peculiar origin of reproductive cells in Catenulida: they are derived from indifferent cells in ectoderm or entoderm, and have been reported by Borkott (1970) to get their male determination under the inductive influence of the skin. In *R. thalia* we found a few large nuclei in the posterior part of the body right underneath the epidermis. These nuclei, which are considered as belonging to glandular cells (see p. 68), could also be early stages of germinative cells. Unfortunately, our material was not sufficient to clarify this point.

SYSTEMATIC POSITION

From the systematic point of view our marine group merits the status of a new family, owing to the following characters:

1 Lack of paratomy.
2 Lack of ciliated pits or furrows.
3 "Testis" not short and compact, but extending into posterior part of the body. Oocytes probably developing from the posterior region of the "testis" (mixed gonad).
4 "Sperm" undergoing a histogenesis usually leading to an elongated shape.
5 In most species, paired granular strands of parenchymous or secretory nature are present between gut and skin.

The new family Retronectidae is related to Stenostomidae especially with regard to brain anatomy. Within the Stenostomidae, *Rhynchoscolex*, a rather aberrant representative of the family, shares several trends with Retronectidae, such as the lack of an adult paratomy, and a polylithophorous statocyst. The main difference, however, lies in the complete lack of male organs in *Rhynchoscolex*. It may become necessary, in the future, to establish a separate family for this genus. As already mentioned, the Tyrrheniellidae must be considered a fairly distant group within the Catenulida. This is especially true with regard to the filiform sperm, the ventral position of the male gonad, and the position of the male efferent duct (ventral to the brain) and the male opening (dorsofrontal).

In summary, the discovery of two genera with fifteen species of marine Catenulida led to the establishment of the new family Retronectidae and contributed some new insights into the morphology of this turbellarian order. From the standpoint of sperm morphology (see Chap. 7) and general morphology (Chap. 1) as well as the peculiar fine structure of the cyrtocytes (Brandenburg, 1966; Reisinger, 1964), it is evident that the Catenulida are quite separate from the main evolutionary line in Turbellaria. We therefore strongly endorse alternative 2 in Karling's phylogenetic tree (Chap. 1). Owing to the isolated position of the group, we should not shrink from unorthodox approaches. To mention just two: Do we really know what is dorsal and what is ventral in this group, when it is a fact that quite a few Catenulida regularly crawl with their "dorsal" side toward the substratum? And after all, isn't it justified to have doubts as to whether Catenulida are Turbellaria, or even Platyhelminthes at all?

DESCRIPTIONS AND DIAGNOSES

Retronectidae fam. n.

Catenulida without paratomy, without ciliated pits or furrows, and without refractile bodies. Brain with two posterior and a varying (two to five) number of anterior gangliar lobes; mixed gonads elongated, usually hourglass-shaped, mostly in male condition: sometimes the posterior part develops into an ovary. Sperm nucleus usually of very distinct shape (conglomerate, rod, ribbon, spicule, banana). Paired granular strands of parenchymous nature usually present between gut and skin. Generally with a statocyst with four nuclei in the ventral wall. Marine.

The family comprises two genera:

Retronectes gen. n.

Retronectidae with mouth, pharynx, and gut lumen. With or without protonephridia. Usually with a statocyst with one, occasionally several stato-liths.

Type species: *R. thalia* sp. n.

Paracatenula gen. n.

Retronectidae without mouth, pharynx, or gut lumen. Usually without protonephridia but with a dorsal cord. With or without a statocyst; statocyst usually with one statolith.

Type species: *P. urania* sp. n.

Species Diagnoses and Descriptions

Genus *Retronectes*

Retronectes thalia sp. n. (Figs. 2*h–j*, 3*a–c*, 4*a–b*, 5, 6, 7, 8, 10*e–h*, 11*a*, *b*, *d*)

DIAGNOSIS: *Retronectes* with granular strands and a statocyst with one statolith. Without protonephridia. Rostrum fairly thick. Mixed gonad, hourglass-shaped. Nucleus of sperms elongated, 18 μm long with irregular edges. The granular strands cause a slightly yellowish orange colored strip in front of the statocyst.

MATERIAL: Seven adult specimens, three of which were sectioned for light microscopy, and two for electron microscopy.

HOLOTYPE: One series of transverse sections, AMNH #769.

PARATYPES: One series of transverse sections, AMNH #770.

TYPE LOCALITY: North American Atlantic coast, off Beaufort, North Carolina; R/V *Eastward* cruise E-4-70, station 14208 at 34°12.5′N and 76°14′W at a depth of 28 m. Fairly coarse, heterogeneous sand rich in organic matter.

DESCRIPTION: Mature specimens (Figs. 2*h*, 3*a*) are about 1400 μm in length and 70 to 80 μm in width. The rostrum is about 200 to 220 μm long and club-shaped. The body, in the region of the statocyst, is slightly constricted.

The epidermis contains small (1 μm) granules. In the anterior and posterior body part some clusters of elongated adenal rhabdites have been found. However, these structures did not show the typical staining in the sections which one would expect from true rhabdites.

The brain (Fig. 5) has five anterior (one mediodorsal, two lateral, two ventrolateral) and two posterior gangliar lobes.

The statocyst (Figs. 2*i*, 3*b*) located 15 μm behind the brain between the two posterior gangliar lobes always contained only one statolith (7 μm). Four nuclei lie in the ventral part of the statocyst (Fig. 5).

The mouth opening (U18) is about 40 μm long and surrounded by glands which produce small (about 2 μm) oval to round granules. The pharynx (Fig. 5) is ciliated, and this is also true for the gut. The latter extends to the very rear end of the animal.

Behind the brain the granular strands are restricted to the sides of the body (Fig. 4*b*). They contain very small (1 μm) elongated granules.

The male opening (U21) is surrounded by gland cells which contain cyanophilous granules (Fig. 5). It leads into a short penis which can be protruded. Posteriorly this penis widens into an oval sack (vesicula granulorum and vesicula seminalis) which, in its distal part, is lined with cuboid cells containing cyanophilous granules. The proximal part is filled with sperms.

The whole copulatory organ is surrounded by a membrane and numerous circular muscles. Right behind this copulatory organ, which is about 100 μm long in fixed animals, one finds two dorsolateral lobes of germinative cells of the testis (Fig. 4*a*, *b*: at). Another germinative zone is clearly separate and situated behind these two lobes (pt). The germinative cells of the latter are especially large and, in one specimen, seemed to be developed into oocytes (Fig. 8).

The nucleus of the sperms in sections is spindle-shaped (Fig. 11*a*, *b*). The plasma is also elongated (48 μm in length), but with an irregular contour line (Figs. 2*j*, 3*c*, 11*a*, *b*). In spermatogenesis we could only observe a transformation of the nucleus from germinative cells (with a large nucleolus) to the final elongated shape (Fig. 10*e*–*h*). No meioses could be observed.

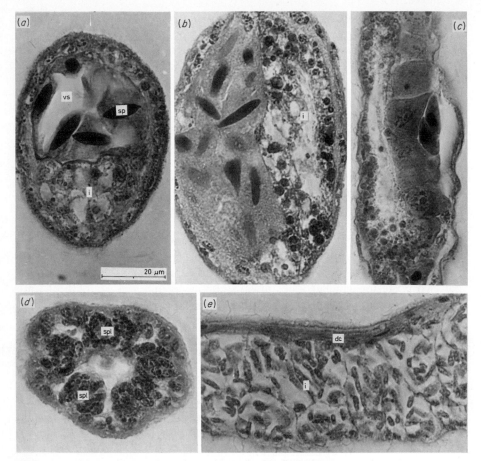

Figure 11 Photomicrographs of sections. (*a–b*) *Retronectes thalia* sp. n.: (*a*) transverse section through vesicula seminalis; (*b*) transverse section through posterior part of "testis." (*c*) *Retronectes clio* sp. n., longitudinal section through posterior part of "testis." (*d*) *Retronectes thalia* sp. n., transverse section in front of the brain. (*e*) *Paracatenula urania* sp. n., sagittal section through mid-body region. All in same scale.

Retronectes clio sp. n. (Figs. 2*a–c*, 10*i*, 11*c*, 13*f*)

DIAGNOSIS: *Retronectes* with granular strands and a statocyst with one statolith. Without protonephridia. Sperm irregular, with nuclei of the conglomerate type.

MATERIAL: Twelve specimens, of which four are adults, in squeeze preparation.

HOLOTYPE: Serial section, AMNH #771.

TYPE LOCALITY: Fine, fairly clean sand in 1.5 m depth off Klubban Beach, Kristineberg (Atlantic coast of Sweden).

DESCRIPTION: Mature animals (Fig. 2a) measure about 1250 μm in length and 60 μm in width. The rostrum is pointed but slightly club-shaped (Fig. 13f) and measures about 130 μm (U10).

A few specimens that were lacking reproductive organs were exceedingly long (1700 to 2800 μm), but agreed in all other features with the smaller specimens. It is possible that this phenomenon results from incomplete paratomy, as suggested by Borkott (1970). The species is brown in transmitted light, owing to the granular strands that originate in the rostrum in fairly compact form, then pass under the brain (Fig. 13f), laterally to the pharynx, and branch out to fill the space between gut and epidermis throughout the whole body.

The mouth opening is situated at U 15. It is 20 μm long and oval-shaped. The pharynx is completely ciliated and opens into a ciliated gut.

R. clio has a statocyst at U10; its diameter is 10 μm (Fig. 2b). There is only one rounded statolith (5 μm). The latter is smooth and does not show any plasmatic cover.

The male opening (Fig. 10i) situated behind the pharynx at U18 is surrounded by granular glands, and opens into a fairly spacious vesicula filled with sperm. The elongated hourglass-shaped testis extends between the male opening and U89 (i.e., 150 μm in front of the posterior end of the body). Whereas the anterior part of the testis is filled with stages of spermatogenesis, the posterior part often contains large granular cells that look rather like oogenesis stages (Fig. 11c).

The sperm, as found in the vesicula, is irregular to triangular, its shape being determined by the surrounding tissue (Fig. 2c). Its diameter is approximately 10 to 15 μm. The nucleus appears as a conglomerate of several particles; its diameter is 3 μm.

Structures found in the posterior part of the testis resemble the sperm in shape, but are larger (25 μm) and do not have the conglomerate-type nucleus. Toward the end of the testis there are slightly smaller structures (17 μm by 7 μm) which are filled with granules of 2 μm diameter.

Retronectes melpomene sp. n. (Figs. 2k–n, 13d, e)

DIAGNOSIS: *Retronectes* without granular strands; with a statocyst with one or more statoliths. Thin rhabdoids in the rostrum. Without protonephridia; sperm with a ribbon-shaped nucleus.

MATERIAL: Four specimens, one adult.

TYPE: No type material available.

TYPE LOCALITY: Fine sand with detritus 4 m depth, off Klubban Beach near Kristineberg, Atlantic coast of Sweden (cf. Sterrer, 1968).

DESCRIPTION: The only adult (Fig. 2k) was 850 μm long and 60 μm wide. The species has a pointed rostrum which tapers gradually into the outline of the body. A very short (20 μm) tail is usually developed. The species is color-less. Bundles of thin, long rhabdoids (8 μm) are present in the rostrum.

Three specimens had a statocyst, in a fourth it was lacking. The number of statoliths was one in two specimens, and five in one specimen (Figs. 2m, 13d). The diameter of the statocyst is 10 to 12 μm. The five statoliths were rounded to oval, between 2 and 5 μm in diameter. They were loosely grouped but not connected, resting on the bottom of the statocyst and slightly vibrating.

The mouth opening extends from U16 to U19, having a length of 20 μm. The pharynx is ciliated, and so is the gut.

A male opening is found dorsally behind the pharynx. The sperm (Fig. 15, 68) is of oval shape, with a coiled ribbonlike nucleus of about 50 μm (un-coiled) length and 3 to 4 μm width.

Retronectes euterpe sp. n. (Fig. 2d–g)

DIAGNOSIS: *Rectronectes* with a statocyst containing one statolith. With protonephridia. Testis hourglass-shaped. Sperm rounded, with nuclei of ir-regular shape.

MATERIAL: Three specimens, two adults, in squeeze preparation.

TYPE: No type material available.

TYPE LOCALITY: Coarse shell above high water, 5 cm sediment depth, intracoastal waterway at Wrightsville Lido, near Wilmington, North Carolina.

DESCRIPTION: The "male" specimen (Fig. 2d) was 1750μm long and 60 μm wide, the "female" specimen (Fig. 2g) 1250 μm long and 60 μm wide. The species is yellowish in color.

In the "female" specimen a protonephridial canal was found. Although the position of its pore could not be ascertained, it is certain that the excretory canal runs mediodorsally through the whole body. Protonephridia flames could most easily be observed in the neighborhood of the statocyst (Fig. 2e).

The mouth opening, situated 100 to 140 μm behind the anterior tip of the body and about 40 μm long, is bordered by rhabdoid-like inclusions of the dimensions 6 by 2 μm. These originate in sac-shaped glands situated laterally to the pharynx.

The testis is hourglass-shaped, and in the "male" specimen it was 1100 μm long. Its anterior part contained densely packed sperm (Fig. 2f) of irregu-lar rounded shape, 20 μm in diameter, with a granular plasma and a nucleus that seems to consist of several agglomerated particles.

The "female" specimen (Fig. 2g) did not contain any sperm, but a mature egg of 140 μm length was situated between U72 and U83.

The gut of the "female" specimen contained a statolith of probably the same species.

Retronectes terpsichore sp. n. (Figs. 2o–r, 12a–c)

DIAGNOSIS: *Retronectes* with a statocyst with one to two statoliths. Rostrum divided into slender anterior and plump posterior part. A tail is

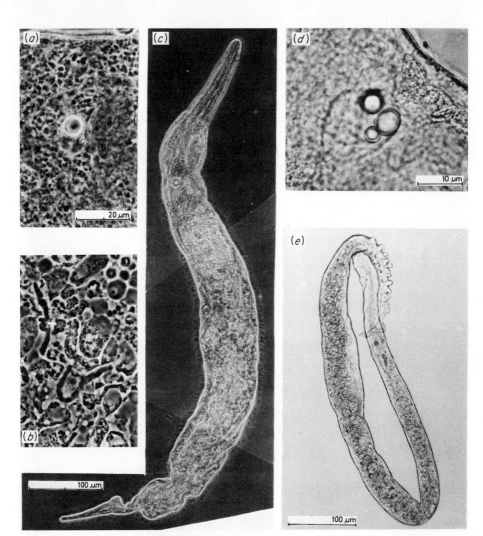

Figure 12 Photomicrographs of live specimens. (a–c) *Retronectes terpsichore* sp. n.: (a) statocyst; (b) sperm; (c) slightly squeezed specimen. (d–e) *Retronectes* sp. III: (d) statocyst; (e) slightly squeezed specimen. Parts a and b are in the same scale, a–c with phase contrast.

present. Thin rhabdoids in the rostrum. Sperm irregular, nucleus ribbon-shaped.

MATERIAL: Two specimens, one adult, in squeeze preparation.

TYPE: No type material available.

TYPE LOCALITY: Coarse sand with detritus at low water level, near Green Island, Portaferry (Northern Ireland).

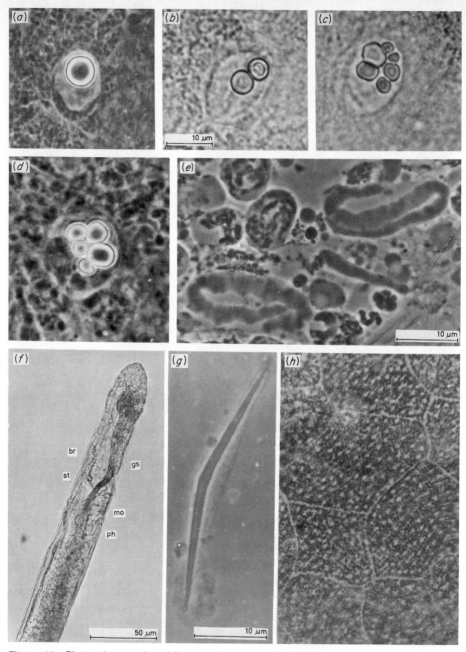

Figure 13 Photomicrographs of live specimens. (a) *Retronectes* sp. from Kristineberg, statocyst; (b) *Retronectes* sp. I, statocyst; (c) *Retronectes* sp. V, statocyst; (d, e) *Retronectes melpomene* sp. n.; (d) statocyst; (e) sperm; (f) *Retronectes clio* sp. n., anterior body region; (g) *Paracatenula polyhymnia* sp. n., sperm; (h) *Retronectes* sp. from Kristineberg, epidermis. Parts *a–d* in the same scale. The scale of part *e* applies also to part *h*.

DESCRIPTION: The specimen (Fig. 2*o*) that furnished most of the data was 650 μm long, and 80 μm wide at U18. The rostrum is divided into a slender anterior part (80 μm long), and a barrel-shaped posterior part (70 μm long). A tail is present (200 μm long). The species is colorless and fairly transparent. One of the specimens, for unknown reasons, was wrinkled all over. The rostrum contains bundles of thin rhabdoids of 5 to 10 μm length (Fig. 2*p*).

The statocyst, situated at U25, is 13 μm in diameter. It contains one statolith in one specimen. The other specimen (Fig. 2*q*) has two statoliths, one of 8 μm, and another of 1.5 μm diameter.

The sperm (Fig. 2*r*) is of irregular shape, with a ribbonlike nucleus (length 25 to 35 μm, width 3 to 4 μm).

Genus *Paracatenula*

Paracatenula urania sp. n. (Figs. 2*v, w*, 11*e*)

DIAGNOSIS: *Paracatenula* without statocyst. 3 to 7 mm long. Dark greenish in color. Sperm with short (10 μm), spindle-shaped nucleus.

MATERIAL: Six specimens, of which four have been sectioned. Three specimens had a male opening. One of the latter had a developing male opening and one developed sperm.

HOLOTYPE: One serial section through a coiled specimen. AMNH #772.

PARATYPES: Two serial sections. AMNH #773, 774.

TYPE LOCALITY: Wrightsville Beach north of Wilmington, North Carolina, at Harbor Island Banks channel side near the bridge of Highway 76, in a fine sand flat, June 1970. The specimens were found on the tubes of an onuphid polychaete, collected by Dr. Charles Jenner, University of North Carolina.

DESCRIPTION: Body length (Fig. 2*v*) 3 to 7 mm, body width about 30 μm. The animals are white in reflected light and olive green in transmitted light.

The rostrum (anterior tip to the beginning of the intestine) is 150 μm long, and filled with the anterior part of the granular strands which extend backward in two main stems.

The cellular epidermis shows normal features. It contains scattered spindle-shaped granules (5 μm long, 2 μm wide). No basement membrane and no subepidermal muscle sheaths could be found in the section. The only fibrillous elements found are in the dorsal cord. Since we observe circular constrictions in the living worms, we assume that fine circular fibers must be present, although they could not be seen in the sections.

The dorsal cord extends throughout the whole body length. It originated between the two posterior lobes of the brain.

The gangliar lobes of the brain are less developed than in other species (e.g., *P. erato*), with two posterior and two anterior ones observed.

No statocyst could be found. A mouth opening and a pharynx are completely lacking in the specimens studied. The gut seems to consist of big (25 μm long, 5 to 10 μm wide) turgescent cells which lie in a row. They contain round to oval bodies (5 to 10 μm) which are filled with small, strongly light-reflecting granules. The latter always lie in a little vacuole.

Three specimens had a dorsal male opening, just in front of the anterior end of the gut. But only in one sectioned specimen was a developed sperm found (Fig. 2w). It was attached to the muscle surrounding the dorsal cord (as in *P. erato*) and contained a 10-μm long, spindle-shaped nucleus.

No female organs have been observed.

Paracatenula erato sp. n. (Figs. 2s–u, 9, 10a–d, 10j, 14b, c)

DIAGNOSIS: *Paracatenula* with a statocyst and one statolith. Sperm with a straight, spindle-shaped nucleus (length 45 μm, width 5 μm), with longitudinal ridges. The species is usually greenish in color.

MATERIAL: Two adult specimens.

HOLOTYPE: Transverse serial section, AMNH #775.

TYPE LOCALITY: Intertidal of "Ann McCrary's mud flat," in 10 cm sediment depth, near Wilmington, North Carolina.

DESCRIPTION: Since both specimens obtained were anterior fragments, the body length is unknown. The fragment to which the following description refers (Fig. 2s) was 1900 μm long and had a maximum width of 100 μm, close to the posterior end. The anterior part of the body (to about 400 μm) is colorless and transparent, whereas the rest is greenish.

The statocyst (Fig. 2t), situated 150 μm behind the anterior end, is 15 μm in diameter; its single statolith is more or less round and measures 7 μm.

The male pore is situated 80 μm behind the statocyst (Fig. 9). It opens into a short copulatory organ (length 45 μm, width 25 μm), which seems connected with the "dorsal cord." Appearing as a narrow canal in live preparation, the latter runs through the whole length of the body.

Sperm and spermatogenesis stages were found along the "dorsal cord." The sperm (Figs. 2u, 10d, 14c) consists of a spindle-shaped nucleus embedded in a granular plasma. The nucleus of the mature sperm shows fine longitudinal ridges. Dimensions are 45 by 5 μm for the nucleus and 45 by 25 μm for the matrix.

The spermatogenesis stages observed are shown in Fig. 10a to d and Fig. 14b and c. In the process, both nucleus and plasma stretch. Granular inclusions in the plasma, first arranged in loose spirals (Fig. 10a), subsequently become denser (Fig. 10b) and aggregate to groups (Fig. 10c). In the final sperm, however, granular inclusions are equally distributed again.

No female organs were observed.

Paracatenula polyhymnia sp. n. (Figs. 2*x*, *y*, 13*g*)

DIAGNOSIS: *Paracatenula* without a statocyst. Sperm with a spicule-shaped nucleus (length 35 μm, width 1.5 μm). Rostrum and tail region are delimited.

MATERIAL: One specimen in squeeze preparation.

TYPE: No type material available.

TYPE LOCALITY: Carbonate sand with detritus from beneath a patch reef, 4 m depth, off Big Pine Key (southern Florida).

DESCRIPTION: The only specimen found (Fig. 2*x*) measured 830 μm in length and maximally 100 μm in width. It had a rather transparent rostrum of 200 μm length, whereas the body, irregular in shape, was dark in transmitted light. The species is provided with a short (100 μm) tail region. A statocyst is lacking.

The sperm (Figs. 2*y*, 13*g*) is arranged in groups of two to eight along the axis of the body. It consists of a spicule-shaped nucleus and a thin layer of surrounding plasma. The nucleus is 34 to 36 μm long and only 1.5 μm thick in its median part, from where it tapers to the finely pointed ends. In one type the ends appeared to be sheathed (Fig. 2*y*, left). The nucleus is bent in the middle to form two symmetrical arms of 160° spread. It can be serrated on the inner edge. The plasma is often thickened around one end of the nucleus where it contains granular inclusions.

No data are available on the female system.

Paracatenula kalliope sp. n. (Figs. 2*z*–*ff*, 14*a*)

DIAGNOSIS: *Paracatenula* with a statocyst and one statolith. Sperm with an elongated nucleus (length up to 65 μm, width 6 μm). Crescent-shaped refractile inclusions (length 25 μm, width 5 μm) are present in the parenchyma. Skin with rows of small epidermal inclusions in groups.

MATERIAL: Six specimens, one adult, in squeeze preparation.

TYPE: No type material available.

TYPE LOCALITY: Carbonate sand with detritus from beneath a patch reef, 4 m depth, off Big Pine Key (southern Florida).

DESCRIPTION: The biggest mature specimen, an anterior fragment, measured 2200 μm in length and 170 μm in width. A complete specimen (Fig. 2*z*) was 1700 μm in length, of which 400 μm were taken by the rostrum and 200 μm by the tail. Occasionally, the animal adheres to the substratum with the tip of its tail. Rostrum and tail are rather transparent, whereas the body appears dark in transmitted light, and white in incident light owing to the gut's containing great numbers of brownish, highly refractile granules of 1 μm diameter, which are arranged in longitudinal rows, 3 μm apart, from each other (Fig. 2*bb*).

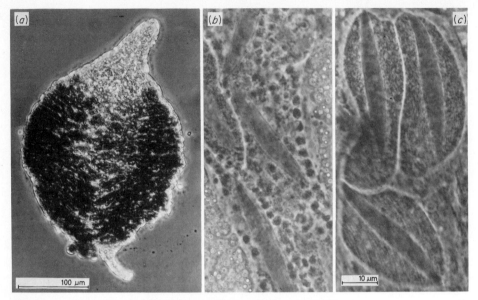

Figure 14   Photomicrographs of live specimens. (a) *Paracatenula kalliope* sp. n., juvenile specimen, squeezed; (b, c) *Paracatenula erato* sp. n., spermatogenesis; (b) late spermatid, (c) sperm. Parts b and c in the same scale.

The statocyst (Fig. 2cc) is round or oval and up to 15 μm in diameter. In one specimen, its anterior wall contained small vacuoles. The single statolith is round or pear-shaped and 5 to 7 μm in diameter. In one specimen the statolith was constricted, as if consisting of two aggregated particles.

In two of the specimens large inclusions (Fig. 2ff) were found in the anterior body region. 25 μm in length and 5 μm in width, their crescent-shaped contour yielded an internal structure of five to nine round and highly refractile bodies.

Sperm was observed in only one specimen. Several shapes were noted, the largest (Fig. 2dd) having a ribbon-shaped nucleus with a length of 65 μm, and 6 μm width. It is not known whether the other structures shown in Fig. 2ee represent sperm or spermatogenesis stages.

No female organs were observed.

Addendum

Unidentified species (Figs. 13a–c, 13h, 15a–r)

In addition to the species of which sufficient material and diagnostic features were available to justify a description, a number of specimens have been recorded that did not meet those criteria. In order to give a more complete picture of the new family, however, we add these fragmentary data represented in Fig. 13*b*, *c*, *h* and Fig. 15*a* to *q*. It may very well be that one or the other of these specimens will later on turn out to be within the variability of one of the described species.

ACKNOWLEDGMENTS

We are indebted to Drs. R. Riedl, C. Jenner, J. Ott (Chapel Hill), and B. Swedmark (Kristineberg) for unpublished data and suggestions. Invaluable help has been received from Dr. Gunde Rieger, who did the preparations for

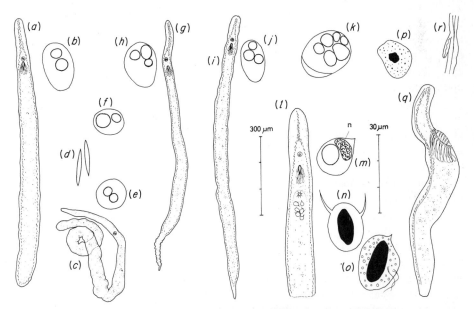

Figure 15 Unidentified species. (*a*) *Retronectes* sp. I from Kristineberg (Sweden). (*b*) Statocyst of the same specimen. (*c*) *Retronectes* sp. II from Kristineberg. (*d*) Rhabdoids from rostrum of the same specimen. (*e*) Statocyst of the same specimen. (*f*) Statocyst of another specimen from the same locality. (*g*) *Retronectes* sp. III from Kristineberg. (*h*) Statocyst of the same specimen. (*i*) *Retronectes* sp. IV from Kristineberg. (*j*) Statocyst of the same specimen. (*k*) *Retronectes* sp. V from Kristineberg, statocyst. (*l*) *Retronectes* sp. VI from Bogue Bank, near Morehead City, N.C. (*m*) Statocyst of the same specimen. (*n*, *o*) Sperm of the same specimen. (*p*) Sperm of retronectid sp. from 40 m, off Morehead City, N.C. (*q*) *Retronectes* sp. VIII from Bogue Inlet, Morehead City, N.C. (*r*) Part of the excretory system of the same specimen. All organizations in one scale, all details in the other.

microanatomy and electron microscopy, and Mrs. Christiane Sterrer, who executed the figures. Our work was supported by the National Science Foundation grants XA120-09 and GA-29592; further by the Hochschul-Jubilaeumsstiftung Wien, and various other Austrian granting agencies. Contribution No. 518 from the Bermuda Biological station.

ABBREVIATIONS USED IN FIGURES

al   anterior gangliar lobes of the brain
at   anterior germinative zone of the "testis"
br   brain
c   cilia
cm   circular muscles
co   male copulatory organ
dc   dorsal cord
ep   epidermis
epc   epidermis cell
f   fibrillous structures in the epidermis
gc   gland cells in the gut
gl   opening of the glands around the mouth
gs   granular strands
i   gut
lm   longitudinal muscles
mi   microvilli
ml   male opening
mo   mouth opening
n   nucleus
ne   nerve tissue
oc   oocyte in the posterior part of the "testis"
o   outer layer of the skin
os   opening of the sensory plate
pe   penis
ph   pharynx
pl   posterior gangliar lobes of the brain
pt   posterior germinative zone of the "testis"
rh   rhabdites
rl   rootlets of the cilia
sp   sperm
spl   sensory plate of the anterior gangliar lobes
spg   spermatogonia
st   statocyst

vg   prostatic vesicle (vesicula granulorum)
vs   seminal vesicle (vesicula seminalis)
x   indigestible material

BIBLIOGRAPHY

Antonius, A. 1965. Methodischer Beitrag zur mikroskopischen Anatomie und graphischen Rekonstruktion sehr kleiner zoologischer Objekte. *Mikroskopie,* **20**(5/6): 145−153.

Borkott, H. 1970. Geschlechtliche Organisation, Fortpflanzungsverhalten und Ursachen der sexuellen Vermehrung von *Stenostomum sthenum* nov. spec. (Turbellaria Catenulida). *Z. Morphol. Tiere,* **67**:183−262.

Brandenburg, J. 1966. Die Reusenformen der Cyrtocyten. Eine Beschreibung von fünf weiteren Reusengeisselzellen und eine vergleichende Betrachtung. *Zool. Beitr.,* **12**(3):345−417.

Bush, L. 1968. Characteristics of Interstitial Sand Turbellaria: the Significance of Body Elongation, Muscular Development, and Adhesive Organs. *Trans. Am. Microscop. Soc.,* **87**(2):244−251.

Dörjes, J. 1968. Die Acoela (Turbellaria) der Deutschen Nordseeküste und ein neues System der Ordnung. *Z. Zool. Syst. Evolutionsforschung,* **6**:56−452.

Fenchel, T. M., and R. J. Riedl. 1970. The Sulfide System: a New Biotic Community underneath the Oxidized Layer of Marine Sand Bottoms. *Marine Biol.,* **7**:255−268.

Hulings, N. C., and J. S. Gray. 1971. A Manual for the Study of Meiofauna. *Smithsonian Contrib. Zool.,* **78**:1−83.

Luther, A. 1960. Die Turbellarien Ostfennoskandiens. I. Acoela, Catenulida, Macrostomida, Lecithoepitheliata, Prolecithophora, und Proseriata. Soc. F. Fl. Fenn., F.F., **7**:1−155.

Marcus, E. 1945a. Sobre Catenulida Brasileiros. *Bol. Fac. Filosof. Ciências Letras Univ. São Paulo, Zool.,* **10**:3−133.

———. 1945b. Sobre Microturbelários do Brasil. *Com. Zool. Museo Hist. Nat. Montevideo,* **1**(25):1−104.

———. 1950. Turbellarios Brasileiros (8). *Bol. Fac. Filosof Ciências Letras Univ. São Paulo, Zool.,* **15**:5−192.

Pullen, E. W. 1957. A Histological Study of *Stenostomum virginianum. J. Morphol.,* **101**:579−621.

Reisinger, E. 1924. Die Gattung *Rhynchoscolex. Z. Morphol. Ökol. Tiere,* **1**:1−37,

———. 1964. Zur Feinstruktur des paranephridialen Plexus und der Cyrtocyten von Codonocephalus (Trematoda Digenea: Strigeidae). *Zool. Anz.,* **172**:16−22.

Riedl, R. 1959. Turbellarien aus submarinen Höhlen, I. Archoophora. *Pubbl. Staz. Zool. Napoli, Suppl.* **30**:178−208.

Rieger, R., and W. Sterrer. 1968. *Megamorion brevicauda* gen. nov. spec. nov., ein Vertreter der Turbellarienordnung Macrostomida aus dem Tiefenschlamm eines norwegischen Fjords. *Sarsia,* **31**:75−100.

Rixen, J. U. 1961. Klein-Turbellarien aus dem Litoral der Binnengewässer Schleswig-Holsteins. *Arch. Hydrobiol.*, **57**:464–538.

Southward, A. J., and E. C. Southward. 1971. Observations on the Role of Dissolved Organic Compounds in the Nutrition of Benthic Invertebrates. *Sarsia*, **45**:69–96.

Sterrer, W. 1966. New Polylithophorous Marine Turbellaria. *Nature*, **210**:436.

———. 1969. Beiträge zur Kenntnis der Gnathostomulida. I. Anatomie und Morphologie des Genus *Pterognathia* Sterrer. *Arkiv Zool.*, **22**(1):1–125.

———. 1971a. On the Biology of Gnathostomulida. *Vie et Milieu, Suppl.*, **22**:493–508.

———. 1971b. Gnathostomulida: Problems and Procedures. *Smithsonian Contrib. Zool.*, **76**:9–15.

Chapter 5

A New Species of the Genus *Pericelis*, a Polyclad Flatworm from Hawaii

Jean L. Poulter
Department of Physiology—Anatomy
University of California, Berkeley, California

During an investigation of the polyclad fauna of the island of Oahu, Hawaii, this interesting and initially perplexing polyclad flatworm was discovered. The outstanding features of the living specimens are more representative of the Acotylea (Fig. 1): a long ruffled pharynx, anteriorly directed uteri, the location of the copulatory complex in the posterior body quarter. The only visible clue in living specimens to indicate that this might be a cotylean is the presence of tiny marginal tentacles. Further investigation did in fact substantiate that this is a cotylean, and it is assigned to the genus *Pericelis* Laidlaw, 1902. The inclusion of this species in the family Pericelidae requires an emendation of the family definition:

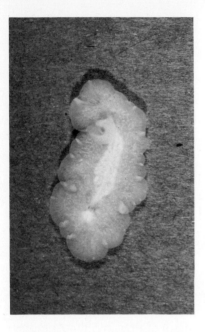

Figure 1 Photograph of dorsal aspect of living holotype specimen.

Family Pericelidae Laidlaw (1902, p. 291) emend. Hyman (1955, p. 262). With a pair of marginal tentacles, each bearing a cluster of eyes.

This emendation conforms with the original definition given by Laidlaw (1902).

MATERIAL, LOCATION, AND TYPES

The material consists of twenty-eight specimens collected at Swanzy Beach Park (Kaaawa), Hauula Beach Park, and Kupikipikio Point (Black Point), Oahu, Hawaii.[1] The specimen designated as the holotype of *Pericelis hymanae* is specimen 23, collected at Swanzy Beach Park, Kaaawa (157°51′14″W and 21°33′30″N) December 15, 1961. Nineteen other specimens from the same location are designated paratypes: nine collected May 13, 1961, four collected July 25, 1961, and six collected December 15, 1961. Eight other specimens assignable to this species are from the other areas. The holotype (USNM 45278), anterior region alcohol-preserved and posterior region frontally sectioned (38 slides), along with three paratypes (USNM 45279, 45280, 45281) including a precipitated borax-carmine-stained whole mount preparation

---

[1]Topographic map of the island of Oahu, United States Department of the Interior, Geological Survey, Washington, D.C. 1954.

mounted on a slide, is deposited in the United States National Museum. Other paratypes are deposited in The American Museum of Natural History, (AMNH 756) the California Academy of Sciences, Bishop Museum, and the British Museum of Natural History. Most of the other specimens assigned to this species are deposited in the California Academy of Sciences.

Since the material to describe a polyclad is often gleaned from several specimens, as is the case in the present study, for the purpose of brevity, the features of the holotype will be considered in a general account of the specimens collected. However, the condition of the holotype will be elucidated where there is variation and the pertinent information is available.

Suborder Cotylea Lang, 1884

Family Pericelidae Laidlaw, 1902

Genus *Pericelis* Laidlaw, 1902

Pericelis hymanae[2] sp. n.

FEATURES OF LIVING MATERIAL (Fig. 1, holotype): The size spans a broad range (Fig. 2); the largest and holotype is 48 mm long and 16 mm wide.

The shape is variable. At rest the worms assume an almost circular form with a mildly ruffled margin, or they assume a pyriform shape with the posterior region broadest. In motion they usually extend to a length-breadth ratio of 3:1 (large specimens may attain a ratio of 4:1) in which case they assume an anteriorly truncated elongated oval form with negligible marginal ruffling.

The body is thin and delicate. The tentacles are widely separated (1.5

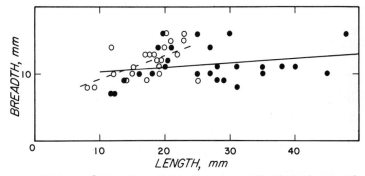

Figure 2 Comparison of length and breadth of specimens collected (living usually but not always crawling extended): •, living; o, alcohol-preserved; trend lines by method of least squares: ——, living; - -, alcohol-preserved.

[2]This species is named for Dr. Libbie Henrietta Hyman who graciously criticized the manuscript of which this was part, and consented to the suggested name.

to 4.5 mm in the holotype; mean, 2.8 mm) tiny V-shaped extensions of the anterior margin with a slight upward crease down the center. They are not "infolded." Midway between the tentacles is the conspicuous marginal indentation reported for other Pericelidae.

The specimens tend toward opacity but are somewhat translucent along the margin beyond the termination of the gut. The color of the dorsal surface is off-white, owing to internal structures and mesenchyme rather than epidermal pigmentation. A light gray, pink, or orange hue may be imparted by material in the gut. Along the dorsal midline, a narrow brown stripe consisting of minute brown pigment granules is seen. It begins lightly at various levels over the pharynx, runs anteriorly, and is discontinuous or extremely light between the paired cerebral eye groups. Anterior to the cerebral eyes, the stripe becomes more vivid and terminates just short of the marginal indentation. Some specimens, including the holotype, exhibit an irregular smoked appearance along the anterior margin between the tentacles. When gravid, as is the holotype, a subepidermal gray-brown pigmentation is seen between the branches of the pharynx and the ova-filled uteri and also in the copulatory region. A few specimens have an isolated patch of tiny orange-brown epidermal pigment granules on the dorsal surface which possibly represents a genetic mosaic. The color of the ventral surface is similar to that of the dorsal, but no pigment marks are noted.

The long (3.8 to 28.5 mm in holotype; mean, 12.2 mm; see Fig. 3) white opaque ruffled pharynx is a most outstanding feature. The anastomosing pat-

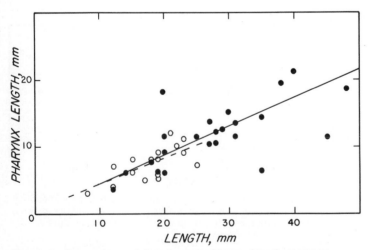

Figure 3 Comparison of length and pharynx length measurements of specimens collected: ●, living; o, alcohol-preserved; trend lines by method of least squares: ——, living; - -, alcohol-preserved.

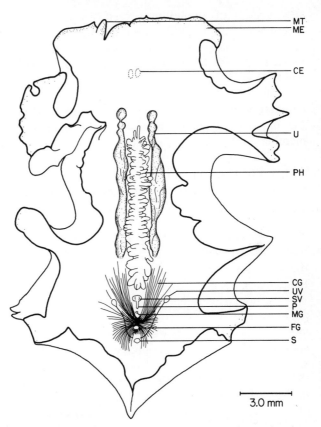

Figure 4 Ventral aspect of cleared holotype specimen showing general anatomical features: location of female gonopore schematically shown: CE, cerebral eyes; CG, cement glands; FG, female gonopore; ME, marginal eyes; MG, male gonopore; MT, marginal tentacle; P, penis; PH, pharynx; S, sucker; SV, seminal vesicle; U, uterus; UV, uterine vesicle.

tern of the gut is easily seen. In gravid specimens, including the holotype, the ova-filled uteri lie on each side of the pharynx extending anteriorly the entire length of the pharynx. Behind the pharynx, the extensive development of the cement glands marks the copulatory region.

The paired elongated oval groups of cerebral eyes are easily located in living material behind the anterior margin (2.3 to 6.0 mm; holotype, 3.8 mm; mean, 3.3 mm). The frontal, anterior marginal and tentacular eyes are more difficult to discern. The tentacular eyes are more easily seen from the ventral surface. The lateral and posterior marginal eyes are not seen in living specimens.

FEATURES OF PRESERVED MATERIAL (Fig. 4, from holotype): The length and breadth of alcohol-preserved specimens (Fig. 2) are, for the smallest, length 8 mm and breadth 8 mm; for the largest and holotype, length 23 mm and breadth 16 mm. The extent of shrinkage due to fixation and alcohol preservation is demonstrated in a comparison of living length and alcohol-preserved length (Fig. 5).

Because of excessive distortion and ruffling of the margin, the anterior indentation and tentacles may be difficult to locate. Also, in a few specimens the tentacles appear to be infolded; this is the condition reported for all other members of the family Pericelidae.

The color of both the dorsal and ventral surfaces is dirty yellow, slightly darker along the margin, and specimens are notably opaque. The more vivid portion of the stripe of brown pigments is retained after fixation.

The paired clusters of cerebral eyes (Fig. 6 of holotype) are roughly triangular in shape in alcohol-preserved specimens. Each of the groups contains approximately 35 to 50 eyes. The variably sized frontal eyes are scattered in front of the cerebral eyes and fan out anteriorly and laterally toward the tentacles. Few frontal eyes are noted in the region of the midline. The tentacular eyes are crowded in the mid-region of the tentacles.

Although marginal eyes completely encircle the margin, the lateral and posterior eyes were discovered in a precipitated borax-carmine-stained whole mount (paratype) and were subsequently found in the cleared holotype specimen (Fig. 4). The marginal eyes are prevalent along the anterior margin, sparse along the lateral and posterior margin, and sometimes located a considerable distance from the margin. However, they are rarely located interior to the perimeter of the gut. The pseudorhabdites of the epithelium, especially of the dorsal surface, obscure the eyespots and they are more easily seen from

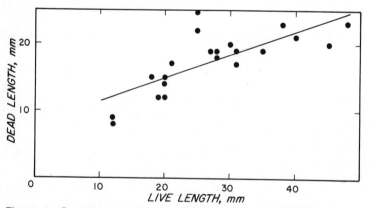

Figure 5 Comparison of living length and alcohol-preserved length. Trend line by method of least squares.

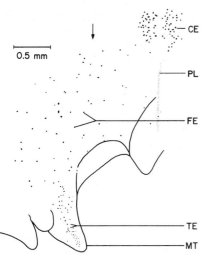

0.5 mm

— CE

— PL

— FE

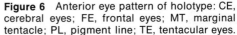

— TE

—MT

Figure 6 Anterior eye pattern of holotype: CE, cerebral eyes; FE, frontal eyes; MT, marginal tentacle; PL, pigment line; TE, tentacular eyes.

the ventral aspect. Marginal eyes are also found in sectioned material. They appear as melanic spheres and have a diametric range of 0.008 to 0.025 mm; mean, 0.014 mm. They demonstrate no apparent nerve connections with the stains employed.

The body is thin, 0.3 mm thick 3 mm from the margin and 1 mm thick in the posterior region of the pharynx.

The dorsal epithelium is thin, composed of cuboidal cells (mean thickness, 0.015 mm). It is well supplied with rhabdites and with another cell type that may represent pseudorhabdites.[3] The latter cell type is most prevalent. In the region of the copulatory apparatus and pharynx, the epithelium is about twice as thick (mean, 0.032 mm) and is profusely supplied with rhabdites and calyciform cells. The poorly organized circular and longitudinal muscle layers under the basement membrane are only 0.012 mm thick. The ventral epithelium is 0.012 mm thick and is less well supplied with calyciform cells and few rhabdites are found. Rhabdites are absent in the ventral epithelium in the region of the copulatory apparatus, where the number of calyciform cells is reduced. The underlying circular and longitudinal muscle layers are more organized than the corresponding dorsal layers. The mean thickness of the ventral muscle layers is 0.014 mm. Inside the body walls, the mesenchyme is loosely packed, and there are many dorsoventral muscle fibers not usually aggregated in bundles.

The mouth is central to the middle of the body and posterior to the middle of the pharynx. The length of the pharynx (smallest, 3 mm; longest, 12 mm; holotype, 9 mm; mean, 7 mm; see Fig. 3) is one-third to one-half (generally

[3]Calyciform cells of Marcus and Marcus (1968).

around 40 percent) of the total body length. The ruffled pharynx (Fig. 4) is slightly displaced toward the anterior. The pharyngeal pouch is deeply branched with blunt pockets whereas the branches of the pharynx are pointed. Over the pharynx lies the main intestine, which is slightly longer than the pharynx. The epithelium of the main intestine (mean thickness, 0.07 mm) is packed with tear-shaped cells with extremely long necks. The cells are filled with numerous basophilic granules, and the histology is unlike that found in the Pseudoceridae. The main intestine terminates just posterior to the pharynx and bifurcates to form two major gut branches which pass around the copulatory region (Fig. 7). In the precipitated borax-carmine-stained whole mount (paratype), the anastomosing pattern of the gut is seen. The network continues to within 2 to 3 mm of the body margin.

The copulatory region (Figs. 7 of holotype, 8) lies just posterior to the termination of the pharynx at a level of the posterior one-fourth of the body. Frontal sections were prepared of the posterior region of the sexually mature holotype specimen. Previously, sagittal sections had been prepared of a sexually mature specimen (number 17, collected at Hauula Beach Park; alcohol-preserved length 18 mm and breadth 13 mm).

The numerous scattered testes are ventrally situated below the level of

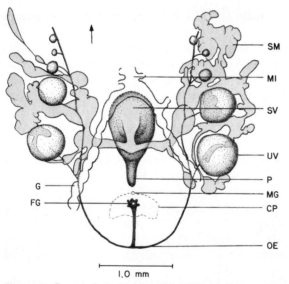

1.0 mm

Figure 7 Dorsal view of a reconstruction of the copulatory apparatus from the frontally sectioned holotype: CP, cement pouch; FG, female gonopore; G, gut; MG, male gonopore; MI, main intestine; OE, oviduct entrance; P, penis; SM, spermiducal vesicle; SV, seminal vesicle; UV, uterine vesicle.

the gut. Sperm ductules of fine caliber lead from the testes to spermiducal vesicles. The sperm ductules may enter the spermiducal vesicles in either a constricted or enlarged portion. The spermiducal vesicles ramify extensively, and the pattern is asymmetrical and tortuous. There are wide openings between some vesicles, and narrow sperm ducts of considerable length connect others. The vesicles may be large or small diametric enlargements of the sperm duct. The outlying vesicles are filled with an acidophilic secretion and few sperm while the vesicles more proximal to the male organ are packed with sperm in the same type of secretion. An anterior or major region of vesicle development joins with a posterior or lesser region of vesicle development. However, this is not the only connection between the anterior and posterior regions of spermiducal vesicle development (see Fig. 7). From this junction of the anterior and posterior spermiducal vesicles, a short transverse sperm duct joins with a similar transverse sperm duct from the opposite side. At this point, they form a short anteriorly directed common sperm duct. This common sperm duct enters the seminal vesicle in a dorsal posterior position. The large seminal vesicle is a slightly elongated orbicular structure surrounded by a thick muscle layer. This muscle layer also extends into the penis. The epithelium of the seminal vesicle is thin (0.004 mm). The lumen (length 0.7 mm; breadth 0.5 mm) contains a moderate quantity of sperm in a secretion similar to that found in the spermiducal vesicles. The sperm demonstrate an orderly orientation lying in concentric rows perpendicular to the wall of the seminal vesicle. The ejaculatory duct leaves the seminal vesicle in a ventral posterior position and soon enters the posteriorly directed penis which descends obliquely toward the posteriorly located male gonopore. The lumen of the ejaculatory duct is irregularly widened in its mid-region. This portion has been assigned a prostatic function by several authors; however, there is no evidence of a secretory epithelium in either of the specimens sectioned. The penis is cylindrical (0.7 mm long) and distally is bluntly pointed. The male antrum is not much larger than the penis itself. The male gonopore is small.

In a reconstruction of the sagittally sectioned specimen the extent of spermiducal vesicle development is reduced, the anterior and posterior regions of vesicle development are connected by sperm ducts of narrow caliber, and the transverse sperm ducts, also of narrow caliber, enter the seminal vesicle separately in a dorsolateral but posterior position. The difference in the arrangement of the sperm duct(s) entering the seminal vesicle is explained by the more advanced sexual maturity of the holotype. The general morphology of the proximal portion of the male system (Fig. 8) is similar to that of the holotype. The size of the seminal vesicle lumen (length 0.5 mm; depth 0.3 mm) and penis (length 0.6 mm) is smaller than the measurements given for the holotype.

The scattered ovaries are dorsally situated above the level of the gut.

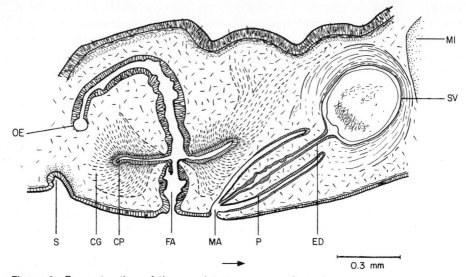

Figure 8 Reconstruction of the copulatory apparatus from the sagittally sectioned specimen: CG, cement glands; CP, cement pouch; ED, ejaculatory duct; FA, female antrum; MI, main intestine; OE, oviduct entrance; P, penis; S, sucker; SV, seminal vesicle.

However, owing to the thinness of the body, a developing oocyte may easily extend to the ventral or testicular region. As far forward as sectioned, the two anteriorly directed ova-filled uteri (see Fig. 4), one lying on each side of the pharynx, appear to be composed of a number of linearly arranged, separate large ova vesicles, closely apposed dorsally but ventrally separated. In the living specimen (Fig. 1) the vesicular nature of the uterus may be noted. The epithelium of these vesicles is in general thin but ventrally, nearing the region of the entrance of the oviduct, it becomes undulated and thicker. Each vesicle is connected ventrally by a branch of the oviduct which in turn joins the main oviduct. The oviduct passes posteriorly, close to the ventral body wall and en route gives rise to a number of uterine vesicles, over 10 in the holotype. The more anterior vesicles may be but tiny buds on the oviduct without sperm, or small thin-walled vesicles almost completely filled with sperm and with short connecting branches to the main oviduct. Other more posterior vesicles are about half filled with sperm, all oriented in a parallel configuration, the most ventral row with their tails toward the mouth of the vesicle. The dorsal half of the vesicle is filled with a granular mass which rests on the sperm heads (Fig. 9) as noted by Meixner (1907) in *P. byerleyana*. The more posterior uterine vesicles are large. The most posterior vesicle is the largest (length 0.5 mm;

breadth 0.7 mm). These posterior vesicles have long branches of the oviduct connecting them to the main oviduct. The vesicles are ovoid and have a stout ciliated columnar epithelium, similar to the histological construction of the oviduct. They are moderately filled with sperm scattered at random in an acidophilic secretion. The heads of the sperm appear swollen and in poor histological condition. The main oviduct continues posteriorly, turns toward the median line, and joins with the main oviduct from the opposite side of the body just anterior to the level of the sucker. From this point of juncture, the vagina courses upward, forward, and downward. Its lumen is irregularly widened and is lined with a tall ciliated epithelium. The vagina enters the wide but dorsoventrally compressed cement pouch, into which the cement glands empty. The cement pouch joins the female antrum, and at this point the antrum is constricted. It immediately widens into a chamber with a number of wide lateral pockets. The antrum is lined with an epithelium of variable height. The antrum narrows, forming the gonopore, which lies 0.03 mm behind the male gonopore.

In the sagittally sectioned specimen, an ova vesicle measures 0.5 mm in

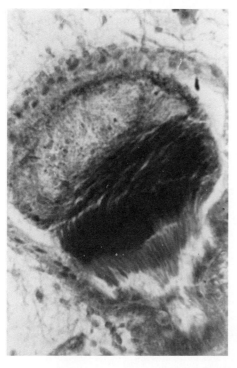

Figure 9 Uterine vesicle in sagittal section, containing sperm and granular acidophilic mass: × 422.

height and 0.45 mm in length. The number of uterine vesicles lying posterior to the ova vesicles is reduced to four or five, but general morphology and histology are similar. The most posterior vesicle (Fig. 10) is 0.6 mm in length and 0.4 mm in height. The more proximal portions of the female system are similar, and the gonopore lies 0.2 mm behind the male gonopore.

The sucker lies about 0.5 mm behind the female gonopore and is weakly developed with an indentation of only 0.03 mm. The epithelium is slightly thicker than that which adjoins it. Beneath the epithelium the muscular system is not developed much further than that of the body wall. The region is somewhat glandular, especially in the anterior portion.

DIFFERENTIAL CHARACTERS

P. hymanae appears to be most closely related to *P. ernesti* (Hyman, 1953) but differs from it and the three other members of the genus: *P. byerleyana* (Collingwood, 1876), redescribed by Laidlaw (1902), Meixner (1907), and Kato (1943); *P. orbicularis* (Schmarda, 1859), reinvestigated by Stummer-Traunfels (1933) and redescribed by Hyman (1955) and Marcus and Marcus (1968); and *P. cata* Marcus and Marcus, 1968: by the combination of color pattern (all others cream or buff with a reticulated brown pattern), tentacular structure (all others infolded), poorly developed sucker and internally by the general location of the sperm duct(s) entrance into the seminal vesicle, general morphology of the proximal portion (seminal vesicle and penis) of the male system, the minor mid-region enlargements of the ejaculatory duct, the lack

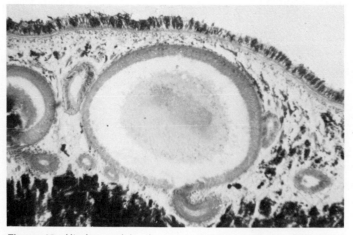

Figure 10 Uterine vesicle, the most posterior and largest; from sagittal section; part of next anterior vesicle seen to left: × 92.

of a secretory epithelium of the ejaculatory duct, and the presence of lateral pockets of the female antrum.

DISCUSSION

The family Pericelidae is clouded by the lack of a definitive description of *P. byerleyana* (Collingwood, 1876), the type of the family and genus from the type locality, Pulo Barundum, Borneo. The anatomical differences reported by subsequent investigators, chiefly Laidlaw (1902) and Meixner (1907), may be due to a number of factors, and the reinvestigation of specimens from the type locality could clarify the situation. The recent addition of *P. cata* Marcus and Marcus, 1968 from the Caribbean, where *P. orbicularis* is also reported, only underscores the necessity of obtaining material from the type locality to elucidate *P. byerleyana*.

P. hymanae is the third Pacific *Pericelis*; *P. byerleyana* is reported from the Indian and West Pacific Oceans, and *P. ernesti* is reported from the East Pacific.

In color and tentacular structure, *P. hymanae* differs distinctly from the other four reported members of the genus. Other features of interest are the general lack of frontal eyes in the anterior midline also noted for *P. ernesti*, and the poorly developed sucker, perhaps similar to that found in *P. byerleyana*.

It is apparent that the terminal portion of the male copulatory organ is of paramount importance in distinguishing the Pericelidae. In *P. hymanae*, the penis is similar in general morphology but shorter than that of *P. byerleyana*. *P. hymanae*, along with *P. byerleyana* and *P. cata*, lacks the large luminal enlargements of the ejaculatory duct demonstrated by *P. orbicularis* and *P. ernesti*. However, *P. byerleyana*, *P. cata*, and possibly *P. orbicularis* (see Stummer-Traunfels, 1933, and Hyman, 1955) demonstrate a glandular secretory epithelium along the ejaculatory duct while in *P. hymanae* and *P. ernesti* this is absent. The bulb-shaped seminal vesicle of *P. hymanae* is also found in *P. ernesti* and *P. orbicularis*.

In the present investigation of *P. hymanae*, it is clear that depending on maturity, wide variation may be found in the distal portion of the male system. Narrow sperm ducts in the less mature become large sinuses in the more mature. This development may be so pronounced that the two lateral sperm ducts may join to form a wide common sperm duct.

A feature of the female system of *P. hymanae* that warrants comment is the enormous size of the most posterior uterine vesicle. This feature, although reduced in size, is also apparent in the smaller immature specimens, always lying on a level parallel with the base of the seminal vesicle. These posterior vesicles are often easily seen in alcohol-preserved specimens.

NOTES

This species was found in large numbers through all seasons in 1961. During the spring and summer of 1962, it showed a tremendous population increase at Kupikipikio Point (Black Point), where only occasional specimens had previously been taken.

P. hymanae is usually found under "clean" basalt rocks and occasionally on broken coral on a substratum of rock on broken coral, broken coral, or coral sand. They can be found beyond 40 m from shore beginning at depths of 15 to 45 cm at low tide. They are usually found near the reef front, in areas of heavy wave action. Patches of *Sargassum* are found in these areas, and the mollusk *Isognomon* (*Meline*) *costellatum* Conrad is common on the underside of the rocks where this *Pericelis* is found. *P. hymanae* uses the underside of the shell of *Isognomon* as a refuge. There is no evidence to indicate that *P. hymanae* is a predator of *Isognomon*, as observations in the field and laboratory do not support this.

P. hymanae never swims. It is gregarious, and several may be found under the same rock. They are especially difficult to remove owing to the thin fragile construction of the body, and the rarely found mature individuals with ova-filled uteri are most subject to damage. The low incidence of sexually mature specimens is a puzzling situation; many more specimens than the number collected were examined, and yet only four or five mature specimens with ova-filled uteri were taken.

Not only the ova, but also the egg capsules are very large (0.5 mm and over). The egg mass is one capsule layer deep and translucent cream in color. The egg mass pattern is roughly square. Unfortunately, those laid in the laboratory failed to mature.

In the fixing of *P. hymanae* with a hot mercuric chloride solution, not only do specimens distort and crumple badly, but unlike other polyclads similarly treated, they adhere to glass and are badly damaged during removal. This can be avoided by coating the glass of the killing vessel with a silicon lubricant. Upon fixation, specimens turned bright yellow.

ACKNOWLEDGMENTS

The author wishes to thank Dr. Sidney J. Townsley for generously making space and equipment in his laboratory available, and also Drs. E. Alison Kay and George W. Chu for making necessary equipment available.

BIBLIOGRAPHY

Collingwood, C. 1876. On Thirty-one Species of Marine Planarians, Collected Partly by the Late Dr. Kelaart, F.L.S., at Trincomalee, and Partly by Dr. Collingwood, F.L.S., in the Eastern Seas. *Trans. Linnean Soc. London*, ser 2. Zool., 1:83–98.

Hyman, L. H. 1953. The Polyclad Flatworms of the Pacific Coast of North America. *Bull. Am. Museum Nat. Hist.*, **100**:265–392.

———. 1955. A Further Study of the Polyclad Flatworms of the West Indian Region. *Bull. Marine Sci. Gulf Caribbean*, **5**(4):259–268.

Kato, K. 1943. Polyclads from Palao. *Bull. Biogeogr. Soc. Japan*, **13**(12):79–90.

Laidlaw, F. F. 1902. "The Marine Turbellaria, with an Account of the Anatomy of Some of the Species." In J. S. Gardiner (ed.), *The Fauna and Geography of the Maldive and Laccadive Archipelagoes*, vol. 1, pp. 282–312. Cambridge University Press.

Lang, A. 1884. "Die Polycladen (Seeplanarien) des Golfes von Neapel und der angrenzenden Meeresabschnitte." *Fauna und Flora des Golfes von Neapel.* Monog. XI, ix + 688 pp.

Marcus, Ev. du B.-R., and E. Marcus. 1968. "Polycladida from Curaçao and Faunistically Related Regions." In P. W. Hummelinck (ed.), *Studies on the Fauna of Curaçao and Other Caribbean Islands*, **26**(101):1–133.

Meixner, A. 1907. Polycladen von der Somaliküste, nebst einer Revision der Stylochinen. *Z. Wiss. Zool.*, **88**(3):385–498.

Schmarda, L. K. 1859. *Neue wirbellose Thiere beobachtet und gesammelt auf einer Reise um die Erde 1853 bis 1857.* 1 Bd. Turbellarien, Rotatorien und Anneliden. Leipzig: Wilhelm Engelmann. Erste Hälfte, xviii + 66 pp.

Stummer-Traunfels, R. von. 1933. "Polycladida." In H. G. Bronn (ed.), *Klassen und Ordnung des Tierreichs.* Bd. IV, Abt. 1c, Leif. **179**: 3485–3596.

Chapter 6

Fine Structure of the Turbellarian Epidermis

Celina Bedini
Floriano Papi
Istituto di Biologia generale dell'Universitá,
Pisa, Italy

The existing information concerning the structure of turbellarian epidermis is to be found in monographs which treat the morphology of particular groups or from scattered notes in which the epidermis was mentioned in the course of describing new or poorly known species.

As is well known, the turbellarian body is clothed with a one-layered, generally ciliated epithelium which is devoid of any secreted protective layer or structures. In many groups it lies on a basement membrane of varying thickness. Notwithstanding the large number of species thus examined, the epithelium has shown little peculiarity for single groups, so that there have not generally resulted characteristics that can be used for phyletic evaluation. Extant treatises, even when from the hand of an eminent specialist of this class such as Hyman (1951), limit themselves to hinting at a few variations in the character of the epithelium.

The data of greatest interest are perhaps derived from the rhabdoids, but these structures, owing to their small dimensions at the level of resolution of the light microscope, have given the impression of rather uniform structure and uncertain classifiability. With the employment of the electron microscope, however, these structures have supplied important data, interesting also from a comparative point of view (Kelbetz, 1962; Reisinger and Kelbetz, 1964). Previously published electron-microscopic studies of the turbellarian epidermis have concerned single species or a few species within the limits of a given group (Török and Röhlich, 1959; Skaer, 1961; Pedersen, 1964; Dorey, 1965).

In order to contribute to a comparative study of turbellarian epidermis at the ultrastructure level, we report in the present paper our observations on a rather large number of species belonging to various orders. In certain cases, the study of related species has served as a control for the constancy of given characteristics within a group. Only normal epithelial cells and, when present, the basement membrane were considered; sensory cells and epidermal glands were not objects of investigation. Particular attention was paid to those structures, such as ciliary rootlets, which showed different characteristics from group to group, so that we might supply new elements for the study of the phylogenesis of turbellarians.

LIST OF SPECIES STUDIED AND METHODS

We have studied twenty-three species[1] belonging to eight different orders. The marine forms (Acoela, Polycladida and Seriata Otoplanidae) were collected along the coast of the Ligurian Sea. All the others are freshwater forms gathered in the vicinity of Pisa, with the exception of *Mesostoma* sp., which was collected by Professor G. Mancino at Tamanrasset in the southern part of Algeria and which we are rearing in our laboratory.

Acoela. Convolutidae: *Convoluta psammophila* Beklemischev, *Mecynostomum* sp.
Catenulida. Catenulidae: *Catenula lemnae* Ant. Dugès
Macrostomida. Macrostomidae: *Macrostomum retortum* Papi, *Promacrostomum gieysztori* (Ferguson)

---

[1]A short abstract has already been published (Bedini and Papi, 1970) with some preliminary data based on the examination of nineteen of these species. Unfortunately, owing to errors in the transcription, an insunk epidermis was attributed to *Macrostomum* although, in fact, we had found this only in *Convoluta, Prorhynchus*, and *Otoplana*; the syncytial structure of the epidermis, mentioned for *Prorhynchus, Gyratrix*, and *Rhynchomesostoma*, is found only in *Gyratrix*.

Polycladida. Pseudoceridae: *Thysanozoon brocchii* Grube
Lecithoepitheliata. Prorhynchidae: *Prorhynchus stagnalis* M. Schultze
Proseriata. Otoplanidae: *Otoplana truncaspina* Lanfranchi, *Notocaryotur-bella bigermaria* Lanfranchi, *Parotoplana macrostyla nomen nudum* Lanfranchi[2]
Tricladida. Dendrocoelidae: *Dendrocoelum lacteum* (O. F. Müller)
Rhabdocoela. Dalyelliidae: *Microdalyellia armigera* (O. Schmidt), *Dalyellia viridis* (G. Shaw), *Gieysztoria diadema* (Hofsten)
  Typhloplanidae: *Castrada viridis* Volz, *Castrada cristatispina* Papi, *Rhynchomesostoma rostratum* (O. F. Müller), *Mesostoma productum* (O. Schmidt), *Mesostoma craci* (O. Schmidt), *Mesostoma ehrenbergi* Focke, *Mesostoma* sp., *Opistomum pallidum* O. Schmidt
  Polycistidae: *Gyratrix hermaphroditus* Ehrenberg

For brevity, we shall give only the genus name in the text when only one species of that genus was examined.

The animals were fixed in 1% OsO_4 in phosphate buffer 0.1 M, 0.2 M, and 0.07 M, and in cacodylate buffer 0.05 M. With the different buffers, one obtains different results depending on the particular species. The material was embedded in the usual manner in Epon 812 or in an Epon-araldite mixture. The sections were stained with uranyl acetate and lead citrate and were examined with an Elmiskop 1/A Siemens electron microscope.

The study has been more or less detailed according to the species, because some of them presented technical difficulties and it was not always possible to get good pictures of some structures.

RESULTS

General Remarks

The epidermis usually shows distinct cell boundaries. In different species, however, the epithelium when viewed with a microscope appears syncytial, but one may presume that, for the most part, one is dealing with an appearance which is due to the absence of glicocalyx between the strictly adjacent membranes. Accordingly, Pedersen (1964) has been able to show with the electron microscope the cellular structure of the epithelium and of the parenchyma of some species of Acoela, a group in which the syncytial structure was believed to be particularly diffuse. In our material, the epithelium has a real syncytial structure only in *Gyratrix* (Fig. 1*j*), and the epidermis shows numerous infoldings of the internal membrane, but not cell boundaries (Fig. 3*a*).

In species belonging to various orders, the epithelium is described as

---

[2]This species will be described by Dr. A. Lanfranchi, whom we wish to thank at this point for having furnished the Otoplanidae.

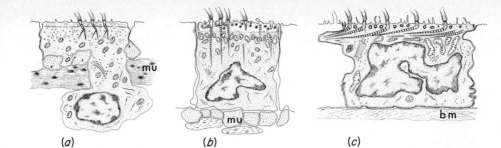

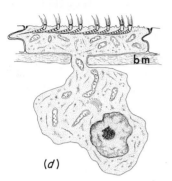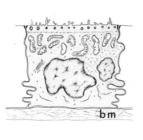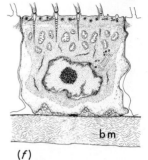

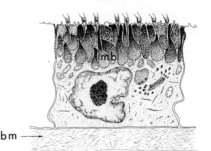

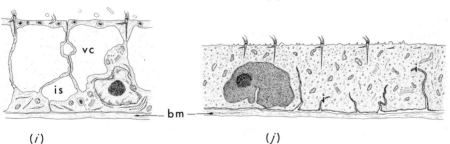

Figure 1 Schematic view of representative types of turbellarian epithelial cells. (*a*) *Convoluta*; (*b*) *Macrostomum*; (*c*) *Parotoplana* (ventral epithelium); (*d*) *Otoplana* (ventral epithelium); (*e*) *Noto-caryoturbella* (dorsal epithelium); (*f*) *Dalyellia*; (*g*) *Mesostoma*; (*h*) *Castrada*; (*i*) *Rhynchomesostoma*; (*j*) *Gyratrix*. bm, basement membrane; is, intercellular space; mb, multigranular bodies; mu, muscle; rh, rhabdites; vc, vacuoles.

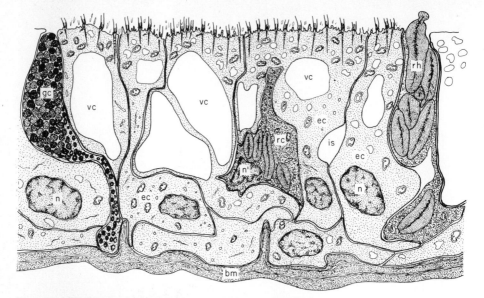

Figure 2 Schematic view of epithelium of *Thysanozoon brocchii.* bm, basement membrane; ec, epithelial cells; gc, gland cell; is, intercellular space; n and n′, nuclei of the epithelial and rhabditogen cells; rc, rhabditogen cell; rh, rhabdites; vc, vacuoles.

"insunk," that is, the nuclei accompanied by some cytoplasm have descended into the parenchyma internal to the muscle fibers. We find this condition in *Convoluta* (Fig. 1a), *Mecynostomum*, *Prorhynchus*, and *Otoplana* (Fig. 1e). In all other species in our studies, the epithelium is cellular with intraepithelial nuclei (Figs. 1b–c, e–i, 2).

The epithelium of turbellarians very often has the characteristics of glandular epithelium because the constituent elements produce rhabdites or bodies similar to rhabdites, granules, and secretions of various appearance. These secretions join themselves to those products of the glandular cells which are interposed between the epithelial elements (*Thysanozoon*; Fig. 4) or are located in the parenchyma.

The nucleus, which very often is polymorphic, is generally displaced to the basal part of the cell, also when it is intraepithelial. In *Gyratrix*, which has a thin epidermis, the nucleus is thereby very flattened and occupies almost the entire thickness (Fig. 3b). In some species in which the epithelium has particularly intense secretory activity (e.g., in *Rhynchomesostoma*) the nucleoli are well developed and clearly show the differentiation between fibrous and granular components (Fig. 5: nu).

The membrane of the epithelial cells does not show any unique features.

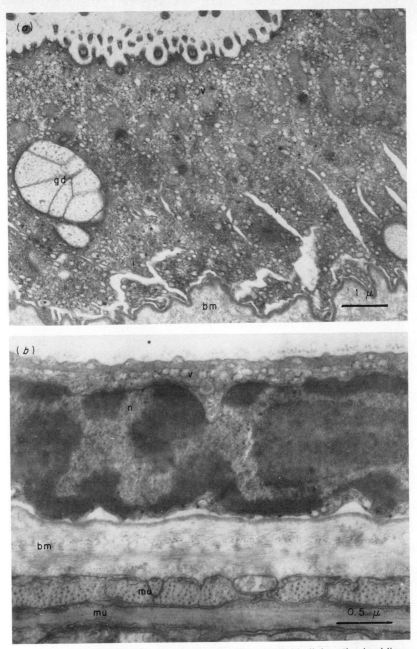

Figure 3 *Gyratrix hermaphroditus*. (*a*) Epithelial portion in oblique section showing infoldings (i) of the inner plasma membrane and numerous secretory vesicles (v) in the cytoplasm; (*b*) epithelial portion showing a large nuclear part (n) and the basement membrane (bm). gd, glandular cell duct; mu, muscle; v, vesicles.

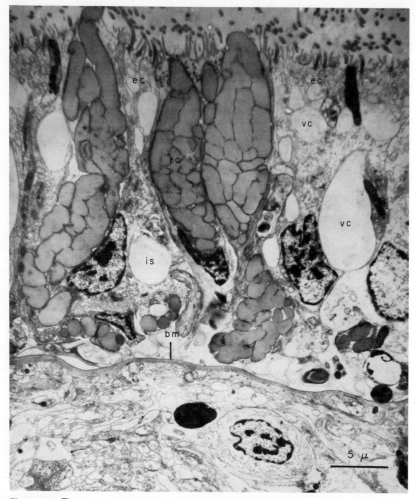

Figure 4 *Thysanozoon brocchii.* Epithelial portion at low resolution. bm, basement membrane; ec, epithelial cells; is, intercellular space; n and n', nucleus of the epithelial and rhabditogen cell; rc, rhabditogen cell; vc, vacuoles.

On the free surface there are microvilli of varying dimensions, more or less packed depending on the species. These are intermingled among cilia, which in our material are missing only on the dorsal surface of the three species of Otoplanidae (Figs. 1*e*, 6).

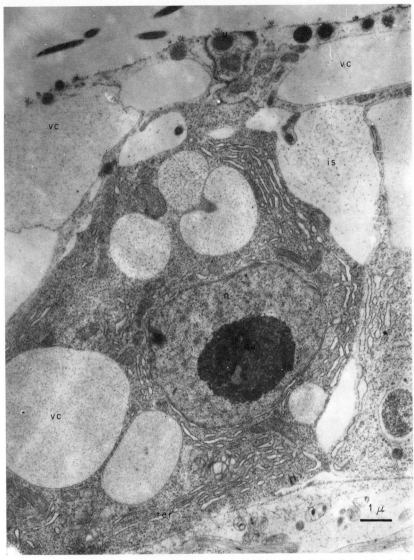

Figure 5 *Rhynchomesostoma rostratum.* Epithelial cell of the posterior body region. The thin cytoplasmic apical layer contains ultrarhabdites (u); note the large vacuoles (vc) in the cytoplasm. is, intercellular spaces; n, nucleus; nu, nucleolus; rer, rough endoplasmic reticulum.

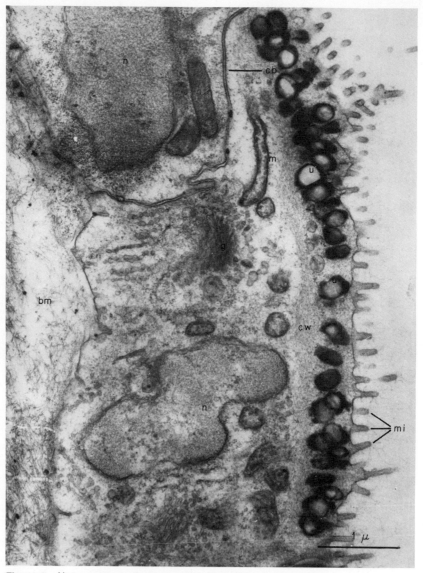

Figure 6 *Notocaryoturbella bigermaria.* Epithelial cell of the dorsal surface showing the cell web (cw) and ultrarhabdites (u). bm, basement menbrane; cb, cell boundary; g, Golgi apparatus; m, mitochondria; mi, microvilli; n, nucleus.

Cell Junctions

In all the species with cellular epithelia we have observed the presence of septate desmosomes between the plasma membrane of neighboring cells. In the Acoela the membranes are separated by 150 to 200 Å at the level of the septate desmosomes, while in all the other groups one records values within the range 200 to 250 Å. Both in *Convoluta* and in *Mecynostomum* there are septate desmosomes which end approximately 0.4 μm before the free surface of the cell. They have septa which are not always evident, giving the impression, in certain parts, of intermediate junctions (Fig. 7a). More evident are the septa between epithelial cells and sensory or glandular cells.

In the other groups, we can distinguish between (1) those cases in which the septate desmosome toward the external border is accompanied by a filamentous dismembranal junction which lies between it and the free surface (Macrostomidae, Fig. 7b; Dalyelliidae, Figs. 7d, 10; *Opistomum*); and (2) those in which the septate desmosomes reach almost to the free surface of the cell and are not accompanied by a junction of any other type (Polycladida; Seriata, Fig. 9a–b; Typhloplanidae, except *Opistomum*, Fig. 9c). The mediocre outcome of our preparations for *Catenula* and *Prorhynchus* has not allowed us to establish whether or not filamentous dismembranal junctions exist in these forms.

The filamentous dismembranal junction extends for 0.4 to 0.7 μm in Macrostomida (Fig. 7b), for 0.7 to 1.0 μm in Dalyelliidae (Fig. 7c, d), and for 1.0 μm in *Opistomum*. It is characterized by the presence of dense material along the inner face of the plasma membrane. The intercellular space (250 to 300 Å) is generally occupied by a disk of moderately dense material. Only in *Dalyellia* (Fig. 7d) is a central lamina very evident. The filaments of the cell web (see below) spread fanlike to the zone of the junction (Fig. 8, insert). The septate desmosome has a range about equal to that of the dismembranal junction, but is more resistant to the action of separation owing to the technical manipulation, as one can see, for example, in *Macrostomum* (Fig. 8), where the cells remain jointed only at the level of the septate desmosomes. The dismembranal junctions seem likewise more resistant in Dalyelliidae and in *Opistomum*, where one rarely sees separations. In the forms lacking filamentous dismembranal junctions, the distance between the septate desmosomes and the free surface is always less than 0.1 μm and shows a thickening on the sides of the plasma membrane (Fig. 9), which in addition in *Thysanozoon* may sometimes show a fibrous material (Fig. 14b). The intercellular space is occasionally occupied by a moderately dense material. It is of note that all the species which show only junctions of the septate type, lack the cell web or have a poorly developed one.

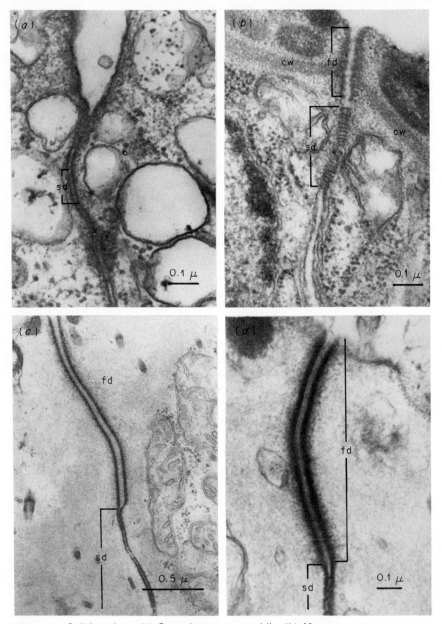

Figure 7 Cell junctions. (a) *Convoluta psammophila*; (b) *Macrostomum retortum*; (c) *Gieysztoria diadema*; (d) *Dalyellia viridis*. In b, c, and d septate desmosomes (sd) and filamentous dismembranal junctions (fd) are visible. In a only a septate junction is present. cw, cell web.

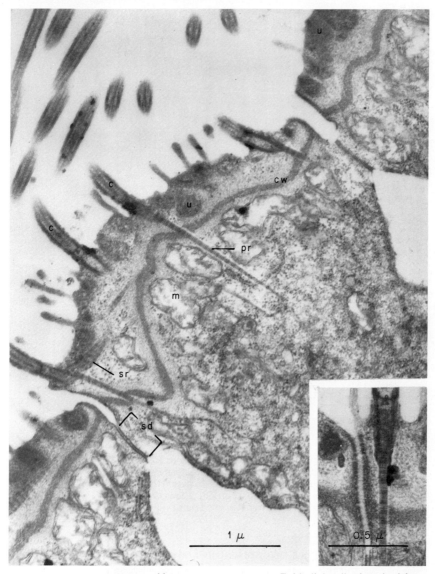

Figure 8 *Macrostomum retortum.* Epithelial cells detached from one another (artifact) and joined together only at the level of the septate desmosome (sd). Insert: cell web near the cell junction. c, cilia; cw, cell web; m, mitochondria; mi, microvilli; pr, principal ciliary rootlets; sr, secondary ciliary rootlets; u, ultrarhabdites.

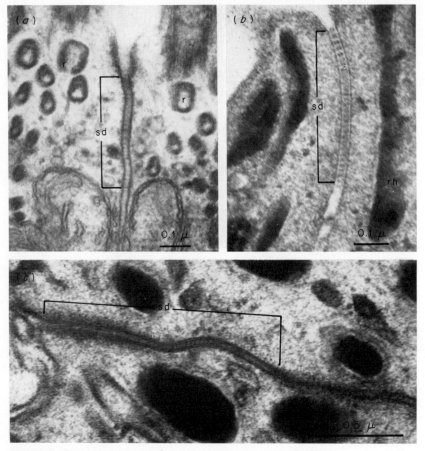

Figure 9 Cell junctions. (a) *Otoplana truncaspina*; (b) *Dendro-coelum lacteum*; (c) *Mesostoma* sp. The septate desmosomes (sd) reach almost to the apical region of the cells. In *a* note the thickening of the plasma membrane inner borders above the septate desmosome. r, ciliary rootlets in transverse section; rh, rhabdites.

Hemidesmosomes are present in Dalyelliidae between the internal surface of the epithelial cell and the basement membrane (Fig. 10:h; see below).

The Cytoplasmic Organelles

In the species in which the cytoplasm is subdivided into an apical portion and a basal one by the cell web, the cytoplasmic organelles, except of course the ciliary basal bodies, are always confined to the basal region.

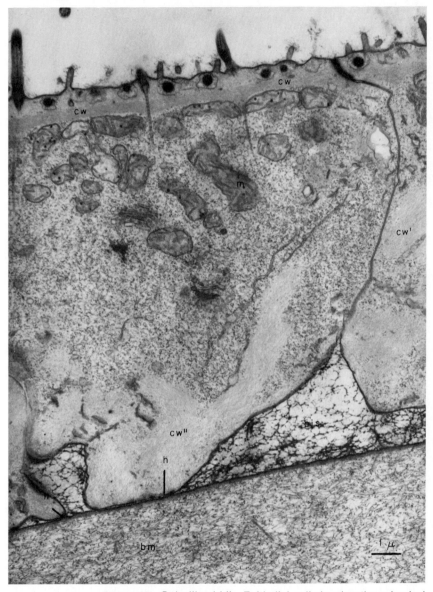

Figure 10 *Dalyellia viridis.* Epithelial cell showing the subapical (cw), lateral (cw′), and basal (cw″) cell web. bis, basal intercellular space; bm, basement membrane; h, hemidesmosome; m, mitochondria; u, ultrarhabdite.

In Otoplanidae, the mitochondria are very elongated, cylindrical, and bent. They are much more numerous in the ventral than in the dorsal epithelium and, in any case, they are more abundant in this family than in all the others. This may be related to the remarkable energy development of the ventral cilia of Otoplanidae, which are known to be capable of very fast movement (Ax, 1956). An orientation of the mitochondria prevalently perpendicular to the free surface is particularly evident in *Otoplana* (Fig. 11a). The length of mitochondria is greater than in other families; in *Notocaryoturbella* they reach about 3 μm in length (Fig. 11b).

As a consequence of its location near the nucleus, the Golgi apparatus of the insunk epithelium is, as a rule, situated deep in the cytoplasm. Only rarely does one see it also in the surface cytoplasm. Often the Golgi apparatus appears to be in clear secretory activity [*Castrada* (Fig. 12), *Rhynchomesostoma*, and *Mesostoma* (Fig. 13)].

The endoplasmic reticulum may be more or less developed, depending on the species; in Acoela it is mostly smooth whereas in *Rhynchomesostoma* and *Mesostoma* it is mostly rough. In the forms with insunk epithelium the reticulum is present almost only in the deep zone of the cytoplasm, but in the species with intraepithelial nuclei it is distributed throughout the cell.

In the cases in which the rough endoplasmic reticulum is poorly developed, one observes the presence of ribosomes scattered throughout the cytoplasmic matrix. Lysosomes and multilamellar bodies are sometimes present in epithelial cells.

Cell Web

In numerous species, the epithelial cells show continual layers of thin, closely packed filaments (about 70 Å in thickness). We call this structure "cell web" despite the fact that it is not in the usual location immediately under the external cellular membrane, as seen in the epithelial cells of vertebrates and some invertebrates (see among others Puchtler and Leblond, 1958; Leblond, Puchtler, and Clermont, 1960; Farquhar and Palade, 1963; Clermont and Pereira, 1966; Satir and Gilula, 1970).

In the majority of cases, the cell web runs parallel to the surface of the epithelium at a distance of 0.5 to 1.0 μm. Thus it subdivides the cytoplasm into a thin apical portion and a more extensive basal portion (Figs. 6, 7b, 8, 10, 14a). The thickness of this subapical cell web varies according to the species. It is extremely thin in *Convoluta* (600 Å) and in *Mecynostomum* (200 Å), and attains the maximal thickness in Macrostomidae (0.25 to 0.4 μm) and in Dalyelliidae (more than 0.4 μm). Intermediate thicknesses are recorded for *Catenula*, *Opistomum*, and *Notocaryoturbella*, where it is present only in the dorsal epithelium. The fanlike spreading of the cell web in the vicinity of the dismembranal junction has already been mentioned above.

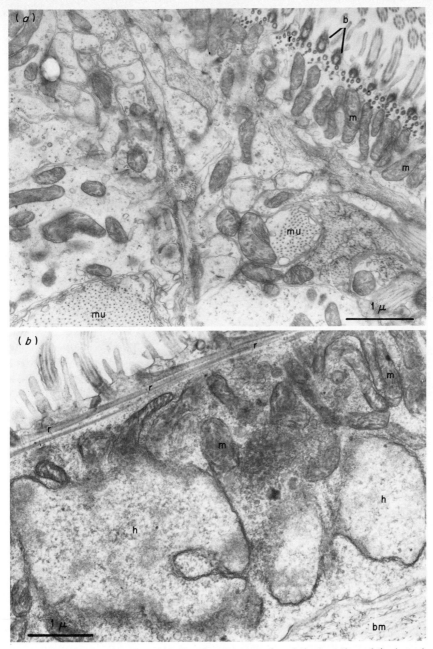

Figure 11 (a) *Otoplana truncaspina*. Apical portion of the insunk epithelium in transversal section. Note the arrangement of mitochondria (m). (b) *Notocaryoturbella bigermaria*. Epithelial cell in longitudinal section. Very large and bent mitochondria (m) are visible above the nucleus (n). bb, basal bodies; bm, basement membrane; mu, muscle; r, ciliary rootlets.

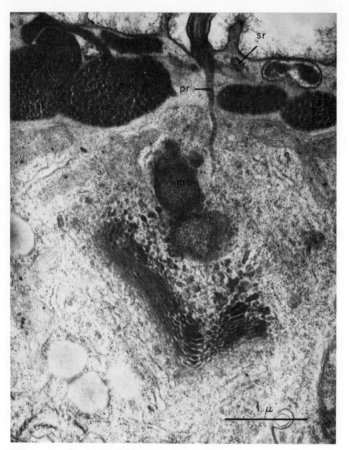

Figure 12 *Castrada viridis.* Epithelial cell portion showing ciliary rootlets (r), multigranular bodies (mb), and Golgi apparatus (g).

In *Gieysztoria* and *Dalyellia* (Fig. 10), the cell web is developed also along the lateral boundary of the cell and in its basal portion. The filaments, however, are less densely packed, and more easily resolvable than those in the subapical portion. In *Thysanozoon* the cell web consists of widely spaced filaments localized under the external wall; in this species, therefore, there is a real "terminal web" (Fig. 14*b*). In *Parotoplana*, as in *Notocaryoturbella*, the cell web is present only in the dorsal epithelium. In *Parotoplana*, however, it is not developed as a subapical cell web, but occurs only along the internal wall and the proximal portion of the lateral wall of the cell (Fig. 15). It appears to consist of more thickly packed fibrils which form a continuous layer almost adhering to the cellular membrane, and which follows the membrane regularly. Its thickness is about 0.15 μm.

We have not found a cell web in *Prorhynchus, Otoplana, Dendrocoelum, Gyratrix*, or Typhloplanidae (with the sole exception of *Opistomum*).

Vesicles, Vacuoles, and Secretory Products

Vesicles and vacuoles of various sizes are present in the epithelial cells of numerous species. Small, empty vesicles (approximate diameter of 0.2 μm) fill almost all the cytoplasm of the epithelium of *Gyratrix* (Fig. 3*a*). Vesicles of the same diameter and appearance are present in *Mecynostomum* and *Convoluta* (Fig. 16). Here they are located mostly in the apical cytoplasmic layer and are only rarely observed in the insunk portions of the cytoplasm.

Very large vacuoles, full of fluid and sometimes attaining a diameter of

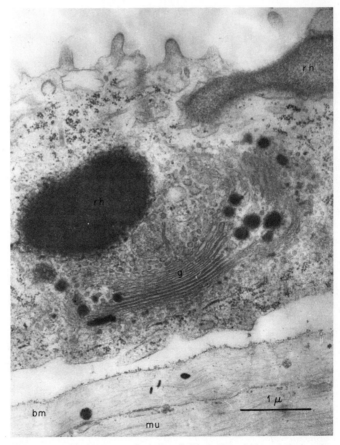

Figure 13 *Mesostoma ehrenbergi.* Rhabdites (rh) and their relation to the Golgi apparatus (g). bm, basement membrane; mu, muscle.

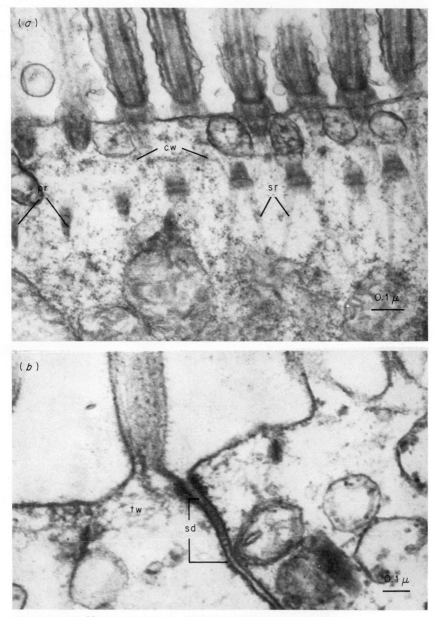

Figure 14 (a) *Mecynostomum* sp. Epithelial cell portion in transverse section, showing the thin cell web (cw). (b) *Thysanozoon brocchii*. Septate desmosome (sd). Note the thin filaments of the terminal web (tw) and the thickening of fibrous material above the septate desmosome. pr, principal rootlets; sr, secondary rootlets.

10 μm, are present in the epithelium of *Thysanozoon*. These may sometimes be confused with the large spaces interposed between cells (Fig. 4).

The apical region of the epithelial cells in both of the *Castrada* species is filled with inclusions which we call multigranular bodies. These bodies measure up to 3 μm and are prevalently pyriform. They pile up one by the other and leave only a very thin cytoplasmic layer interposed (Fig. 17). The inside is occupied by very dense homogeneous granules, which measure up to 0.16 μm. It should be pointed out that some bodies contain mostly large granules, while others contain only small granules. The granules in both cases are immersed in a finely granular matrix of medium density (Fig. 18).

The discharge of the bodies takes place after their membranes fuse

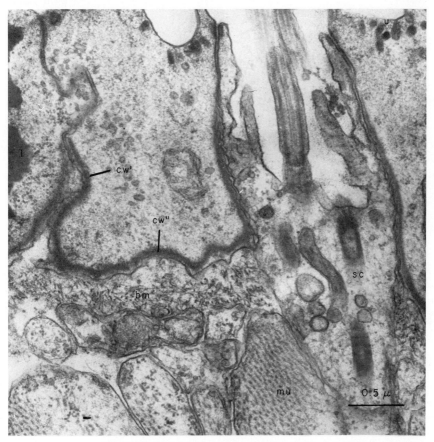

Figure 15 *Parotoplana truncaspina.* Epithelial cell showing the laterobasal cell web (cw′, cw″). bm, basement membrane; l, lipidic droplet; mu, muscle; sc, sensory cell; u, ultrarhabdites.

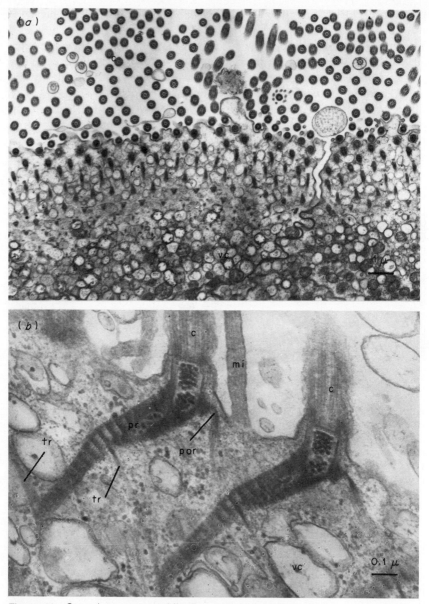

Figure 16 *Convoluta psammophila.* Tangential (*a*) and longitudinal (*b*) sections of epithelial cells. c, cilia; mi, microvilli; pr, principal rootlet; por, posterior rootlet; r, rootlet; tr, twin rootlet; vc, vacuoles.

apically with the cellular membrane; the successive disappearance of the wall interposed between the secretion and the outside permits the escape of the granules. The emptied bodies contain only sparse fibers and have minimal densities.

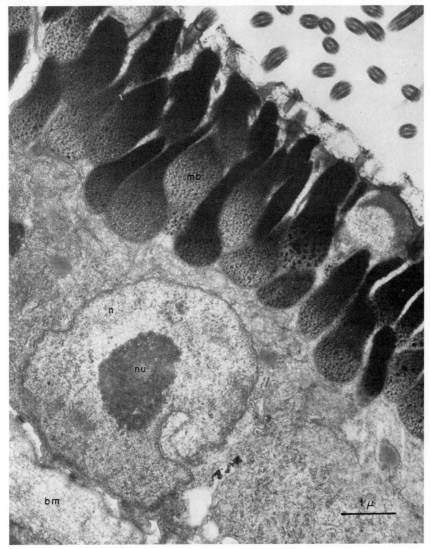

Figure 17 *Castrada cristatispina*. Epithelial cell portion showing the multigranular bodies (mb). bm, basement membrane; n, nucleus; nu, nucleolus.

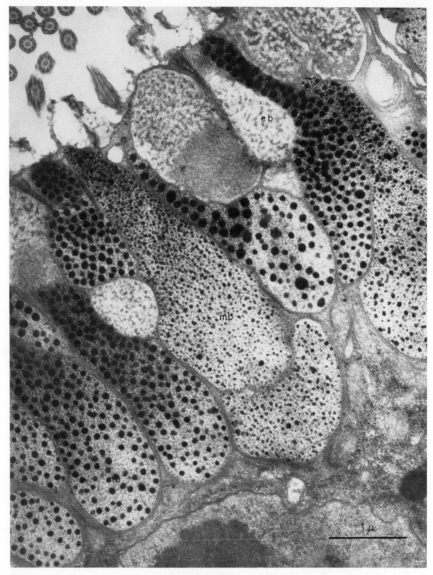

Figure 18 *Castrada cristatispina*. Multigranular (mb) and empty
bodies (eb). Note the variable size of granules.

The multigranular bodies appear to originate in a deeper zone of cytoplasm. This is suggested by some micrographs which show probable precursors of the multigranular bodies in the vicinity of the Golgi apparatus,
which is in intense secretory activity (Fig. 12).

This layer of multigranular bodies corresponds to the *Alveolarschicht* described by Luther (1904, p. 8) in the epithelium of *Castrada* and other Typhloplanidae. Since this author found these bodies empty, it was reasonable to suppose that the granules of secretion would contain lipid materials. Indeed, we observed that they had disappeared after treatment with hydrogen peroxide.

In *Prorhynchus* the cytoplasm of the not insunk portion of the epithelium is almost totally full of vacuoles which contain fibrous material of medium density. These vacuoles remind us of the discharged multigranular bodies of *Castrada*.

In *Rhynchomesostoma* the epithelium, as already noted by light microscope (Luther, 1904), has different appearances in different regions of the body. Except for the contractible snout, the epithelial cells are almost wholly occupied by large vacuoles which are empty in our preparations. There are also wide intercellular spaces, equally empty. The cytoplasm is reduced to a thin apical layer and to a basal layer containing the nucleus, the Golgi apparatus, and a rough endoplasmic reticulum (Figs. 5, 19*a*).

In the region of the snout one distinguishes a terminal cone which, as is well known, may be telescopically retracted, and a region farther back which is folded in during the contraction. In this latter region the epithelium is more compact than in the former, and in the apical region it shows vacuoles which are empty or which contain only loose filaments (Fig. 19*b*). On the terminal cone the vacuoles become looser as we progress toward the tip of the snout. At the tip, the epithelium is characterized by the presence of numerous sensory cells and ducts of rhabditogen cells (*Stäbchenstrassen*) and of glandular cells. The epithelial cells, rich with microvilli, contain very numerous electron-dense granules (Fig. 20*a*).

Among the secretory products of epithelial cells are the dermal rhabdites. Those of *Dendrocoelum* and *Mesostoma* appear like dense oval or rod-like bodies surrounded by a membrane. Peripherally, they seem to consist of a homogeneous layer of material which is very electron-opaque, while the central zone has a granular appearance. In *Mesostoma* a material very similar to the rhabdite is contained in the vesicles of the Golgi apparatus, indicating that the material forming the rhabdite is probably produced by this apparatus (Figs. 13, 21:dv).

In *Rhynchomesostoma*, in the region of the snout immediately behind the apical cone, one sees rounded rhabdites of about 2 μm diameter which are surrounded by a layer of fibrous cytoplasm (Fig. 20*b*). They are much more abundant dorsally than ventrally and may be discharged, as is evident in some micrographs in which the fibrous layer has ascended to the external limit of the cell (Fig. 20*c*).

In a series of species which may not exhibit dermal rhabdites under the light microscope, we find rounded electron-dense granules of 0.4 to 0.8 μm

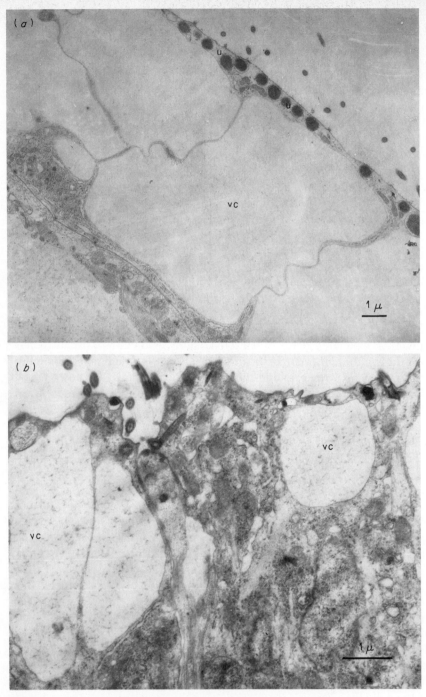

Figure 19 *Rhynchomesostoma rostratum.* Epithelial portion (*a*)
through the posterior region of the body (*b*) in the region just back
of the terminal cone. Note the presence in part *a* of large vacuoles
(vc) limiting the outer and inner cytoplasmic layer of the epithelial
cell. In (*b*) the cytoplasm is more compact and the vacuoles are
scattered in it. u, ultrarhabdites.

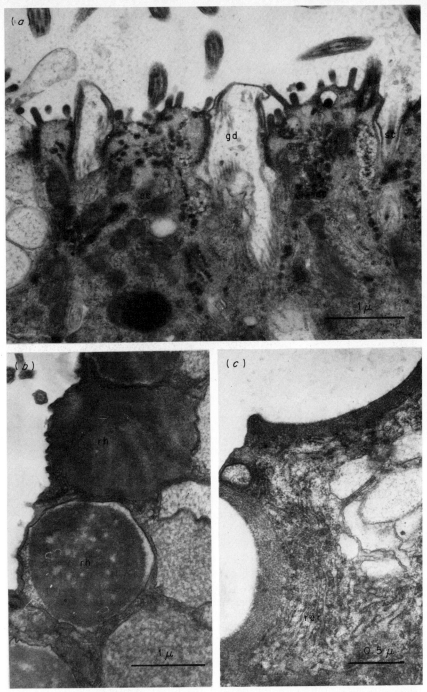

Figure 20 *Rhynchomesostoma rostratum.* (a) Epithelium of the tip of the terminal cone; (b) epithelium of the basal and dorsal part of the terminal cell cone containing rhabdites (rh); (c) the same region after the rhabdite discharge. ec, epithelial cells; gd, glandular duct; rer, rough endoplasmic reticulum; sc, sensory cell ending.

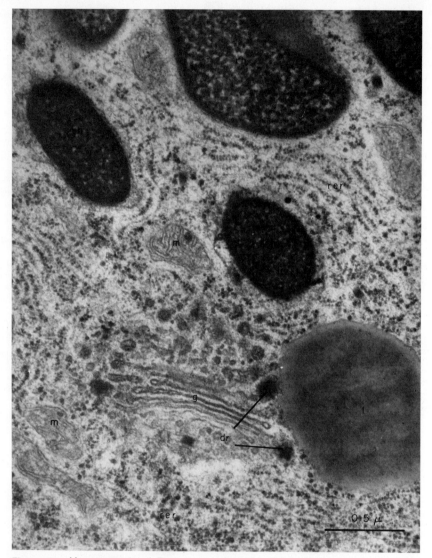

Figure 21 *Mesostoma craci.* Dermal rhabdites (rh), Golgi apparatus (g), and dense vesicles (dv) are present near a lipid droplet (l). m, mitochondria; rer, rough endoplasmic reticulum.

diameter in the epithelium. Except for their form and dimensions, these granules are very similar to the dermal rhabdites. Until these bodies have been tested histochemically and have been shown to have a different nature, they may be regarded as homologous to the dermal rhabdites. Since the presence or absence of dermal rhabdites is taxonomically employed, it might generate confusion to use the name "rhabdite" for structures which are invisible or difficult to identify with the light microscope. We, therefore, propose to call the bodies described above "ultrarhabdite." It would seem less appropriate to use the term "pseudorhabdite," which has been employed by Reisinger and Kelbetz (1964) for similar bodies in *Megalorhabdites inermis*.

Ultrarhabdites are present in the more apical portion of the cytoplasm (outside the cell web, if a cell web is present) in the Macrostomidae (Fig. 8), in *Catenula*, in the Dalyelliidae, in *Opistomum*, and in *Rhynchomesostoma* [in the vacuolized epithelium behind the snout region (Figs. 5, 19a)]. In the Otoplanidae the ultrarhabdites are present only in the dorsal epithelium, and they show a core of less density (Fig. 6). In *Dalyellia* they are contained in vacuoles (Fig. 10).

In the Acoela studied as well as in *Prorhynchus*, *Gyratrix*, and *Castrada*, dermal rhabdites and ultrarhabdites are lacking. In the latter genus the function of the rhabdites is probably assumed by the multigranular bodies.

In the dorsal epithelium of the Otoplanidae, which completely lacks cilia, the ultrarhabdites may be easily mistaken, when seen with the light microscope, for basal granules of cilia. This has induced Hofsten (1918) and Ax (1956) to say that in some species there is a degeneration of the dorsal ciliation. This interpretation was made even more likely by the fact that the microvilli may simulate residues of cilia (Fig. 6).

Cilia

In all the species studied, the free surface of the epithelial cells is supplied with cilia, with the exception of the dorsal surface of the Otoplanidae. Cilia are regularly arranged in rows as is evident in tangential or oblique sections. Particularly clear is the regular arrangement of the cilia in *Convoluta* (Fig. 16).

The structure of the ciliary shaft does not present any outstanding characteristics; in every species there is the same pattern with $9 + 2$ fibrils. The basal body protrudes above the free surface of the cell. In *Convoluta* and *Macrostomum* one sees inside the basal body a mass of granules which push also into the rootlet (Fig. 16b).

According to the number and disposition of the rootlets, the cilia may be classified into three types: the first type is that of the Acoela, and we can confirm the observations of Dorey (1965). Their cilia have four rootlets (Fig. 16b): a principal rootlet (pr) directed obliquely to the front and below, a very small posterior rootlet departing from the posterior part of the basal body, and two twin rootlets which depart from a point of flexion of the major rootlet. These latter diverge one from the other, pointing obliquely backwards and below, and each one makes contact with the tip of the major rootlet of another cilium (Figs. 14a, 16b, 22a).

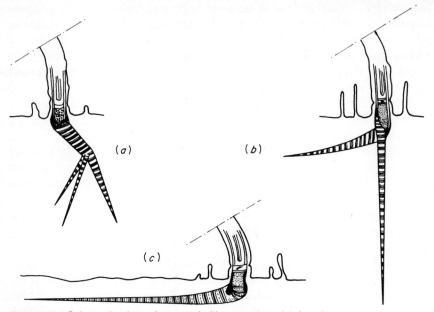

Figure 22 Schematic view of types of ciliary rootlets. (a) Acoela;
(b) other groups; (c) Otoplanidae.

The second type of rootlet is found in Macrostomida, Catenulida, Lecithoepitheliata, Polycladida, Tricladida, and Rhabdocoela. Here there are
only two rootlets. The one which we call the "principal" rootlet runs per-
pendicularly to the cell surface. The "secondary" rootlet is directed caudally,
going more or less parallel to the free surface of the cell (Figs. 22b, 23).

The third type is that of Otoplanidae. In the three species of our study,
the cilia are provided with a single rootlet, about 5 μm long, which runs toward the back, parallel to the cell surface (Fig. 22c). Owing to their length,
the rootlets exceed the basal bodies of some cilia which lie behind on the same
longitudinal row, so that more rootlets run closely packed to each other (Fig.
11b). As a consequence of this arrangement, there is a particular morphological adaptation of the epithelial cell which sends backwards and immediately
below the apical region of the contiguous cell, a wedge-shaped process which
contains the rootlets of the cilia located toward the caudal margin of the cell.
These rootlets converge toward the tip of this wedge-shaped process (Figs.
24, 25a). We have not observed junctions between ciliary rootlets.

The rootlets are composed of longitudinal filaments, packed side by
side and arranged in such a way as to form a pattern which is striated because

of the repetition of periodical structures. The rootlets of the Otoplanidae are hollow and, with the exception of those of sensory cilia, do not allow recognition of longitudinal filaments. These appear therefore as made up of superimposed rings of various densities (Fig. 25b).

The striation of the rootlets is particularly evident after staining with uranyl acetate. Dark bands appear alternated with clear interbands, in both of

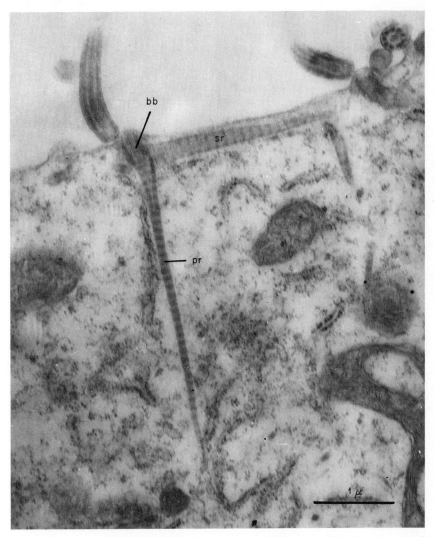

Figure 23 *Mesostoma productum*. Rootlets of a cilium. bb, basal body; pr, principal rootlet; sr, secondary rootlet.

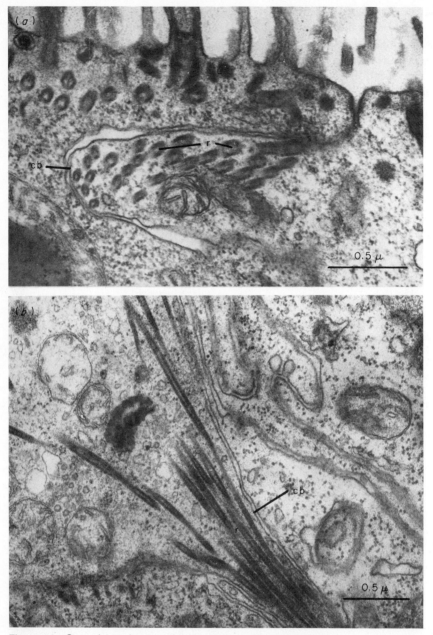

Figure 24 Cross (*a*) and tangential (*b*) sections through the epithelial cell of *Parotoplana macrostyla*. Long wedge-shaped processes of the epithelial cell containing ciliary rootlets (r) are visible. cb, cell boundary; u, ultrarhabdites.

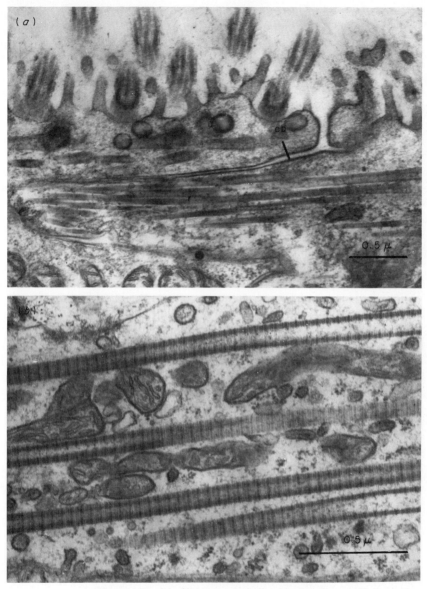

Figure 25 (a) *Notocaryoturbella bigermaria*: sagittal section through the epithelium showing wedge-shaped process of the cell. cb, cell boundary; r, ciliary rootlets; u, ultrarhabdites. (b) *Otoplana truncaspina*: ciliary rootlets. Note their central hollow core.

which we may distinguish darker stripes (subbands) and clearer stripes (inter-subbands). The most common pattern consists of subdivisions of a four- or five-subband period, and as many intersubbands. It is noteworthy that the period pattern may change either between different rootlets of the same cilium or in different parts of the same rootlet. According to Dorey (1965) this latter difference may be attributed to the presence of two distinct regions with a different pattern superimposed one on the other within the thickness of the same rootlet.

The length of the repetition period varies from one preparation to another even in the same species, but there are real differences from species to species and also between main and secondary rootlets of the same species. The extreme values go from a minimum of 490 Å in some preparations of *Mecynosto-mum* to a maximum of about 800 Å in the secondary rootlets of *Mesostoma productum* and *Macrostomum*. The diversity in the striation of the rootlets of different species can be appreciated in Fig. 26, where there are photographs of the species which afford the clearest picture.

Our findings relative to the arrangement of the rootlets explain the real nature of the supposed system of fibrils discovered by Luther (1904) in the Typhloplanidae which were thought to connect the basal bodies. This system has been regarded as a general characteristic of the ciliary system of turbellarians (Hyman, 1951). It is now evident that these presumed fibrils are in reality rootlets of cilia which run horizontally and align according to longitudinal files. In this way, for example, the *Plasmafaden* described by Hein (1928, p. 491) in *Dalyellia viridis* are none other than the secondary rootlets of the cilia. In *Castrada* [which is referred to in fig. 14 table I of Luther (1904)] these presumed fibrils in reality result from the secondary rootlets and from thin cytoplasmic layers interposed between the multigranular bodies, which are inserted between the longitudinal rows of cilia (Fig. 27). Luther also described transverse fibrils which join basal bodies of cilia belonging to contiguous longitudinal rows. In our opinion these presumed transverse filaments are sections of cytoplasmic layers interposed between the multigranular bodies because rootlets oriented in this way are lacking.

Basement Membrane

A basement membrane is missing in the Acoela, Macrostomida, and Catenulida which we studied. In the other groups it has varying degrees of development and measures from 0.2 to 4.0 μm. Its thickness depends also on the region of the body. In *Gieysztoria* and *Dalyellia* ample spaces, filled with a material which seems to have resulted from the coagulation of a fluid, are interposed between epithelial cells and the basement membrane, In these forms therefore the epithelial cells come directly in contact with the basement membrane

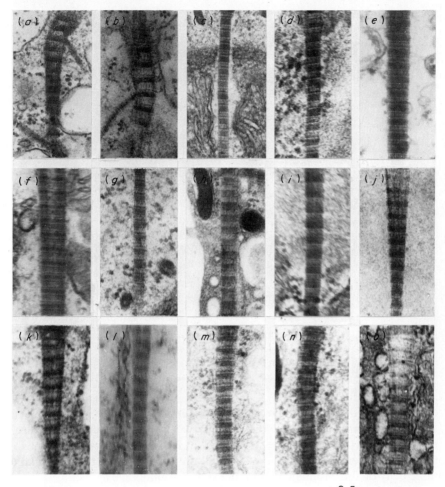

0.5 μ

Figure 26 Striation of ciliary rootlets. (a) *Convoluta psammophila*: principal rootlet showing a twin rootlet departing at top and the contact with a twin rootlet of an adjacent cilium (bottom). (b) *Mecynostomum* sp.: principal rootlet and twin rootlet. (c) *Macrostomum retortum*: principal rootlet. (d) *Macrostomum retortum*: secondary rootlet. (e) *Thysanozoon brocchii*: principal rootlet. (f) *Otoplana truncaspina*. (g) *Parotoplana macrostyla*. (h) *Dendrocoelum lacteum*: secondary rootlet. (i) *Dalyellia viridis*: principal rootlet. (j) *Gieysztoria diadema*: secondary rootlet. (k) *Rhynchomesostoma rostratum*: secondary rootlet. (l) *Mesostoma productum*: principal rootlet. (m) *Mesostoma* sp.: secondary rootlet. (n) *Mesostoma* sp.: principal rootlet. (o) *Gyratrix hermaphroditus*: secondary rootlet.

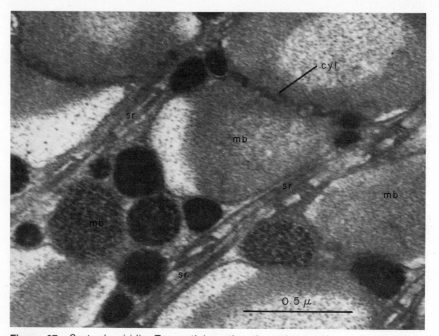

Figure 27 *Castrada viridis.* Tangential section through an epithelial cell showing the arrangement in longitudinal rows of the secondary rootlets (sr) and the thin cytoplasmic layer (cyl) between the multigranular bodies (mb).

only in limited areas, which show hemidesmosomes (Fig. 10). Few small spaces filled with fluid and interposed between the basement membrane and epithelial cells are found also in *Microdalyellia*. Here, however, we have not been able to see hemidesmosomes. In all the other species the epithelial cells lie directly on the basement membrane and lack junctional areas.

The basement membrane consists of fibrils varying between 80 and 100 Å in thickness, embedded in a homogeneous ground substance. The fibrils do not show any periodicity and are likely collagen of a primitive type. The inside and outside of the basement membrane is frequently bordered by a dense zone consisting probably of a condensation of the ground substance. The fibrils mostly lie parallel to the surface of the body. On the basis of their orientation in this plane one may distinguish three different cases:

1   The fibrils do not have a prevalent orientation (*Castrada cristatispina*, Fig. 28*a*).

2   The fibrils are arranged for the most part according to two directions. By analogy with the muscle fibers of the subepidermal musculature, we can

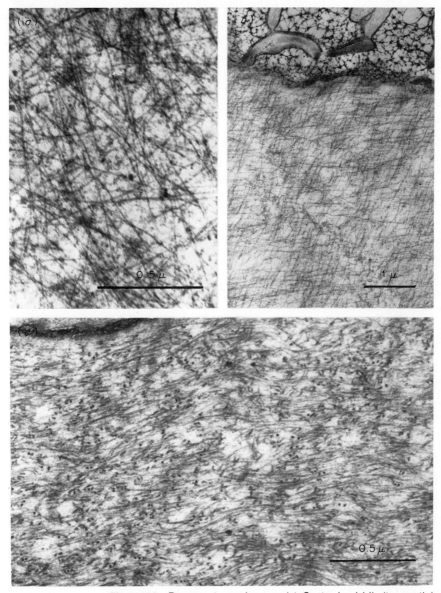

Figure 28 Basement membranes. (a) *Castrada viridis* (tangential section); (b) *Gieysztoria diadema* (tangential section); (c) *Dendrocoelum lacteum* (longitudinal section).

distinguish longitudinal and circular fibrils. They are mixed together and not separated into layers (*Otoplana, Dendrocoelum, Gieysztoria*, Fig. 28*b, c*).

3 The arrangement of fibrils shows a complex tridimensional pattern. Both transverse and longitudinal sections show alternating layers of circularly and longitudinally coursing fibrils (*Gyratrix* and *Thysanozoon*, Fig. 29). From many sections of different orientation, one can conclude that the same fibril passes from one layer to another by bending upward or downward and then by a lateral turn of 90°.

CONCLUSIONS

The series of morphological findings shown above are interesting chiefly in relation to the phylogenetic value of the new characteristics brought to light. At our present state of knowledge, the functional significance of the structures remains obscure, in large part.

Within single systematic groups, the ultramicroscopic characteristics of the epithelium reveal a noteworthy constancy which confirms in general, at least for the taxa which we have studied, the validity of the present system. For example, we find that in all the Acoela the membranes of contiguous cells are closer to one another than in any of the species of the other groups.

Equally notable is the fact that in the seven species of Acoela, belonging to five different genera investigated by Dorey (1965) and by us, there is a single type of ciliary rootlet which appears exclusive to the group. Also in the Otoplanidae we find an equally exclusive type of ciliary rootlet; it is not improbable that the strong development of the only rootlet may be connected with the presence of a creeping sole which can produce a noteworthy thrust. In addition there is probably a functional relationship between the intense ciliary labor and the large and numerous mitochondria.

A characteristic limited to Dalyelliida is the presence of large spaces filled with fluid interposed between the basement membrane and epithelial cells. In *Dalyellia* and *Gieysztoria* this characteristic is paired with the presence of hemidesmosomes between epithelial cells and basement membrane. It is also associated with the development of the cell web which occurs along the lateral and interior walls of the epithelial cells.

Other characteristics are common to more groups. Formed secretions of various types, but probably homologous (dermal rhabdites, ultrarhabdites, and multigranular bodies), are absent only in the two Acoela which we studied.

The cell web is present in various species where it assumes various levels of development and different positions in the cell. When it is subapical and well developed, there are at the same time filamentous dismembranal desmosomes. Of note is the absence of cell web in all the Typhloplanidae except *Opistomum*.

With the light microscope it has already been observed that the basement membrane is missing in Acoela, while conflicting data exist concerning

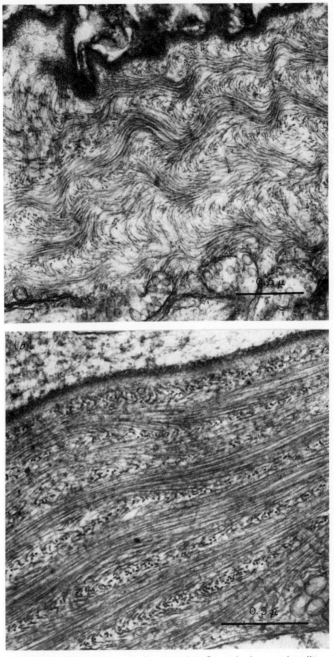

Figure 29 Basement membranes. (a) *Gyratrix hermaphroditus*; (b) *Thysanozoon brocchii*.

the Macrostomida and the Catenulida. The pictures from the electron microscope confirm this absence for the Acoela, and we may at this point consider the problem of verifying at the ultrastructural level the presence of the basement membrane in representatives of the other two orders. It would be very interesting to find out whether in both of these groups there exist forms with and without basement membrane, because this may indicate that this structure appeared independently at least two times. Our data on the arrangement of the filaments of the basement membrane enhance the interest of further study of this structure's evolution. Noteworthy is the complicated disposition of the filaments of Polycladida, where it is perhaps connected with the large size of these animals. One notes also that according to MacRae (1965) the filaments of *Notoplana acticola* would even show a periodicity, thus suggesting a further step in evolution.

Thysanozoon and *Gyratrix* show interesting peculiarities in the structure of the epithelium. Unfortunately these cannot be discussed further because we cannot offer comparisons with related species.

We hope that the electron micrographs we have presented suggest the value of extending to other structures of the turbellarian body the comparative approach at the ultrastructural level.

BIBLIOGRAPHY

Ax, P. 1956. Monographie der Otoplanidae (Turbellaria). *Akad. Wiss. Lit. Mainz Abhandl. Math. Nat. Kl.,* 1955, **13:**499–796.

Bedini, C., and F. Papi. 1970. Fine Structure of the Turbellarian Epidermis. *Am. Zool.,* **10**(4):545.

Clermont, Y., and G. Pereira. 1966. The Cell Web in Epithelial Cells of the Rat Kidney. *Anat. Record,* **156:**215–228.

Dorey, A. E. 1965. The Organization and Replacement of the Epidermis in Acoelous Turbellarians. *Quart. J. Microscop. Sci.,* **106:**147–172.

Farquhar, M. G., and G. E. Palade. 1963. Junctional Complexes in Various Epithelia. *J. Cell. Biol.,* **17:**375–412.

Hein, C. 1928. Zur Kenntnis der Regenerationsorgane bei den Rhabdocoelen. Mit Angaben über den feineren Bau und die Lebensäusserungen. *Z. Wiss. Zool.,* **130:**469–546.

Hofsten, N. (v). 1918. Anatomie, Histologie und systematische Stellung von *Otoplana intermedia* Du Plessis. *Zool. Bidr. Uppsala,* **7:**1–74.

Hyman, L. H. 1951. *The Invertebrates,* vol. II. New York: McGraw-Hill. 550 pp.

Kelbetz, S. 1962. Zum Feinbau der Rhabditen. *Naturwiss. Deut.,* **49:**548.

Leblond, G. P., H. Puchtler, and Y. Clermont. 1960. Structures Corresponding to Terminal Bars and Terminal Web in Many Types of Cells. *Nature,* **186:**784–788.

Luther, A. 1904. Die Eumesostominen. *Z. Wiss. Zool.,* **77:**1–273.

MacRae, E. K. 1965. Fine Structure of the Basement Lamella in a Marine Turbellarian. *Am. Zool.*, **5:**247.

Pedersen, K. J. 1964. The Cellular Organization of *Convoluta convoluta*, an Acoel Turbellarian: A Cytological, Histochemical and a Fine Structural Study. *Z. Zellforsch.*, **64:**655–687.

Puchtler, H., and C. P. Leblond. 1958. Histochemical Analysis of Cell Membranes and Associated Structures as Seen in the Intestinal Epithelium. *Am. J. Anat.*, **102:**1–32.

Reisinger, E., and S. Kelbetz. 1964. Feinbau und Entladungsmechanismus der Rhabditen. *Z. Wiss. Mikr.*, **65:**472–508.

Satir, P., and N. B. Gilula. 1970. The Cell Junction in a Lamellibranch Gill Ciliated Epithelium. *J. Cell. Biol.*, **47:**468–487.

Skaer, R. J. 1961. Some Aspects of the Cytology of *Polycelis nigra. Quart. J. Microscop. Sci.*, **102:**295–317.

Török, L. J., and P. Röhlich. 1959. Contribution to the Fine Structure of the Epidermis of *Dugesia lugubris* O. Schm. *Acta Biol. Acad. Sci. Hung.*, **10:**23–48.

Chapter 7

Spermiogenesis, Sperm Morphology, and Biology of Fertilization in the Turbellaria

Jan Hendelberg
Zoological Institute, University of Uppsala,
Uppsala, Sweden

The present work is intended to give an outline of our current knowledge of turbellarian spermatology. It will summarize my own results (Hendelberg, 1965, 1969; light, including phase-contrast, and electron microscope studies) and also take into consideration results arrived at by other authors.

HIGHLY SPECIALIZED SPERMATOZOA AND INTERNAL FERTILIZATION IN THE TURBELLARIA

First a few words on turbellarian spermatozoa and the general relation between the morphology of the spermatozoon and the biology of fertilization.

The turbellarian spermatozoa are of many different types. Three of them are shown by Hyman (1951, fig. 41C, D). In the spermatozoa of triclads, for example, we find two lateral flagella or *Nebengeisseln*. The spermatozoa of some macrostomids are characterized by two stiff bristles. Other spermatozoa have no visible flagella. Ax (1961, fig. 12; 1963) illustrates some other types, too.

All these types of turbellarian spermatozoa are very different from the type of spermatozoon which, according to Franzén (1956), is the primitive type of spermatozoon in the Metazoa (Fig. 1*d*). This primitive type, characterized by a short middle piece containing a number of mitochondrial spheres and a tail of a long, free flagellum, has a widespread distribution among the Metazoa. It is found in species with external fertilization, that is, the type of fertilization following the shedding of eggs and spermatozoa freely into the water, as a rule seawater. Franzén found the spermatozoa of species with internal fertilization to be more or less modified (cf. Fig. 1*e*).

The great differences between the primitive metazoan spermatozoon and the spermatozoa within the Turbellaria have been considered by Franzén (1956) to be due to the strongly changed biology of propagation in the Turbellaria, all of which have internal fertilization. This theory has been strongly supported by Jägersten (1959, 1968) and others, among them Ax (1961, 1963), who considers the "Turbellarian Archetype" to have had internal fertilization. Also the turbellarian archetype introduced in this volume by Karling is assumed by him to have had internal fertilization.

It is true that *Xenoturbella bocki* Westblad, described as a turbellarian, has spermatozoa of the primitive type (Westblad, 1949, fig. 7; Franzén, 1956, p. 366). However, the turbellarian nature of this species has been doubted and even denied (Remane, 1958; Jägersten, 1959; Reisinger, 1960; Ax, 1961, 1963). Thus, for the time being, the theory of the origin of the turbellarian spermatozoa from the mentioned primitive type lacks direct support in the morphology of the spermatozoa in the Turbellaria.

In other groups of animals the modifications following the change into internal fertilization often, according to Franzén (1956), resulted in spermatozoa with a more or less elongated middle piece enveloping a great part of the flagellum (Fig. 1*e*). Spermatozoa of a similar appearance have been described from many turbellarians. A head, a middle piece containing a central axial filament, and a tail have been discerned. In fact, the structure of such turbellarian spermatozoa is much more complicated (see below). Nothing has been found in the morphology of these spermatozoa which supports the theory that they have evolved from the primitive type mentioned above. It must be stressed, however, that this does not mean that they have not evolved from that type. As has been demonstrated by Franzén (1956, 1967a, b), the alterations of the sperm morphology in other groups with internal fertilization

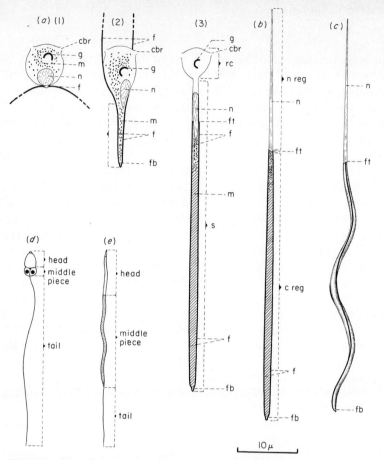

Figure 1 (a–c) Spermatids and spermatozoa of an acotylean polyclad, *Cryptocelides loveni*: (1) Young spermatid with two flagella
growing freely. (2) Spermatid. The basal parts of the two flagella
are united with the spermatid shaft. (3) Spermatid. The flagella are
united in the whole of their length with the spermatid shaft. (b)
Spermatozoon drawn schematically in the same fashion as the
spermatids in the preceding figures. (c) Spermatozoon. If a spermatozoon is studied from the edge, the undulating movements may
appear momentarily as shown in the figure. cbr, cytoplasmic
bridge; c reg, cytoplasmic region; f, flagellum; fb, basal end of
flagellum; ft, tip of flagellum; g, Golgi complex; m, mitochondria,
mitochondrial material; n, nucleus; n reg, nuclear region; rc,
residual cytoplasm; s, spermatid shaft. (d) The primitive type of
spermatozoon in the Metazoa. *(After Franzén, 1956, p. 462, somewhat simplified.)* (e) An example of a spermatozoon evolved from
the primitive type. The modifications following the change into
internal fertilization as a rule result in an elongation of the middle
piece. A great part of the (single) flagellum is incorporated into the
middle piece. *(After Franzén, 1956.)* Many turbellarian spermatozoa
have been described earlier to be of this type. In fact their morphology is much more complicated; e.g., see parts b and c.

have often led to very specialized spermatozoa. This has been found in groups in which some subgroups have retained the external fertilization and the primitive type of spermatozoon.

THE DEVELOPMENT OF DIFFERENT TYPES OF TURBELLARIAN SPERMATOZOA FROM SPERMATIDS WITH TWO FLAGELLA

The Spermiogenesis in the Polycladida

Many of the acotylean polyclads have spermatozoa of the so-called thread-shaped type, i.e., filiform spermatozoa without *Nebengeisseln* or undulating membranes. I have studied the spermiogenesis in five of these species, representing five genera (Hendelberg, 1965). They were found to be very similar in respect of spermiogenesis and sperm morphology. As an example I will give an account of the spermiogenesis of *Cryptocelides loveni* Bergendal (Figs. 1*a–c*, 3, 7), and also make some comparisons with that of *Leptoplana tremellaris* (O. F. Müller) (Fig. 2*c*, *d*).

In these species, as in the other turbellarians I have studied (Hendelberg, 1969), cytoplasmic bridges (Fig. 1*a*: cbr; cf. Fig. 4*d*) unite the spermatids in groups. These bridges are the result of incomplete cell cleavage.

Two flagella grow from the young, still spherical spermatid [Figs. 1*a*, 2*c*(1), (2)]. They grow opposite each other from a small protuberance of the cell, distal of the still spherical nucleus. (In Fig. 2*c*, as in Figs. 2*a* and 4*a*, the cytoplasm bridges are not marked; the proximal ends of the spermatids, which are united by cytoplasm bridges, are always oriented upwards, however.)

When the flagella have reached about their full length, the spermatid elongates and becomes club-shaped. As the shaft of the club, the spermatid shaft, grows, the two flagella become united with it [Figs. 1*a*(2), (3), 2*c* (3)–(7)]. Mitochondria [Fig. 1*a*(2): m] and other cytoplasm components move out into the shaft. At the same time the nucleus elongates and takes up its final position in the proximal part of the spermatid shaft.

The spermatid shaft is transformed into the spermatozoon [Figs. 1*a*(3)–1*b*, 2*c*(7)–2*d*] Part of the proximal lump of cytoplasm is cast off as residual cytoplasm.

The species ("turbellarian B") studied by Retzuis (1905, figs. XIII:6–14) is not determined but belongs, according to the descriptions, most probably to the Polycladida. Like many others, Retzius drew a single flagellum in the early spermatid, but he could not explain the transition between this stage and the next, club-shaped one. I think he had seen one of the two flagella, for very often only one is in the focal plane, and the reason why he could not find an axial filament in the spermatid shaft is that there is no such structure.

In the polyclad species with spermatozoa of the so-called thread-shaped type I found the two flagella to be situated along opposite sides of the sperma-

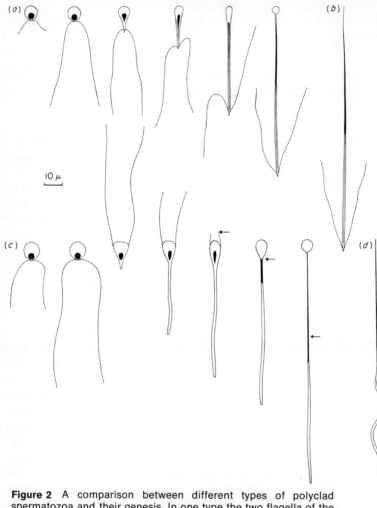

Figure 2 A comparison between different types of polyclad spermatozoa and their genesis. In one type the two flagella of the young spermatid remain free; in the other type they are united with the spermatid shaft which is transformed into the spermatozoon. (a) Consecutive stages in spermatid development of *Stylostomum ellipse*; (b) spermatozoon of *S. ellipse*; (c) consecutive stages in spermatid development of *Leptoplana tremellaris*; (d) spermatozoon of *L. tremellaris*. Nucleus and nuclear region black; the arrows indicate the position of the tips of the flagella. *(From Hendelberg, 1965.)*

tid shaft [Fig. 1*a*(3)]. The distal part of this shaft becomes the cytoplasm region of the spermatozoon (Fig. 1*b*: c reg). The other, narrow, tapering part of the spermatozoon, earlier regarded as the tail, is the nuclear region. The flagella border the cytoplasm region and in some of the species, *Leptoplana tremellaris*, for example, part of the nuclear region, too (Fig. 2*c*). In transverse sections studied by electron microscopy the two flagella are seen to be very superficially situated (Figs. 3*b*, 7).

The observations mean that the so-called thread-shaped polyclad spermatozoa develop in about the same way as the turbellarian spermatozoa with undulating membranes studied by Koltzoff (1909). For further details see Hendelberg (1965).

The occurrence in the Polycladida of spermatozoa with two *undulating membranes*, each bordered by a filament of flagellum type, has been confirmed (Hendelberg, 1965).

In the cotylean polyclads the spermatozoa studied are characterized by two free flagella (*Nebengeisseln*). The two flagella grow from the young, still almost spherical spermatid and remain free during spermiogenesis (Fig. 2*a*, *b*).

Spermiogenesis in the Acoela

Koltzoff (1909) describes two undulating membranes in the spermatozoa of acoels he studied. However, Westblad (1948) found only one undulating membrane, or spermatozoa without undulating membranes. Westblad's figures, and not Koltzoff's, have been reproduced usually since then.

I have reinvestigated most of the spermatozoa studied by Westblad, among them the one of *Haploposthia viridis* (An der Lan), which, according to Westblad, is equipped with one undulating membrane. My studies revealed *two* undulating membranes, easily discernible by phase-contrast microscopy (Fig. 4*e*; cf. Hendelberg, 1969, fig. 56). Two flagella grow freely from the young spermatid (Fig. 4*d*) and then become incorporated into the spermatid shaft, which becomes the spermatozoon. In electron micrographs we find each of the incorporated flagella near the margin of each undulating membrane (Fig. 4*f*).

Henley, Costello, and Ault (1968) report a similar morphology of the spermatozoa of the acoels *Childia groenlandica* and *Polychoerus carmelensis*. In another paper (Costello *et al.*, 1969) they mention that two free flagella are given off by the young spermatid of *Childia*. The flagella are then incorporated into the parts which become the two undulating membranes of the mature spermatozoon.

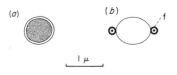

Figure 3 Transverse sections of the spermatozoon of *Cryptocelides loveni* (schematically drawn) showing the superficial position of the two flagella bordering the cytoplasmic region. (*a*) The nuclear region; (*b*) the cytoplasmic region.

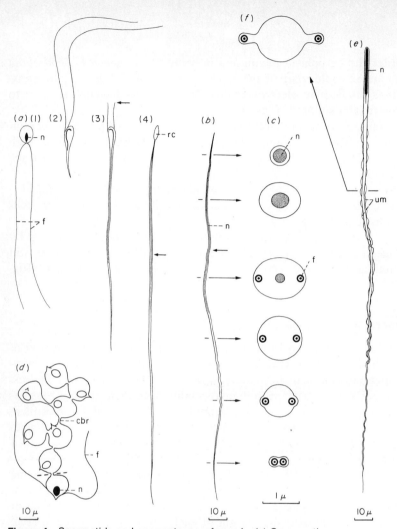

Figure 4 Spermatids and spermatozoa of acoels. (*a*) Consecutive stages in spermatid development, showing the growth and incorporation of two flagella, of *Convoluta saliens*; (*b*) spermatozoon of *C. saliens*; (*c*) schematic drawings of transverse sections of the spermatozoon of *C. saliens* showing the position of the incorporated flagella; (*d*) part of smeared spermatid cluster of *Haploposthia viridis*; the flagella are omitted from all but one spermatid, the nucleus of which is marked black; (*e*) spermatozoon of *H. viridis*; this spermatozoon is characterized by two undulating membranes; (*f*) transverse section of spermatozoon of *H. viridis*, showing the position of the flagella in the undulating membranes. cbr, cytoplasmic bridge; f, flagellum, free or incorporated; n, nucleus; rc, residual cytoplasm; um, undulating membranes. The arrow in Fig. 4*a*(4) and the corresponding arrow in *b* indicate the supposed positions of the tips of the incorporated flagella. (Cf. Hendelberg, 1969)

In respect of the flagella and the undulating membranes the recent results are then quite in line with those arrived at by Koltzoff.

However, not only spermatozoa with undulating membranes but also *other types of spermatozoa of the Acoela* develop from spermatids with two flagella. The thread-shaped spermatozoa of *Convoluta saliens* (Graff, 1882) offer one example (Fig. 4a–c) of several I studied (Hendelberg, 1969). We can see how the flagella become incorporated (Fig. 4a). In electron micrographs they are found in the cytoplasm region and also partly along the nucleus (Fig. 4c). Note that the flagella are found inside the body of the spermatozoon (Figs. 4c, 6) and not in the superficial position found in the acotylean polyclads (Figs. 3b, 7). According to my observations, by phase-contrast microscopy, the tips of the flagella ought to be in the positions marked by the arrows in Figs. 4a(4), 4b. It must be mentioned, however, that the position of the two incorporated flagella is reported by Bedini and Papi (1970) to be different in a species of the Acoela they studied.

The spermiogenesis of the acoel *Paraphanostoma submaculatum* Westblad was studied, too. The spermatozoon of this species is characterized by a bundle of terminal filaments. However, also this type of spermatozoon develops from spermatids with two flagella (Hendelberg, 1969).

GENERAL CONSIDERATIONS ON THE FLAGELLA IN TURBELLARIAN SPERMATOZOA

The results of the studies of polyclads and acoels imply that the difference between spermatozoa with free flagella, *Nebengeisseln*, and spermatozoa of the thread-shaped types without undulating membranes or *Nebengeisseln* is not very great. In all these types we find a development of the spermatozoon from a spermatid with two flagella. The basic difference is that these flagella either remain free (Fig. 2a, b) or are united with (Fig. 2c, d) or are incorporated into (Fig. 4a, b) the body of the spermatozoon.

In some species studied *no flagella* were found, e.g., the species of Prolecithophora and Macrostomida I studied (Hendelberg, 1969). In the spermatozoa of some macrostomids there are, as mentioned above, two bristles. According to Bedini and Papi (1970) the bristles are of a structure different from flagella.

In the Typhloplanoida some species have very short flagella and others have no flagella at all. Thus the flagella may be more or less reduced, and when there are no flagella at all, I think they have been reduced completely, that is, eliminated.

The most obvious function of the flagella is to transport the spermatozoon. When there are no flagella, alternatives are necessary, as, for example, in swimming species with spermatophores, or other structures cause movement of the spermatozoon.

Thus microtubular filaments underlaying the cell membrane (cf. Fig. 7) have been suggested by Christensen (1961) to cause the undulations of the nonflagellate spermatozoon of a *Plagiostomum* species. Such microtubular filaments are suggested by Silveira and Porter (1964) to be a possible explanation of the movements of the cell body of the type of spermatozoon found in the Tricladida. In this type of spermatozoon there are two free, motile flagella, too.

In the polyclad spermatozoa of the thread-shaped type I studied, only the part of the spermatozoon bordered by the two flagella was observed to undulate (Fig. 1c). This refers to, among others, spermatozoa in which the cell membrane is underlain by microtubular filaments for the whole of the length of the spermatozoon. Thus the microtubular filaments occur also in that part of the nuclear region which does not undulate. Flagella removed from the cell body could sometimes be observed to continue undulating. Thus, in this type of turbellarian spermatozoa, I think the flagella cause the undulations.

Further information on motion and motility of turbellarian spermatozoa is given by Hendelberg (1965, 1969); see also Costello and Costello (1968) and Costello *et al.* (1969).

The results regarding the occurrence of flagella in spermatids and spermatozoa have been plotted (Fig. 5) in the scheme of evolution of the Turbellaria suggested by Ax (1961, 1963). The symbol to the left of each (horizontal) arrow stands for the spermatid which is either biflagellate or nonflagellate. In the turbellarian species studied hitherto (by modern methods) no observation of spermatids or spermatozoa with a single flagellum was made. The symbols to the right of the (horizontal) arrows stand for the different types of spermatozoa. The different positions of the flagella in the spermatozoa of acoels and acotylean polyclads is shown in schematically drawn transverse sections. (The derivation of the Trematoda and the Cestoda from rhabdocoels, suggested by many authors, is indicated in the scheme. These groups are discussed below, under the heading "Sperm Morphology and Taxonomy.")

As can be seen in the scheme (Fig. 5) the biflagellate type of spermatid occurs in many groups, in the Acoela, in both cotylean and acotylean polyclads, in the Proseriata, in the Tricladida, and in many of the studied rhabdocoels. In the spermatozoa of the Acoela the incorporated flagella are found inside the cell body. In the acotylean polyclads they are superficially situated, and in some other groups they remain free. But also the spermatozoa with free flagella are of different types. The mitochondria, for instance, form quite different patterns in the spermatozoa of cotylean polyclads and triclads studied hitherto; see Hendelberg (1965) and Silveira and Porter (1964), respectively.

I think that this widespread occurrence of similar spermatids which give rise to spermatozoa of different types may be explained most easily by as-

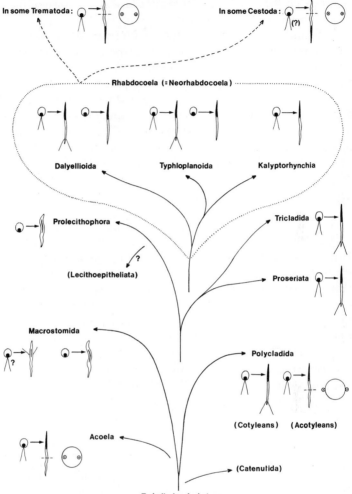

Figure 5 The results arrived at hitherto regarding the occurrence of flagella in spermatids and spermatozoa of the Turbellaria, plotted in the scheme of the evolution of the group suggested by Ax (1961, 1963). The derivation of the Trematoda and the Cestoda from rhaodocoels (of dalyellioid type?) suggested by many authors is indicated in the scheme. The development from the young spermatid to the spermatozoon is indicated by an arrow. The nuclei are marked black. The transverse sections of the spermatozoa show the position of the two incorporated flagella. (Cf. Hendelberg, 1969.)

suming that the presence of two flagella was a feature of the common ancestor of the turbellarians in which they are found.

It must be stressed that the occurrence of two flagella in the turbellarian spermatozoa does not imply that these spermatozoa have not evolved from the primitive type of spermatozoon within the Metazoa. It is true that the theory was advanced at the beginning of the present century that the occurrence of two flagella in many turbellarian spermatozoa is a primitive feature, surviving from the period before the metazoan stage. However, a doubling of the flagellum has been found in the spermatozoa of many other groups of animals, often in single species only, but in some groups—for instance, the Insecta—in quite a few species; see references in Hendelberg (1969, p. 41).

THE ULTRASTRUCTURE OF THE FLAGELLA

In the spermatids and spermatozoa of many turbellarians the ultrastructure of the flagella has been found to be of the so-called "9 + 1" pattern (Fig. 7). Instead of the central pair of microtubules generally found in cilia and flagella we have here in the center of the flagellum a cylindric core which has a rather complex structure, recently analyzed by Silveira (1969) and Henley *et al.* (1969). This "9 + 1" pattern has been found only in spermatids and spermatozoa of flatworms, but not in the somatic cells, the cilia of which have been found to be of the common "9 + 2" pattern.

Hiterto the "9 + 1" pattern of spermatozoan flagella has been found in the following groups of the Turbellaria: in the Polycladida, both the acotyleans and cotyleans studied, by Hendelberg (1965), Silveira (1967a, according to Silveira, 1969), and Thomas (1970); in the Proseriata by Lanfranchi (pers.

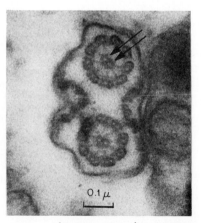

Figure 6 Transverse section of spermatozoon of *Convoluta saliens,* Acoela, showing the two tubular filaments (arrows) in the center of each flagellum complex. X 80,000. *(From Hendelberg, 1965.)*

0.1 μ

(0.1 μ = 8 mm)

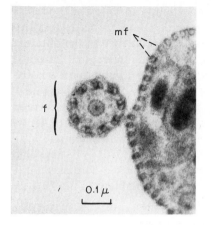

Figure 7 Transverse section of part of a spermatozoon of *Cryptocelides loveni,* Polycladida, showing one of the flagella (f) adhering to the body of the cell. The aberrant ultrastructure, the "9 + 1" pattern, is known from the flagella of different flatworm spermatozoa. Microtubular filaments (mf) are situated just beneath the cell membrane. X 80,000. *(From Hendelberg, 1969.)*

comm.); in the Tricladida by Klima (1961), Silveira and Porter (1964), and Silveira (1968, 1969); and in the Typhloplanoida by Henley *et al.* (1969). This pattern has been reported also from an acoel, *Mecynostomum auritum,* by Afzelius (1966).

However, this "9 + 1" pattern is not found in all turbellarian spermatozoa. In the two acoels I studied by high-resolution electron microscopy, *Mecynostomum lutheri* and *Convoluta saliens,* I found a "9 + 2" pattern (Fig. 6); cf. Hendelberg (1969). The spermatozoa of the latter species were sectioned at many different levels. In all these sections the "9 + 2" pattern was found.

The "9 + 2" pattern is reported from another acoel, *Convoluta psammophila,* by Bedini and Papi (1970). However, they found also a "9 + 0" pattern, that is, parts of the flagella were found to be devoid of central microtubules.

From other acoels, *Childia, Polychoerus,* and *Anaperus,* a "9 + 0" pattern is reported by Henley *et al.* (1968), Costello *et al.* (1969), and Henley and Costello (1969); cf. Silveira's (1967a) reporting of *Anaperus,* according to Silveira (1969).

I have suggested earlier that the "9 + 1" pattern of spermatozoan flagella might be explained most easily by assuming that it occurred in the common ancestor of the turbellarians in which it is found. Now that we have found a "9 + 2" pattern in some acoels, that is, a pattern which is very reminiscent of and perhaps identical with the one commonly found in cilia and flagella, one cannot presume that the ancestors of the Acoela had a "9 + 1" pattern in their spermatozoan flagella. On the contrary, it does not seem unlikely that the "9 + 2" pattern found in the Acoela is the original one. If so, the "9 + 1" pattern in one or more acoels is of great interest from a phylogenetic point of view. Has this "9 + 1" pattern evolved independently in different groups of the Turbellaria, or are the Acoela a group from which other turbellarians have evolved, or is there some other explanation of the occurrence of the different patterns in the Acoela? These questions are of such great interest that Afzelius, who ascribed the "9 + 1" pattern to an acoel, *Mecynostomum auritum*, and I have started a reinvestigation of this species. Preliminary results from new, controlled material suggest that the old determination of species was correct.

The "9 + 1" pattern certainly differs greatly from the "9 + 2" pattern. However, as has been demonstrated by Henley *et al.* (1969) and Silveira (1969), two longitudinal, coiled structures occur in the core of the "9 + 1" flagellum of turbellarian spermatozoa. It is not said whether we have here structures which may be homologous with the central pair of microtubules in the "9 + 2" flagellum, but this seems to be a problem worth investigating.

SPERM MORPHOLOGY AND TAXONOMY

The results arrived at by Franzén (1956) and others clearly show the interdependence between sperm morphology and function. Nevertheless, our knowledge of turbellarian spermatology can be of value from a phylogenetic and taxonomical point of view. I will mention here a few examples.

The derivation of the Trematoda and the Cestoda from the Turbellaria, perhaps the Rhabdocoela, has been suggested in consequence of other morphological characters. This derivation is strongly supported by the sperm morphology. Both trematode and cestode spermatozoa have flagella with the "9 + 1" pattern; see references in Hendelberg (1969, 1970) or Silveira (1969); see also Morseth (1969). In the trematode and in some cestode spermatids and spermatozoa studied by modern methods two flagella have been found; see references in Hendelberg (1970). This is so in trematode spermatozoa earlier regarded as monoflagellate (Hendelberg, 1962). The reported occurrence in some Cestoda of only one flagellum or axial filament may, I think, be explained as being due to reduction (Hendelberg, 1970).

The relationship between the Gnathostomulida and the Turbellaria has been discussed by Ax (1961, 1963) and others. The ultrastructure of the spermatozoa of three *Gnathostomula* species has been studied by Graebner (1969). These spermatozoa, characterized by many small feet (membranous protrusions), are quite unlike the turbellarian spermatozoa known hitherto. Thus, the ultrastructure of these gnathostomulid spermatozoa provides no evidence of a close relationship between turbellarians and gnathostomulids. It will be of interest in this context to study the fine structure of the more filiform spermatozoa found by Sterrer (1966) in another gnathostomulid genus, *Pterognathia*.

The occurrence of some general differences between spermatozoa from different turbellarian groups is of interest from a taxonomical point of view. Such differences imply possibilities to use the sperm morphology as a taxonomic character, among others, e.g., when trying to place a species in its correct position in the turbellarian system.

CURRENT PROBLEMS
OF TURBELLARIAN SPERMATOLOGY

One of the many interesting problems of turbellarian spermatology is to determine in which part of the spermatozoon the *acrosome* function is located. Is it a tip reaction in the head filament as Henley (1968) suggested might be possible? Or is it a lateral reaction like the one found in trematodes, as described by Burton (1967)?

In the spermatozoa of other groups of animals the acrosome is produced by the Golgi complex. We find a prominent Golgi complex in many turbellarian spermatids. In the Polycladida I studied I found the structure of flattened Golgi vesicles (Fig. 1*a*: g) to be cast off together with the residual cytoplasm (Hendelberg, 1965). However, its products most probably move out into the spermatid shaft which is transformed into the spermatozoon. Silveira (1967b) has studied the secretion of a special kind of granules from the Golgi complex in an acoel, *Anaperus gardineri*. Perhaps the refractile bodies found by Henley (1968) in the cytoplasmic region of the spermatozoa of the acoel *Childia groenlandica* are of the same origin? Special cytoplasmic structures other than mitochondria are found in the spermatids and spermatozoa of polyclads, too (Hendelberg, 1965). However, as yet none of these structures have been demonstrated to have an acrosome function. Much work remains to be done on this problem.

Another problem is how the spermatozoon is supplied with nourishment when waiting for the ovum. Is there any *reserve nourishment* in the turbellarian spermatozoa?

The general function of the *nucleus* is well known. It supplies what is to become the zygote with one set of chromosomes. An interesting aberration is the pseudogamy found in polyploid forms of triclads. The pseudogamy implies that the spermatozoon first enters the ovum, but that it (or at least its nuclear material) is then eliminated in some way or other; see Benazzi Lentati (1970).

Another aberration is the absence of nuclear material, or at least of DNA, in one of the types of spermatozoa in species with sperm dimorphism. This has been reported from the acoel *Heterochaerus australis* by Haswell (1905) and was found by me (Hendelberg, 1969) in the acoel *Otocelis rubropunctata*. However, the function of the different types of spermatozoa in these acoels is not yet known.

In some turbellarian spermatozoa the nucleus is very aberrant; see Ferguson (1940, p. 249) and Karling (1940, pp. 33–34). What are these aberrant structures and what purpose do they serve?

Still another problem is the role played by the spermatozoan centrioles in the fertilized ovum. This is a general problem of spermatology; cf. Afzelius (1970).

As can be seen, our knowledge of turbellarian spermatology is in many respects still incomplete. The ideas presented here of the significance of recent results are not to be regarded as definitive, but will lead, I hope, to further investigation in this field.

BIBLIOGRAPHY

Afzelius, B. 1966. *Anatomy of the Cell*. Chicago and London: University of Chicago Press. 127 pp.

———. 1970. "Thoughts on Comparative Spermatology." In B. Baccetti (ed.), *Comparative Spermatology*, pp. 565–573. New York: Academic Press.

Ax, P. 1961. Verwandtschaftsbeziehungen und Phylogenie der Turbellarien. *Ergeb. Biol.*, 24:1–68.

———. 1963. "Relationships and Phylogeny of the Turbellaria." In E. C. Dougherty (ed.), *The Lower Metazoa*, pp. 191–224. Berkeley and Los Angeles: University of California Press.

Bedini, C., and F. Papi. 1970. "Peculiar Patterns of Microtubular Organization in Spermatozoa of Lower Turbellaria." In B. Baccetti (ed.), *Comparative Spermatology*, pp. 363–366. New York: Academic Press.

Benazzi Lentati, G. 1970. Gametogenesis and Egg Fertilization in Planarians. *Intern. Rev. Cytol.*, 27:101–179.

Burton, P. R. 1967. Fine Structure of the Reproductive System of a Frog Lung Fluke. II. Penetration of the Ovum by a Spermatozoon. *J. Parasitol.*, 53(5):994–999.

Christensen, K. 1961. Fine Structure of an Unusual Spermatozoan in the Flatworm *Plagiostomum*. *Biol. Bull.*, 121:416.

Costello, D. P., and H. M. Costello. 1968. Immotility and Motility of Acoel Turbellarian Spermatozoa, with Special Reference to *Polychoerus carmelensis*. *Biol. Bull.*, 135:417.

———, C. Henley, and C. R. Ault. 1969. Microtubules in Spermatozoa of *Childia* (Turbellaria, Acoela) Revealed by Negative Staining. *Science*, 163:678–679.

Ferguson, F. F. 1940. A Monograph of the Genus *Macrostomum* O. Schmidt 1848. Part VIII. *Zool. Anz.*, 129:244–266.

Franzén, Å. 1956. On Spermiogenesis, Morphology of the Spermatozoon, and Biology of Fertilization among Invertebrates. *Zool. Bidr. Uppsala*, 31:355–482.

———. 1967a. Spermiogenesis and Spermatozoa of the Cephalopoda. *Arkiv Zool.*, (2)19:323–334.

———. 1967b. Remarks on Spermiogenesis and Morphology of the Spermatozoon among the Lower Metazoa. *Arkiv Zool.*, (2)19:355–342.

Graebner, I. 1969. Vergleichende elektronenmikroskopische Untersuchung der Spermienmorphologie und spermiogenese einiger Gnathostomula-Arten: *Gnathostomula paradoxa* (Ax 1956), *Gnathostomula axi* (Kirsteuer 1964), *Gnathostomula jenneri* (Riedl 1969). *Mikroskopie*, 24:131–160.

Haswell, W. A. 1905. Studies on the Turbellaria. *Quart. J. Microscop. Sci.*, 49:425–467.

Hendelberg, J. 1962. Paired Flagella and Nucleus Migration in the Spermiogenesis of *Dicrocoelium* and *Fasciola* (Digenea, Trematoda). *Zool. Bidr. Uppsala*, 35:569–587.

———. 1965. On Different Types of Spermatozoa in Polycladida, Turbellaria. *Arkiv Zool.*, 18:267–304.

———. 1969. On the Development of Different Types of Spermatozoa from Spermatids with Two Flagella in the Turbellaria with Remarks on the Ultrastructure of the Flagella. *Zool. Bidr. Uppsala*, 38:1–50.

———. 1970. "On the Number and Ultrastructure of the Flagella of Flatworm Spermatozoa." In B. Baccetti (ed.), *Comparative Spermatology*, pp. 367–374. New York: Academic Press.

Henley, C. 1968. Refractile Bodies in the Developing and Mature Spermatozoa of *Childia groenlandica* (Turbellaria: Acoela) and Their Possible Significance. *Biol. Bull.*, 134:382–397.

——— and D. P. Costello. 1969. Microtubules in Spermatozoa of Some Turbellarian Flatworms. *Biol. Bull.*, 137:403.

———, ———, and C. R. Ault. 1968. Microtubules in the Axial Filament Complexes of Acoel Turbellarian Spermatozoa, as Revealed by Negative Staining. *Biol. Bull.*, 135:422–423.

———, ———, M. B. Thomas, and W. D. Newton. 1969. The "9 + 1" Pattern of Microtubules in Spermatozoa of *Mesostoma* (Platyhelminthes, Turbellaria). *Proc. Natl. Acad. Sci.*, 64:849–856.

Hyman, L. H. 1951. *The Invertebrates*. II. *Platyhelminthes and Rhynchocoela. The Acoelomate Bilateria*. New York: McGraw-Hill. 550 pp.

Jägersten, G. 1959. Further Remarks on the Early Phylogeny of the Metazoa. *Zool. Bidr. Uppsala*, 33:79–108.

———. 1968. *Livscykelns evolution hos Metazoa. En generell teori*. Lund: Scandinavian University Books. 295 pp. (With English summary.)

Karling, T. G. 1940. Zur Morphologie und Systematik der Alloeocoela Cumulata und Rhabdocoela Lecithophora (Turbellaria). *Acta Zool. Fenn.*, 26:1–260.

Klima, J. 1961. Elektronenmikroskopische Studien über die Feinstruktur der Tricladen (Turbellaria). *Protoplasma*, 54:101–162.

Koltzoff, N. K. 1909. Studien über die Gestalt der Zelle. II. Untersuchungen über das Kopfskelett des tierischen Spermiums. *Arch. Zellforsch.*, 2:1–65.

Morseth, D. J. 1969. Spermtail Finestructure of *Echinococcus granulosus* and *Dicrocoelium dendriticum*. *Exptl. Parasitol.*, 24:47–53.

Reisinger, E. 1960. *Was ist Xenoturbella?* *Z. Wiss. Zool.*, 164:188–198.

Remane, A. 1958. Zur Verwandtschaft und Ableitung der niederen Metazoen. *Verhandl. Deut. Zool. Ges.*, 1957, 179–196.

Retzius, G. 1905. Zur Kenntnis der Spermien der Evertebraten. II, III. Die Spermien der Würmer. 1. Die Turbellarien. *Biol. Untersuch.*, N. F. 12(9):84–85.

Silveira, M. 1967a. "Ultraestruturas ciliares de Turbelários e suas implicações fisiológicas." Doctoral thesis, University of São Paulo.

——. 1967b. Formation of Structured Secretory Granules within the Golgi Complex in an Acoel Turbellarian. *J. Microscopie*, **6**:95–100.

——. 1968. Action de la pepsine sur un flagelle du type "9 + 1." *Experientia*, **24**: 1243–1245.

——. 1969. Ultrastructural Studies on a "Nine plus One" Flagellum. *J. Ultrastruct. Res.*, **26**:274–288.

—— and K. R. Porter. 1964. The Spermatozoids of Flatworms and Their Microtubular System. *Protoplasma*, **59**:240–265.

Sterrer, W. 1966. *Gnathostomula paradoxa* Ax und Vertreter von *Pterognathia* (ein neues Gnathostomuliden-Genus) von der schwedischen Westküste. *Arkiv Zool.*, (2)**18**(16):405–412.

Thomas, M. B. 1970. Transitions between Helical and Protofibrillar Configurations in Doublet and Singlet Microtubules in Spermatozoa of *Stylochus zebra* (Turbellaria, Polycladida). *Biol. Bull.*, **138**:219–234.

Westblad, E. 1948. Studien über skandinavische Turbellaria Acoela. V. *Arkiv Zool.*, **41A**(7):1–82.

——. 1949. *Xenoturbella bocki* n.g., n.sp., a Peculiar, Primitive Turbellarian Type. *Arkiv Zool.*, (2)**1**(3):11–29.

On the Male Copulatory Organ of some Polycystididae and Its Importance for the Systematics of the Family

Ernest R. Schockaert
State University of Ghent, Belgium

Since 1956 an important diagnostic character for the family Polycystididae Graff, 1905 has been the fact that the seminal duct opens separately from the prostatic vesicle into the male genital canal (so-called *divisa* type). *Koinocystella inermis* Karling, 1952 and *Phonorhynchoides flagellatus* Beklemischev, 1927 were the only exceptions, but neither species was recognized as Polycystididae at the time of their description. Karling (1953, p. 361), studying the genus *Rogneda* Uljanin, 1870, modified very slightly the family diagnosis. In his work on the male atrial organs of the Kalyptorhynchia, Karling (1956) described several new species and genera of polycystidids in which the organization of the male copulatory organ deviated in one point or another from that described in the original diagnosis, and he wrote (loc. cit., p. 217): "Ein für Polycystididen gemeinsames diagnostisches Merkmal hinsichtlich des Baues der männliche Atrialorganen kann heute noch nicht formuliert werden."

An important fact that required an adaptation of the old diagnosis (Karling, 1964) was the presence in many newly described species of a copulatory organ of the *conjuncta* type, i.e., with an interposed prostatic bulb. At this moment the anatomy of the proboscis and of the pharynx gives us the main characters to recognize an eukalyptorhynchid turbellarian as a polycystidid.

For our discussion we have mainly been guided by the ideas of Karling (1956). The most important points of his theory on which our conclusions are based are the following:

  1   All glands of the male genital canal (prostatic glands, accessory glands) are derived from differentiated cells of the epithelium of the male genital canal (Karling, 1956, pp. 216–217).
  2   The stylet in Polycystididae (stylet of the prostatic vesicle and accessory stylets) are derivations of cirrus spines (Karling, 1956, pp. 204–206, 216–218) (*Hakenstilett*).
  3   A stylet can also originate from the cuticularization of a penis papilla (Karling, 1940, pp. 192–193; 1956, pp. 206–207) (*Papillenstilett*).

COPULATORY ORGANS WITH SECONDARY INTERPOSED PROSTATIC VESICLE

In the genera *Annulorhynchus* Karling, 1956, *Psammopolycystis* Meixner, 1938, *Phonorhynchella* Karling, 1956, and *Gallorhynchus* Schockaert and Brunet, 1971 the male copulatory organ is composed of an unpaired seminal vesicle, an interposed prostatic vesicle, and a double-walled stylet. In *Annulorhynchus adriaticus* Karling, 1956 and in *Gallorhynchus mediterraneus* Schockaert and Brunet, 1971 the sperm-conducting stylet is accompanied by an accessory cuticular organ; in *Psammopolycystis* and in *Phonorhynchella* this accessory cuticular organ is provided with an accessory secretion reservoir.

On the other hand, a rather similar situation has been described for the *Scanorhynchus* species Karling, 1955, *Danorhynchus gösoeensis* Karling, 1955, and *Neopolycystis tridentata* Karling, 1965. In these species there is an interposed prostatic part, but a free prostatic vesicle occurs as well, and the seminal duct enters freely the male genital canal (but see below).

In both his publications of 1955 and 1956 Karling discussed the matter whether the interposed prostatic vesicle in the species known at that time was a primary or a secondary acquisition. In all above-mentioned polycystidids the interposed prostatic part is evidently a secondary acquisition. Its relative importance differs from species to species: in *Neopolycystis tridentata* the interposed prostatic part is weakly developed; in the *Scanorhynchus* species interposed and free prostatic vesicles have the same importance, while the free prostatic vesicle is rudimentary in *Danorhynchus gösoeensis*. The stylet is associated with the free prostatic vesicle and accompanied by an accessory cuticular organ.

We call this kind of organization of the copulatory organ the Scanorhynchus type (Fig. 1*a–c*).

Although the stylet in this type is associated with the free prostatic vesicle, it also helps to conduct sperm, e.g., as in *Polycystis riedli* Karling, 1956, *Polycystis dolichocephala* (Pereyaslawseva, 1895), *Gyratrix hermaphroditus* Ehrenberg, 1831, *Phonorhynchus helgolandicus* (Mecznikov, 1865). (See also Karling, 1956, pp. 205, 212.)

When the free prostatic vesicle has now completely disappeared, and the stylet (with double wall!) is used only for transport of sperm and secretion of the interposed prostatic bulb, we have an organization of the male copulatory organ that can be called the Gallorhynchus type (Fig. 1*d–f*). It is found in its typical form in *Gallorhynchus mediterraneus* and *Annulorhynchus adriaticus*. Two modifications occur: the accessory cuticular organ is lost (*Gallorhynchus simplex* Schockaert and Brunet, 1971) or provided with an accessory secretion reservoir (*Psammopolycystis, Phonorhynchella*). The organization of the male copulatory organ of the latter two genera is termed the Psammopolycystis type. The occurrence of an accessory secretion reservoir can be observed in other, not necessarily closely related Polycystididae such as *Phonorhynchus helgolandicus, Typhlopolycystis coeca* Karling, 1956, *Polycystis nägelii* Kölliker, 1845.

What are the conclusions for the taxonomy? We are convinced that *Neopolycystis, Scanorhynchus, Danorhynchus gösoeensis, Annulorhynchus,* and *Gallorhynchus* are closely related to each other. In addition to the similarity of the organization of the male copulatory organ, they all possess a single ovary (most of them a single testis as well), a common genital pore in terminal position, and a comparable organization of the female atrial organs. This group can be connected with *Gyratrix* and *Gyratricella* Karling, 1955 through *Danorhynchus duplostylis* Karling, 1955. *Psammopolycystis* and *Phonorhynchella* are probably not so closely related to the above-mentioned group: the genital pore is not terminal in position and the organization of the female atrial organs is different.

COPULATORY ORGANS WITH PRIMARILY INTERPOSED PROSTATIC BULB

In the genus *Parachrorhynchus* Karling, 1956 and in *Koinocystella inermis* Karling, 1952 the interposed prostatic vesicle is evidently a primary situation. The same is true for the Duplacrorhynchinae Schockaert and Karling,

(a)   (b)   (c)

(d)   (e)   (f)

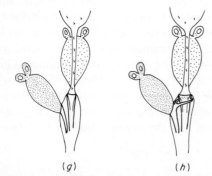

(g)   (h)

Figure 1 (a) *Neopolycystis*; (b) *Scanorhynchus*; (c) *Danorhynchus gösoeensis*; (d) *Annulorhynchus*; (e) *Gallorhynchus mediterraneus*; (f) *Gallorhynchus simplex*; (g) *Phonorhynchella*; (h) *Psammopolycystis*.

1970. In this subfamily a muscle septum is differentiated around the male duct to form a copulatory organ of the duplex type.

Recently another polycystidid has been described with a *conjuncta* copulatory organ that is to be considered as primitive: *Djeziraia pardii* Schockaert, 1971. The interposed prostatic vesicle contains two kinds of prostatic secretions and bears a simple single-walled stylet. It is difficult to see a relation of such a stylet with the so-called *Hakennatur* of the stylet of Polycystididae, and it is therefore considered as a *Papillenstilett*. Moreover, the similarity to the copulatory organ of *Yaquinaia microrhynchus* Schockaert and Karling, 1970 is striking. In this species the stylet is rudimentary and lies in an eversible cirrus, but here the copulatory organ is clearly of the duplex type, for the cirrus is enclosed in a cirrus sac. For a copulatory organ with a stylet it is often impossible to state whether it is built according to the duplex or simplex system (Karling, 1956, p. 192), and this is indeed the case for *Djeziraia pardii*. However, a copulatory organ of the type of *Djeziraia* can easily be derived from one as found in *Yaquinaia*, the stylet protractors being the incompletely formed muscle septum.

Within the group of species with a *conjuncta* copulatory organ, and where this situation is to be considered as a primary one, we can distinguish three categories:

1 The *Parachrorhynchus* type: a *conjuncta* type with a cirrus not enclosed in a cirrus sac (Fig. 2*a*).

2 The *Duplacrorhynchus* type: as the first type, but the cirrus enclosed in a cirrus sac (Fig. 2*b*, *c*). The organization as it is found in *Yaquinaia microrhynchus* can be considered as a transition to the next type.

3 The *Djeziraia* type: a *conjuncta* type with a *Papillenstilett*, the male genital canal not surrounded by a muscle septum (Fig. 2*d*).

It must be stressed that all species with a copulatory organ of the *Duplacrorhynchus* or *Djeziraia* type have been found on the coasts of the Indo-Pacific ocean and that this area is practically unexplored for turbellarians.

There is certainly a close relationship between the Polycystididae with a copulatory organ with primary interposed prostatic vesicle, but at the present state of our knowledge we cannot draw definitive conclusions on the taxonomy.

THE GENUS Phonorhynchoides

A special situation occurs in the genus *Phonorhynchoides* Beklemischev, 1927, with two species known so far: *P. flagellatus* Beklemischev, 1927 and *P. somaliensis* Schockaert, 1971. The organization resembles very much that of *Phonorhynchus helgolandicus*, but the sperm-conducting apparatus is of the *conjuncta* type.

 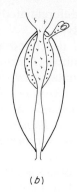 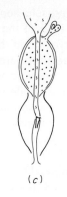

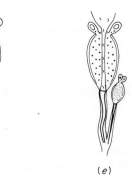

Figure 2 (a) *Parachrorhynchus*; (b) *Duplacrorhynchus*; (c) *Ya-quinaia*; (d) *Djeziraia*; (e) *Phonorhynchoides*.

Typologically the male apparatus of *Phonorhynchoides* is comparable with those of *Psammopolycystis* and *Phonorhynchella*. For these two genera we have good evidence to consider the *conjuncta* condition as derived, but this is not the case for *Phonorhynchoides*. With the discovery of *Djeziraia* and *Yaquinaia* the possibility of a primary interposed prostatic vesicle with stylet must be taken into consideration. The stylet is apparently single-walled in *Phonorhynchoides somaliensis*, and it might be a penis derivation as it is in *Djeziraia*. However, an accessory cuticular stylet with secretion reservoir is also present.

We only include *Phonorhynchoides* in the discussion to have an all-round survey of Polycystididae with a male copulatory organ with interposed prostatic bulb. We call this kind of organization the Phonorhynchoides type, but for the moment we are unable to give indications about the origin of this type. Taking other features into consideration, such as proboscis, female atrial organs, and number of gonads (paired in *Phonorhynchoides*), it becomes quite difficult to indicate any relationship of *Phonorhynchoides* with other representatives of the family with similar organization of the male atrial organs.

CONCLUSIONS

A male copulatory organ with interposed prostatic bulb is not at all an exceptional condition in the family Polycystididae. It is very difficult, however, and sometimes impossible to say whether this *conjuncta* condition is a primary or secondary one. Closely related to this problem is the origin of the stylet: can this be considered as derived from a cuticularized penis or from cirrus spines? A thorough morphological study and the discovery of new forms will bring us the solution for at least some cases.

BIBLIOGRAPHY

Karling, T. G. 1940. Zur Morphologie und Systematik der Alloeocoela Cumulata und Rhabdocoela Lecithophora. (Turbellaria). *Acta Zool. Fenn.*, 26:1–260.

———. 1953. Zur Kenntnis der Gattung *Rogneda* Ulianin (Turbellaria, Kalyptorhynchia). *Arkiv Zool.*, N. S., 5(6):349–368.

———. 1955. Studien über Kalyptorhynchia (Turbellaria). V. Der Verwandtschaftskreis von *Gyratrix* Ehrenberg. *Acta Zool. Fenn.*, 88:1–39.

———. 1956. Morphologisch-histologische Untersuchungen an den männlichen Atrialorganen der Kalyptorhynchia (Turbellaria). *Arkiv Zool.*, N.S., 9(3):187–279.

———. 1964. Über einige neue und ungenügend bekannte Turbellaria Eukalyptorhynchia. *Zool. Anz.*, 172(3):159–183.

Literature with Special References to Some Cited Species
(See also references mentioned above)

Beklemischev, W. N. 1927. Über die Turbellarienfauna des Aralsees. Zugleich ein Beitrag zur Morphologie und zum System der Dalyelliida. *Zool. Jahrb. (Syst.),* 54:87–138.

Karling, T. G. 1952. Kalyptorhynchia (Turbellaria). *Further Zool. Results Swed. Antarctic Expedition, 1901–1903,* 4, 9:1–50.

Schockaert, E. R. 1971. Turbellaria from Somalia. I. Kalyptorhynchia (part 1). *Monitore Zool. Ital.*, N.S. Suppl. **4:** 101–122.

—— and M. Brunet. 1971. Turbellaria Polycystididae (Kalyptorhynchia) from the Marseille-area. II. *Gallorhynchus simplex* n. sp. and *G. mediterraneus* n. sp. *Ann. Soc. Roy. Zool. Belg.*, **101**(1):65–75.

—— and T. G. Karling. 1970. Three New Anatomically Remarkable Turbellaria Eukalyptorhynchia from the North American Pacific Coast. *Arkiv Zool.*, N.S., **23**(2):237–253.

Chapter 9

Digestive Physiology
of the Turbellaria

J. B. Jennings
Department of Zoology, University of Leeds,
England

The majority of the Turbellaria are free-living predators, and the prey utilized by different members of the class include protozoa, coelenterates, annelids, mollusks, nematodes, rotifers, arthropods, and tunicates (Percival, 1925; Estham, 1933; Hyman, 1951; Johri, 1952; Froehlich, 1955; Jennings, 1957, 1959; Pfitzner, 1958; Reynoldson and Young, 1963). A number of turbellarians have adopted ecto- or entocommensal habits, but these species, whilst supplementing their diet to varying extents with particles of the host's food, generally feed on the same types of food organisms as are favored by their free-living relatives (Gonzales, 1949; Jennings, 1968a, 1971). These organisms occur by chance, or as epizoites, co-commensals, or parasites upon or within the flatworm's host. The few truly parasitic Turbellaria utilize the host's own ingesta, its tissues or body fluids, or, where the mouth or gut is lost, presumably absorb predigested metabolites across the body wall in some manner akin to that used by the Cestoda.

Thus in most turbellarians the diet is relatively unspecialized, in biochemical terms, and there is no particular emphasis upon foods which contain excessive proportions of protein, fat, or carbohydrate. It would be expected, therefore, that the digestive physiology will be equally unspecialized and that the full complement of digestive enzymes will be present. As will be seen later, this has been found to be the case in those species which have been intensively studied, and the indications are that these species demonstrate a general pattern of digestive physiology characteristic of the Turbellaria as a class.

Before examining this pattern in detail, it is convenient to review briefly the form of the turbellarian alimentary system. Basically this consists of a mouth, pharynx, and blind saccate intestine. The pharynx, derived in part from the stomodaeal invagination, has been elaborated to varying degrees of complexity in the different orders and forms the basis of the feeding mechanism. It is in the pharynx, in fact, that the Turbellaria show the greatest degree of regional specialization within the alimentary system. Three basic types of pharynges can be recognized; these are, respectively, the simple, bulbous, and plicate. The bulbous pharynx occurs in two forms, the rosulate and dolioform, and the plicate also has two variants, the cylindrical and ruffled plicate. The structure and methods of use of these various types of pharynx have been fully described elsewhere (Hyman, 1951; Jennings, 1957, 1968b); fuller discussion of this organ and of turbellarian feeding mechanisms is not appropriate here. It is relevant, though, to point out that this elaboration of the forepart of the gut, and the versatility of the organ so developed, confers upon the Turbellaria their ability to utilize such an extraordinarily wide range of food organisms, and hence it has had a considerable influence upon the digestive physiology. Depending upon its form, the pharynx allows the turbellarian to swallow the prey intact (simple, bulbous, and some plicate types), or pierce the integument to suck out the body contents (bulbous and cylindrical plicate), or to envelop the prey and subject it within the extruded pharynx to extracorporeal digestion (ruffled plicate). Thus, despite the lack of segmental appendages, jaws, or sclerotized or otherwise hardened parts which form the basis of feeding devices in other animals, the Turbellaria prey upon almost any type of invertebrate including forms such as mollusks and arthropods which are protected by tough exoskeletons.

The intestinal region of the alimentary system, in contrast to the pharyngeal region, shows no regional specialization, and specialization of function, when it occurs, is entirely at the cellular level. The intestine in the Acoela remains as a syncytial undifferentiated mass of endoderm, containing numerous vacuoles in which digestion occurs, but in all other Turbellaria apart from the few aberrant parasitic species it is well organized and contains a permanent central lumen. The intestine is essentially a blind endodermal sac, which may be a simple tubular structure with or without lateral pouches as in the rhabdocoels and many alloeocoels, or it may be much subdivided as in the Tricladida and Polycladida.

The intestinal wall, or gastrodermis, consists of a single layer of cells standing on a thin basement membrane. In lower rhabdocoels, a few alloeocoels, and some cotylean polyclads, the gastrodermal cells are all alike morphologically and are columnar with ciliated distal margins. In most other turbellarians, though, the cells are differentiated both morphologically and physiologically into two types (Fig. 1). The first is pyriform to spherical, loaded with proteinaceous spheres and now known to be entirely secretory in function, whilst the second is clavate to columnar, phagocytic, and the site of intracellular digestion. A variation of this type of gastrodermis occurs in the temnocephalid rhabdocoels, where some of the pyriform gland cells have sunk below the gastrodermis and lie in the parenchyma but continue to discharge their secretions into the gut lumen via elongated necks running between the columnar cells. A second variation is seen in the umagillid rhabdocoels, where the gland cells are either few in number or absent.

The first detailed investigation of digestion in the Turbellaria was that of Metschnikoff (1878), and one consequence of his studies was his discovery of the phenomena of phagocytosis and intracellular digestion. He described intracellular digestion in the acoelan *Convoluta* and in a variety of rhabdocoels and triclads. Metschnikoff's findings were verified in rhabdocoels by Jacek (1916), using *Stenostomum*, and later by Westblad (1922) for this species and

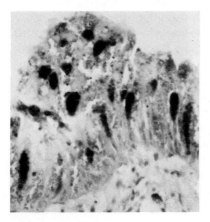

Figure 1 Transverse section through a portion of the gastrodermis of the freshwater triclad *Polycelis felina* (= *P. cornuta*), treated by the Hess and Pearse (1958) method for endopeptidase. The gland cells which secrete the endopeptidase responsible for extracellular proteolysis appear black, indicating a strong positive reaction to this technique. The columnar cells are not well preserved, owing to the method of fixation necessary for demonstration of enzymic activity in the gland cells, but are seen to be devoid of food vacuoles, the triclad having been starved for 7 days before fixation. Scale: 1 cm = 20 μm.

also for *Microstomum* and *Macrostomum*. Westblad claimed that there is also some extracellular digestion in *Stenostomum*.

The occurrence of phagocytosis and intracellular digestion in the triclads was confirmed by Arnold (1909), Saint-Hilaire (1910), Löhner (1916), and Westblad (1922). Arnold was the first to ascribe a secretory function to the pyriform nonphagocytic cells of the gastrodermis and stated that they go through a cycle of growth and secretory activities, and produce enzymes responsible for some degree of extracellular digestion. His views were based entirely on histological preparations and they tended to be ignored by subsequent investigators, who either disregarded these cells or believed them and their contents to be protein reserves.

Until the 1920s, apart from Arnold's paper, studies on digestion in the Turbellaria tended simply to confirm Metschnikoff's findings and extend them to other species. The significant advance upon Metschnikoff's work came with the publication by Hyman, Willier, and Rifenburgh (1924) and Willier, Hyman, and Rifenburgh (1925) of their classic studies on phagocytosis, food vacuole formation, intracellular digestion, fat utilization, and oxygen consumption in the triclad *Dugesia dorotocephala*. These gave the first detailed account of these processes in flatworms, and all subsequent investigations into the nutritional physiology of the Turbellaria and many comparable ones in the Trematoda, Rhynchocoela, Nematoda, and, indeed, invertebrates in general owe a considerable debt to these papers, to which Libbie H. Hyman contributed.

Hyman and her associates showed that food particles in vacuoles newly formed in the distal region of the columnar gastrodermal cells retain their identity for a while and stain only lightly. The vacuoles slowly move deeper within the cell, water is absorbed from them, and the contents gradually condense into homogeneous spheres which have a strong affinity for eosin and other acidic stains. The spheres decrease in size as they approach the base of the cell and eventually disappear, apart from indigestible residues, as digestion and absorption are completed. The residues are eventually returned to the free distal surface of the cell and voided into the gut lumen, but the vacuoles and residues do not follow any definite course, and there is no cyclosis such as occurs, for example, in many ciliate Protozoa.

The columnar cells normally contain large deposits of triglyceride fat, but during the early stages of vacuole formation and intracellular digestion the fat content drops rapidly and there is a simultaneous increase in the rate of oxygen consumption by the entire flatworm, indicating the use of the fat to provide energy for digestive processes. During the later stages of digestion the fat content of the columnar cells returns to normal, presumably from the synthesis of fat from the products of digestion, and oxygen consumption also returns to the normal level. Kelley (1931) substantiated these findings, again in the triclad *Dugesia dorotocephala*, by using calf thymus as a test food and applying the Feulgen technique to trace the progressive intracellular breakdown of nucleoprotein.

Hyman (1951), on the basis of these earlier studies, considered digestion in the triclad to be entirely intracellular and believed that the pyriform gland cells, or "granular clubs" as she termed them, represented protein reserves. Jennings (1957, 1959) supported these views but in further studies (1962a), using histochemical methods which permit identification and localization of the major groups of digestive enzymes within tissue sections, showed that the pyriform cells are in fact secretory and produce an enzyme which is responsible for some degree of extracellular digestion. These histochemical studies have revealed the basic pattern of triclad digestive physiology and it has become increasingly apparent that this is also, with some exceptions, the basic pattern for the Turbellaria. In an examination of this pattern, therefore, it is convenient to consider the Tricladida first and to use the situation in this group as the basis of a comparative survey of digestion throughout the class.

ORDER TRICLADIDA

It is now established that digestion in the triclad Turbellaria is achieved by a combination of extra- and intracellular processes, in which proteases, carbohydrases, lipases, and phosphatases act in specific sequences to bring about complete hydrolysis of the food into soluble constituents (Jennings, 1962a, 1968b).

The cylindrical plicate pharynx of the triclad contains numerous acidophilic and basophilic gland cells (Hyman, 1951). A large proportion of the acidophils produce a secretion which shows a strong positive reaction to the Hess and Pearse (1958) method for endopeptidases (proteases which initiate proteolysis by cleaving peptide bonds in the inner regions of the protein molecule, to produce short-chain proteins or polypeptides). The gland cells producing this secretion are flask-shaped and open onto the outer surface of the pharynx, never into the lumen, and they are particularly concentrated around the free distal end (Fig. 2). During feeding the pharynx penetrates the prey's integument very quickly (Jennings, 1957, 1959), and penetration, therefore, would appear to be largely mechanical, but it may well be supplemented to some extent by histolysis effected by the secretions of these pharyngeal glands. Once within the prey, the pharynx moves about freely in the body cavity, withdrawing the body contents, and since this part of the feeding process generally occupies several minutes, there can be little doubt that the endopeptic secretions from the pharynx contribute significantly to the disruption of the prey's body contents before these are sucked through the pharynx into the intestine. The fact that the glands open onto the external surface of the

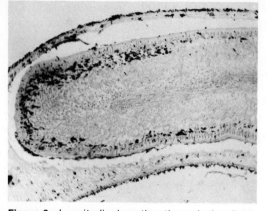

Figure 2 Longitudinal section through the distal half of the pharynx in *Polycelis felina*, showing the endopeptidase-producing gland cells (black) which discharge onto the outer surface of the pharynx. The gland cells are especially concentrated distally. Rhabdites in the epidermal cells on the dorsal and ventral body surfaces also appear black; this does not represent endopeptidase activity but is due to their extreme acidophilic nature and consequent affinity for the eosin counterstain. Hess and Pearse method for endopeptidase. Scale: 1 cm = 125 μm. *(From Jennings, 1962a, by permission of the Biological Bulletin.)*

pharynx, and not into its lumen, supports this interpretation of the role of their secretions during ingestion. In this connection it is significant that triclads always feed by inserting the pharynx into the prey, even when the latter is small enough to be swallowed intact, as when oligochaetes of a smaller diameter than the resting pharynx are captured. In these instances the pharynx is extended until it is slender enough to enter the prey and thus, presumably, allow the pharyngeal secretions to attack the body contents. This typical feeding pattern is followed even with homogeneous masses of food, such as homogenized liver or clotted blood, when the pharynx is inserted into the food mass to withdraw material, rather than being applied to the surface.

The optimum pH for visualizing endopeptic activity in the pharyngeal glands is 5.0, demonstrating the acidic nature of this initial proteolysis, but no evidence for the production of acidic secretions in the pharynx has been found. The gland cells, however, do show, consistently, a weak reaction for acid phosphatase as demonstrated by the Gomori (1952) and Burstone (1958) methods.

The proteinaceous spheres contained within the pyriform gland cells of the gastrodermis show an intense positive reaction for endopeptidase (Fig. 1). Newly ingested material lying in the intestinal lumen shows at first only faint traces of endopeptic activity, derived, presumably, from the pharyngeal secretions, but this increases in amount up to a maximum reached about 4 hours after feeding (Fig. 3). During this time the gland cells discharge their spheres of endopeptidase and then slowly synthesize more over the next 24 to 48 hours. The optimum pH value for demonstration of endopeptic activity in both the gland cells and gut lumen is 5.0, as with the pharyngeal secretion.

No other enzymic activity has been demonstrated in the gut contents of the triclad, and it is concluded that the extracellular stage of digestion is entirely endopeptic. This, however, does not cause complete homogenization of the food, and even at the peak of activity, as shown by the intensity of the histochemical reaction in the gut contents, distinct components of the food such as cell nuclei may often be recognized. The columnar cells of the gas-

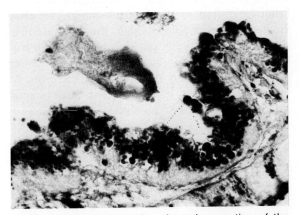

Figure 3 Transverse section through a portion of the intestine in *Polycelis felina*, fixed 4 hours after ingestion of boiled liver. Control sections of uningested liver showed no enzymic activity, but the portion seen here, lying in the gut lumen, shows some reaction for endopeptidase especially in the darker righthand region. Gland cells are not evident in the gastrodermis, having discharged their spheres of endopeptidase, and the columnar cells are becoming loaded with food vacuoles which are passing down into their basal regions. The contents of the vacuoles show a positive reaction for endopeptidase, which is stronger than that apparent in the lumen contents owing to intracellular secretion of more enzyme. Hess and Pearse method. Scale: 1 cm = 40 μm. *(From Jennings, 1962a, by permission of the Biological Bulletin.)*

trodermis (Fig. 1) commence phagocytosis of the food immediately as it enters the intestine, and small food particles rapidly become enclosed in food vacuoles without being subjected to extensive extracellular proteolysis. The function of the latter appears to be, primarily, to facilitate phagocytosis by softening and breaking up the larger pieces of food which have survived the disintegrating action of the pharynx during ingestion. Lumen proteolysis does not achieve or even approach complete hydrolysis of ingested protein into polypeptides, and, indeed, discharge of endopeptidase appears to be restricted to the first 4 hours after ingestion, so that there is not a continued discharge for as long as food remains in the lumen. This accounts for the earlier inability to detect the extracellular stage of digestion by purely histological techniques, and also for the fact that large food particles occasionally persist unchanged in the gut lumen until eventually expelled through the pharynx around 48 hours after feeding.

This somewhat limited amount of extracellular proteolysis contrasts sharply with the comparable situation in the closely related Rhynchocoela, where lumen proteolysis is extensive and does achieve a great degree of homogenization of the food (Jennings, 1962b; Jennings and Gibson, 1969). In the Rhynchocoela, though, the gastrodermis contains a much higher proportion of gland cells per unit area than in the triclad or, indeed, any other turbellarian and, significantly, the food is generally swallowed intact without any fragmentation by the feeding mechanism. A comparison of the Tricladida and Rhynchocoela, in this context, affords an excellent demonstration of the influence a particular method of feeding may have on subsequent digestive processes.

Food vacuoles formed distally in the columnar cells move down the cells, with their contents condensing, becoming homogeneous, and finally disappearing, as described by Willier, Hyman, and Rifenburg (1925). During their movement down the cells, a number of different digestive enzymes, working under different pH conditions, exert their effect.

Material in the newly formed vacuole shows endopeptic activity carried over from the intestinal lumen. As the vacuole moves back into the cell, however, the intensity of the reaction increases and eventually all the vacuoles in the cell show an extremely strong reaction (Fig. 3). This clearly indicates a continuation and extension of the endopeptic activity initiated in the gut lumen, but effected by endopeptidases secreted into the vacuole from the cytoplasm of the columnar cells. Again, optimum pH value for visualization of enzymic activity is 5.0, as with the endopeptidases originating in the pharyngeal and gastrodermal gland cells. Thus these three sets of enzymic activities are all strictly comparable and can be regarded, physiologically, as sequential parts of the same process, namely, the initial acidic and endopeptic phase of digestion. This phase, although apparently uniform throughout in physiological terms, is divided spatially into extracellular and intracellular stages.

Acid phosphatase activity is demonstrable in the phagocytic cells during food vacuole formation and the early stages of intracellular digestion (Rosenbaum and Rolon, 1960; Osborne and Miller, 1963). The reaction increases in intensity and the peak of activity coincides with that of the endopeptidase (Jennings, 1962a). During this time the cytoplasm surrounding the food vacuoles shows strong acid phosphatase activity, and this often extends into the vacuolar contents also. The function of this enzyme is not clear, but it may well be a manifestation of the energy-releasing mechanisms associated with vacuole formation and secretion of the endopeptidase, involving utilization of the fat stored in the cell as described by Willier, Hyman, and Rifenburgh (1925). Alternatively, it may well be concerned with the establishment and maintenance of the necessary low pH conditions in the vacuolar fluid.

Endopeptidase activity persists in the food vacuoles for 8 to 12 hours after feeding. By this time the vacuolar contents are reduced to compact homogeneous masses (see Willier, Hyman, and Rifenburgh, 1925), and the endopeptic and acid phosphatase activities begin to decline, first within vacuoles lying deep in the columnar cells and then in the remainder. This disappearance of endopeptic activity indicates completion of the first stage of proteolysis, with proteins having been hydrolyzed to peptones and polypeptides. The vacuolar contents then pass into the second phase of digestion, in which exopeptidases (proteases removing terminal amino acids from polypeptide chains) complete proteolysis, and lipases and carbohydrases attack fats and carbohydrates exposed by the digestion of cell walls and other limiting membranes.

As endopeptic activity declines in the food vacuoles, it is replaced, progressively, by exopeptidases of which one, leucine aminopeptidase, is easily demonstrated by the Burstone and Folk (1956) technique. This technique is not specific for leucine aminopeptidase, as was at first believed, and it is very likely that the reaction obtained with it within triclad vacuoles at this stage of digestion represents the combined activities of a number of aminopeptidases and possibly of some carboxypeptidases also. Nevertheless, the use of the technique in this particular context is quite valid, irrespective of its lack of specificity, since it does demonstrate *exopeptidase* activity as distinct from *endopeptidase*.

The time of onset of exopeptidase activity varies with the size of the original meal, which determines the number of food vacuoles formed in the columnar cells. If only a small meal has been taken and each cell forms only a few vacuoles, exopeptidase activity may appear as early as 2 hours after feeding. Alternatively, if a large meal has been taken and the cells are packed with vacuoles, the endopeptic phase is considerably extended and exopeptidases may not appear until 18 or more hours after feeding. Generally, though, endopeptidase activity is replaced by exopeptidase between 8 and 12 hours after ingestion of a meal.

Exopeptic activity overlaps endopeptic, therefore, to a degree determined by the amount of food originally ingested, and vacuoles in the distal portion of a columnar cell may show various degrees of endopeptidase activity whilst those deeper within the cell show increasing amounts of exopeptidase. Eventually all the vacuoles show a strong exopeptic reaction (Fig. 4), and this activity persists for as long as the vacuoles continue to have visible contents.

Optimum pH value for demonstration of exopeptic activity is 7.2, indicating that there is a significant change in the pH conditions attending this second and terminal phase of proteolysis.

During replacement of endopeptic activity by exopeptic there is a simultaneous replacement of acid phosphatase in the cytoplasm and vacuolar contents by alkaline phosphatase. The pH value for optimum visualization here is 9.0, confirming the alkaline nature of this phase of digestion.

Simultaneously with the onset of exopeptic digestion lipases appear in the food vacuole; they can be demonstrated by the Gomori (1952) Tween 80 technique. As with the exopeptidase activity, the optimum pH for histochemical demonstration is around 7.2, indicating a similar pH value *in vivo*. Lipolytic activity has never been found in the gut lumen.

Carbohydrases also operate during this digestive phase but are extremely difficult to demonstrate by specific histochemical techniques. Incorporation of starch in test meals and subsequent treatment of histological sections or squash preparations of living triclads by Lugol's iodine solution reveal progressive conversion and disappearance of starch within the food vacuoles, whilst unphagocytosed starch remaining in the gut lumen is quite unchanged.

Virtually no information is available regarding the movement of products of digestion from the gastrodermis to the rest of the body. In the absence of a

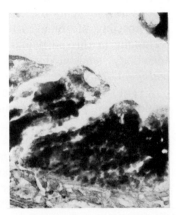

Figure 4 Transverse section of a portion of the gastrodermis in *Polycelis felina*, fixed 12 hours after a meal of boiled liver. The food vacuoles fill the entire length of the columnar cells and show a positive exopeptidase reaction. Burstone and Folk (1956) method for exopeptidases of the leucine aminopeptidase type. Scale: 1 cm = 20 μm.

vascular system transport of metabolites must presumably depend either on diffusion along concentration gradients or on some mechanism involving active transfer from cell to cell. The absence of a vascular transport system is probably compensated for to a large extent by the ramifying nature of the intestine, which extends into most parts of the body so that no tissues or organs are very far removed from the site of primary digestion and assimilation despite the expanded somewhat leaflike body form characteristic of most triclad Turbellaria. The same situation is found, of course, in the Polycladida, and many of the Trematoda Monogenea and Digenea. Conversely, where the gut is simple and unbranched, the body tends to be unexpanded and somewhat cylindrical, as in the Rhabdocoela and those trematodes possessing only a bifid intestine. In the great majority of these animals, triclads included, the distance of any part of the body from the digestive tissue is very similar to that from the surface of the body. Gaseous exchange is presumably by diffusion, also, and these distances may well prove to be of a critical value which is determined by the rates of diffusion through living tissues of metabolites such as oxygen, carbon dioxide, monosaccharides, or amino acids.

Food is stored in the triclad principally as fat, which is laid down in the columnar cells of the gastrodermis and in the mesenchyme. Glycogen is also stored at these sites but to a lesser extent.

Ultrastructural studies reveal that the columnar cells possess scattered microvilli on their free distal surface, and the phagocytic lamellae which extend out into the gut to engulf material and form a food vacuole develop between these (Jennings, unpublished work). Microvilli are a characteristic feature of most cells possessing an absorptive function, but where they are dense and regular in their distribution on cells in the alimentary system, there is now considerable evidence that they play an important part in digestion as well as absorption. Ugolev (1960a, b, 1965) has shown that in a variety of vertebrates the microvilli act as a porous reactor and bring substrates and enzymes into intimate contact, thereby considerably accelerating hydrolysis. This "membrane" or "contact digestion," as opposed to "cavital" or "luminal digestion" in the gut lumen, probably also occurs in many invertebrates and has certainly been suggested for a number of digenetic trematodes (Halton, 1966; Jennings, 1968b). Membrane digestion, however, depends on the presence of densely packed regular microvilli, and those of the triclad intestine are not of this nature. Their function remains unknown, but it is probably concerned with absorption of soluble materials from the gut lumen, perhaps immediately after ingestion of a meal.

The sequence of digestive processes outlined above is summarized in Table 1. The basic pattern foreshadows the situation found in higher animals, particularly in the separation of distinct phases characterized by wide differences in attendant pH conditions.

Table 1   A Summary of Digestive Physiology in the Turbellaria: Tricladida

Site	Type of enzyme	Source	Function	pH phase
Extracellular	Endopeptidase	Gland cells in pharynx and gastrodermis	Initiation of proteolysis	Acidic (around pH 4.0)
Intracellular	Endopeptidase, acid phosphatase	Cytoplasm of phagocytic cells	Initiation of proteolysis	Acidic (around pH 4.0)
	Exopeptidases (amino-, carboxy-, and presumably dipeptidases), alkaline phosphatase	Cytoplasm of phagocytic cells	Completion of proteolysis	Alkaline (in the range pH 7.2–9.0)
	Lipase, alkaline phosphatase	Cytoplasm of phagocytic cells	Lipolysis	Alkaline (in the range pH 7.2–9.0)
	Carbohydrases, alkaline phosphatase	Cytoplasm of phagocytic cells	Carbohydrate digestion	Alkaline (in the range pH 7.2–9.0)

Source: After Jennings, 1972.

ORDER ACOELA

Relatively little is known of the digestive processes of the Acoela. Feeding in this order is both macrophagous and microphagous (Jennings, 1957), but whatever the size of the prey, it is eventually enclosed within vacuoles in the endodermal syncytium which performs the functions of an intestine in these animals. Functionally, it is not clear whether these vacuoles are homologous with the truly intracellular vacuoles of the triclad gastrodermis, or whether they should be regarded as temporary gut lumina. When relatively lårge prey are ingested, such as copepod larvae, the primary vacuole enclosing the food sometimes acquires one or more satellite vacuoles which eventually fuse with it, and the resultant structure certainly performs the same function as an organized gut lumen, despite its lack of a cellular wall. Subsequent formation of satellite vacuoles, containing parts of fragmented food, has never been observed, however, and so there is nothing comparable to the intracellular phase in the triclad if these large vacuoles are to be considered as representing extracellular digestion.

Isolated observations (Jennings, unpublished work) indicate that the sequence of events within the vacuole is much the same as in the triclad. In *Convoluta convoluta*, for example, newly ingested copepod larvae and ciliate protozoa rapidly develop intense endopeptic activity, often within a few minutes of ingestion. The enzyme or enzymes responsible are presumably secreted from the cytoplasm surrounding the vacuole. They are subsequently replaced by exopeptidases, and these remain demonstrable until copepod larvae, for instance, are entirely digested and only the indigestible exoskeleton remains. It is not known whether these stages of peptic digestion are attended by appropriate types of phosphatase activity, nor have carbohydrases or lipases been demonstrated. The last two types of enzymes are presumably secreted, though, as the food will normally contain carbohydrates and fats. Acoelans which take a large proportion of diatoms in their diet, or which can be maintained exclusively upon these organisms (Kozloff, 1970), may well show a considerable shift of emphasis in their complement of digestive enzymes toward production of carbohydrases, and this is a point which would probably repay further study.

One difference between digestion in the acoel and the triclad is that whereas in the latter adjacent food vacuoles generally show similar or only slightly different enzymic activities at any one time, in the acoel the contents of immediately adjacent vacuoles may be in widely separate stages of digestion. One vacuole, for example, may contain a recently ingested copepod which is being subjected to maximum endopeptic hydrolysis, whilst another separated from it by a cytoplasmic strand only a few microns in thickness may contain either material showing intense exopeptic activity or merely exoskeletal remains. The food vacuoles do show a considerable amount of movement within the digestive syncytium, and it may be that they acquire different types of enzymes as they pass through different areas. This implies a significant amount of physiological differentiation within the syncytium, and although there is no evidence to support this, it is a possibility for future examination. Alternatively, the submicroscopic organelles of the cytoplasm in all parts of the syncytium may be capable of producing the entire range of digestive enzymes and secreting various types as required into adjacent vacuoles. Both mechanisms must operate in the phagocytic cell of the triclad gastrodermis, with one or the other being brought into action, depending upon how rapidly the food vacuole passes down the cell (which in turn depends on the size of the meal and the amount of food available in the gut lumen and in the correct condition for phagocytosis at any one time). It would be interesting to examine specialization of function at this level in the two types of turbellarians, using such histochemical techniques as are currently applicable at the ultrastructural level.

Food reserves in the Acoela, as in the Tricladida, consist mainly of fat which is deposited in the mesenchyme and digestive tissue. Small amounts of glycogen, also, are found at these sites.

ORDER RHABDOCOELA *sensu lato*

The gastrodermis is not uniform in structure throughout the Rhabdocoela and, as a consequence, the digestive physiology varies somewhat between the different suborders.

Where the rhabdocoel gastrodermis is morphologically uniform in structure and apparently undifferentiated into glandular and other components (Catenulida or suborder Notandropora), its cells are columnar and generally ciliated. Westblad (1922) stated that in *Stenostomum* digestion is both extra- and intracellular, but Jennings (1957) claimed that the observable extracellular breakdown of food organisms such as ciliate protozoa and rotifers is largely mechanical, and is a consequence of the food being driven up and down the intestine by contractions of the body musculature. Both authors state that subsequent to this breakdown food particles are phagocytosed by the columnar cells and digestion is completed intracellularly. Details of gastrodermal structure and the mechanism of entry of materials into the cells are not available, but the presence of a ciliated distal border is not necessarily incompatible with the phagocytosis of discrete particles. In the related phylum Rhynchocoela, for example, the gastrodermis is uniformly ciliated, but phagocytic lamellae develop between the cilia and engulf material from the gut lumen to form typical food vacuoles (Jennings, 1969). It would be of considerable interest if this was found to be the case in these lower rhabdocoels also. Nothing is known of the sequence of enzymic hydrolysis which follows phagocytosis in *Stenostomum*, but the contents of the food vacuoles undergo the same visible changes as in the triclad.

In the Macrostomida or suborder Opisthandropora, the gastrodermis in most species resembles that of the Catenulida in being ciliated and lacking differentiation into distinct components, but nothing is known of the digestive physiology in these instances. In a few cases, the gastrodermis is very similar to the triclad type and digestion follows the same pattern (Jennings, 1968b). *Macrostomum*, for example, feeds on minute crustaceans, annelids, and other similar-sized organisms, which it seizes and swallows intact. Gastrodermal gland cells produce an endopeptidase which effects partial extracellular digestion. Phagocytosis then occurs and digestion is completed by endopeptidases, exopeptidases, lipases, and carbohydrases acting in sequence exactly as in the triclad.

In the suborder Temnocephalida the gastrodermis is of the triclad type and the digestive physiology is of the expected pattern and virtually identical with that of the triclad (Gonzales, 1949; Jennings, 1968a). One point of difference from the triclad, however, is that the extracellular endopeptic stage is somewhat extended and the food is rendered much more homogeneous before it is phagocytosed. A number of the gastrodermal gland cells are subgastrodermal, lying in the parenchyma immediately below the gastrodermis and discharging into the gut lumen by long necks which extend between the columnar phagocytes. This allows an increase in the number of gland cells (linked with the increased emphasis on extracellular digestion) without decreasing the relative number of columnar cells and hence the phagocytic capacity of the gastrodermis.

A characteristic feature of the temnocephalid alimentary system is the close association with it of part of the reproductive system, namely, the vesicula resorbiens or seminal bursa. The function of this structure is not fully understood, but it is believed to be concerned somehow with the utilization and reincorporation into the general metabolism of excess sexual products. It usually contains degenerating spermatozoa (derived, perhaps, from its partner in copulation rather than from the male part of its own reproductive system) and yolk globules. In *Temnocephala brenesi* the vesicula resorbiens is embedded in the gastrodermis and in many other species, e.g., *T. novaezealandiae*, it is pressed closely against it. In both *T. brenesi* and *T. novaezealandiae* the walls of the vesicle and of the duct connecting it to the rest of the reproductive system at all times show intense reactions to techniques for acid phosphatase and exopeptidase (Jennings, 1968a). The contents, however, show little enzymic activity of any kind, so that it does not appear that hydrolytic enzymes are actively secreted into the vesicle to digest its contents. It may be that autolysis alone is sufficient to produce soluble breakdown products which are absorbed (with the acid phosphatase being concerned in this) and then perhaps digested further in the walls of the vesicle and duct by cytoplasmic hydrolases such as the exopeptidase. The exopeptidase is identical in properties, so far as can be determined histochemically, with that produced in the gastrodermis during the second, alkaline phase of digestion. There is no direct open connection between the vesicula resorbiens and the gut, comparable to the genitointestinal canal which occurs in some trematodes, and there is no apparent reason for its close association with the gut unless the degradation products of the contained spermatozoa and yolk are passed to the nearby gut cells for completion of digestion. The association would appear to have some functional significance, though, for it is a consistent feature of the temnocephalid organization.

The diet of the Temnocephalida is essentially the same as that of freeliving rhabdocoels (protozoa, rotifers, minute annelids, and crustaceans), except for occasional supplementation by particles of the host's food, which for temnocephalids living on crustacean gills consists of shreds of animal tissue dropped by the crustacean's mouthparts and swept back over the gills by the respiratory current. The food reserves in *T. brenesi* and *T. novaezealandiae* similarly show no significant differences from those of free-living flatworms, so that the ectocommensal mode of life seems to have exerted little influence upon the general nutritional physiology. Two species from Ceylon, however, are reported to store considerable amounts of glycogen (Fernando, 1945).

In the remaining suborder of rhabdocoels, the Lecithophora, only the typhloplanoid genus *Mesostoma* amongst the free-living forms has been studied to any extent as regards digestive processes. *Mesostoma* seizes annelids, crustaceans, or insect larvae and either ingests them intact or uses the rosulate pharynx to suck out the body contents (Jennings, 1957). Ingestion of intact prey implies a considerable degree of extracellular digestion, but the nature and source of the enzyme or enzymes involved are not known. The gastrodermis often appears to be syncytial and does not have specific secretory cells. It is generally loaded with food vacuoles, whose contents show the same visible changes as in the triclad, but again nothing is known of the enzymes involved.

A number of rhabdocoels from the suborder Lecithophora have adopted entocommensal or, rarely, parasitic modes of life. These are mainly dalyellioids and most are members of the Umagillidae or Fecampiidae, but include a few members from other families (Jennings, 1971).

Of the entocommensals only two species, *Syndesmis antillarum* and *S. franciscana* from the coelom and gut of echinoids, have been studied in detail as regards nutrition and digestion (Jennings and Mettrick, 1968; Mettrick and Jennings, 1969). Both species feed on the ciliate protozoa which abound in their habitat, and *S. antillarum* occasionally supplements this diet by ingesting host coelomocytes. This may be purely by chance, or it may represent a tendency to utilize host tissues as food and may, therefore, illustrate a very early stage in the development of a parasitic mode of life.

The gastrodermis in both species differs from the type found in other rhabdocoels, with the possible exception of *Mesostoma*. It is unciliated and phagocytic, but lacks specific gland cells (Fig. 5). Clusters of gland cells occur around the proximal end of the dolioform pharynx, at its junction with the intestine, but these show no reaction for hydrolases and seem to be concerned with secreting the cuticlelike lining of the pharynx lumen. The gastrodermis often appears syncytial when loaded with food vacuoles, and cell boundaries are difficult to distinguish in many preparations.

Digestion occurs both extra- and intracellularly, but in *Syndesmis* these terms do not have the same precise connotations as in other Rhabdocoela and the Tricladida. Digestion in *S. antillarum* (misidentified as *S. franciscana* by Jennings and Mettrick, 1968) was considered to be entirely intracellular, as it was also in *S. franciscana* until better methods of fixation and examination were employed. These showed that a number of food vacuoles which earlier would have been considered to be intracellular are in fact extracellular. Soon after feeding opposing faces of the gastrodermis come together at irregular

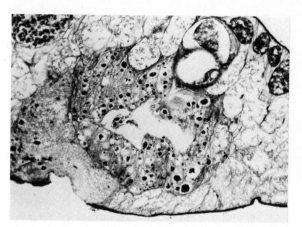

Figure 5 Transverse section through the intestine of the umagillid rhabdocoel *Syndesmis antillarum*, showing absence of gland cells from the gastrodermis. Numerous food vacuoles are present, with their contents undergoing intracellular digestion. In this specimen the gut lumen is well organized and not occluded by temporary fusions of opposing faces of the gut wall. Giemsa. Scale: 1 cm = 40 μm. *(From Jennings and Mettrick, 1968, by permission of the Caribbean Journal of Science.)*

intervals down the length of the intestine, and the lumen becomes divided off into a number of large vacuoles each of which contains digesting food. Unfortunately the demonstration of the true extracellular nature of these "vacuoles" requires fixation in buffered glutaraldehyde and postfixation in osmic acid, i.e., standard fixation for ultrastructural studies. This type of fixation precludes histochemical study of the digestive process as the active enzymes are not preserved, whilst, conversely, fixation permitting demonstration of enzymic activities in the food vacuoles does not give optimum preservation of cell walls. Histochemical studies show that digestion in many vacuoles follows the pattern typical of the triclad, with acidic endopeptic hydrolysis accompanied by acid phosphatase activity (Fig. 6) being followed by development of exopeptic, lipolytic, diastatic, and alkaline phosphatase activities. Because of the inherent difficulties due to fixation it is not possible to determine whether the full sequence of events occurs only within truly intracellular vacuoles, as in other Turbellaria, or whether it occurs also in some or all of the extracellular "vacuoles" which are really temporarily isolated portions of the gut lumen. In all the Turbellaria so far studied by these methods, and in the related Rhynchocoela (Jennings and Gibson, 1969) the extracellular phase of digestion in the gut lumen is restricted to endopeptic hydrolysis. The only exception to this is the admittedly anomalous situation in the Acoela. It would

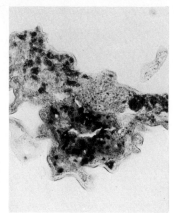

Figure 6 Transverse section of *Syndesmis antillarum*, showing strong acid phosphatase activity (black) in food vacuoles within the gastrodermis. Burstone (1958) method for acid phosphatase. Scale: 1 cm = 80 μm. *(From Jennings and Mettrick, 1968. by permission of the Caribbean Journal of Science.)*

be interesting to discover whether the departure from the usual gut morphology in *Syndesmis* during digestion is accompanied by a corresponding variation in physiology, with enzymes which are entirely intracellular in other Turbellaria acting here extracellularly.

Both *S. antillarum* and *S. franciscana* vary from the typical turbellarian pattern as regards food reserves, also. It has already been shown that most turbellarians possess food reserves of the type to be expected of free-living predators; i.e., they store food principally as fat (but not in excessive amounts), and glycogen is stored in only small quantities. In contrast, *S. antillarum* and *S. franciscana* both form extensive reserves of glycogen (15 to 19 percent of their dry weight), and the latter species also stores an extraordinary amount of fat (approximately 25 percent of the dry weight). These values are more reminiscent of those reported for endoparasitic rather than free-living species (Brand, 1966) and their significance is not clear. One explanation often put forward to account for the high carbohydrate content of entoparasitic helminths is that their metabolism depends on anaerobic glycolysis and hence they tend to store large amounts of glycogen. This explanation can hardly be justified here, since the habitats of *S. antillarum* and *S. franciscana* are not anaerobic. One factor common to the syndesmids and entoparasitic helminths such as cestodes, though, is that both live within a host organism and overcome the problems of dispersal and establishment of the next generation in new hosts partly by producing large numbers of eggs. The umagillids certainly produce more eggs than free-living rhabdocoels (Hyman, 1951), and their extraordinary amounts of food reserves may be set aside either for eventual inclusion in the eggs, or for use as energy-releasing substrates during egg production.

Members of the Fecampiidae are entirely entoparasitic, but unfortunately virtually nothing is known of their nutritional physiology beyond the obvious inference from their mode of life that they must obtain all their nutrients from the host. *Fecampia*, in decapods and isopods (Giard, 1886; Caullery and Mesnil, 1903; Baylis, 1949), possesses a normal gut in the free-swimming larval stage, but the intestine loses all connection with the exterior in the adult parasitic stage. *Kronborgia*, which parasitizes amphipods, however, lacks a gut at all stages of the life history (Christensen and Kanneworff, 1964, 1965). Thus food must be absorbed in both genera across the body surface, as in cestodes. *Fecampia* does not appear to harm its host by its presence, but *Kronborgia* causes parasitic castration presumably owing to its successful competition with the gonads for food materials.

ORDER ALLOEOCOELA

Nothing is known of the digestive physiology of the Alloeocoela. The structure of the gut, as described by Hyman (1951), indicates that the feeding mechanisms will be similar to that of rhabdocoels or triclads depending on the type of pharynx present. Digestion, too, probably follows the triclad pattern in those genera where the gastrodermis has a similar structure to that of the triclads.

ORDER POLYCLADIDA

The two suborders of the Polycladida, the Cotylea and Acotylea, differ in their alimentary systems in the form of the pharynx and the structure of the gastrodermis. In many cotyleans the pharynx is of the cylindrical plicate type, and in those species whose feeding habits are known, its use is similar to that in the triclads but with the important exception that the food is not extensively disorganized during ingestion. *Cycloporus papillosus*, for example, feeds on colonial tunicates such as *Botryllus* by protruding the pharynx and pushing it down into the colony to suck out individual zooids (Jennings, 1957). These are swallowed virtually intact, in contrast to the situation in the triclad. The gastrodermis in *Cycloporus*, and probably in most cotyleans, consists entirely of tall columnar cells which are uniformly and quite densely ciliated at their distal border. There is no morphological differentiation into glandular and absorptive or phagocytic cells. Histological examination reveals progressive homogenization of the food organisms, and the homogenate eventually disappears from the intestinal lumen, presumably by absorption into the gastrodermis. The gastrodermal cells never show any traces of food vacuoles and it is concluded, therefore, that digestion is entirely extracellular. There have been no histochemical studies on cotylean digestion, and nothing is known of either the source or the nature of the enzymes responsible for this complete

extracellular digestion. The cotyleans, so far as is known, are the only turbellarian group with this type of digestive physiology, and it is radically different from the basic pattern of extra- and intracellular processes discernible
elsewhere in the class.

The acotylean polyclads possess the ruffled plicate type of pharynx, and
with its aid they are able to feed on annelids, crustaceans, and occasionally
mollusks. In *Leptoplana* the food is swallowed intact if it is small enough, but
animals too large for this are enveloped in the protruded and expanded pharynx
and subjected within it to extracorporeal digestion (Jennings, 1957). The digestive juices poured onto the food whilst it is held in the pharynx are strongly
proteolytic, with a pH value of 4.5, and they reduce the food into fragments
small enough for ingestion. Observation of prey swallowed intact confirms
the occurrence of extracellular digestion in the intestinal lumen.

The gastrodermis, as in the cotyleans, is composed of tall columnar cells,
all apparently of the same type, but these are not ciliated. They phagocytose
material from the intestinal lumen as it becomes available, and the resultant
food vacuoles, and their contents, progress down the cell and undergo the
same visible changes as do those in the triclad gastrodermis.

Nothing is known of the enzymes responsible for these extra- and intracellular digestive processes. The pharynx is well equipped with acidophilic
gland cells, some of which are probably homologous with those of the triclad's
pharynx, and these, no doubt, are the source of the proteases responsible for
the extracorporeal digestion. The source of the proteases which act in the
gut lumen is unknown, unless the proteolysis here depends entirely upon enzymes carried in with the food from the pharynx. The absence of gland cells
from the gastrodermis is unexpected, in view of the extensive extracellular
proteolysis. In *Leptoplana* the gastrodermal cells are often indistinct from each
other, particularly when full of food vacuoles, and it may be that gland cells
are present but difficult to detect histologically. A histochemical investigation
would be well worth while, to check this point and also to determine whether
the sequence of enzymic events in food vacuoles is the same as in those of the
triclads and many other Turbellaria.

DISCUSSION

The available evidence indicates that the basic pattern of digestive physiology
seen in the Tricladida is repeated, to varying extents, in most other Turbellaria.
Where the gastrodermis is of the type found in the triclads and is fully differentiated into glandular and phagocytic components, as in the temnocephalid
and some macrostomid rhabdocoels, there is the complete pattern of extra-
and intracellular digestion and acidic and alkaline phases. The only difference
between digestion in these rhabdocoels and in the triclad is that they show relatively greater emphasis on the extracellular stage of endopeptic hydrolysis,

and this is linked with their habit of swallowing a proportion of the food intact.

The basic pattern can also be discerned in those turbellarians whose gastrodermis lacks visibly differentiated gland cells but where the gastrodermal cells are still phagocytic. This has been demonstrated in histochemical studies on the entocommensal lecithophoran rhabdocoel *Syndesmis*, and it is strongly indicated in other turbellarians which so far have been examined only by histological methods, namely, free-living lecithophorans such as *Mesostoma*, catenulid rhabdocoels like *Stenostomum*, and the acotylean polyclads. The situation in the catenulids is complicated by the presence of a ciliated gastrodermis but, as pointed out earlier, this is not as unusual as it might seem, since similar circumstances prevail throughout the related phylum Rhynchocoela.

The Acoela are anomalous in that the structure of their gut is not directly comparable with that of other turbellarians, but in this order too a semblance of the basic pattern occurs in the food vacuole, with initial acidic endopeptidase activity being followed by alkaline exopeptic hydrolysis.

So far as is known, then, only the cotylean polyclads show a fundamental departure from the general turbellarian pattern of digestion. Their gastrodermis lacks visible gland cells and is ciliated, but it is also nonphagocytic, for food vacuoles have never been observed. Digestion in this suborder, therefore, appears to be entirely extracellular, but the source of the enzymes involved, and the nature and sequence of their actions, are unknown. The absence of visibly differentiated gland cells is somewhat surprising, in view of the total lack of intracellular digestion, but this feature is by no means uncommon in invertebrate alimentary systems. Apart from the other examples in the Turbellaria, absence of gastrodermal gland cells has been reported from the Trematoda Monogenea (Halton and Jennings, 1965), Trematoda Digenea (Halton, 1967), Nematoda (Jennings and Colam, 1970; Colam, 1971a, b, c), and Hirudinea (Jennings and van der Lande, 1967). In many of these instances, though, digestion is not entirely extracellular and there is considerable intracellular breakdown of materials taken in from the gut lumen in a partially digested state.

The division of digestive processes into acidic and alkaline phases is a fundamental feature of almost all animal digestive systems (Jennings, 1972) and occurs even in the Protozoa (Mast, 1947). In the protozoan, though, the two phases occur within the same vacuole and consequently are separated from each other only in time. In most other animals, however, evolution of the alimentary system in the form of specialization of function at the cellular and then regional levels has led to varying degrees of spatial as well as temporal separation of the two phases of digestion. The climax is seen in the Chordata, where there generally is complete separation in both time and space as digesting food moves through consecutive regions of the gut.

In the turbellarian digestive system, as seen in the majority of the class, there is fairly strict temporal separation of acidic and alkaline phases, but there is also the beginning of spatial separation in those instances where an extracellular stage has been separated off. Biochemically, though, this seems to be identical with the first part of intracellular digestion, and it probably arose, originally, as an extension of this and basically as a device for preparing the food for phagocytosis. This view is suggested by the somewhat limited nature of extracellular digestion in the Tricladida, where the feeding mechanism effects a very large proportion of the necessary reduction in the particle size of the food, and the greater emphasis placed on it in the temnocephalid rhabdocoels, where the prey is often swallowed intact. Support for this interpretation of the primitive role of extracellular digestion comes from the phylum Rhynchocoela, where most of the members swallow their prey intact and where there is a very large amount of extracellular digestion (Jennings, 1962b; Jennings and Gibson, 1969). Further, it is probably significant in this connection that in both the Turbellaria and Rhynchocoela the extracellular breakdown, irrespective of the emphasis placed upon it relative to the remainder of digestion, is exclusively of the acidic endopeptic type and is concerned only with initiation of proteolysis. In the Turbellaria there is further secretion of endopeptidase within the gastrodermal cell, and endopeptic activity is still an integral part of the intracellular stage of digestion. In the Rhynchocoela, though, the evidence indicates absence of such secretion and it is believed that the endopeptidase activity seen within the food vacuoles has been carried in with the food from the gut lumen; it would therefore appear that the separation of the acidic and alkaline phases is more clear-cut and represents an advance on the turbellarian condition. The logical conclusion to further development along these lines, of course, would be the establishment of a pattern of digestion consisting of an entirely extracellular acidic endopeptic phase, succeeded by intracellular completion of proteolysis, and the initiation and completion of lipolysis and carbohydrate digestion, under alkaline conditions. So far as is known, this pattern has not been evolved in the Turbellaria or Rhynchocoela, but there are indications that it occurs in the Nematoda (Jennings and Colam, 1970).

The situation in the cotylean polyclads would clearly repay further study. Their digestive physiology appears to be radically different from that of other Turbellaria, in that an intracellular stage is absent, and it is not known whether the extracellular breakdown incorporates both acid and alkaline phases. The intestine is of the ramifying type characteristic of triclads and polyclads, and it is difficult to see how there can be anything but purely temporal separation of the two phases, if they occur, with the entire contents of the gut being of the same pH value at any one time.

In the majority of the Turbellaria, though, a pattern of digestive physiology is discernible which clearly foreshadows that found in the higher animals, provided due recognition is given to the influence of particular types of feeding mechanisms and the inherent limitations imposed by the turbellarian level of organization. The turbellarian alimentary system as a whole, therefore, can be regarded as representing an early stage in the evolution of more highly differentiated and physiologically sophisticated systems for the digestion and assimilation of food.

BIBLIOGRAPHY

Arnold, G. 1909. Intracellular and General Digestive Processes in Planariae. *Quart. J. Microscop. Sci.*, **54**:207–220.

Baylis, H. A. 1949. *Fecampia spiralis*, a Cocoon-forming Parasite of the Antarctic Isopod *Serolis schytei*. *Proc. Linnean Soc. London*, **161**:64–71.

Brand, T. von. 1966. *Biochemistry of Parasites*. New York: Academic Press. 429 pp.

Burstone, M. S. 1958. Histochemical Demonstration of Acid Phosphatase with Naphthol AS-phosphates. *J. Natl. Cancer Inst.*, **21**:523–539.

—— and J. E. Folk. 1956. Histochemical Demonstration of Aminopeptidase. *J. Histochem. Cytochem.*, **4**:217–226.

Caullery, M., and F. Mesnil. 1903. Recherches sur les *Fecampia* Giard, Turbellariés Rhabdocoeles, parasites internes des Crustacés. *Ann. Fac. Sci. Marseille*, **13**:131–168.

Christensen, A. M., and B. Kanneworff. 1964. *Kronborgia amphipodicola* gen. et sp. nov., a Dioecious Turbellarian Parasitizing Ampeliscid Amphipods. *Ophelia*, **1**:147–166.

—— and ——. 1965. Life History and Biology of *Kronborgia amphipodicola* Christensen and Kanneworff (Turbellaria, Neorhabdocoela). *Ophelia*, **2**:237–251.

Colam, J. B. 1971a. Studies on Gut Ultrastructure and Digestive Physiology in *Rhabdias bufonis* and *R. sphaerocephala* (Nematoda: Rhabditida). *Parasitology*, **62**:247–258.

——. 1971b. Studies on Gut Ultrastructure and Digestive Physiology in *Cyathostoma lari* (Nematoda: Strongylida). *Parasitology*, **62**:259–272.

——. 1971c. Studies on Gut Ultrastructure and Digestive Physiology in *Cosmocerca ornata* (Nematoda: Ascaridida). *Parasitology*, **62**: 273–284.

Eastham, L. E. S. 1933. Morphological Notes on the Terrestrial Triclad *Rhynchodemus brittanicus* Percival. *Proc. Zool. Soc. London*, 1933, 889–895.

Fernando, W. 1945. The Storage of Glycogen in the Temnocephaloidea. *J. Parasitol.*, **31**:185–190.

Froehlich, C. G. 1955. On the Biology of Land Planarians. *Bol. Fac. Filosof. Ciências Letras Univ. São Paulo, Zool.*, **20**:263–272.

Giard, M. A. 1886. Sur un rhabdocoele nouveau, parasite et nidulant (*Fecampia erythrocephala*). *Compt. Rend. Acad. Sci. Paris*, **103**:499–501.

Gomori, G. 1952. *Microscopic Histochemistry*. Chicago: University of Chicago Press. 273 pp.

Gonzales, M. D. P. 1949. Sobre a digestao e a respiraçâo das Temnocephalas (*Temnocephalus bresslaui* spec. nov.). *Bol. Fac. Filosof. Ciências Letras Univ. São Paulo, Zool.*, **14**:277–323.

Halton, D. W. 1966. Occurrence of Microvilli-like Structures in the Gut of Digenetic Trematodes. *Experientia.* **22**:828–829.

———. 1967. Observations on the Nutrition of Digenetic Trematodes. *Parasitology*, **57**:639–660.

——— and J. B. Jennings. 1965. Observations on the Nutrition of Monogenetic Trematodes. *Biol. Bull.*, **129**:257–272.

Hess, R., and A. G. E. Pearse. 1958. The Histochemistry of Indoxylesterase of Rat Kidney with Special Reference to Its Cathepsin-like Activity. *Brit. J. Exptl. Pathol.*, **39**:292–299.

Hyman, L. H. 1951. *The Invertebrates.* II. *Platyhelminthes and Rhynchocoela.* New York: McGraw-Hill. 550 pp.

———, B. H. Willier, and S. A. Rifenburgh. 1924. Physiological Studies on Planaria. VI. A Respiratory and Histochemical Investigation of the Source of Increased Metabolism after Feeding. *J. Exptl. Zool.*, **40**:473–494.

Jacek, S. 1916. Untersuchungen über den Stoffwechsel bei rhabdocoelen Turbellarien. *Bull. Intern. Acad. Sci. Cracovie. Classe Sci. Math. Nat.*, **B**:241–261.

Jennings, J. B. 1957. Studies on Feeding, Digestion and Food Storage in Free Living Flatworms. *Biol. Bull.*, **112**:63–80.

———. 1959. Observations on the Nutrition of the Land Planarian *Orthodemus terrestris* (O. F. Müller). *Biol. Bull.*, **117**:119–124.

———. 1962a. Further Studies on Feeding and Digestion in Triclad Turbellaria. *Biol. Bull.*, **123**:571–581.

———. 1962b. A Histochemical Study of Digestion and Digestive Enzymes in the Rhynchocoelan *Lineus ruber* (O. F. Müller). *Biol. Bull.*, **122**:63–72.

———. 1968a. Feeding, Digestion and Food Storage in Two Species of Temnocephalid Flatworms (Turbellaria: Rhabdocoela). *J. Zool. London*, **156**:1–8.

———. 1968b. "Platyhelminthes: Nutrition and Digestion." In M. Florkin and B. T. Scheer (eds.), *Chemical Zoology*, vol 2, pp. 303–326. New York: Academic Press, 639 pp.

———. 1969. Ultrastructural Observations on the Phagocytic Uptake of Food Materials by the Ciliated Cells of the Rhynchocoelan Intestine. *Biol. Bull.*, **137**:476–485.

———. 1971. "Parasitism and Commensalism in the Turbellaria." In B. Dawes (ed.), *Advances in Parasitology*, vol. 9, pp. 1–32. New York and London: Academic Press.

———. 1972. *Feeding, Digestion and Assimilation in Animals*, 2d ed. London: Macmillan.

——— and J. B. Colam. 1970. Gut Structure, Digestive Physiology and Food Storage in *Pontonema vulgaris* (Nematoda: Enoplida). *J. Zool. London*, **161**:211–221.

——— and R. Gibson. 1969. Observations on the Nutrition of Seven Species of Rhynchocoelan Worms. *Biol. Bull.*, **136**:405–433.

——— and V. M. van der Lande. 1967. Histochemical and Bacteriological Studies on

Digestion in Nine Species of Leeches (Annelida: Hirudinea). *Biol. Bull.*, **133**: 166-183.

—— and D. F. Mettrick. 1968. Observations on the Ecology, Morphology and Nutrition of the Rhabdocoel Turbellarian *Syndesmis franciscana* (Lehman, 1946) in Jamaica. *Caribbean J. Sci.*, **8**:57-69.

Johri, L. N. 1952. A Report on a Turbellarian *Placocephalus kewense*, from Delhi State, and Its Feeding Behaviour on the Live Earthworm *Pheretima posthuma*. *Sci. Cult. Calcutta*, **18**(6):291.

Kelley, E. G. 1931. The Intracellular Digestion of Thymus Nucleo-protein in Triclad Flatworms. *Physiol. Zool.*, **4**:515-542.

Kozloff, E. N. 1970. Selection of Food, Feeding and Digestion in an Acoel Turbellarian. *Am. Zool.*, **10**:553.

Löhner, L. 1916. Zur Kenntnis der Blutverdauung bei Wirbellosen. *Zool. Jahrb. Abt. Allgem. Zool.*, **36**:1-10.

Mast, S. O. 1947. The Food Vacuole in *Paramecium*. *Biol. Bull.*, **92**:31-72.

Metschnikoff, E. 1878. Über die Verdauungsorgane einiger Süsswasserturbellarien. *Zool. Anz.*, **1**:387-390.

Mettrick, D. F., and J. B. Jennings, 1969. Nutrition and Chemical Composition of the Rhabdocoel Turbellarian *Syndesmis franciscana* (Lehman, 1946), with Notes on the Taxonomy of *S. antillarum* Stunkard and Corliss, 1951. *J. Fisheries Res. Board Can.*, **26**(10):2669-2679.

Osborne, P. J., and A. T. Miller. 1963. Acid and Alkaline Phosphatase Changes Associated with Feeding, Starvation and Regeneration in Planarians. *Biol. Bull.*, **124**:285-292.

Percival, E. 1925. *Rhynchodemus brittanicus*, n. sp. A New British Terrestrial Triclad, with a Note on the Excretion of Calcium Carbonate. *Quart. J. Microscop. Sci.*, **69**:344-355.

Pfitzner, I. 1958. Die Bedingungen der Fortbewegung bei den deutschen Landplanarien. *Zool. Beitr.*, **3**:235-310.

Reynoldson, T. B., and J. O. Young. 1963. The Food of Four Species of Lake-dwelling Triclads. *J. Animal Ecol.*, **32**:175-191.

Rosenbaum, R. M., and Carmen I. Rolon. 1960. Intracellular Digestion and Hydrolytic Enzymes in the Phagocytes of Planarians. *Biol. Bull.*, **118**:315-323.

Saint-Hilaire, C. 1910. Beobachtungen über die intracelluläre Verdauung in den Darmzellen der Planarien. *Z. Allgem. Physiol.*, **11**:177-248.

Ugolev, A. M. 1960a. The Existence of Parietal Contact Digestion. *Bull. Exptl. Biol. Med. U.S.S.R..*, **49**:1-12.

——. 1960b. Influence of the Surface of the Small Intestine on Enzymatic Hydrolysis of Starch by Enzymes. *Nature London*, **188**:588-589.

——. 1965. Membrane (Contact) Digestion. *Physiol. Rev.*, **45**:555-595.

Westblad, E. 1922. Zur Physiologie der Turbellarien. I. Die Verdauung. II. Die Exkretion. *Lunds Univ. Arsskr. N.F. Avd. 2*, **18**(2):1-212.

Willier, B. H., L. H. Hyman, and S. A. Rifenburgh. 1925. A Histochemical Study of Intracellular Digestion in Triclad Flatworms. *J. Morphol.*, **40**:299-340.

Chapter 10

Some Aspects of the Physiology and Organization of the Nerve Plexus in Polyclad Flatworms

Harold Koopowitz
Developmental and Cell Biology,
University of California, Irvine, California

According to modern phylogenetic speculations the turbellarians occupy a strategic position with regard to metazoan evolution. These animals are still at the tissue level of organization and have a number of characteristics which are intermediate between the cnidaria and higher protostomes. The composition and organization of the nervous system is particularly interesting. There is an anterior median brain from which radiate a number of nerve trunks, which subdivide and anastomose repeatedly. These may form either a regular ladderlike pattern of nerves, as in many of the freshwater triclads, or a netlike reticulation like that found in the polyclads. The peripheral plexiform system of the polyclads suggests a primitive organization somewhat akin to a condensed coelenterate nerve net, although the brain may appear to be as complex as some found in the higher annelids (Bullock and Horridge, 1965). The phyletic position occupied by the polyclads is by no means clear, but it seems much closer to the other protostomous phyla than the more commonly studied triclads. In the latter case it is rather difficult to distinguish the contribution that the peripheral nervous system makes and how this might interact with the centralized cords and brain. Polyclads afford preparations which allow one to distinguish, to a certain extent, the role played by the peripheral nervous system. This paper deals with the physiological organization of polyclad nervous systems with the main emphasis on the plexiform peripheral system. A number of different species have been used, and these indicate that the organization of the peripheral parts varies from species to species. Further, the peripheral plexus possesses a good deal of autonomy.

MATERIALS AND METHODS

A small notoplanid-type polyclad used in this study has been provisionally described as *Notoplana acticola*, although its identity is uncertain. Other species which can be more definitely classified and have been used in this study are *Planocera gilchristi*, *Thysanozoon brocchii*, and *T. californica*. *P. gilchristi* and *T. brocchii* are South African and the others are Californian. All the animals were collected in the upper intertidal zone. They were maintained in trays of seawater kept at room temperature ±20°C. Seawater was changed every day and the animals were fed their preferred foodstuffs when these were known. *Notoplana* will accept frozen brine shrimp. Generally animals were used within 1 week of collecting, but those which would feed could be maintained in good condition from 4 to 6 weeks. A large number of experimental approaches and techniques have been used, and these will be detailed when the specific experiments are discussed.

RESULTS

Anatomy of the Plexus

Most of the nervous system in the polyclad flatworms is in the form of a net-like reticulation of branching and anastomosing nerve trunks. This form of organization is comparatively rare, and the functional significance of this type of arrangement is not at all clear (Horridge, 1968). On a gross level the system might be considered as an intermediate stage between the anatomically diffuse cnidarian system and the more condensed, centralized nervous systems of most higher invertebrates.

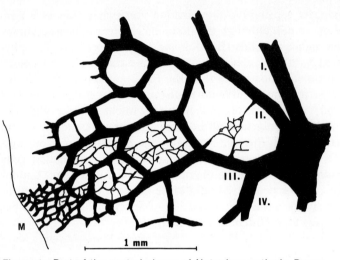

Figure 1 Part of the ventral plexus of *Notoplana acticola*. Drawn from material stained for cholinesterase. Only a portion of the finer meshwork of nerves between the coarse strands of the plexus has been drawn. The roman numerals refer to the major trunks leaving the brain where I is closest to the midline at the anterior portion of the brain. M = margin of the animal.

The peripheral plexus of most polyclads occurs in two parts, a ventral and a dorsal portion. The ventral submuscular plexus is well developed and consists of stout anastomosing branches (Fig. 1), between which lies a much finer reticulation of fibers. At the edges of the animal the ventral plexus branches to form a dense reticulation. The dorsal submuscular plexus is made up from much finer nerves and resembles the coelenterate configuration more closely. The resemblance is fortuitous, for the strands of the polyclad plexus are made up from numerous neurons whereas the coelenterate nerve net is made up from single naked nerve cells. The polyclad plexus, however, does contain numerous multi- and bipolar neurons and in this respect resembles the cnidaria more closely than most other invertebrates.

At the ultrastructural level the branches of the plexus appear to be composed of parallel fibers, at right angles to which run other axons (Fig. 2*a*). Scattered along the length of the plexus strands are clusters of neuropil (Fig. 2*b*). These are probably places where integration of information takes place but could also be involved with some of the netlike properties which will be discussed later. Synaptic vesicles are common and three types have been found: small clear vesicles, large irregular clear ones, and a smaller dense one. Mixed synapses with two kinds of vesicles are common, and in one case all three vesicle types in the same axon have been found (Fig. 2*c*, *d*). These appear similar to those in planarian central nervous systems (Morita and Best, 1966). Occasionally gap junctions have been found between adjacent axons (Fig. 3a). These may be electrical *en passant* synapses. In numerous cases axons, while not displaying clear gap junctions, may be very closely appressed (Fig. 3b), and often axons within axons are found (Fig. 3c). Presumably these indicate some calyxlike configuration between the two cells. A subepithelial plexus has not been found in any of the polyclads that have been investigated so far. The presence of a subepidermal plexus resembling a diffuse nerve net has only been demonstrated in a few flatworms, and it is not clear whether this represents a primitive condition or the result of reduction. The acoela are generally considered to possess such a plexus, and a diffuse nerve net has also been demonstrated in freshwater planarians (Lentz, 1968). In a number of polyclad genera including *Planocera* (Ewer, 1965), *Notoplana* and *Stylocho-plana* (Hadenfeldt, 1929), and *Thysanozoon* (Lang, 1884; Levetzow, 1936) the plexuses are so netlike that one is forced to assume some important functional significance for the anastomoses of adjacent branches.

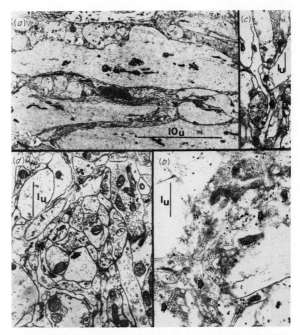

Figure 2 Ultrastructure of the peripheral part of the plexus of *N. acticola*. (a) Tangential section to show the main orientation of the nerve axons; (b) a knot of neuropil with at least three kinds of synapses, dense core (1), large irregular (2) and small clear (3) vesicles; (c, d) mixed synapses which contain a number of different vesicles.

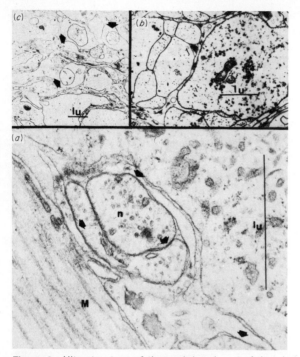

Figure 3 Ultrastructure of the peripheral part of the plexus of *N. acticola.* (*a*) Gap junctions between nerves close to the muscle fibers; (*b*) closely packed axons; (*c*) axons within axons; often one appears to be at right angles to the axis of the other.

Physiology of the Plexus in *Planocera*

The polyclad which has received most attention in recent years has been *Planocera gilchristi* (Gruber and Ewer, 1962; Ewer, 1965; Koopowitz and Ewer, 1970). This is a large oval animal with a diameter up to 70 mm. Previous studies (Koopowitz and Ewer, 1970) indicate that the animals have inhibitory as well as excitatory neurons. Figure 4 shows a partially hemisected preparation with intact brain and major nerve trunks. Stimulation of the control side causes a contraction of that as well as the contralateral witness half. Changing to a higher frequency of stimulation brings the inhibitory system into action with concurrent relaxation in the witness as well as control side. The contractions measured in this manner result mainly from the large parapharyngeal muscles which lie alongside the pharynx (Gruber and Ewer, 1962), and these are innervated by the main nerve trunks. It was shown (Koopowitz and Ewer, 1970) that transmission of information from one side to the other involved the main nerves and brain. A lesion through a major tract between the site of stimulation and the brain impedes the transmission of both excitation and inhibition. Inhibition, however, is not confined to central parts of the nervous system. If one cuts out the brain, large nerve trunks, and most of the parapharyngeal muscles, one ends up with a ring of margin. It is possible to demonstrate both excitation and inhibition (Fig. 5) in this strip of tissue.

Because of the netlike appearance of the peripheral parts of the nervous system, an experiment was designed which we hoped might give some information about conduction in the strip preparation. The strip was set up in such a way that a beam of light could be bounced off a tiny mirror lying on part of the preparation so that the light was focused on a revolving drum covered with photographic paper. In this way movement at any particular place on the preparation could be monitored (Fig. 6). The results, in Fig. 6c, indicate that the strip is continuously active. Whether this is due to pacemaker neurons in the net or small reflexes is unclear. This background activity, however, makes it rather difficult to interpret the results of stimulation at the distant end of the preparation (Fig. 6d). In fact it is not until the electrodes are placed within 2.5 cm from the preparation that a recordable response is obtained (Fig. 6e). It is likely that the record is one of direct muscle stimulation evoked by some kind of electrotonic spread. When the strips of margin were stimulated electrically, one occasionally found localized contractions which appeared at unpredictable sites. They could be distinguished from spontaneous activity, for they lasted much longer and only occurred after electrical stimulation. This type of conduction can be demonstrated by another method. If strip preparations are made and then cut across at either posterior or anterior midline, then sticking a pin into either an anterior or posterior end causes writhing movements to occur. These, however, are not in the form of a progressive wave of

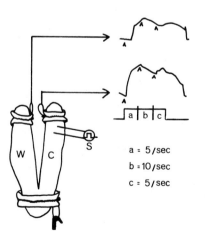

a = 5/sec

b = 10/sec

c = 5/sec

Figure 4 Responses by both sides of *Planocera gilchristi* when only one side is stimulated. W = witness side of the preparation; C = control side which received the stimulus (S); a, b, and c = trains of stimuli; each train lasted 5 sec. Contraction is shown as an upward deflection on the trace. Levers are attached to the two sides by foam plastic collars. The anterior end is attached to a fixed hook.

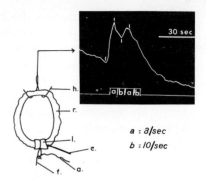

Figure 5 Contraction and relaxation elicited from a ring of margin (r) from *P. gilchristi*. Electrodes, e and a, are at the posterior end of the ring. The ring was ligatured with foam plastic (l) and attached to a fixed hook (f). A double hook (h) was used to support the preparation's weight. Trains of stimuli at different frequencies, a and b, were delivered.

activity but appear randomly in the strip. An unusual feature is that these writhing movements never cross the midline of the animal; they are confined to the stimulated half of the strip.

From the above observations on *Planocera*, one could tentatively conclude that the peripheral plexus contains both excitatory and inhibitory fibers. Pacemaker cells may be scattered throughout. Coordination between the two halves of a strip preparation seems limited, if it exists at all. This bilateral symmetry is not unusual, for it is found also in *Thysanozoon*.

Coordination of Papillae in *Thysanozoon*

The *Thysanozoon* species are large cosmopolitan polyclads which have retractile papillae on their dorsal surface. The retractile nature of the papillae has been known for a long time, but no one seems to have made the observation that if one papilla is touched, then a number of surrounding papillae also retract and bend toward the stimulated organ. If stimulated papillae are close to the median long axis of the animal, then only adjacent papillae on the ipsilateral side of the stimulated area contract. This has been observed with both *T. brocchii* and *T. californica*; however, conduction across the midline was found in one rather small individual of the latter. The response is rather reminiscent of that found in a colony of coelenterate polyps; Ewer (1965) postulated that the underlying organization could be either in the form of a nerve net that connected the papillae or in the form of direct connections between close papillae. I have evidence that suggests that the latter may be the case.

With a glass stylus, lightly touching the tip of a papilla caused a surrounding field to contract. The average diameter of the field was about seven papillae in width. On the other hand, touching the base of the structure resulted in a field with an average diameter of nine papillae. In order to determine

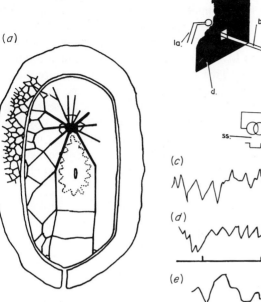

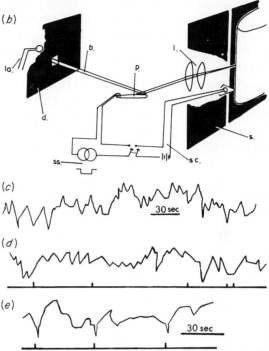

Figure 6   Recording from isolated portions of a strip preparation. (a) That part of the animal used to make the strip preparation; some of the ventral plexus has been drawn. (b) Apparatus used, a beam of light, b, is reflected off a mirror on the preparation, p, and focused, I, on a kymograph drum covered with photographic paper. d and s are screens to cut out background illumination; la = light source; ss = electrical stimulator; and sc = signal marker. (c) Spontaneous activity recorded from part of the strip, movement in the vertical plane could mean either contraction or relaxation. (d) Activity from one end of the strip while stimulating at the other. (e) Responses from the strip when the electrodes are close to the recording site.

whether the stimulus was relayed from papilla to papilla through a field, a row of papillae between the stimulated and an observed one were excised as indicated in Fig. 7a. This did not interfere with transmission. Figure 7b shows a further experiment. Here stimulation of a papilla at A causes those at positions B and C to contract too. When a cut between A and C is made, then stimulation of A only results in contraction of B and not C, although stimulation of B results in a contraction of both A and C. Physiologically a nerve net could be defined by its ability to conduct around lesions (Bullock, 1965);

this is clearly not the case here. More evidence that there may be direct
neural connections between papillae comes from the following experiment,
the results of which are shown diagrammatically in Fig. 7c. When a papilla
is cut off at its base, then those in its field bend toward the injured site and
maintain a tetanic contraction for 10 to 15 min. Severing another papilla on
the other side of one already bent toward an injured site will cause the bent
one to both straighten and shorten. This suggests that there are direct connections between a papilla and the musculature of other papillae in its field,
and that the connections are with the musculature which is closest to the
origin. Bullock (1965) has presented similar evidence for this kind of organization between sea urchin spines. Here where the innervated field is much
larger than in *Thysanozoon*, he finds that the number of neurons necessary to
provide direct links between spines is well within the limits set by the number
of nerve cells in the epithelium.

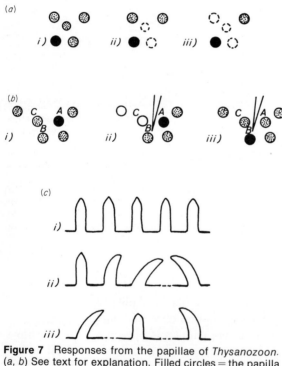

Figure 7 Responses from the papillae of *Thysanozoon*.
(*a, b*) See text for explanation. Filled circles = the papilla
which is stimulated, stippled circles = papillae contract-
ing, clear circles refer to papillae not responding, and
broken clear circles refer to the position of papillae which
have been excised. (*c*) See the text for explanation.

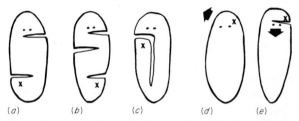

Figure 8 *a, b* and *c* = lesions which still allow ditaxic locomotion to be elicited at the anterior end of *N. acticola* when stimulated at position x. *d* and *e* refer to directions of movement when stimulated at the anterior lateral edge before and after a lesion in front of the brain. The brain lies between the position of the eye clusters.

Organization of the Sensory System in *Notoplana*

The sensory system of *Notoplana acticola* appears to have properties more similar to those of a classic nerve net than any other polyclad studied so far. Like many polyclads this species moves by means of ditaxic locomotion, i.e., alternate waves of extension and contraction moving out of phase along the two sides of the animal's body. Here ditaxic locomotion requires the presence of the brain, and hence its initiation means that information must have reached as far as this structure. It is possible to make a variety of cuts and lesions (Fig. 8) which demonstrate that sensory information can spread around and between cuts. Figure 8c is of major interest because it indicates that the information passes posteriorly as well as anteriorly, suggesting that the sensory messages are spread out diffusely from their origin. It seems likely that the response is initiated by sensory neurons. The animals move if they are prodded with a mechanical probe; if one applies suction electrodes to the same area, one can give the animal a shock so that a large portion of the worm's body contracts, but this will not lead to ditaxic locomotion. It is probable that in this case the muscles are stimulated directly and not the sensory nerves. This acts as a double control because the electrical stimulation rules out direct muscle-muscle propagation as a means of conducting around the lesions, and also that the prodding which initiates locomotion might do so by pushing the anterior end of the body over the substrate and so stimulating receptors at the front of the animal. During electrical stimulation the contractions jerk the entire body, but this does not lead to locomotion. The overlapping lesions shown in Fig. 8a and b cut through the main nerve trunks, and it seems that information must arrive at the brain in a rather circuitous fashion. The system acts as a physiological nerve net, and presumably the plexus is where conduction takes place.

The entire sensory system is not organized like a nerve net. In those cases where information about the location of the sensory cell is required, specific pathways are used. If an animal is touched at the anterior side in front of the brain, it moves away by twisting to the contralateral side (Fig. 8d), but if the main nerve trunks between the point of stimulation and the brain are severed, then the stimulation results in an entirely different kind of response. The animal performs a "reverse ditaxis" and backs away (Fig. 8e). Thus it appears that there are at least two types of sensory information elicited the same way: *specific* information which travels to the brain along set pathways and *nonspecific* information which may travel a circuitous route to the brain. Under normal circumstances *specific* information takes precedence over *nonspecific* and the latter might only be of functional significance to the animal if it is injured.

The organization of the motor system, in contrast to the sensory system, requires direct lines to those muscles which take part in locomotion. A lesion anterior to the brain renders the innervated muscle field flaccid and it is no longer used in locomotion. Likewise, those lesions posterior to the brain stop muscles behind the lesion from being used in locomotion.

DISCUSSION

These observations highlight two points which are worthy of further discussion. First, the organization in the peripheral parts of the nervous system appears to differ in those species tested, and second, the relationship between the peripheral and central parts of the nervous system suggests some ideas about the evolution of centralization.

It is of interest that one cannot clearly demonstrate nerve-net-like properties in *Planocera*, and this makes the functional arrangement of the plexiform organization obscure. The experiments reported indicate that there may be conducting pathways in the plexus which occur without reference to the central nervous system, but the ability to conduct around lesions appears to be missing (Gruber and Ewer, 1962). There are a number of possible explanations which could account for these findings. It may be that such a nerve net actually exists, but we have not as yet found the correct means for stimulating it. The ability to demonstrate the presence of a net in *Notoplana*, with mechanical but not electrical stimulation, also points to this possibility. Alternatively there may not be netlike properties in this plexus and it could merely be physically organized in a plexus as some secondary arrangement brought about by asymmetric growth involved with the flattening and relatively large size of the animals. The basic plan might well have been a simple ladderlike arrangement as is found in the polyplacophora, annelids, and even triclad flatworms (Bullock and Horridge, 1965). Even though the plexus is so netlike in appearance, one would not expect that the motor side of the nervous system would be arranged as a physiological net because much of the activity of the animals entails fairly stringent control of particular muscle fields. Levetzow (1936) thought that such a motor nerve net was involved with locomotion in *Thysano-zoon*. Most locomotion in this animal occurs by means of ciliary rather than muscle movement except during swimming, when waves of activity pass along the sides of the animal in an anterior-posterior direction. Lesions through the body wall did not appear to interfere with the rhythmicity of these contraction waves. In *Planocera*, most of the experiments set out to determine the net-like properties of the system have involved conduction in an anterior direction (Gruber and Ewer, 1962), but it is possible that a polarized net which conducts only in a posterior direction exists and this possibility should be examined. The sensory netlike properties which are found in *Notoplana* are explicable in terms of either netlike properties of the nervous system or some form of neuroid conduction in the epithelium (Mackie, 1970), but it is difficult to visualize how this latter sort of information might be fed to the brain. At all events this system appears to be of secondary importance, and it is tempting to view it as a vestigial holdover from an ancestral form where netlike properties were more important, and as such the "net" is merely a supplementary warning system. This idea, however, must be rationalized in light of the fact that the anastomoses of the plexus are well developed and one would expect a major functional significance for this arrangement. More data will be required before this conflict can be reconciled.

There is no clear morphological distinction between the peripheral parts of the plexus and the cords. One merges into the other and it may be that the only differences are those of quantity rather than quality. This suggests that the evolution of centralized nervous systems merely entailed a migration of the component parts and their functions into what was to become the main nerve trunks. Evidence for the possession of large fragments of behavior patterns in the plexus have been presented elsewhere (Koopowitz, 1970), but an example may stress this, as well as illustrate the interactions between central and more peripheral parts of the system. If a piece of periwinkle is offered to a hungry *Planocera gilchristi* at a point on its margin about two-thirds from the anterior end, the animal will twist around and extend its anterior margin over the food. When the brain is removed and the procedure repeated, then that lateral portion in contact with the food will extend and engulf it. These observations suggest that information about the presence of food is sent centrally, and this evokes both inhibitory feedback to the local reflexes and also commands activity in the anterior of the animal. These and other observations indicated that local reflexes may be spread throughout the plexus. It is possible that the activity recorded from strip preparations could be due to this kind of organization rather than some inherent pacemaker activity. Pacemaker activity, however, is probably present throughout the system because pieces of animals may have bursts of activity (Gruber and Ewer, 1962). Action potentials of irregular frequency can also be recorded from isolated nerve trunks (unpublished data). Pacemaker activity in flatworms tends to be irregular, and there is little evidence for rhythmicity. The latter might not be expected if the pacemakers are scattered throughout the system and have little interaction with each other. Earlier workers (Olmsted, 1922; Moore, 1923) reported that spontaneity resided in the cerebral cortex and the decerebrate animals remained motionless and did not react unless stimulated. Even *Planocera* will not appear to move after decerebration if one merely observes the animal in a casual manner and does not actually record activity. It is possible that the earlier workers did not observe their animals for long continuous periods.

One of the most striking features of the anatomical organization of the nervous system is that it does appear to be intermediate between the primitive diffuse nerve net of the coelenterates and the more advanced ganglionated central nervous systems of higher protostomes. The plexus with its meshwork of finer branches and anastomoses is reminiscent of a nerve net in the process of concentration. It is as if fibers have become clumped together to form trunks which increase in diameter as they approach the major nerve cords and brain, and yet continue to retain the essential anastomosing and branching plan. Another intermediate character is the submergence of the cell bodies into the general tissue of the nerve trunks. Generally in protostomes the cell bodies form a rind on the outside of neural tissue (Bullock and Horridge, 1965). The cerebral ganglia of polyclads do have a rind of unipolar somata, but the cell bodies of neurons in the plexus are multi- or bipolar and resemble those of coelenterate preparations (Robson, 1965; Jha, 1965; Leghissa, 1965). Further evidence for uncentralized organization is to be found in knots of neuropil scattered throughout the plexus and the considerable volume of brain occupied with fibers of passage between the two halves of the animal (Hadenfeldt, 1929).

The presence of pacemaker cells, local reflexes, and the netlike properties (*Notoplana*) of the plexus also suggest that physiologically the organization is to a large part uncentralized. Therefore at both anatomical and physiological levels, polyclad nervous systems appear intermediate in nature between diffuse and centralized arrangements. It is not unreasonable to expect that this information about polyclad nervous systems is applicable to hypotheses concerning the evolution of centralization. Passano (1963) suggested that during the course of evolution a hierarchy of pacemakers arose with the more centralized systems overriding and controlling those which were placed more peripherally. He notes that such a system requires points where integration can occur. Present knowledge regarding those polyclad systems which have been studied indicates a number of features which correlate well with Passano's hypothesis, e.g., knots of scattered neuropil and activity in strip preparations. More evidence for this point of view can be obtained from studies on the pharynx of *Enchiridium punctatum*, a polyclad (Koopowitz, unpublished data). Here the isolated pharynx contracts rhythmically. If the organ is sliced into a number of smaller segments, each section starts to contract. However, the rhythms of the individual sections may differ both from each other and from the original contraction frequency. These observations do indicate some kind of pacemaker hierarchy. Autonomy in flatworm pharynges is also known from the triclad planarians (Wulzen, 1917).

The key factor which led to the evolution of the brain has long been thought to be either an accumulation of nerves around an anterior statocyst or an anterior accumulation of many types of receptors (see Bullock and Horridge, 1965, for further review). Another possibility (Koopowitz, 1970) is that it is important for a bilaterally symmetrical animal to be able to coordinate the two sides of the body, and hence the evolution of the apparatus to do this. There are two ways by which the brain could exercise this control: first, by inhibiting individual pacemaker centers which can be clearly demonstrated in many polyclads where decerebration leads to eversion of the pharynx, premature egg laying (Gruber and Ewer, 1962), and rhythmical contractions in the genitalia (Koopowitz, unpublished data); second, the brain could route information from one side of the body to the other (Fig. 3) where the witness half of the animal follows the control with surprisingly little modification. The ability of decerebrate animals to respond to food stimuli indicates that sensory stimuli may not proceed directly to the brain.

It is possible to construct a plan of the centralization that evolved after a "polyclad" stage. The next important step was formation of a rind of cells with migration of cell bodies toward the periphery of the nerve trunks. This stage is evident in certain nemertines like *Cerebratulus marginatus* (Bürger, 1891), where the lateral nerve cords which are not ganglionated already possess a rind. Rind formation would be followed by the confinement of cell bodies to specific parts of the nervous system, i.e., in ganglia.

Molluscan organization still retains many features which are reminiscent of the polyclad plan, of which the ventral plexus in the sole of the foot is the most similar. Annelids, on the other hand, appear to have only branching nerves and not anastomosing ones. This arrangement is most easily derived from a ladderlike organization which occurs in both turbellarians and nemertines. Finally there is a trend toward consolidation of ganglia which occurs in many of the higher invertebrates.

BIBLIOGRAPHY

Bullock, T. H. 1965. Comparative Aspects of Superficial Conduction Systems in Echinoids and Asteroids. *Am. Zool.*, **5**:545-562.

—— and G. A. Horridge. 1965. *Structure and Function in the Nervous Systems of Invertebrates*. San Francisco: W. H. Freeman and Co.

Bürger, O. 1891. Beiträge zur Kenntnis des Nervensystems der Wirbellosen. Neue Untersuchungen über das Nervensystem der Nemertinen. *Mitt. Zool. Sta. Neapel*, **10**:206–254.

Ewer, D. W. 1965. Networks and Spontaneous Activity in Echinoderms and Platyhelminthes. *Am. Zool.*, **5**:563–572.

Gruber, S. A., and D. W. Ewer. 1962. Observations on the Myo-neural Physiology of the Polyclad *Planocera gilchristi*. *J. Exptl. Biol.*, **39**:459–477.

Hadenfeldt, D. 1929. Das Nervensystem von *Stylochoplana maculata* und *Notoplana atomata*. *Z. Wiss. Zool.*, **133**:586–638.

Horridge, G. A. 1968. Chap. 6 in *Interneurones*. London: W. H. Freeman and Co.

Jha, R. K. 1965. The Nerve Elements in Silver Stained Preparations of *Cordylophora*. *Am. Zool.*, **5**:431–438.

Koopowitz, H. 1970. Feeding Behaviour and the Role of the Brain in the Polyclad Flatworm, *Planocera gilchristi*. *Anim. Behav.*, **18**:31–35.

—— and D. W. Ewer. 1970. Observations on the Myo-neural Physiology of a Polyclad Flatworm: Inhibitory Systems. *J. Exptl. Biol.*, **53**:1–8.

Lang, A. 1884. Die Polycladen des Golfes von Neapel und der angrenzenden Meeresabschnitte. *Fauna und Flora des Golfes von Neapel*. Monog. XI, ix + 688 pp.

Leghissa, S. 1965. Nervous Organization and the Problem of the Synapse in *Actinia equina*. *Am. Zool.*, **5**:411–424.

Lentz, T. L. 1968. *Primitive Nervous Systems*. New Haven, Conn.: Yale University Press.

Levetzow, K. G. von. 1936. Beiträge zur Reizphysiologie der Polycladen Strudelwürmer. *Z. Vergleich. Physiol.*, **23**:721–726.

Mackie, G. O. 1970. Neuroid Conduction and the Evolution of Conducting Tissues. *Quart. Rev. Biol.*, **45**:319–332.

Moore, A. R. 1923. The Function of the Brain in Locomotion of the Polyclad Worm, *Yungia aurantiaca*. *J. Gen. Physiol.*, **6**:73–76.

Morita, M., and J. B. Best. 1966. Electron Microscopic Studies of Planaria. III. Some Observations on the Fine Structure of Planarian Nervous Tissue. *J. Exptl, Zool.*, **161**:391–412.

Olmsted, J. M. D. 1922. The Role of the Nervous System in the Locomotion of Certain Marine Polyclads. *J. Exptl. Zool.*, **36**:57–66.

Passano, L. M. 1963. Primitive Nervous Systems. *Proc. Natl. Acad. Sci.*, **50**:306–313.

Robson, E. A. 1965. Some Aspects of the Structure of the Nervous System in the Anemone *Calliactis*. *Am. Zool.*, **5**:403–410.

Wulzen, R. 1917. Some Chemotropic and Feeding Reactions of *Planaria maculata*. *Biol. Bull.*, **33**:67–69.

Chapter 11

Ecological Separation in British Triclads (Turbellaria) with a Comment on Two American Species

T. B. Reynoldson
Zoology Department, University College of North
Wales, Bangor, Caernarvonshire, U.K.

A taxonomic group invariably shows adaptive radiation due in part to the pressure of numbers, itself a function of the potential to increase geometrically, possessed by all species. However, there is also a corresponding restriction from several processes of which competition for limited resources is one, and ultimately a physiological barrier becomes operative. The distribution and abundance of a species within narrow confines of space is often the result of the interaction of these opposing forces. The outcome will differ not only according to historical events but also as a result of the basic attributes of the type of organisms with which we are concerned. For example, the size of the organisms, the type of organization, the physiology and genetics, etc., of the taxon may be decisive in this context. With any given taxon the properties of the environment will also contribute significantly. In the case of the Tricladida they have spread into the freshwater, marine, and terrestrial environments. Since these three major environments are contiguous, the spread of triclads among them must be regarded as the first main step in ecological separation but such a macroscale is not the focus of this paper; it is the separation within these major ecosystems with which I am concerned. In considering this topic, it is helpful to bear in mind that most, if not all, triclads are predators.

SEPARATION IN TERRESTRIAL (TERRICOLA) AND MARINE (MARICOLA) TRICLADS

Under British conditions of climate, topography, and geology and their complex interactions, triclads have not been very successful in colonizing land, for there is only one endemic terrestrial species, *Orthodemus terrestris* (O. F. Müller), usually referred to under *Rhynchodemus* by European workers (e.g., Luther, 1961; Hartog, 1962a). Although this species is widespread in Britain, it is never abundant locally and its habitat requirements are difficult to define although soil type and moisture are important (Hartog, 1962b). *R. bilineatus* (Mecznikow) was not recorded in Britain until 1944 (Pantin, 1944). It is probably widespread and occurs in the same habitat as *O. terrestris* (Pantin, 1950) but the ecological relationships of these two species in Britain are unknown. The validity of the two species *Rhynchodemus scharffi* (von Graff) and *Rhynchodemus brittanicus* (Percival) mentioned in Southern (1936) is a question I am not competent to discuss. Three other species of land triclads have been recorded sporadically from the British Isles, namely, *Dolichoplana feildeni* von Graff, *Bipalium kewense* Moseley, sometimes recorded under *Placocephalus*, and *Artioposthia triangulata* (Dendy). The first two are recorded from Ireland by Southern (1936), and the third is a more recent discovery (Anonymous, 1963, 1964) which has been identified by Mr. Stephen Prudhoe of the British Museum (Natural History). One feature which they have in common is that all are introductions from distant places; *D. feildeni* was described originally from Java, Ceylon, and Barbados; *B. kewense* is widespread in tropical and semitropical areas, but its original distribution is unknown (Hyman, 1951); and *A. triangulata* is a common species in New Zealand (S. Prudhoe, pers. comm.). There can be little doubt that they have been transported to Britain in soil around plant roots. While records of the first two refer to their occurrences in greenhouses, usually artificially heated, *A. triangulata* seems to be adapting to British conditions outdoors and several populations have been recorded from garden soils in Northern Ireland (Anonymous, 1963, 1964) and also from a garden in Carlisle, England (T. B. Reynoldson, 1965, unpublished data). It is a large triclad measuring some 10 cm in length and is reported to feed voraciously on our native earthworms and therefore may be filling a vacant niche, since earthworms are abundant and have few invertebrate predators. It would appear to be ecologically separated from *O. terrestris* on the basis of the size of food it takes, since the native species is small (10 to 35 mm in length). Although the latter's natural diet is not known, it feeds on small earthworms and slugs in the laboratory, and probably does so in the field (Jennings, 1959).

The marine species of triclad are not well known in Britain, but recently Hartog (1968) has discussed the status of four species, two of which are new records, including a species new to science. A fifth species, *Sabussowia dioica* Claparède according to Wilhelmi (1908), was recorded by Gamble (1893) under *Fovia affinis* Stimpson. It was not reported by Hartog, who mentions that he did not search driftweed, where it lives. Most of the habitat descriptions of marine triclads, apart from those of the commonly recorded *Procerodes littoralis* (Strom) previously called *Procerodes* (or *Gunda*) *ulvae* in the European literature, refer to the Plymouth area although they must be more widespread. These species are usually separated spatially from both the Terricola and Paludicola, but Naylor and Slinn (1958) found the freshwater species *Crenobia alpina* and *Procerodes littoralis* intermingling in a stream on the Isle of Man. All the marine species except *S. dioica* are found where salinity fluctuations are likely to be wide; there is also evidence that they may be spatially separated. For example, Hartog makes the point that *P. littoralis* rarely exists with the other species but sometimes overlaps slightly with *P. lobata*. However, the latter at Plymouth lives among pebbles at a depth of 30 to 50 cm in a belt immediately above the ridge formed by wave action at the level of mean high-water spring tides whereas *P. littoralis* occurs in streams running over beaches. *Uteriporus vulgaris* has been described only from salt marshes in the Plymouth area, and it is clear from Hartog's account that no other triclad species were present. However, Southern (1936) recorded this species from a different habitat, namely, the *Fucus spiralis* zone to the Orange Lichen zone on the shore at Galway Bay in Ireland; he makes no mention of other species. *Procerodes ireneae* Hartog, the new species, was found on one small pebble beach at Wembury, near Plymouth, between high-water neaps and mean sea level. There was some intermingling of it and *P. littoralis* from a nearby stream. How far restriction of these marine triclads to the upper shore is due to physiological limitations (e.g., Pantin, 1931a, b; Beadle, 1934) and how far to competition with marine animals is not known, but Hartog expresses the opinion that the seaward limitation of *P. littoralis* to mean sea level may be due to competition with the polyclad *Notoplana atomata*. This is not the case, however, with a population of *P. littoralis* in the Bangor (North Wales) area (T. B. Reynoldson, Jr., pers. comm.). Definitive conclusions on the nature of ecological separation in these few marine species must await further study; at the moment it appears to be spatial. At this point it is relevant to consider the freshwater species in relation to the marine habitat. It seems probable that physiological limitations, as opposed to ecological, restrict such triclads to freshwater habitats or to those with very dilute seawater of fairly constant salinity. Their ability to tolerate competition and predation in the marine habitat is suggested by the occurrence of *Polycelis tenuis*, *Dugesia polychroa* (and *lugubris* Reynoldson and Bellamy, 1970), *Dendrocoelum lacteum*, and *Planaria torva*, all typical freshwater species, in the littoral zone of the Baltic Sea at Tvarmine (Finland) (Luther, 1961; Reynoldson, 1958a), where the salinity is 3‰ Cl⁻ (Hartog, 1968). They can be collected along with typical marine animals on seaweed and stones immersed permanently in the sea.

SEPARATION IN THE FRESHWATER TRICLADS (PALUDICOLA)

Eleven species of freshwater triclad are recognized in Britain at present, including *Dugesia tigrina*, a relatively recent introduction from the Americas (Dahm, 1955; Reynoldson, 1956). These can be divided into groups (Table 1) which are characteristic of (1) ponds and lakes, (2) streams and rivers, while a third (3) contains species equally common in still and moving waters. Thus there is spatial separation on a macroscale, but like most biological classifications it is not absolute. Lake species may be found in the quiet areas of rivers while stream species occur in the colder lakes of northern Britain (Reynoldson, 1953). The question arises whether or not there is spatial separation on a smaller scale within a stream or lake system. There is a large literature on this subject for streams on the continent of Europe (see Dahm, 1958), and in Britain Carpenter (1928) and Beauchamp and Ullyott (1932) have described a spatial separation with *Crenobia alpina* occupying the cooler, more stenothermal headwaters, *Polycelis felina* living in the warmer zone below, and *P. nigra* occupying the warm slower-flowing regions on the coastal plain.

Table 1   The Typical Habitat of British Triclads

Lake	Stream	Lake and stream
Polycelis tenuis Ijima	*Polycelis felina* (Dalyell)	*Polycelis nigra* (Müller)
Dugesia lugubris (Schmidt)	*Crenobia alpina* (Dana)	*Phagocata vitta* (Dugès)
Dugesia polychroa (Schmidt)		
Dendrocoelum lacteum (Müller)		
Bdellocephala punctata (Pallas)		
Planaria torva (Müller)		

Such a classical distribution has been attributed to specific contrasts in the optimum temperature regime with interspecific competition also playing a significant role. However, several ecologists have expressed doubt about the importance of such competition, most recently Dahm (1958) and Pattee (1965). Wright (1968) working in my laboratory has made a detailed study of the distribution of these species in streams flowing northwards from the Snowdonia massif in North Wales and also in the less steep Anglesey streams. He confirmed the occurrence of spatial separation in these three species but found it to be related to the slope of the terrain over which the stream flowed, especially in the case of *C. alpina* and *P. felina*. Most of the Snowdonian streams flow rapidly to the sea over a narrow coastal plain, and Wright showed that the temperature regime from source to mouth was suitable for the existence of both species. A close agreement was shown between gradient and species distribution irrespective of distance from the source (Fig. 1). He concluded that *C. alpina* was abundant in streams with a steep gradient while *P. felina* rarely maintained a population at a gradient in excess of 27 percent. The position is likely to be more complicated than this, but sufficient has been said to demonstrate the occurrence of spatial separation. The causes of such a distribution, whether behavioral, a result of different tolerance to flow, competition, or some combination of these, remains to be shown.

 Phagocata vitta has been less well studied, but it appears to differ in its ecology from the others. It is more characteristic of groundwater (Gislén, 1946; Dahm, 1949, 1958). There is evidence showing that this species may have the bulk of its population in the subsurface water and moreover may be restricted to special soil conditions. In British lakes it is confined to high-altitude places usually with a surrounding terrain having *Sphagnum* moss dominant in the flora (Reynoldson, 1958b). Dr. H.B.N. Hynes (pers. comm.) has reported that when an excavation is made reaching the water table in these Snowdonian soils, large numbers of *Ph. vitta* appear in the seepage although it may be scarce in the nearby stream. This species would appear to be spatially separated to a large degree from the more usual stream and lake species.

 Considering now typical lake species (Table 1), the evidence points convincingly to the conclusion that they are not separated spatially. In our experience, the bulk of the populations of the four common species occurs in shallow water, from the lake edge to a depth of some 25 cm. There are records, however, of these species living at greater depths (e.g., Chodorowski, 1959, 1960), and Macan and Maudsley (1969) reported peak populations of *Dugesia polychroa* and *Dendrocoelum lacteum* at 1.8 m depth in Windermere. The depths at which these triclads occur varies with the type of substratum, the amount of silting, the distribution of prey, etc., and it also varies temporarily in the same lake, depending on wind conditions, an onshore wind causing them

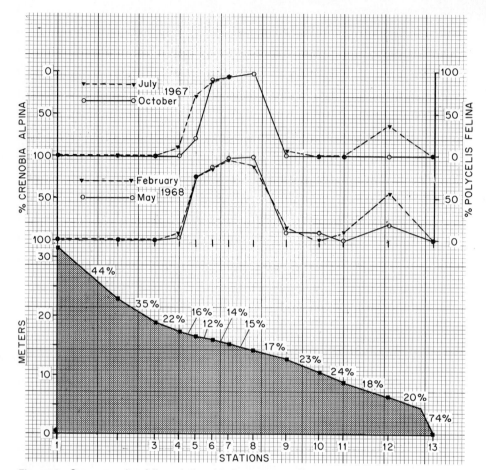

Figure 1 One example of the relationship between gradient and the proportion of *Crenobia alpina* and *Polycelis felina* in the Anafon Stream A, Snowdonia, Wales (*after Wright, 1968*). The upper diagram records the triclad faunas on four occasions at 13 stations along its length from the source; the lower gives the gradient between stations. The distance from station 1 to station 13 is 111 m.

to go deeper for a short period. The ability of these species to coexist is confirmed by finding them together in small pools only some 20 m² in area and not exceeding a depth of 30 cm. It is also common to find all four species on the same stone or leaf in a lake. Since they are able to coexist, competition is likely to be minimized in some way. Hutchinson (1959) has suggested that where there are size differences among potential competitors of the predator type, the food resource may also be partitioned on a size basis. This has been shown to be so for birds (Schoener, 1965) and also in part for lizards (Schoener, 1968; Pianka, 1969). Although there are contrasts in adult size of these triclad species—e.g., *Polycelis nigra* and *P. tenuis* are 8 to 12 mm, *Dugesia polychroa* 11 to 17 mm, and *Dendrocoelum lacteum* 14 to 25 mm in length—they do not appear to use size differences in prey to reduce the competition which undoubtedly occurs (Reynoldson, 1966). Rather, they have partitioned the food resource on a taxonomic basis. This aspect has been examined fully by using a serological technique (Davies, 1970), and the results have shown (Reynoldson and Davies, 1970) that despite much overlap among the four species each feeds to a greater extent on different prey. Thus the *Polycelis* species eat more oligochaetes, *Dugesia polychroa* more gastropods, and *Dendrocoelum lacteum* more *Asellus*. If the differences between the genera are measured as a proportion of total food eaten they fall into two groups, one with a small value of 10.4 percent (95 percent limits, 0 to 21.8 percent) and one with a larger value of 46.2 percent (95 percent limits, 31.4 to 61.0 percent) of the food. The former are not of prime importance from the viewpoint of coexistence, but the latter provide what have been termed "food refuges" and allow coexistence by reducing competition for food. To date, the identity of the food refuges of the two *Polycelis* species has not been discovered, but there is some evidence that they might be separated ecologically by feeding on different families of Oligochaeta.

Planaria torva is an interesting species in the present context, and its biology has been studied by Dr. A. D. Sefton. The species is not widespread in Britain (Ball, Reynoldson, and Warwick, 1969; Sefton, 1969), but its distribution does not fit that of a relict species. There seems to be a real concentration in the vicinity of ports or associated canals and a large number of populations occur in the Glasgow area. It has been suggested (Sefton, 1969) that *P. torva* is a recent introduction via the timber trade, possibly from Scandinavia, where it is an abundant and widespread species. There would seem to be good opportunities for cocoons to be deposited and survive on round logs. A similar idea has been proposed with regard to the recent discovery of *Dugesia polychroa* in Canada (Ball, 1969). The food niches of *Planaria torva* and *Dugesia polychroa*, taken from Sefton (1969) and Reynoldson and Davies (1970) respectively, are similar in respect of the main item, gastropods, as shown below, and they may be expected to compete severely.

Percentage of Meals Taken

	Gastropoda	*Asellus*	Oligochaeta (1)	(2)	Chironomidae	*Gammarus*	Tests
Planaria torva	68.6	21.0	4.7	. . . .	5.6	. . .	326
Dugesia polychroa	57.0	16.0	. . .	22.0	. . .	4.0	176

The reason for the two sets of data under Oligochaeta is due to the inability of the antiserum used for *P. torva* to detect *Lumbriculus variegatus*, a common aquatic worm. When both triclads occurred together, there was evidence that they had fed to some extent on different gastropod species. In this context it is significant that *P. torva* is at present almost restricted to very productive lakes with a varied gastropod fauna, and in these ecosystems ecological separation on a food basis might be sufficiently great to allow coexistence; in less productive lakes with a smaller and more restricted gastropod fauna any separation might not be large enough. If this hypothesis is valid, *Dugesia polychroa* would be expected to influence the future spread of *P. torva*.

Bdellocephala punctata is a challenging species from the viewpoint of ecological separation. While it occurs in less than 10 percent of lakes in northern Britain, it is able to inhabit both unproductive lakes such as Llyn Geirionydd and Mymbyr in Snowdonia with a calcium concentration of 1 to 2 mg/l and productive lakes like Johnston Loch (Lanarkshire, Scotland) and Comber Mere (Shropshire) with 57.5 and 48.4 mg/l of calcium respectively. There seems to be no characteristic pattern of distribution such as shown by the common species. *Bdellocephala punctata* is the largest of the triclads, reaching a length of 35 mm. It belongs to the same family as *Dendrocoelum lacteum*, and they show a similar feeding behavior in taking active prey much more commonly than the planariid species (Reynoldson and Young, 1963). A priori, it would be anticipated that *B. punctata* would compete most severely with *D. lacteum* although we know that these two species coexist in about half of the *B. punctata* habitats. In Comber Mere, where there is a relatively large *Bdellocephala* population, *Dendrocoelum* is uncharacteristically absent. Preliminary observations on the food of field specimens of *B. punctata* (Davies, unpublished data) show it to prey on *Asellus* (65 percent of total food), *Gammarus* (17 percent), and oligochaetes (22 percent), i.e., in much the same proportions as *D. lacteum*, but only 23 specimens were tested giving 24 positive reactions to the antisera. Since *B. punctata* is the largest species, there is the possibility of separation, especially from *D. lacteum*, by taking larger prey, but this would not explain its sparse distribution over a wide spectrum of lake types. It is also tempting to consider *Gammarus* as its food refuge, but their distributions do not coincide, as do those of *D. lacteum* and *Asellus* (Reynoldson and Young, 1966). Spatial separation does not seem likely either since Berg (1938) and Beauchamp (1932) reported *Bdellocephala* as occurring most abundantly in shallow water, a pattern which I can confirm. This problem will be solved only when a suitably large population is found to study. *Dugesia lugubris* (O. Schmidt) is a species which has been recognized definitively only recently (Reynoldson and Bellamy, 1970) in Britain. It is not widespread and its ecology remains to be studied, especially potential competition with *D. polychroa*.

It is appropriate to conclude by considering, in a preliminary way, the ecological separation of two American species in a Californian habitat. I am grateful to Miss Sandra M. Harris for permission to use some of her preliminary observations on *Dugesia dorotocephala* (Woodworth) and *D. tigrina* (Girard) in North Lake, Golden Gate Park, San Francisco, made while she studied in the Zoology Department at the University of California, Berkeley. These are two species on which Dr. Libbie Hyman worked extensively and successfully (e.g., 1925, 1931, 1939, 1941). *Dugesia tigrina* is one of the commonest species in the United States living in both streams and lakes, but it is absent from cold springs and spring-fed ponds. *Dugesia dorotocephala* is more localized and an inhabitant of springs, spring-fed marshes, cool unpolluted creeks, and spring-fed lakes (Hyman, 1951; Kenk, 1944); it is reported to show a positive rheotaxis (Chandler, 1966). The two species seem to be spatially separated, possibly by behavioral mechanisms. The situation in the productive North Lake is interesting in this context. An asexual population of *D. dorotocephala* inhabited the waterfall area and several meters of an inflowing stream, living on the undersurface of stones. It was rarely found in the lake, forming less than 1 percent of several hundreds of triclads collected. A sexual population of *D. tigrina* occupied the lake where it lived in the axils of water flag; there were no stony areas. In a short channel connecting stream and lake, with a slight flow, both species intermingled, but *D. tigrina* did not extend far into the stream (Fig. 2). While the general habitat of *D. tigrina* is typical, that of *D. dorotocephala* is not, since the stream connected North Lake with other lakes and was not spring-fed or cool; it was also mildly polluted. During May–June 1968, the temperature of the stream at 20°C was marginally higher than that of the lake, which was heavily shaded by trees. The causes of this ecological separation are intriguing, especially the restriction of *tigrina* to the lake, since the stream, on physicochemical grounds, would appear suitable for it. Preliminary studies of the food of these triclads was made from the viewpoint of food refuges. It was examined by three methods, searching squashes of field specimens for the remains of prey (Reynoldson and Young, 1963), using a serological technique (Young *et al.*, 1964; Davies, 1970; Reynoldson and Davies, 1970), and laboratory experiments with normal and damaged prey. Sampling from these two habitats showed that an amphipod, *Hyalella azteca* (Saussure), a snail, *Physa virgata* and oligochaetes were abundant in both; chironomids were common on the stones and presumably in the organic mud of the lake. Attention was confined to the first three diverse types of potential prey. Examining squashes provided information only on oligochaetes as food since chaetae were the only skeletal remains found. Of 93 *D. tigrina* and 90 *D. dorotocephala* examined, 22 percent and 23 percent respectively contained chaetae. Approximately 100 specimens of each species were squashed onto filter paper in the field and these were tested serologically by Dr. R. W. Davies in Bangor, North Wales, for positive reactions to the three

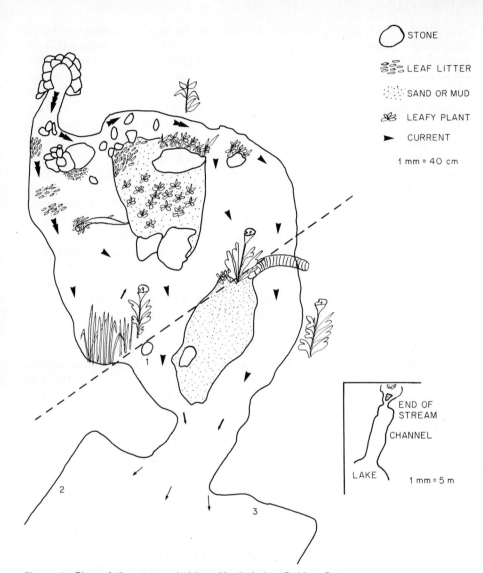

STONE

LEAF LITTER

SAND OR MUD

LEAFY PLANT

CURRENT

1 mm = 40 cm

END OF
STREAM

CHANNEL

LAKE   1 mm = 5 m

Figure 2 Plan of the stream habitat, North Lake, Golden Gate Park, San Francisco. Current speed is proportional to the size of the arrows, which also indicate direction of flow. Above the broken line all the triclads collected were *Dugesia dorotocephala*. At stations 1, 2, and 3 the following numbers of the two species were collected: station 1, 20 *Dugesia dorotocephala*, 9 *D. tigrina*; 2, 3 *Dugesia dorotocephala*, 28 *D. tigrina*; 3, 22 *Dugesia dorotocephala*, 1 *D. tigrina*. The inset plan shows the relation of stream, channel, and lake.

types of prey. It had been shown previously that each of the prey types reacted well with our antisera prepared against gastropods, oligochaetes, and amphipods for testing British triclads (Reynoldson and Davies, 1970). The results (Table 2) show a low proportion of total positive reactions, which may suggest that the triclads were finding food scarce or alternatively that an important item of food was not included in the antisera. The detailed results confirm that both species were feeding equally on oligochaetes. The higher proportion of feeding suggested by the search for chaetae is probably due to the retention of these beyond the period when a positive serological reaction is given. *Dugesia tigrina* seems to eat more gastropods and *D. dorotocephala* significantly more amphipods. The failure to find gastropod radulae in squashes of *D. tigrina* is interesting, for it clearly eats *Physa*; perhaps, in contrast to *D. polychroa*, it ingests only the softer parts.

When starved triclads were given access to undamaged *Physa* and *Hyalella*, neither fed on *Physa* whereas $^{28}/_{30}$ *Hyalella* were eaten by *D. dorotocephala* and only $^{3}/_{30}$ by *D. tigrina*. This certainly reflects the higher activity and greater speed of movement of *D. dorotocephala*, but it probably also involves a different basic response. Experiments were also made using damaged *Physa* and *Hyalella* by exposing them in the same bowl to the two triclad species separately. The experiments were repeated five times and the total period spent feeding on each prey by 100 triclads of each species was noted. *Dugesia tigrina* spent similar periods feeding on *Physa* and *Hyalella* ($P > 0.05$), but *D. dorotocephala* spent 1.7 times longer feeding on *Hyalella* ($P < 0.01$). The several experiments suggest that *Hyalella* is both more easily captured and a more attractive prey for *D. dorotocephala*; *D. tigrina* is not frequently stimulated to capture intact prey, as Libbie Hyman had noted (1925), but it is attracted equally to both prey types when they are damaged and exuding body fluids. This is a result in broad agreement with the serological data. The situation cannot be examined in terms of food refuges with the same relevance as for British triclads since the samples were necessarily obtained from different habitats, are smaller, and were not repeated at different seasons. However, there is some validity in doing so since the populations of *Hyalella* and *Physa* were not obviously disparate numerically in the stream and lake.

Table 2   Food of Field Specimens of *Dugesia tigrina* and *D. dorotocephala* Detected by a Serological Method

Food type	Dugesia tigrina		Dugesia dorotocephala	
	No. tested	No. positive	No. tested	No. positive
Oligochaetes	96	11	100	9
Physa virgata	96	19	100	11
Hyalella azteca	96	9	100	25

When this is done, a difference value of 36 percent for *Hyalella* in favor of *D. dorotocephala* is obtained and a corresponding value of 21 percent for *Physa* in favor of *D. tigrina*. These results suggest that *D. dorotocephala* may have a safer food refuge from triclad competition in the stream than *D. tigrina* has and may explain the inability of *D. tigrina* to inhabit the stream. The absence of *D. dorotocephala* from the lake cannot be explained in terms of interspecific competition and seems more likely to be an evolved behavioral response. The channel is most probably to be regarded as a boundary habitat supplied continuously from both populations. I am aware that there are many gaps in this argument, but if the ideas provoke further study on such lines, it will have served its purpose, in my view.

DISCUSSION

Species diversity is a central theme in ecology because it has implications for evolution, it influences production and energy flow in the ecosystem, and it is also concerned with population dynamics. Ecological separation is another way of looking at species diversity, and Hutchinson (1959) was one of the first of the present-day ecologists to examine quantitative aspects. MacArthur and his school (MacArthur, 1964; MacArthur and MacArthur, 1961; Mac-Arthur and Pianka, 1966; MacArthur and Wilson, 1967) have taken this further and according to them, spatial separation is the commonest strategy, a suggestion supported by this review of triclads. Whether such separation is the result of behavioral responses or maintained by competitive exclusion will have to be determined in each case. Opportunities for ecological separation are a function of the physicochemical complexity of the environment enhanced by the flora and, in the case of predators and parasites, by the diversity of the fauna. Other bases relate to size and type of food and temporal separation of several kinds. Total diversity attained by any taxon is, in theory, the product of these several parameters. The extent to which a particular taxon diversifies in the ecosystem will be a function of the inherent ability of the organisms to respond to selective pressures, the heterogeneity of the environment in space and time, the mode of life of the taxon, and the size of the organisms contained in it. In general, smaller organisms are likely to evolve greater diversity than larger organisms, parasites greater diversity than free-living animals; a nice recent example to illustrate this is the considerable diversity in a genus of small ectoparasitic mites (Lanciani, 1970). In the case of the British freshwater triclads, the restriction to one species in unproductive lakes and the presence of up to six species in productive lakes reflects the interaction of these processes within the framework of historical events. In contrast to such limited diversity we have the case of the large ancient lakes of Europe and Asia such as Ohrid (Stankovic, 1960) and Baikal (Kozhov, 1963), where 20 and 80 triclad species, respectively, are reported to coexist. Some of the Lake Baikal species reach a relatively huge size with a *Polycotylus* species reported as measuring 30 cm in length. One of the bases of such diversity is probably the corresponding diversity of prey organisms, but it would be a fascinating experience to investigate the details.

ACKNOWLEDGMENTS

I wish to thank the Natural Environment Research Council for financial support and Drs. J. F. Wright and A. D. Sefton for permission to use some of their research data. I am also grateful to the chairman of the Zoology Department, University of California, Berkeley, for the invitation to spend a year in the department as visiting professor.

BIBLIOGRAPHY

Anonymous. 1963. Annual Progress Report on Research and Technical work. Minist. of Agric., N. Ireland, 1963, 13–14.
Anonymous. 1964. Annual Progress Report on Research and Technical Work. Minist. of Agric., N. Ireland, 1964, 17–18.
Ball, I. R. 1969. *Dugesia lugubris* (Tricladida, Paludicola). A European Immigrant into North American Fresh Waters. *J. Fisheries Res. Board Can.*, **26**:221–228.
———, T. B. Reynoldson, and T. Warwick. 1969. The Taxonomy, Habitat and Distribution of the Freshwater Triclad *Planaria torva* (Platyhelminthes: Turbellaria) in Britain. *J. Zool. London*, **157**:99–123.
Beadle, L. C. 1934. Osmotic Regulation in *Gunda ulvae. J. Exptl. Biol.*, **11**:382–396.
Beauchamp, R. S. A. 1932. Some Ecological Factors and Their Influence on Competition between Stream and Lake-living Triclads. *J. Animal Ecol.*, **1**:175–190.
——— and P. Ullyott. 1932. Competitive Relationships between Certain Species of Freshwater Triclads. *J. Ecol.*, **20**:200–208.
Berg, K. 1938. Studies on the Bottom Animals of Esrom Lake. *Kgl. Danske Videnskab. Selskab Skrifter*, 8, 255 pp.
Carpenter, K. 1928. On the Distribution of Freshwater Turbellaria in the Aberystwyth District, with Special Reference to Two Ice-Age Relicts. *J. Ecol.*, **16**:105–122.
Chandler, C. 1966. Environmental Factors Affecting the Local Distribution and Abundance of Four Species of Stream-dwelling Triclads. *Invest. Indiana Lakes Streams*, 7(1):1–56.
Chodorowski, A. 1959. Ecological Differentiation of Turbellarians in Harz Lake. *Polsk. Arch. Hydrobiol.*, **6**:33–73.
———. 1960. Vertical Stratification of Turbellaria Species in Some Littoral Habitats of Harz Lake. *Polsk. Arch. Hydrobiol.*, **8**:153–163.
Dahm, A. G. 1949. *Phagocata* (= *Fonticola*) from South Sweden (Turbellaria Tri-

cladida, Paludicola) Taxonomical, Ecological and Chorological Studies. *Kgl. Fysiograf. Sällsk, Handl.*, N.F. **60**(7), 32 pp.

———. 1955. *Dugesia tigrina* (Girard) an American Immigrant into European Waters. *Verhandl. Int. Ver. Limnol.*, **12**:554–561.

———. 1958. *Taxonomy and Ecology of Five Species Groups in the Family* Planariidae (*Turbellaria Tricladida Paludicola*). Malmö: Nya Litografen. 241 pp.

Davies, R. W. 1970. The Production of Antisera for Detecting Specific Triclad Antigens in the Gut Contents of Predators. *Oikos*, **20**:248–260.

Gamble, F. W. 1893. Contributions to a Knowledge of British Marine Turbellaria. *Quart. J. Microscop. Sci.*, **34**:433–528.

Gislén, T. 1946. About the European Species of the Genus *Fonticola* with Some Notes concerning the Distribution and Ecology of *F. vitta. Biol. Jaarboek Dodonaea*, **13**:174–183.

Hartog, C. den. 1962a. De Nederlandse Platwormen (Tricladida). *Mededel. K.N.N. V.*, 42, 40 pp.

———. 1962b. The Distribution of the Land Planarian *Rhynchodemus terrestris* in the Netherlands. *Proc. Koninkl. Ned. Akad. Wetenschap.*, **65**(C):369–379.

———. 1968. Marine Triclads from the Plymouth Area. *J. Marine Biol. Assoc. U.K.*, **48**:209–223.

Hutchinson, G. E. 1959. Homage to Santa Rosalia, or Why Are There So Many Kinds of Animals. *Am. Naturalist*, **93**:145–159.

Hyman, L. H. 1925. The Reproductive System and Other Characters of *Planaria dorotocephala* Woodworth. *Trans. Am. Microscop. Soc.*, **44**:51–89.

———. 1931. Studies on Morphology, Taxonomy and Distribution of North American Triclad Turbellaria. IV. Recent European Revisions of the Triclads and Their Application to the American Forms, with a Key to the Latter and New Notes on Distribution. *Trans. Am. Microscop. Soc.*, **50**:316–335.

———. 1939. North American Triclad Turbellaria. IX. The Priority of *Dugesia* Girard 1850 over *Euplanaria* Hesse 1897 with Notes on American Species of *Dugesia. Trans. Am. Microscop. Soc.*, **58**:264–275.

———. 1941. Environmental Control of Sexual Reproduction. *Anat. Record*, **81**(4): Suppl., p. 108.

———. 1951. *The Invertebrates:* II. *Platyhelminthes and Rhynchocoela, the Acoelomate Bilateria.* New York: McGraw-Hill. 550 pp.

Jennings, J. B. 1959. Observations on the Nutrition of the Land Planarian *Orthodemus terrestris* (O. F. Müller). *Biol. Bull.*, **117**:119–124.

Kenk, R. 1944. The Fresh-water Triclads of Michigan. *Misc. Publ. Museum Zool. Univ. Mich.*, no. 60, 44 pp.

Kozhov, M. 1963. Lake Baikal and Its Life. Monogr. biol., 11, 344 pp. Ed. W. W. Weisbach and P. van Oye, The Hague: W. Junk.

Lanciani, C. A. 1970. Resource Partitioning in Species of the Water Mite Genus *Eylais. Ecology*, **51**:338–342.

Luther, A. 1961. Die Turbellarien Ostfennoskandiens II Tricladida. *Soc. Fauna Flora Fennica, Fauna Fennica* 11, 42 pp.

Macan, T. T., and M. Maudsley. 1969. Fauna of the Stony Substratum in Lakes in the English Lake District. *Verhandl. Int. Ver. Limnol.*, **17**:173–180.

MacArthur, R. H. 1964. Environmental Factors Affecting Species Diversity. *Am. Naturalist*, **98**:387–398.

—— and J. W. MacArthur. 1961. On Bird Species Diversity. *Ecology*, **42**:594–598.

—— and E. R. Pianka. 1966. On Optimal Use of a Patchy Environment. *Am. Naturalist*, **100**:603–609.

—— and E. O. Wilson. 1967. *The Theory of Island Biogeography*. Princeton, N.J.: Princeton University Press. 199 pp.

Naylor, E., and D. J. Slinn. 1958. Observations on the Ecology of Some Brackish-water Organisms in Pools at Scarlett Point, Isle of Man. *J. Animal Ecol.*, **27**:15–25.

Pantin, C. F. A. 1931a. The Adaptability of *Gunda ulvae* to Salinity I. The Environment. *J. Exptl. Biol.*, **8**:63–72.

——. 1931b. The Adaptability of *Gunda ulvae* to Salinity III. The Electrolyte Exchange. *J. Exptl. Biol.*, **8**:82–94.

——. 1944. Terrestrial Nemertines and Planarians in Britain. *Nature London*, **154**:80.

——. 1950. Locomotion in British Terrestrial Nemertines and Planarians: with a Discussion on the Identity of *Rhynchodemus bilineatus* (Mecznikow) in Britain and on the Name *Fasciola terrestris* O. F. Müller. *Proc. Linnean Soc. London*, **162**:23–37.

Pattee, E. 1965. Stenothermie et eurythermie. Les invertébrés d'eau douce et la variation journaliére de temperature. *Ann. Limnol.*, **1**:281–434.

Pianka, E. R. 1969. Sympatry of Desert Lizards. *Ecology*, **50**:1012–1030.

Reynoldson, T. B. 1953. Habitat of *Polycelis felina* (= *cornuta*) and *Crenobia alpina* in the British Isles. *Nature London*, **171**:660.

——. 1956. The Occurrence in Britain of the American Triclad *Dugesia tigrina* (Girard) and the Status of *D. gonocephala*. *Ann. Mag. Nat. Hist.*, S12, **9**:102–105.

——. 1958a. Observations on the Comparative Ecology of Lake-dwelling Triclads in Southern Sweden, Finland and Northern Britain. *Hydrobiologia*, **12**:129–141.

——. 1958b. The Quantitative Ecology of Lake-dwelling Triclads in Northern Britain. *Oikos*, **9**:94–138.

——. 1966. The Distribution and Abundance of Lake-dwelling Triclads—towards a Hypothesis. *Advances Ecol. Res.*, **3**:1–71.

—— and L. S. Bellamy. 1970. The Status of *Dugesia lugubris* and *D. polychroa* (Turbellaria, Tricladida) in Britain. *J. Zool. London*, **162**:157–177.

—— and R. W. Davies. 1970. Food Niche and Co-existence in Lake-dwelling Triclads. *J. Animal Ecol.*, **39**:599–617.

—— and J. O. Young. 1963. The Food of Four Species of Lake-dwelling Triclads. *J. Animal Ecol.*, **32**:175–191.

—— and ——. 1966. The Relationship between the Distribution of *Dendrocoelum lacteum* (Müll.) and *Asellus* in Britain and Fennoscandia. *Verhandl. Int. Ver. Limnol.*, **16**:1633–1639.

Schoener, T. W. 1965. The Evolution of Bill Size Differences among Sympatric Congeneric Species of Birds. *Evolution*, **19**:189–213.

——. 1968. The *Anolis* Lizards of Bimini: Resource Partitioning in a Complex Fauna. *Ecology*, **49**:704–726.

Sefton, A. D. 1969. "The Biology of *Planaria torva* (Müll.)." Ph.D. thesis, University of Wales. 149 pp.

Southern, R. 1936. Turbellaria of Ireland. *Proc. Roy. Irish Acad.*, **43B**:43–72.

Stankovic, S. 1960. *The Balkan Lake Ohrid and Its Living World*. Monogr. biol. 9, 357 pp. Ed. F. S. Bodenheimer and W. W. Weisbach, The Hague: W. Junk.

Wilhelmi, J. 1908. Seetricladen von Plymouth. *Zool. Anz.*, **38**:618–620.

Wright, J. F. 1968. "The Ecology of Stream-dwelling Triclads." Ph.D. thesis, University of Wales. 186 pp.

Young, J. O., I. G. Morris, and T. B. Reynoldson. 1964. A Serological Study of *Asellus* in the Diet of Lake-dwelling Triclads. *Arch. Hydrobiol.*, **60**:366–373.

Chapter 12

Salt-marsh Turbellaria

C. den Hartog
Rijksherbarium, Leiden, Netherlands

Interest in the salt-marsh Turbellaria is rather recent. The first special study on them was published by Ax (1960). This initial publication contained a list of 33 species, collected along the Baltic and North Sea coasts of Germany. In fact the paper was a taxonomic precursor of the ecological studies of Bilio (1963a, b, 1964, 1965, 1967a). Bilio did not restrict himself to the study of the German salt marshes, but made investigations also in the deltaic area of the rivers Rhine, Meuse, and Scheldt in the southwestern part of the Netherlands, and in southwestern Finland (Bilio, 1966). In most of the works of Bilio the stress is laid on the peculiar environment or on general tendencies in the ecological and distributional pattern of the aquatic fauna of the salt marshes; the presentation of the data is rather abstract where the fauna is concerned and with the exception of his thesis (Bilio, 1964, 1967a) hardly any information on the species themselves is given.

The general works "Die Turbellarien Ostfennoskandiens" by Luther (1960, 1962, 1963) and Karling (1963) contain considerable information on Baltic salt-marsh Turbellaria. What work I have done on salt-marsh Turbellaria has been included in treatises of taxonomical groups (den Hartog, 1963, 1964a, b, 1965, 1966a, b). My data were obtained from field work in the southwestern part of the Netherlands, supplemented by observations in the Dutch Wadden Zee, along the coast of Normandy, in the surroundings of Plymouth, and along the Firth of Forth.

As a result of the investigations by Bilio and myself the number of species now known to occur on northwestern European salt marshes has increased to 56 (Table 1). This number does not include some species as yet unidentified, e.g., the blind solenopharyngid recorded by Ax and Bilio and three other species found by myself.

THE SALT MARSH

A salt marsh is considered to be the high-littoral belt of a sheltered sandy or muddy shore covered by halophilous angiosperms. It is in fact the habitat in which the conflict between the sea on the one hand and the freshwater and the

Table 1   Turbellaria Found on Northwestern European Salt Marshes

Order Acoela:	
Mecynostomum auritum (Schultze)	B
Order Macrostomida:	
Macrostomum balticum Luther	B*
Macrostomum hamatum Luther	B
Macrostomum hystricinum Beklemischev	B
Macrostomum pusillum Ax	M
Macrostomum spirale Ax	B
Macrostomum tenuicauda Luther	S
Order Lecithoepitheliata:	
Archimonotresis limophila Meixner	M
Order Proseriata:	
Archilopsis unipunctata (Fabricius)	M
Coelogynopora schulzii Meixner	M†
Minona baltica Karling and Kinnander	B†
Monocelis fusca Oersted	M
Monocelis lineata (O. F. Müller)	M
Order Tricladida:	
Uteriporus vulgaris Bergendal	M†

Table 1 (*continued*)

Order Neorhabdocoela:
  Suborder Dalyellioidea:
    Baicalellia brevituba (Luther)   M
    Jensenia angulata (Jensen)   M
    Provortex balticus (Schultze)   M
    Provortex karlingi Ax   M
    Provortex pallidus Luther   B†
    Pseudograffilla arenicola Meixner   M
    Vejdovskya halileimonia Ax   S
    Vejdovskya ignava Ax   B
    Vejdovskya mesostyla ssp. *hemicycla* Ax   S
    Vejdovskya pellucida (Schultze)   B
  Suborder Typhloplanoida:
    Anthopharynx vaginatus Karling   B*
    Castrada subsalsa Luther   S
    Coronhelmis multispinosus Luther   B*
    Haloplanella minuta Luther   B
    Haloplanella obtusituba Luther   B
    Lutheriella diplostyla den Hartog   S
    Maehrenthalia dubia Ax   B
    Pratoplana salsa Ax   S
    Promesostoma caligulatum Ax   M
    Promesostoma marmoratum (Schultze)   M
    Proxenetes britannicus den Hartog   S
    Proxenetes cisorius den Hartog   S
    Proxenetes deltoides den Hartog   B†
    Proxenetes flabellifer Jensen   M
    Proxenetes karlingi Luther   B*
    Proxenetes minimus den Hartog   S
    Proxenetes monotubulus den Hartog   S
    Proxenetes pratensis Ax   S
    Proxenetes puccinellicola Ax   S
    Proxenetes trigonus Ax   M
    Proxenetes unidentatus den Hartog   S
    Ptychopera spinifera den Hartog   S
    Ptychopera tuberculata (von Graff)   S
    Ptychopera westbladi (Luther)   M
    Thalassoplanella collaris Luther   B
    Westbladiella obliquipharynx Luther   B*
  Suborder Kalyptorhynchia:
    Acrorhynchides robustus (Karling)   B
    Gyratrix hermaphroditus Ehrenb.   H
    Parautelga bilioi Karling   S
    Placorhynchus o. octaculeatus Karling   B
    Prognathorhynchus canaliculatus Karling   B
    Zonorhynchus salinus Karling   B?*

Legend of the symbols used: B = brackish-water species; H = holeuryhaline species; M = marine species; S = salt-marsh species; † = littoral species; * = species showing brackish-water submergence.

land on the other hand finds its reflection in the composition of the flora and fauna. Apart from marine and terrestrial organisms on the salt marsh, quite a number of species which are characteristic for conflict situations between sea and freshwater occur. These may be referred to as "brackish-water species." Further, there is a group of species which seems to be confined to the "littoral border environment" (den Hartog, 1968), which the salt marsh has in common with the high-littoral belt of the rocky shore (Table 2). Finally there is a group of species which is completely restricted to the salt-marsh habitat. The freshwater element is almost completely absent.

The flora and fauna of the lower-situated mud and sand flats are almost completely marine.

The vegetation of an ideal, accrescent salt marsh shows a distinct zonation pattern (Fig. 1), and this is in general similar in the whole temperate part of northwestern Europe (Beeftink, 1962, 1965; Corillion, 1953; Gillner, 1952, 1960), on tidal coasts as well as on coasts where the fluctuations of the water level are induced by meteorological conditions, e.g., the Baltic coast. The levels mentioned in this paper relate solely to the tidal coasts.

Table 2   Species Common to the High-littoral Belt of Rocky Shores and Salt Marshes

Turbellaria:
> *Coelogynopora schulzii* Meixner
> *Uteriporus vulgaris* Bergendal

Isopoda:
> *Ligia oceanica* (L.)
> *Sphaeroma rugicauda* Leach

Amphipoda:
> *Orchestia gammarella* (Pall.)
> *Orchestia mediterranea* A. Costa

Gastropoda:
> *Assiminea grayana* Fleming
> *Leucophytia bidentata* (Mont.)
> *Littorina saxatilis* (Olivi)

Chlorophyceae:
> *Blidingia marginata* (J. Ag.) Dangeard
> *Blidingia minima* (Näg. ex Kütz.) Kylin
> *Rhizoclonium riparium* (Roth) Harv.
> *Ulothrix pseudoflacca* Wille
> *Ulothrix subflaccida* Wille

Rhodophyceae:
> *Bostrychia scorpioides* (Huds.) Mont.
> *Catenella repens* (Lightf.) Batt.

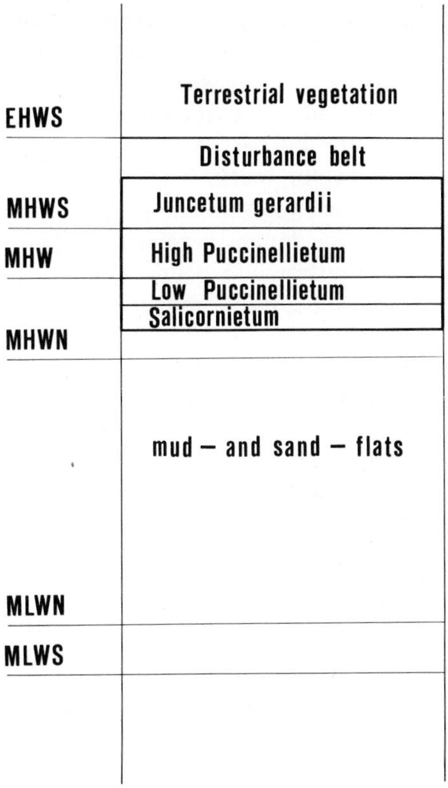

EHWS	Terrestrial vegetation
	Disturbance belt
MHWS	Juncetum gerardii
MHW	High Puccinellietum
	Low Puccinellietum
MHWN	Salicornietum
	mud — and sand — flats
MLWN	
MLWS	

Figure 1 Scheme of the zonation pattern in the intertidal belt along a sheltered shore in temperate northwestern Europe. The belt which represents the salt marsh is delineated by a thick line. MLWS, mean low water at spring tides; MLWN, mean low water at neap tides; MHWN, mean high water at neap tides; MHW, mean high-water mark; MHWS, mean high water at spring tides; EHWS, extremely high high water at spring tides.

The lowest vegetation belt is dominated by annual glassworts of the genus *Salicornia* (*Salicornietum europaeae*). The bottom is either bare or covered with pure or mixed algal growths consisting of diatoms, Cyanophyceae and *Vaucheria* species. The fauna of this vegetation belt is not very characteristic, but is mainly an upshore extension of that of the mud and sand flats. Most species are members of the association of still-water biotopes or the association of detritus-rich fine sand, as defined by Ax (1951) in his study on the Turbellaria of Kiel Bay. A few representatives of the brackish-water fauna and also some typical salt-marsh species may sometimes be found. Turbellaria found in this belt but not in the higher-situated belts have not been incorporated in Table 1. As is indicated in Fig. 1, the lower border of the *Salicornia* belt is situated above the level of mean high-water neap tides, but its upper border is below the level of mean high water.

The *Salicornietum* is followed in the zonation as well as in the succession by a belt dominated by the perennial grass *Puccinellia maritima* (Huds.)

Parl. (*Puccinellietum maritimae*); this belt extends to the level of mean high-water spring tides, where it is abruptly replaced by plant communities dominated by the grass *Festuca rubra* L. or the rush *Juncus gerardii* Loisl.

The *Puccinellia* community can be subdivided into two entities, according to the floristic and faunistic composition; the border between these entities coincides more or less with mean high-water mark. The lower *Puccinellietum* consists mainly of *Puccinellia* cushions, a few annual plants, and a well-developed growth of filamentous algae. In the fauna there are still quite a number of marine euryhaline species, but the brackish-water element as well as the typical salt-marsh element has increased considerably in number of species and of specimens.

The higher *Puccinellietum* consists of a closed mat of *Puccinellia maritima*, and this species is accompanied not only by some annuals but also by quite a number of perennial herbs (e.g., *Plantago maritima* L., *Triglochin maritima* L., *Limonium vulgare* Mill., etc.) and even by a dwarf shrub, *Halimione portulacoides* (L.) Aellen, which forms a community of its own.

This ideal zonation pattern is generally obscured by other factors. When a salt marsh is dissected by creeks, a system of levees and depressions is formed; and this has a profound influence on the hydrology of the salt marsh. This is reflected in the composition of flora and fauna. Other factors which modify the zonation pattern are the texture of the substratum, the rate of

Figure 2 *Salicornia* belt along the estuary of the river Avon near Aveton Gifford (Devon, England). Locality of *Macrostomum spirale*, *M. hystricinum*, and *Monocelis lineata*. July 1970.

Figure 3 Vegetation dominated by *Halimione portulacoides* along the estuary of the river Avon near Aveton Gifford (Devon, England). Habitat of *Proxenetes deltoides, Vejdovskya halileimonia,* and *Ptychopera westbladi.* July 1970.

sedimentation, the occurrence of subterranean freshwater tracks, and grazing by cattle.

A very important disturbance of the zonation pattern has been brought about by the settlement of the cordgrass *Spartina townsendii* H. and J. Groves (a male-sterile hybrid between the native *S. maritima* (Curt.) Fernald and the American *S. alterniflora* Loisl., and including the fertile polyploid *S. anglica* Hubbard, which has arisen from the hybrid) on many salt marshes. Since its first discovery in 1870 this new *Spartina* has become extremely common along the European Atlantic coast, partly by natural dispersal and partly by human interference. As a result of its very dense growth it has in very many places completely superseded the plant communities of the lower salt marsh, and converted the belt between the level of mean high water of neap tides and mean high-water mark into a monotonous *Spartina* marsh. The bottom of these *Spartina* marshes is poor in animals including Turbellaria.

The level of mean high water of spring tides is an important ecological border (Bilio, 1965) not only because of the sudden change in the vegetation, but also because the marine element disappears completely above this line and the brackish-water species as well as the salt-marsh species show an enormous quantitative and qualitative decrease. The vegetation above this level is dominated by *Festuca rubra* and *Juncus gerardii*; the most conspicuous companion species are *Artemisia maritima* L., *Armeria maritima*

Figure 4 General view of the Balgzand salt marsh near Den Helder (the Netherlands). June 1970.

(Mill.) Willd., and *Glaux maritima* L. (*Artemisietum maritimae*; *Juncetum gerardii*).

The population density of animals in the *Festuca rubra* and *Juncus gerardii* communities is usually extremely low. Often samples of about 200 cm³ of ground give a yield of one or two flatworms, and sometimes even none. The upper border of the *Festuca rubra* community is indistinct; this vegetation gradually changes into a vegetation of salt-tolerant, nonhalophilous plants. The latter vegetation is characterized by some "disturbance indicators," e.g., *Oenanthe lachenalii* Gmel., *Apium graveolens* L., *Bupleurum tenuissimum* L., *Elytrigia pungens* (Pers.) Tutin, *Carex extensa* Good., or *Blysmus rufus* (Huds.) Link, depending on ecological circumstances. The disturbance element is brought about by the fact that this vegetation is flooded at irregular intervals, usually for short periods in the winter season only, and sometimes comes under the influence of slightly saline groundwater. The width of this disturbance belt is highly variable, depending on the relief. Where there is a slope, the belt may be a few decimeters wide; on the extensive beaches along the south coasts of the Frisian Islands (the Netherlands) this belt may extend for a mile or more. Although the turbellarian fauna of the disturbance belt has not yet been investigated systematically, some preliminary samples appeared to contain a still-unidentified species of the genus *Macrostomum*.

THE TURBELLARIA OF THE SALT MARSH

The Turbellaria occurring on the salt marshes can be arranged into four groups.

1 The euryhaline marine group consisting of 17 species (i.e., 30.3 percent)
2 The brackish-water group consisting of 22 species (i.e., 39.3 percent)
3 The typical salt-marsh group consisting of 16 species (i.e., 28.6 percent)
4 The holeuryhaline group consisting of 1 species (i.e., 1.8 percent)

1. The euryhaline marine species are those species which have their main distribution in the marine environment but are sufficiently tolerant with respect to salinity fluctuations to penetrate into less stable brackish environ-

Figure 5 Salt-marsh vegetation on the Balgzand near Den Helder (the Netherlands). Along the creeklet *Salicornia* species form a narrow belt, at a slightly higher level the grass *Puccinellia maritima* dominates. A rich fauna mainly consisting of *Proxenetes puccinellicola, P. deltoides, P. minimus, Ptychopera westbladi*, and *Macrostomum balticum* was obtained from the latter vegetation. June 1970.

ments. Most of these species are inhabitants of the lower-situated sand and mud flats. Some of them are only casuals (*Irrgäste*) in the salt-marsh habitat and have been found there once or a few times, e.g., *Promesostoma marmoratum, P. caligulatum, Proxenetes flabellifer, P. trigonus, Provortex balticus,* and *Macrostomum pusillum.* Among these *Promesostoma marmoratum* and *Provortex balticus* are often extremely abundant at lower levels and in salt-marsh creeks and ponds. Species fairly frequent in the salt-marsh habitat, and which must be regarded as permanent or seasonal inhabitants of at least the lower part of the *Puccinellietum* are *Monocelis lineata, M. fusca, Archilopsis unipunctata, Archimonotresis limophila, Provortex karlingi, Baicalellia brevituba, Pseudograffilla arenicola,* and *Ptychopera westbladi.*

However, there are two species, *Uteriporus vulgaris* and *Coelogynopora schulzii,* which are not invaders from lower levels. They seem to be restricted to the upper part of the eulittoral and are common on salt marshes and under stones of rocky shores. It appears to be an ecological requirement for *Co-*

Figure 6 Vegetation pattern on the Balgzand salt marsh near Den Helder (the Netherlands). The depression is surrounded by a narrow belt of *Salicornia* sp. and a wide belt of *Puccinellia maritima* with much *Salicornia.* On the slight elevations *Festuca rubra* dominates (light patches). September 1970.

elogynopora that such stones be embedded in a sediment, but *Uteriporus* is found also in the spaces between stones and boulders on true rock. Both species penetrate into estuaries where they just cross the average isohaline of 10‰ Cl' at high tide. Along the shores of the Baltic *Uteriporus* reaches more or less the same isohaline, but *Coelogynopora* changes its habitat and with decreasing salinity it becomes an inhabitant of coastal groundwater. This species possibly has to be classified as a brackish-water species (e.g., see Bilio, 1964). *Jensenia angulata* has occasionally been found in the Netherlands together with the two above-mentioned species, but I have no further data on it; it may also belong to the littoral group.

2. The "brackish-water species" are those species which have their main distribution in unstable conflict situations between sea and freshwater. They are found in brackish pools permanently or temporarily cut off from the sea, in lagoons, in estuaries, in coastal groundwater, and on salt marshes and they are generally absent from true marine habitats. Several of these species are quite common on salt marshes, e.g., *Mecynostomum auritum, Macrostomum balticum, M. spirale, Proxenetes deltoides, P. karlingi,* and *Maehrenthalia dubia.*

Although most of the brackish-water Turbellaria that have been found on the salt marshes have a wide distribution and can be found in all kinds of habitats, there are two groups of species which are more restricted in their occurrence. The first group consists of three species which are confined to the eulittoral belt, viz., *Minona baltica, Provortex pallidus,* and *Proxenetes deltoides.* The first two of these species have been found mainly on salt marshes, but also repeatedly in coastal groundwater. The third species is certainly most abundant on salt marshes, but it is too frequent and too numerous on the higher-situated sand and mud flats to be regarded as a true salt-marsh species; moreover, this species has also been found in blocked brackish-water ponds. The second group consists of *Macrostomum balticum, Anthopharynx vaginatus, Coronhelmis multispinosus, Proxenetes karlingi,* and *Westbladiella obliquipharynx.* These species exhibit "brackish-water submergence." Along the Atlantic coast they are strictly bound to the high eulittoral, where they are most abundant on the salt marshes or are completely restricted to them. Along the Baltic shores they descend to considerable depths, in proportion to the salinity decrease. *Anthopharynx vaginatus* is a true salt-marsh species along the Atlantic coast, but in the southwestern part of Finland it has been found even at a depth of 20 m. *Zonorhynchus salinus* may also belong to this group, but it has been recorded too infrequently to be sure of this.

All brackish-water species found on the salt marshes are of marine origin.

3. There are 16 turbellarian species which as yet have been found only on salt marshes or very rarely on the adjacent sand and mud flats and in no

other habitats. As most of these species have quite conspicuous characteristics, it is unlikely that they would have been overlooked if they had occurred in other habitats. The fact that they have been found repeatedly, often in large numbers, in salt-marsh habitats proves in my opinion that they are exclusive to them.

Two of these 16 species have been found only once. These are *Vejdovskya mesostyla* ssp. *hemicycla*, which was found in a *Festuca rubra* vegetation along the German Baltic coast, and *Proxenetes monotubulus*, which was found in a *Puccinellia maritima* vegetation in the southwestern part of the Netherlands. They will not be considered further.

Of the other 14 species, it is possible to give only some general indications. Only three species extend into the *Festuca rubra* and *Juncus gerardii* vegetations, viz., *Proxenetes unidentatus*, *Lutheriella diplostyla*, and *Vejdovskya halileimonia*. Along marine coasts the first-mentioned species is restricted to this belt, but on salt marshes in the low-salinity section of estuaries it inhabits the *Puccinellietum maritimae*. *Vejdovskya halileimonia* and *Proxenetes cisorius* are most abundant and almost confined to the higher parts of the *Puccinellietum* and to the *Halimione portulacoides* association. *Parautelga bilioi* is restricted to muddy substrata of the lower *Puccinellietum*. In contrast, *Castrada subsalsa* occurs in sandy places, where there is a subterranean groundwater flow.

The other species are all confined to the *Puccinellietum*, or equivalent vegetations in the low-salinity section of the estuaries, but their detailed distribution within these communities is not at all clear, partly owing to the lack of knowledge regarding their ecological demands. There also seems to be some antagonism between some of the species. *Proxenetes puccinellicola*, which is most numerous in the lower *Puccinellietum*, has been found regularly in depressions in the high *Puccinellietum* and also in the *Halimione* vegetation. It is absent from the latter two niches when *Proxenetes cisorius* is present. These two species, which are quite common, exclude each other almost completely. So far I have only found a few specimens of these two species in one sample.

If the true salt-marsh Turbellaria are arranged according to their distribution on the marshes along the banks of estuaries, three groups of species can be recognized. Most of them are restricted to the euhaline and the polyhaline section; i.e., they do not pass the average isohaline of 10‰ Cl' at high tide (den Hartog, 1971). This group consists of *Proxenetes britannicus*, *P. cisorius*, *P. puccinellicola*, *P. pratensis*, *Vejdovskya halileimonia*, *Pratoplana salsa*, *Parautelga bilioi*, and *Castrada subsalsa*. The first three of these species are restricted to the Atlantic coast; the others have been recorded also from Kiel Bay in the Baltic, except for *Castrada subsalsa*, which is known only from the Dutch and the Finnish salt marshes. The second group consists of four very

euryhaline species, which have been found in the euhalinicum, the polyhalini-
cum, and the mesohalinicum and which do not pass the average isohaline of
2‰ Cl' at high tide. These species are *Ptychopera tuberculata*, *Proxenetes
minimus*, *P. unidentatus*, and *Lutheriella diplostyla*. Of these only *P. unidentatus* has been recorded from the Baltic, where it was found in southwestern
Finland. The third group consists of two species which have been found only
in the mesohaline and oligohaline sections, thus in the section enclosed by
the average isohalines of 0.3 and 10‰ Cl' at high tide; these species are
Ptychopera spinifera and *Macrostomum tenuicauda*, both also known from the
southwestern coast of Finland. *Ptychopera spinifera* invades even the marginal
freshwater section, where salinity shows seasonal increases but the average
salinity is between 0.1 and 0.3‰ Cl' at high tide (den Hartog, 1966b; 1971).

The typical salt-marsh species are all of marine origin except one. *Castrada subsalsa* is so far the only salt-tolerant species of the genus *Castrada*
which is represented by many species in freshwater. It is remarkable that three
genera, *Parautelga*, *Pratoplana*, and *Lutheriella* are confined to the salt-
marsh habitat.

4. The holeuryhaline species are those which inhabit marine, freshwater,
and brackish-water habitats and seem to be quite indifferent to the salinity
factor. The only species belonging to this group is *Gyratrix hermaphroditus*.
Whether this species is really so indifferent is still an open question, since
Reuter (1961) distinguished three races in Finland on the grounds of cytogenetic characters.

ECOLOGICAL FACTORS AND TURBELLARIAN COENOSES

Although it is at present well known that the salt-marsh Turbellaria are not
evenly spread over the salt marshes, their detailed distribution within this
habitat is still not understood. I have mentioned that the ideal zonation pattern
is obscured by many factors. Seen in this light, it is easy to understand why
Beeftink (1965) was able, within the *Puccinellietum maritimae*, to distinguish
three subassociations and within these eight finer subdivisions (variants,
phases, and facies) each with its own ecological peculiarities. Therefore, I
suggest that future samples from the salt marsh for turbellarian studies should
be taken with closer regard for the finer subdivisions of the plant communities.

The relationship between the plant communities and the Turbellaria is
indirect, of course, but it cannot be denied that the ecological factors which
determine the composition of the plant communities are for the greater part
of importance to the Turbellaria as well. The plants themselves exercise an
important influence on the habitat by stabilizing the substratum, by the production of litter, and by protecting the substratum against direct radiation and
desiccation.

Figure 7 A mixed vegetation of *Puccinellia maritima* and *Salicornia* species in a depression of the Balgzand salt marsh near Den Helder (the Netherlands). Habitat of *Proxenetes britannicus.* September 1970.

The most important factors which determine the composition of the turbellarian coenoses are substratum, hydrology, and salinity.

The substratum of salt marshes may vary from coarse sand via fine sand to mud; it may be humous and can be even peaty. The texture of the substratum determines to a high degree its water-holding capacity and its permeability. Further, the rate of silting is of importance. On many salt marshes along the Dutch and German Wadden Zee the sedimentation process is being accelerated by the construction of low dams in order to extend the land surface. The vegetation of these artificial salt marshes is poor, as is the fauna, including the Turbellaria (Bilio, 1966).

The hydrology of the salt marshes is of paramount importance for the Turbellaria. On a salt marsh where the fluctuations of the water level are dependent on wind force and wind direction (as along the Baltic), hydrological circumstances are quite different from those on a tidal salt marsh. Further, there is quite a difference in hydrology between a salt-marsh depression in which the water stagnates after high tide and a marsh where the water recedes with the tide. The tidal difference itself also plays a part; for example, a marsh where the daily tidal difference exceeds 10 m, as in the Bay of Mont-Saint-Michel along the French coast of the Channel, the draining is much more intensive than on the marshes of the Dutch Wadden Zee, where the tidal difference varies between 1.3 and 2.5 m.

The salinity of the substratum is a very unstable factor; it is dependent on the rate of flooding, the salinity of the floodwater, meteorological conditions (precipitation, evaporation), and the lateral freshwater runoff. Generally the average salinity decreases in upshore direction, but the rate of fluctuation increases to above the level of mean high water of spring tides and then decreases rapidly (see also Bilio, 1967a). The influence of freshwater runoff especially has to be stressed here. Many marshes along the Dutch coast are separated from the mainland by a dike, which is usually bordered by a ditch on the land side. This means that the dike functions as an artificial watershed, and the freshwater drained off from the salt marsh is the freshwater originating from precipitation on the salt marsh itself. Where such a break in the freshwater effluence does not exist—e.g., in salt marshes bordered by dunes or in a rocky

Figure 8 Typical salt-marsh vegetation with *Limonium vulgare*, *Halimione portulacoides*, *Triglochin maritima*, *Suaeda maritima*, and *Spartina townsendii* near Zijpe (Schouwen-Duiveland, the Netherlands). August 1962.

area—the lateral freshwater runoff draining via the salt marshes is considerably more, and this is reflected in the vegetation by the occurrence of *Juncus maritimus* Lamk. In such seepage areas *Coronhelmis multispinosus* and *Castrada subsalsa* are found.

The quality of the freshwater also influences the composition of the salt-marsh fauna. The turbellarian fauna on the salt marshes along the low-salinity sections of the estuaries in the Plymouth area is very poor, and this must be ascribed to the fact that these marshes are flooded in the winter half year by soft water originating from the adjacent moors. The salt marshes in the low-salinity sections of the rivers Rhine and Meuse are often flooded by hard fresh river water in winter, but here the turbellarian fauna is rich (den Hartog, 1970). Some euryhaline Turbellaria penetrate here even into the marginal freshwater section, e.g., *Ptychopera spinifera* and *Provortex pallidus*, and these live there together with freshwater species such as *Prorhynchus stagnalis* Schultze, *Macrostomum rostratum* (Papi), *M. distinguendum* (Papi), *M. finlandense* (Ferguson), and various *Castrada* species in reed swamps with marsh marigolds (*Caltha palustris* L.).

From the foregoing it is obvious that the salt-marsh habitat is differ-

Figure 9   The salt marsh of the Mokbaai (Texel, the Netherlands). Subterranean freshwater tracks are marked by dense growths of *Juncus maritimus*. June 1970.

entiated into vegetation belts, and that each belt comprises a great number of niches, each of which may have its own turbellarian coenoses, i.e., its special set of turbellarian species. However, it is premature to distinguish such turbellarian coenoses on the basis of present knowledge. From the many samples I have investigated so far, only one combination of species has been found that may represent such a coenosis. In all other cases it has appeared impossible to recognize any grouping. It has not even been established whether the large numbers of some species (e.g., *Proxenetes puccinellicola*, *P. britannicus*, *P. deltoides*, *Macrostomum spirale*, *M. balticum*, *Monocelis lineata*) sometimes found in one sample are due to the fact that they live in swarms or that they are really so numerous in the niche which was sampled. In other words, it is not known whether the picture of the turbellarian fauna obtained from the samples is reliable or not.

Another reason why it is premature to distinguish turbellarian coenoses is that little is known of the species on which such coenoses would be based other than their names and morphological features. Hardly anything is known about their way of life, their activity rhythm, the duration of their lifetime, their rate of reproduction, their means for surviving unfavorable periods, etc. Only about their food is something known, thanks to a paper by Bilio (1967b).

In order to obtain a better picture of the distribution of the turbellarian fauna in the salt-marsh habitat, very detailed sampling is needed on selected, relatively simple salt marshes. Samples should be taken at all seasons of the year. Apart from this fieldwork, the life cycle and the preference for, and tolerance to, various ecological factors of the salt-marsh Turbellaria should be studied by means of cultures in the laboratory. In my opinion this is the only way to arrive at a well-based coenological subdivision of the salt-marsh habitat. It is also the only way to find out why the typical salt-marsh Turbellaria are restricted to this habitat.

BIBLIOGRAPHY

Ax, P. 1951. Die Turbellarien des Eulittorals der Kieler Bucht. *Zool. Jahrb.* (*Syst.*), **80**:277-378.

——. 1960. Turbellarien aus salzdurchtränkten Wiesenböden der deutschen Meeresküsten, *Z. Wiss. Zool.*, **163**(1/2):210-235.

Beeftink, W. G. 1962. Conspectus of the Phanerogamic Salt Plant Communities in the Netherlands. *Biol. Jaarboek Dodonaea*, **30**:325-362.

——. 1965. De zoutvegetatie van ZW-Nederland beschouwd in Europees verband. *Mededel. Landbouwhogeschool Wageningen*, **65**(1):1-167.

Bilio, M. 1963a. Die Zonierung der aquatischen Bodenfauna in den Küstensalzwiesen Schleswig-Holsteins. *Zool. Anz.*, **171**(5-8):328-337.

——. 1963b. Die biozönotische Stellung der Salzwiesen unter den Strandbiotopen. *Zool. Anz. Suppl.*, **27**:417-425.

―――. 1964. Die aquatische Bodenfauna von Salzwiesen der Nord- und Ostsee. I. Biotop und ökologische Faunenanalyse: Turbellaria. *Intern. Rev. Ges. Hydrobiol.*, **49**(4):509–562.

―――. 1965. Die Verteilung der aquatischen Bodenfauna und die Gliederung der Vegetation im Strandbereich der deutschen Nord- und Ostseeküste. *Botanica Gothoburg*, **3**:25–42.

―――. 1966. Charakteristische Unterschiede in der Besiedlung finnischer, deutscher und holländischer Küstensalzwiesen durch Turbellarien. *Veröffentl. Inst. Meeresforsch. Bremerhaven* Sonderband, **2**:305–318.

―――. 1967a. Die aquatische Bodenfauna von Salzwiesen der Nord- und Ostsee. III. Die Biotopeinflüsse auf die Faunenverteilung. *Intern. Rev. Ges. Hydrobiol.*, **52**(4):487–533.

―――. 1967b. Nahrungsbeziehungen der Turbellarien in Küstensalzwiesen. *Helgoländer Wiss. Meeresuntersuch.*, **15**:602–621.

Corillion, R. 1953. Les halipèdes du Nord de la Bretagne. *Rev. Gén. Bot.*, **60**:609–658, 707–775.

Gillner, V. 1952. Die Gürtelung der Strandwiesen und der Wasserstandswechsel an der Westküste Schwedens. *Svensk Botan. Tidskr.*, **46**:393–428.

―――. 1960. Vegetations- und Standortsuntersuchungen in den Strandwiesen der schwedischen Westküste. *Acta Phytogeograf. Suec.*, **43**:1–198.

Hartog, C. den. 1963. The distribution of the marine Triclad *Uteriporus vulgaris* in the Netherlands. *Proc. Koninkl. Ned. Akad. Wetenschap.*, **C66**(2):196–204.

―――. 1964a. Proseriate Flatworms from the Deltaic Area of the Rivers Rhine, Meuse and Scheldt I–II. *Proc. Koninkl. Ned. Akad. Wetenschap.*, **C67**(1): 10–34.

―――. 1964b. A Preliminary Revision of the Proxenetes Group (Trigonostomidae, Turbellaria) I–III. *Proc. Koninkl. Ned. Akad. Wetenschap.*, **C67**(5):371–407.

―――. 1965. A Preliminary Revision of the Proxenetes Group (Trigonostomidae, Turbellaria) IV–V. *Proc. Koninkl. Ned. Akad. Wetenschap.*, **C68**(2):98–120.

―――. 1966a. A Preliminary Revision of the Proxenetes Group (Trigonostomidae, Turbellaria) VI–X. *Proc. Koninkl. Ned. Akad. Wetenschap.*, **C69**(2):97–163.

―――. 1966b. A Preliminary Revision of the Proxenetes Group (Trigonostomidae, Turbellaria) Supplement. *Proc. Koninkl. Ned. Akad. Wetenschap.*, **C69**(5):557–570.

―――. 1968. The Littoral Environment of Rocky Shores as a Border between the Sea and the Land and between the Sea and the Fresh Water. *Blumea*, **16**(2):374–393.

―――. 1970. Some Aspects of Brackish-water Biology. *Comment. Biol. Soc. Sci. Fenn.*, **31**(9):1–15.

―――. 1971. The Border Environment between the Sea and the Fresh Water, with Special Reference to the Estuary. *Vie et Milieu, Suppl.*, **22**:739–751.

Karling, T. G. 1963. Die Turbellarien Ostfennoskandiens V. Neorhabdocoela 3. Kalyptorhynchia. *Soc. F. Fl. Fenn., F.F.*, **17**:1–59.

Luther, A. 1960. Die Turbellarien Ostfennoskandiens I. Acoela, Catenulida, Macrostomida, Lecithoepitheliata, Prolecithophora, und Proseriata. *Soc. F. Fl. Fenn., F.F.*, **7**:1–155.

————. 1962. Die Turbellarien Ostfennoskandiens III. Neorhabdocoela 1. Dalyellioideae, Typhloplanoida: Byrsophlebidae und Trigonostomidae. *Soc. F. Fl. Fenn., F.F.*, **12**:1–71.

————. 1963. Die Turbellarien Ostfennoskandiens IV. Neorhabdocoela 2. Typhloplanoida: Typhloplanidae. Solenopharyngidae und Carcharodopharyngidae. *Soc. F. Fl. Fenn., F.F.*, **16**:1–163.

Reuter, M. 1961. Untersuchungen über Rassenbildung bei *Gyratrix hermaphroditus* (Turbellaria Neorhabdocoela). *Acta Zool. Fenn.*, **100**:1–32.

Chapter 13

The Turbellarian Fauna of the Romanian Littoral Waters of the Black Sea and Its Annexes

Valeria Mack-Fira
Faculty of Biology (Invertebrate Zoology),
University of Bucharest, Romania

Romania is situated at the crossroad of the faunal population exchange between the north and the south, the east and the west. Owing to its paleogeographical history and to its relief of different shapes harmoniously combined with the most various types of waters, under favorable climatic conditions, this country is a true natural reservation comprising a fauna of unexpected heterogeneous origin and age. In this area endemic populations meet northern and Siberian elements, Central European, Atlantic-Mediterranean, and Indo-Pacific species, the Ponto-Caspian and glacial relicts, and Palaearctic, Holarctic, and cosmopolitan species.

Such a mosaic structure of the Romanian fauna, demonstrated in many animal groups, may be noted also in turbellarians, which are inhabitants of the inland waters or of the various biotopes in the Black Sea and its littoral annexes (Table 1).

No comprehensive study on the Romanian seashore turbellarians of the

TABLE 1 – THE SYNOPSES OF THE STUDIED SPECIES

Nr.	The species	Sand	Stone	Littoral basins	The brackish area	The freshwater area	The brackish area	The freshwater area	Razelmul Mare Lake (Jurilovca)	Golovitza Lake	Portitza	Sinoë Lake	Tuzla-Duingi Lake	Slut-Ghiol Lake	Tasaul Lake	Agigea Lake	Mangalia Lake	Techirghiol Lake	Permanent swamps	Sources	Zoogeography	Distribution in Rumania
		14-17‰	9.9‰	3-34‰			2.5-4.5‰		17-20‰			15‰	32‰	17‰	80‰							
1	Oligochoerus sp.				•	•															Ponto-caspian relict	
2	Convoluta convoluta (Abildg. 1806)	•																			Atlantic-medit. immigrant	
3	Convoluta albomaculata (Per. 1892)	•																			endemic	
4	Mecynostomum auritum (Schultze 1851)												•								Atlantic-medit. immigrant	
5	Mecynostomum arenarium Ax 1959	•																			Atlantic-medit. immigrant	
6	Archaphanostoma agile (Jensen 1878)	•																			Atlantic-medit. immigrant	
7	Stenostoma leucops (Ant Dugès 1828)				•	•	•	•	•	•	•		•	•							holarctic	in all inner freshwater basins
8	Stenostoma unicolor O. Schmidt 1848				•	•	•	•	•	•			•	•							palaearctic	in all inner freshwater basins
9	Catenula lemnae Ant Dugès 1832						•	•	•	•			•	•							cosmopolitan	Rumanian Plain, alpine Lakes
10	Macrostomum hystricinum Bekl. 1951					•		•	•	•											European	
11	Macrostomum romanicum Mack-Fîră 1968														•						Mediterranean	
12	Macrostomum distinguendum Papi 1951				•	•		•													European	Transylvania and Danube swamps
13	Macrostomum clavistylum Bekl. 1951											•									Ponto-caspian relict	
14	Macrostomum rubrocinctum Ax 1951													•							Northern immigrant	
15	Macrostomum ventriflavum Per. 1892	•																			endemic	
16	Macrostomum peteraxi Mack-Fîră 1971	•																			endemic	
17	Microstomum lineare (Müller 1774)					•	•		•			•									palaeartic	in all inner freshwater basins
18	Plagiostomum ponticum Per. 1892	•																			Atlantic-medit. immigrant	
19	Plagiostomum lemani (Plessis 1874)					•			•												European	
20	Pseudostomum klostermanni (v. Graff 1874)	•																			Atlantic-medit. immigrant	
21	Allostomum pallidum Beneden 1861	•	•																		Atlantic-medit. immigrant	
22	Allostomum catinosum (Bekl. 1927)	•																			Mediterranean	
23	Monocelis lineata (Müller 1774)	•	•										•								Atlantic-medit. immigrant	
24	Monocelis longiceps (Ant Dugès 1830)	•																			Atlantic-medit. immigrant	
25	Archilina endostyla Ax 1959	•																			Atlantic-medit. immigrant	
26	Promonotus ponticus Ax 1959												•								endemic	
27	Promonotus sp.											•									Atlantic-medit. immigrant	
28	Postbursoplana fibulata Ax 1955	•																			Atlantic-medit. immigrant	
29	Coelogynopora sp.																				Atlantic-medit. immigrant	
30	Procerodes lobata (O. Schmidt 1862)	•																			Atlantic-medit. immigrant	
31	Bresslauilla relicta Reisinger 1929	•																			Atlantic-medit. immigrant	Rumanian Plain
32	Pseudoaraffilla hymanae n.sp.	•																			Atlantic-medit. immigrant	
33	Microdalyellia armigera (O. Schmidt 1861)																		•		palaeartic	Rumanian Plain and Carpathian Mt.
34	Microdalyellia fusca (Fuhrmann 1894)					•	•		•												palaeartic	
35	Microdalyellia brevimana (Bekl. 1921)				•							•									North Euro-Siberian immig.	Câtuşa swamps
36	Gieysztoria cuspidata (O. Schmidt 1861)					•	•		•												palaeartic	Danube swamps
37	Gieysztoria ornata maritima Luther 1955											•									Northern immigrant	
38	Gieysztoria macrovariata (Weise 1942)						•														Central-European	
39	Gieysztoria triquetra (Fuhrmann 1894)				•	•	•	•													Central-European	Danube swamps
40	Castrella truncata (Abildg. 1789)					•															palaeartic	in all inner freshwater basins
41	Typhloplana viridata (Abildg. 1789)					•			•												European	Transylvania and Romanian Plain
42	Strongylostoma radiatum (Müller 1774)					•															cosmopolitan	in all inner freshwater basins
43	Strongylostoma elong. spinosum Luther 1950					•		•													Northern immigrant	Danube swamps
44	Strongylostoma cirratum Bekl. 1922					•	•		•												Siberian immigrant	Danube swamps
45	Castrada viridis Volz 1898																		•		European	Transylvania
46	Papiella otophthalma (Plotnicov 1906)					•															Ponto-caspian relict	Danube and Romanian Plain
47	Mesostoma lingua (Abildg. 1789)					•	•	•	•				•								cosmopolitan	in all inner freshwater basins
48	Bothromesostoma personatum (O. Schmidt 1848)					•															holarctic	in all inner freshwater basins
49	Phaenocora unipunctata (Orsted 1843)																		•		cosmopolitan	Rumanian Plain and Carpathian Mt.
50	Phaenocora typhlops (Vejdovsky 1880)																		•		palaearctic	
51	Hartogia pontica Mack-Fîră 1968	•																			Atlantic-medit. immigrant	
52	Proxenetes angustus Ax 1951	•																			Atlantic-medit. immigrant	
53	Trigonostomum mirabile (Per. 1892)	•																			Atlantic-medit. immigrant	
54	Trigonostomum venenosum (Ulj. 1870)	•												•							Atlantic-medit. immigrant	
55	Ptychopera plebeia (Bekl. 1927)	•																			Atlantic-medit. immigrant	
56	Promesostoma bilineatum (Per. 1892)	•	•																		Atlantic-medit. immigrant	
57	Paramesostoma pachidermum (Per. 1892)	•																			Atlantic-medit. immigrant	
58	Polycystis naegeli Kölliker 1845	•																			Atlantic-medit. immigrant	
59	Gyratrix hermaphroditus Ehrenb. 1831	•				•	•	•	•			•									Atlantic-medit. immigrant	ubiqviste
60	Gregyator mamertinus (v. Graff 1874)	•																			Atlantic-medit. immigrant	
61	Phonorhynchus pernix Ax 1959											•									Atlantic-medit. immigrant	
62	Phonorhynchoides flagellatus Bekl. 1927					•															Ponto-caspian relict	
63	Utelga heinckei Attems 1897	•																			Atlantic-medit. immigrant	
64	Itaipusa karlingi Mack-Fîră 1968	•																			Atlantic-medit. immigrant	
65	Utsurus camarguensis Brunet 1965	•																			Atlantic-medit. immigrant	
66	Pontaralia beklemichevi Mack-Fîră 1968					•															Ponto-caspian relict	
67	Pontaralia relicta (Bekl. 1927)					•															Ponto-caspian relict	
68	Torkarlingia euxinica Mack-Fîră 1971	•																			Atlantic-medit. immigrant	
69	Schizorhynchus tataricus v. Graff 1905	•	•																		endemic	
70	Cheliplana euxeinos Ax 1959	•																			endemic	

Pontic basin was made before our investigations (1968–1970). They were known only from the classic writings of Uljanin (1870), Pereyaslawzewa (1892), Graff (1904, 1905, 1911), Jacubova (1909), Beklemischev (1927) concerning the U.S.S.R. Black Sea coast, Valkanov (1936, 1954, 1957) for the Bulgarian shore, and Ax's work on the pre-Bosporus area (1959).

The present paper is a synthesis of my observations on marine Turbellaria and those living in freshwater and brackish coastal lakes, which were originally stream mouths or Black Sea lagoons. I am obliged to include annexes as well as the predeltaic area, for the zoogeographical origin of the Pontic turbellarian fauna to be understandable, since they shelter many relicts surviving the periods of unrest and faunal calamities which represent the early history of the eastern region of the ancient Pliocene brackish sea.

The following is a summary report on the basins and biotopes where the material was collected, the listing of the species studied, and ecological and zoogeographical considerations.

LIST OF SPECIES

1. *Convoluta convoluta* (Abildgaard, 1806)
OCCURRENCE. The Black Sea, stony bottom, 0.5 m depth, algae: Agigea, Costineşti, Vama Veche. Numerous specimens.
GEOGRAPHICAL DISTRIBUTION. The Barents, the Baltic, and the North Seas, the European coast of the Atlantic, the Mediterranean, the Marmara Sea, Bosporus, the Black Sea (the Anatolian coast, Sevastopol, Yalta, Suhumi).

2. *Convoluta albomaculata* (Pereyaslawzewa, 1892)
OCCURRENCE. The same as *C. convoluta*. A few specimens.
GEOGRAPHICAL DISTRIBUTION. The Black Sea (Sevastopol).

3. *Archaphanostoma agile* (Jensen, 1878)
OCCURRENCE. The Black Sea: Agigea, stony bottom, 2.5 m depth, in sand, 20 August, 1970. A few specimens.
GEOGRAPHICAL DISTRIBUTION. The Baltic–North Sea canal, the North Sea, the Adriatic Sea.

4. *Mecynostomum auritum* (O. Schultze, 1851) (Fig. 1)
OCCURRENCE. Mangalia Lake (village Limanu), 12‰ salinity, 0.5 m depth, 17 October, 1968. One specimen.
GEOGRAPHICAL DISTRIBUTION. Finnish Gulf, the Baltic, the North Sea, the Baltic–North Sea canal, the Marmara Sea.

5. *Mecynostomum arenarium* Ax, 1959
OCCURRENCE. The Black Sea: Agigea, stony bottom, in sand, 0.5 to 3 m depth. A few specimens.
GEOGRAPHICAL DISTRIBUTION. The Marmara Sea, the Black Sea (Anatolian coast).

Figure 1 *Mecynostomum auritum* (O. Schultze, 1851), general organization (from life).

6. *Oligochoerus* sp.

OCCURRENCE. The lagoon complex Razelm-Sinoë: Golovitza Lake, among *Enteromorpha*, one specimen; the predeltaic region: in the freshwater area of the "melea" of Sahalin, a few specimens.

7. *Stenostomum leucops* (Ant. Dugès, 1828)

OCCURRENCE. The lagoon complex Razelm-Sinoë: Razelmul Mare (Doloşman), Golovitza, Portitza, among plants in a swamp near the fisheries station, the Sinoë Lake (Grindul Lupilor); Taşaul Lake, in sand; Siut-Ghiol; in the freshwater area of the Northern Bay–Sulina arm (predeltaic region), 0.20 m depth. Numerous specimens.

GEOGRAPHICAL DISTRIBUTION. Europe, Asia, North America

IN ROMANIA. In all inner freshwaters.

8. *Stenostomum unicolor* O. Schmidt, 1848

OCCURRENCE. The same as for *S. leucops.*

GEOGRAPHICAL DISTRIBUTION. Europe and Asia.

IN ROMANIA. In all inner freshwater basins.

9. *Catenula lemnae* Ant. Dugès, 1832

OCCURRENCE. The lagoon complex Razelm-Sinoë. The same stations as for *S. leucops.* Numerous specimens.

GEOGRAPHICAL DISTRIBUTION. Europe, Asia, North America, Brazil.

IN ROMANIA. The Danube (the delta and the flooded area), the Romanian Plain, the alpine lakes (the Retezat Mountains–southern Carpathians).

10. *Macrostomum hystricinum* Beklemischev, 1951

OCCURRENCE. The lagoon complex Razelm-Sinoë: Golovitza, the end of Sinoë Lake; the predeltaic sector: Ciotic-Zaton, the south of Sahalin Island, 3 to 3.4‰ salinity. Numerous specimens.

GEOGRAPHICAL DISTRIBUTION. The brackish basins of the North Atlantic and of the North Sea, the Baltic–North Sea canal (Kiel Bay), the Finnish Gulf, the brackish basins of the Mediterranean (Pisa, the Canet marsh), the Black Sea (Sile, in front of a freshwater flow), the Caspian Sea, and the Aral Lake.

11. *Macrostomum romanicum* Mack-Fira, 1968

OCCURRENCE. Tekirghiol Lake, among *Cladophora,* 0.5 m depth, 80 to 100 ‰ salinity. Very numerous.

GEOGRAPHICAL DISTRIBUTION. La Camargue (Mediterranean basin).

IN ROMANIA. Lake Sărat-Brăila (Romanian Plain), 100 to 200‰ salinity (Mack-Fira, 1968).

12. *Macrostomum distinguendum* Papi, 1951

OCCURRENCE. Predeltaic sector: the Northern Bay and the "melea" of Sahalin (St. George arm), in freshwater areas. Among plants.

GEOGRAPHICAL DISTRIBUTION. Finland, the U.S.S.R., Poland, Italy, Austria.

IN ROMANIA. Danube swamps (the delta and the flooded zone), Transylvania (Mack-Fira, 1968).

13. *Macrostomum clavistylum* Beklemischev, 1951

OCCURRENCE. Taşaul Lake, in sand, 30 September, 1970. Prevailing.

GEOGRAPHICAL DISTRIBUTION. Ai-Dai Lake (U.S.S.R.).

14. *Macrostomum rubrocinctum* Ax, 1951

OCCURRENCE. Mangalia Lake (village Limanu), 0.5 m depth, among plants, 12‰ salinity, 17 October, 1968. Numerous specimens.

GEOGRAPHICAL DISTRIBUTION. Kiel Bay, Skagerrak.

15. *Macrostomum ventriflavum* Pereyaslawzewa, 1892

OCCURRENCE. The Black Sea: Costineşti, in the sand of the stone zone, 21 September, 1969; Agigea, sand, stone, 1 m depth. 18 August, 1970; Cape Midia, stone, in sand, 1 m depth. 5 December, 1970. Five specimens.

GEOGRAPHICAL DISTRIBUTION. Black Sea (Sevastopol).

16. *Macrostomum peteraxi* Mack-Fira, 1971

OCCURRENCE. The Black Sea, in the sand of the stone zone: Agigea, 2.5 m depth; Costineşti, 0.5 m depth; Vama Veche, 1 m depth. Few specimens.

17. *Microstomum lineare* (Müller, 1774)

OCCURRENCE. Lagoon complex Razelm-Sinoë: Golovitza Lake; predeltaic sector: Sahalin Island, freshwater area. Numerous specimens.

GEOGRAPHICAL DISTRIBUTION. Widespread in Eurasia.

IN ROMANIA. In all inner freshwaters.

18. *Plagiostomum ponticum* Pereyaslawzewa, 1892

OCCURRENCE. The Black Sea, stone zone: Agigea, Costineşti, Vama Veche, 1 m depth; Mamaia, among *Enteromorpha* and *Cystoseira*, 3.5 and 5.5 m depth. Numerous specimens.

GEOGRAPHICAL DISTRIBUTION. The Adriatic Sea, the brackish basins of the French coast of the Mediterranean Sea, the Black Sea (Sevastopol).

19. *Plagiostomum lemani* (Plessis, 1874)

OCCURRENCE. Lagoon complex Razelm-Sinoë: Golovitza Lake, 1 to 2‰ salinity; Siut-Ghiol Lake, among plants, September 1965, 1967, 1969. Numerous specimens.

GEOGRAPHICAL DISTRIBUTION. England, Finland, Denmark, Baltic countries, U.S.S.R. (the Caspian Sea), Germany, France, Italy, Switzerland.

20. *Pseudostomum klostermanni* (von Graff, 1874)

OCCURRENCE. The Black Sea: Costineşti, the "littoral basins," among *Cladophora*, 0.2 to 0.3 m depth; Agigea, Costineşti, and Vama Veche, stone zone, among algae, 1 m depth. Numerous specimens.

GEOGRAPHICAL DISTRIBUTION. The North Atlantic, the North Sea, the English Channel, the Mediterranean, the Adriatic Sea, the Black Sea (Sevastopol).

21. *Allostomum pallidum* Beneden, 1861

OCCURRENCE. The Black Sea: Costineşti, the "littoral basins," among *Cladophora*, 0.2 to 0.3 m depth; Agigea, Costineşti, Vama Veche, stone zone, among algae, 1 m depth. Numerous specimens.

GEOGRAPHICAL DISTRIBUTION. The North Sea, the Adriatic Sea.

22. *Allostomum catinosum* (Beklemischev, 1927)

OCCURRENCE. The Black Sea: Agigea, the stone zone, among algae, 0.5 m depth. Rare.

GEOGRAPHICAL DISTRIBUTION. The French Mediterranean marshes (Salses) and the Black Sea (Odessa).

23. *Monocelis lineata* (O. F. Müller, 1774)

OCCURRENCE. The Black Sea, stone zone: Agigea, sand, 0.5 to 3 m depth; Costineşti, the "littoral basins," 0.2 to 0.3 m depth; Constantza, among *Enteromorpha*; Vama Veche, in the sand of the stone zone; Cape

Midia, 5 December, 1970, in sand, 1 m depth. Very numerous. Mangalia Lake (village Limanu), in sand, 12‰ salinity. Many specimens.

GEOGRAPHICAL DISTRIBUTION. Greenland, the White Sea, the North Atlantic, North Sea, the Baltic, the English Channel, the Mediterranean, the Adriatic Sea, the Marmara Sea, the Bosporus, the Black Sea (Odessa, Sevastopol, Suhumi, Yalta, the Anatolian coast).

24. *Monocelis longiceps* (Ant. Dugès, 1830) (Fig. 2)

OCCURRENCE. The Black Sea: Cape Midia (5 December, 1970), Agigea, Costineşti, Vama Veche, in the sand of the stone zone, 0.5 to 3 m depth. Abundant.

GEOGRAPHICAL DISTRIBUTION. The North Atlantic, the North Sea, the Mediterranean, the Adriatic Sea, the Black Sea (Sevastopol, Suhumi).

25. *Archilina endostyla* Ax, 1959

OCCURRENCE. The Black Sea: Agigea, Costineşti, Vama Veche, in the sand of the stone zone, 0.5 to 1 m depth, very numerous; in the "littoral basins" (Costineşti).

GEOGRAPHICAL DISTRIBUTION. The Mediterranean Sea (Banyuls-sur-mer), the Marmara Sea, the Bosporus, the Black Sea (Anatolian coast).

26. *Promonotus ponticus* Ax, 1959

OCCURRENCE. The predeltaic region: the Northern Bay, 3 and 5 m depth, and in the Danube (fog signal), 1 km away from the river mouth. Ten specimens.

GEOGRAPHICAL DISTRIBUTION. The Black Sea (Anatolian coast), the Bosporus, and the Marmara Sea.

27. *Promonotus* sp.

OCCURRENCE. The lagoon complex Razelm-Sinoë: Tuzla-Duingi Lakes. Five specimens.

28. *Postbursoplana fibulata* Ax, 1955 (Figs. 3, 4)

OCCURRENCE. The Black Sea: Agigea, sand zone, "the *Otoplana* zone," November 1968. Two specimens.

GEOGRAPHICAL DISTRIBUTION. The Mediterranean Sea (Banyuls-sur-mer).

29. *Coelogynopora* sp. (Fig. 5)

OCCURRENCE. The Black Sea: Agigea, stone zone, in sand, 2 m depth, 31 July, 1969. One immature specimen.

30. *Procerodes lobata* (O. Schmidt, 1862)

OCCURRENCE. The Black Sea, stone zone: Agigea, Constantza, Vama Veche, 0.5 m depth, among *Enteromorpha* and in the phreatic layer close to the shore (1 August, 1970). Very numerous.

GEOGRAPHICAL DISTRIBUTION. The English Channel (Plymouth), the Meterranean, the Black Sea (Sevastopol, Yalta, Suhumi).

31. *Bresslauilla relicta* Reisinger, 1929

OCCURRENCE. The Black Sea: Agigea, in phytal, 1 m depth, August 1970. Two specimens.

Figure 2 *Monocelis longiceps* (Ant. Dugès, 1830), general organization.

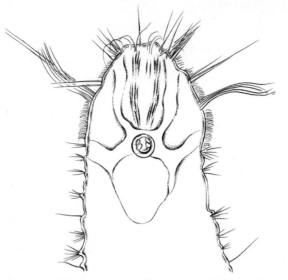

Figure 3 *Postbursoplana fibulata Ax*, 1955, cephalic region (from life).

GEOGRAPHICAL DISTRIBUTION. Finnish Bay, the Baltic Sea, Kiel Bay, Germany, Switzerland, Austria, Italy, France, the Anatolian coast of the Black Sea.

IN ROMANIA. In the sources and the stagnant basins of the Romanian Plain (Mack-Fira, 1970).

32. *Pseudograffilla hymanae* sp. n. (Fig. 6)

OCCURRENCE. Black Sea, stone zone, in sand at the algae base, 2 to 3 m depth, 20 August, 1970, in great numbers.

MATERIAL. Studies on living specimens, six series of sections, several specimens in alcohol. Holotype (one slide, sagittal sections) and two paratypes in the Swedish Museum of Natural History, Section for Invertebrate Zoology, Stockholm; three paratypes in the Faculty of Biology, University of Bucharest, Romania.

DIAGNOSIS. Length of living animals 1 to 1.25 mm. Clumsy little transparent body. Brick-red-yellowish pigment within the parenchyma. Frontal border with a few tactile hairs. Black eyes, with two or three retinal cells, equally spaced to each other and to the body sides. Subterminal mouth. Pharynx about one-third to one-quarter of the animal length. Slightly lobated intestine. Lobated testes in the second body quarter. Largely developed prostatic glands, with cyanophile and erythrophile secretion. Egg-shaped male copulatory organ, with strong muscular walls. Deferent ducts joining within the copulatory bulb. Ductus ejaculatorius very long and winding, with strong muscular walls, crossing the copulatory organ from one end to the other. Copulatory bulb having no male atrium prominence. Large ovary on the right side, with a much-developed germinative portion. Vitellogens dorsally and ventrally extended on the hind-half sides of the animal, surrounding the intestine. The elongated uterus, flattened in the animal's front plane and connected by a short genito-intestinal duct with the intestine, and by an orifice closed by a strong sphincter with a vesicula resorbiens having lacunose walls. The genital atrium, without a diverticulum playing the role of a bursa copulatrix, is followed by a long muscular duct leading to the uterus, from which it is separated by a small sphincter. The genital orifice lies in the middle of the ventral side of the animal.

DISCUSSION. *Pseudograffilla hymanae* sp. n. is very close to *P. arenicola* Meixner, 1938, from which it differs in the form and length of the ductus ejaculatorius, the form of the uterus, and the lack of a distinctive bursa copulatrix as an annex of the atrium.

33. *Microdalyellia armigera* (O. Schmidt, 1861)

OCCURRENCE. The freshwater sources of the Mangalia Lake. 17 October, 1968. Few specimens.

GEOGRAPHICAL DISTRIBUTION. Widespread in Europe.

IN ROMANIA. In the sources and rivulets of the Bucegi Mountains, in glacial

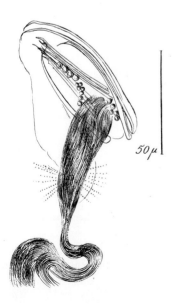

50 μ

Figure 4 *Postbursoplana fibulata* Ax, 1955, male copulatory organ (from life).

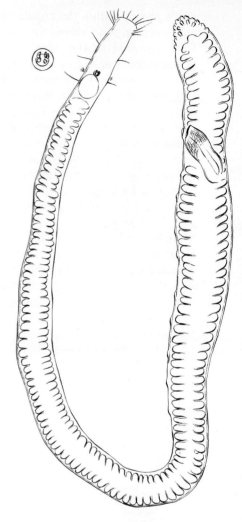

Figure 5 *Coelogynopora* sp., the organization of an immature animal (from life).

lakes (the Retezat Mountains–southern Carpathians), in the sources of
   the Romanian Plain (Mack-Fira, 1967).

34. *Microdalyellia fusca* (Fuhrmann, 1894)

OCCURRENCE. Lagoon complex Razelm-Sinoë: Golovitza Lake (among *Enteromorpha*), Doloşman; Siut-Ghiol Lake, among plants.

GEOGRAPHICAL DISTRIBUTION. Europe and Asia.

IN ROMANIA. In the sources of the Romanian Plain and Carpathian Mountains
   (Mack-Fira, 1967).

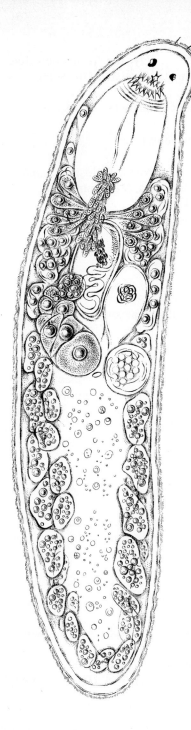

Figure 6   *Pseudograffilla hymanae* sp. n., general organization (from life).

35. *Microdalyellia brevimana* (Beklemischev, 1921)

OCCURRENCE. The predeltaic region: the Northern Bay, the freshwater area, 10 cm depth, among plants, alluvial bottom. 12 August, 1970. Numerous specimens.

GEOGRAPHICAL DISTRIBUTION. Sweden, Finland, Germany (Kurmark), Poland, Italian Alps, northern U.S.S.R., Siberia.

IN ROMANIA. Cătuşa swamp (the Siret River meadow).

36. *Gieysztoria cuspidata* (O. Schmidt, 1861)

OCCURRENCE. Lagoon complex Razelm-Sinoë: Golovitza, Doloşman, among plants, numerous; Siut-Ghiol Lake, among plants. Very numerous.

GEOGRAPHICAL DISTRIBUTION. Widespread in Eurasia.

IN ROMANIA. In the Danube swamps (Mack-Fira, 1967, 1968).

37. *Gieysztoria ornata maritima* Luther, 1955

OCCURRENCE. The Agigea Lake, among plants, 0.5 to 1 m depth, salinity 1.3‰ Very numerous.

GEOGRAPHICAL DISTRIBUTION. Finnish Bay.

38. *Gieysztoria macrovariata* (Weise, 1942)

OCCURRENCE. Lagoon complex Razelm-Sinoë: Portitza, swamp with a rich vegetation, 4 August, 1970. Few specimens.

GEOGRAPHICAL DISTRIBUTION. Italy, Germany.

39. *Gieysztoria triquetra* (Fuhrmann, 1894)

OCCURRENCE. Predeltaic region: the freshwater area of the "melea" of the Sahalin Island and the Northern Bay, 10 to 30 cm depth, among plants; the lagoon complex Razelm-Sinoë: Golovitza, Doloşman. Portitza, (Mack-Fira, 1970). Very numerous.

GEOGRAPHICAL DISTRIBUTION. Germany, Switzerland, Italy, Yugoslavia.

IN ROMANIA. In the Danube swamps (Mack-Fira, 1967, 1968).

40. *Castrella truncata* (Abildgaard, 1789)

OCCURRENCE. The predeltaic region: Sahalin Island, the freshwater area of the "melea." A few specimens.

GEOGRAPHICAL DISTRIBUTION. Greenland, Faeroes, Europe, Siberia.

IN ROMANIA. In the Danube swamps (delta and the flooded zone), the stagnant waters from Transylvania and the Romanian Plain (Paradi, 1882; Mack-Fira, 1967, 1968).

41. *Typhloplana viridata* (Abildgaard, 1789)

OCCURRENCE. The predeltaic region: Sahalin Island, the freshwater area of the "melea," among *Salvinia natans*. A few specimens.

GEOGRAPHICAL DISTRIBUTION. Greenland, Europe, Asia, Africa, North America.

IN ROMANIA. The delta and the flooded zone of the Danube; the stagnant freshwater of the Romanian Plain and Transylvania (Paradi, 1882; Mack-Fira, 1967, 1968, 1970).

Figure 7 *Strongylostoma cirratum* Beklemischev, 1922, male copulatory organ (from life).

42. *Strongylostoma radiatum* (O. F. Müller, 1774)

OCCURRENCE. The predeltaic region: freshwater area of the "melea" Sahalin. Numerous specimens.

GEOGRAPHICAL DISTRIBUTION. Europe, Asia, South America.

IN ROMANIA. The delta and the flooded zone of the Danube; the stagnant freshwater basins of the Romanian Plain; the Cătuşa swamp (Mack-Fira, 1970).

43. *Strongylostoma elongatum spinosum* Luther, 1950

OCCURRENCE. The lagoon complex Razelm-Sinoë: Golovitza Lake. Numerous specimens.

GEOGRAPHICAL DISTRIBUTION. The Finnish Bay.

IN ROMANIA. The flooded zone of the Danube (Mack-Fira, 1968, 1970).

44. *Strongylostoma cirratum* Beklemischev, 1922 (Figs. 7, 8)

OCCURRENCE. The lagoon complex Razelm-Sinoë: Golovitza and Sinoë (Grindul Lupilor) Lakes; predeltaic region: freshwater area of "melea" Sahalin. Among plants. Numerous specimens.

GEOGRAPHICAL DISTRIBUTION. Northern Siberia (Tomsk).

IN ROMANIA. In the Danube swamps (Mack-Fira, 1968, 1970).

45. *Castrada viridis* Voltz, 1898

OCCURRENCE. Mangalia Lake, in the swamp at the lake end, 0.98‰ salinity. 17 October, 1968. Four specimens.

GEOGRAPHICAL DISTRIBUTION. Iceland, Faeroes, Europe.

IN ROMANIA. In the stagnant freshwaters of the Romanian Plain and Transylvania.

46. *Papiella otophthalma* (Plotnicov, 1906)

OCCURRENCE. The predeltaic region: the freshwater area of the "melea" Sahalin. Abundant.

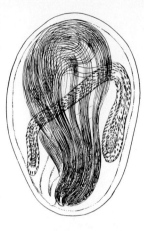

Figure 8 *Strongylostoma cirratum* Beklemischev, 1922, male copulatory organ (from life).

GEOGRAPHICAL DISTRIBUTION. U.S.S.R. (Bologovsk and the Volga steppe).
IN ROMANIA. In the Danube swamps (delta and the flooded zone), Snagov and Herăstrău Lakes (Romanian Plain) (Mack-Fira, 1968, 1970).

47. *Mesostoma lingua* (Abildgaard, 1789) (Figs. 9, 10)

OCCURRENCE. The predeltaic region: in the freshwater sector of the Sahalin Island "melea," among plants, very numerous; the lagoon complex Razelm-Sinoë: Golovitza Lake, Doloşman, Portitza; Agigea Lake, 1.3‰ salinity.

GEOGRAPHICAL DISTRIBUTION. Western Greenland, Europe, Asia, Africa.
IN ROMANIA. Common in all the stagnant freshwaters and in the Danube swamps (Mack-Fira, 1968, 1970).

48. *Bothromesostoma personatum* (O. Schmidt, 1848)

OCCURRENCE. Mangalia Lake: among green and filamentous algae of a freshwater source, 17 October, 1968. Numerous specimens.

GEOGRAPHICAL DISTRIBUTION. Greenland, Iceland, Europe, Asia, North America.
IN ROMANIA. Common in the stagnant freshwater basins and in the Danube swamps (Mack-Fira, 1970).

49. *Phaenocora unipunctata* (Örsted, 1843)

OCCURRENCE. Mangalia Lake: in a holocrene freshwater source without vegetation, 17 October, 1968. Numerous specimens.

GEOGRAPHICAL DISTRIBUTION. Europe, Asia.
IN ROMANIA. In the Carpathian sources and the stagnant freshwater basins of the Romanian Plain (Mack-Fira, 1970).

50. *Phaenocora typhlops* (Vejdovsky, 1880)

OCCURRENCE. Costineşti: in a large permanent swamp with vegetation, close to the beach, 17 June, 1966. Very abundant.

Figure 9 *Mesostoma lingua* (Abildgaard, 1789), specimen with thick-shelled (dormant) eggs, dorsal view (from life).

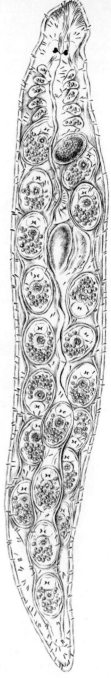

Figure 10 *Mesostoma lingua* (Abildgaard, 1789), specimen with thin-shelled (subitaneous) eggs, dorsal view (from life).

GEOGRAPHICAL DISTRIBUTION. Europe, Asia.

IN ROMANIA. Known in the surroundings of Sibiu (Paradi, 1881).

51. *Hartogia pontica* Mack-Fira, 1968

OCCURRENCE. The Black Sea, the stone zone: Agigea, 2.5 m depth, two specimens; Costineşti, 0.5 m depth, 21 September, 1969, one specimen.

52. *Proxenetes angustus* Ax, 1951 (Fig. 11)

OCCURRENCE. The predeltaic region: the Northern Bay (Sulina arm), 3 m and 5 m depth; Agigea (the Black Sea shore), stone, 1 m depth, in a little quiet bay, 16 August, 1970. Numerous specimens.

GEOGRAPHICAL DISTRIBUTION. The North Sea, the marshes of the French Mediterranean Sea, the Marmara Sea, the Black Sea (Sile).

53. *Trigonostomum mirabile* (Pereyaslawzewa, 1892)

OCCURRENCE. The Black Sea, stony bottom: Mamaia, among *Cystoseira*, 5 m depth; Agigea, 3 m. Rare. Cape Midia, among algae, 1 m depth, 5 December, 1970. Numerous specimens.

GEOGRAPHICAL DISTRIBUTION. The Black Sea (Sevastopol, Odessa, Sile) and Marmara Sea.

54. *Trigonostomum venenosum* (Uljanin, 1870)

OCCURRENCE. The Black Sea, stony bottom: Vama Veche, among *Cystoseira*, 1 m depth. Numerous specimens.

GEOGRAPHICAL DISTRIBUTION. The North Atlantic, the North Sea, the Mediterranean Sea, the Marmara Sea, the Black Sea (Sevastopol, Sile), Bosporus.

55. *Ptychopera plebeia* (Beklemischev, 1927) (Fig. 12).

OCCURRENCE. Mangalia Lake, 12‰ salinity, among algae and sand, few specimens. The Black Sea, stony bottom: Cape Midia, 1 m depth, among algae, 5 December, 1970. Numerous specimens.

GEOGRAPHICAL DISTRIBUTION. The marshes of the French Mediterranean (Salses and Canet), the Marmara Sea, the Black Sea.

56. *Promesostoma bilineatum* (Pereyaslawzewa, 1892)

OCCURRENCE. The Black Sea, stony bottom: Agigea, 1 m depth; Constantza, among *Enteromorpha*, 1 m depth; Costineşti, the "littoral basins." Numerous specimens.

GEOGRAPHICAL DISTRIBUTION. Kiel Bay, the English Channel, the Black Sea (Sevastopol, Odessa).

57. *Paramesostoma pachidermum* (Pereyaslawzewa, 1892) (Fig. 13)

OCCURRENCE. The Black Sea, the stone zone: Vama Veche and Agigea, among *Cystoseira*; Constantza, among *Enteromorpha*. Few specimens.

GEOGRAPHICAL DISTRIBUTION. The Marmara Sea, the Bosporus, the Black Sea (Sevastopol).

58. *Polycystis naegeli* Kölliker, 1845

OCCURRENCE. The Black Sea, the stone zone, among algae: Agigea and

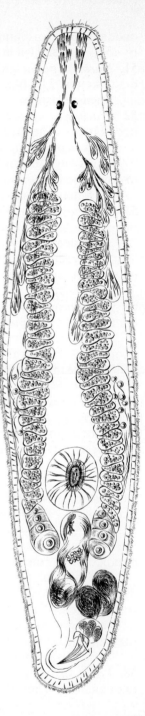

Figure 11 *Proxenetes angustus Ax*, 1951, general organization (from life).

Mamaia, among *Cystoseira* and *Enteromorpha*, 0.5 to 5.5 m depth; Costineşti, Vama Veche, 1 m depth, among algae. Very numerous.

GEOGRAPHICAL DISTRIBUTION. North Atlantic, the North Sea, the Baltic Sea, the English Channel, the Mediterranean, the Black Sea (Sevastopol, Yalta, Suhumi, the Anatolian coast).

59. *Gyratrix hermaphroditus* Ehrenberg, 1831

OCCURRENCE. The Black Sea, stony bottom, Agigea, 0.5 to 1 m depth; the lagoon complex Razelm-Sinoë: Golovitza, Doloşman, Portitza; Siut-Ghiol Lake, among plants; Mangalia Lake: freshwater source without vegetation, 17 October, 1968; predeltaic region: the Sulina and St. George arms, in the freshwater portion of the "melea."

GEOGRAPHICAL DISTRIBUTION. Ubiquitous.

Figure 11 *Proxenetes angustus* Ax (1951), general organization (from life).

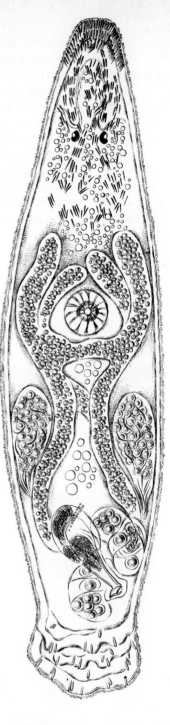

Figure 13 *Paramesostoma pachidermum* (Pereya-slawzewa, 1892), general organization, dorsal view (from life).

IN ROMANIA. In all inner basins except the supersalines.

60. *Progyrator mamertinus* (Graff, 1870)

OCCURRENCE. The Black Sea, stony bottom, Vama Veche, among *Cystoseira* and *Enteromorpha*, 1 m depth. Two specimens.

GEOGRAPHICAL DISTRIBUTION. The North Atlantic, the English Channel, the Mediterranean, the Black Sea (Odessa, the Anatolian coast), the Marmara Sea.

61. *Phonorhynchus pernix* Ax, 1959 (Fig. 14)

OCCURRENCE. The lagoon complex Razelm-Sinoë: Tuzla-Duingi, 4 August, 1970. Three specimens.

GEOGRAPHICAL DISTRIBUTION. The Marmara Sea.

62. *Phonorhynchoides flagellatus* Beklemischev, 1927 (Fig. 15)

OCCURRENCE. Lagoon complex Razelm-Sinoë: Golovitza Lake, 1 to 2‰ salinity.

GEOGRAPHICAL DISTRIBUTION. The Aral Lake.

63. *Utelga heinckei* (Attems, 1897) (Figs. 16, 17)

OCCURRENCE. The Black Sea, stony bottom, Agigea, 2 to 3 m depth, in the sand among algae, August, 1970. Eight specimens.

GEOGRAPHICAL DISTRIBUTION. The North Sea (Helgoland, Blyth), Skagerrak (Kristineberg), the North Atlantic coast of Ireland, the Mediterranean Sea (Marseille).

64. *Itaipusa karlingi* Mack-Fira, 1968

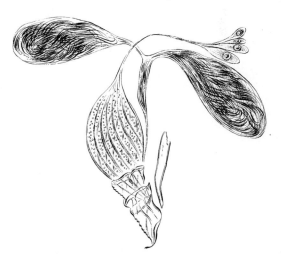

Figure 14 *Phonorhynchus pernix* Ax, 1959, male copulatory organ (from life).

Figure 15 *Phonorhynchoides flagellatus* Beklemischev, 1927, general organization (from life).

OCCURRENCE. Black Sea shore, stone, in the sand under the algae: Agigea, Costineşti, Vama Veche, 0.5 to 3 m depth. Numerous.

GEOGRAPHICAL DISTRIBUTION. The Mediterranean Sea (Marseille, M. Brunet, *in litt.*).

65. *Utsurus camarguensis* Brunet, 1965 (Figs. 18, 19)

OCCURRENCE. The Black Sea shore, stony bottom: Agigea, 1 to 2 m depth, August 1970, in the sand under the algae. Few specimens.

GEOGRAPHICAL DISTRIBUTION. The Mediterranean (Marseille).

66. *Pontaralia beklemichevi* Mack-Fira, 1968

OCCURRENCE. The lagoon complex Razelm-Sinoë: Golovitza, 1 to 2‰ salinity, among algae, 31 July and 20 September, 1969. In great numbers.

IN ROMANIA. The Snagov Lake (Romanian Plain), among plants.

67. *Pontaralia relicta* (Beklemischev, 1927)

OCCURRENCE. The lagoon complex Razelm-Sinoë: Golovitza, 1 to 2‰ salinity, 31 July, 1969, among algae. One specimen.

GEOGRAPHICAL DISTRIBUTION. The Aral Lake, the Caspian Sea.

68. *Torkarlingia euxinica* Mack-Fira, 1971

OCCURRENCE. The Black Sea shore, stone: Costineşti, 0.5 m depth; Agigea, 1 to 3 m depth. Numerous specimens.

69. *Schizorhynchus tataricus* Graff, 1905

OCCURRENCE. The predeltaic region: Sulina arm, the Northern Bay, 5 m depth. One specimen.

GEOGRAPHICAL DISTRIBUTION. The Black Sea (Sevastopol).

70. *Cheliplana euxeinos* Ax, 1959

OCCURRENCE. The Black Sea shore, stone zone: Agigea, in the sand, under the *Cystoseira*, 3 m depth, 19 October, 1968. One specimen.

GEOGRAPHICAL DISTRIBUTION. The Black Sea (Sile).

BASINS AND TURBELLARIAN BIOCENOSES

1. The Black Sea (Fig. 22)

This brackish sea of Sarmatic origin, with a salinity varying vertically from 17 to 18‰ in the superficial layers and up to 22.5‰ in the deep layers (except the narrow pre-Bosporus area) (Zenkevich, 1963), occupies an intermediate position between the mesomixohaline and polymixohaline basins.

The Romanian seashore stretches from Stambulul Vechi (Chilia arm) to the Vama Veche and comprises two parts: a sandy one to the north of Cape Midia, representing 59 percent of the whole length, and another with steep cliffs of loess, limestone, and gritstone, south of Cape Midia. The investigated material came more from the rocky than from the sandy region. In the "Otoplana zone" of the sandy region I collected specimens of *Post-bursoplana fibulata*.

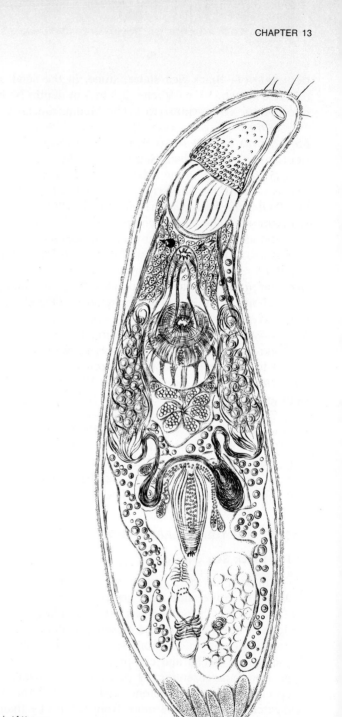

Figure 16 *Utelga heinckei* (Attems,
1879), general organization (from life).

In the rocky supralittoral of Costineşti the "littoral basins" of low depth (15 to 30 cm) with calm water subjected to great temperature and salinity variations are rich in *Cladophora* and *Enteromorpha*. Sometimes these basins may be covered by the waves, but they remain separated when the sea is calm. I found here a great number of *Allostomum pallidum, Pseudostomum klostermanni, Promesostoma bilineatum, Monocelis longiceps,* and *Archilina endostyla*.

In the mediolittoral occur *Monocelis longiceps, M. lineata, Archilina endostyla, Procerodes lobata, Convoluta convoluta,* and *C. albomaculata,* the latter two among vegetation.

The samples from the infralittoral were taken in 0.5 to 5.5 m depth and yielded: *Archaphanostoma agile, Mecynostomum arenarium, Convoluta convoluta, C. albomaculata, Monocelis lineata, M. longiceps, Archilina endostyla, Macrostomum peteraxi, M. ventriflavum, Allostomum pallidum, A. catinosum, Plagiostomum ponticum, Pseudostomum klostermanni, Trigonostomum venenosum, T. mirabile, Ptychopera plebeia, Hartogia pontica, Proxenetes angustus, Promesostoma bilineatum, Paramesostoma pachidermum, Bresslauilla relicta, Pseudograffilla hymanae, Gyratrix hermaphroditus, Progyrator mamertinus, Polycystis naegeli, Utelga heinckei, Utsurus camarguensis, Itaipusa karlingi, Torkarlingia euxinica, Cheliplana euxeinos,* and *Coelogynopora* sp.

2. The Predeltaic Region (Figs. 20, 21)

The Danube delta comprises an area of 251,000 hectares enclosed by the three main arms of the rivers, Chilia, Sulina, and Saint George. It is an early marine gulf, a remaining part of which forms today the lagoon complex Razelm-

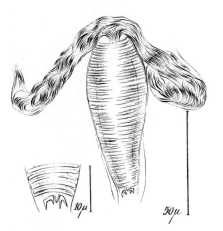

Figure 17 *Utelga heinckei* (Attems, 1879), male copulatory organ (from life).

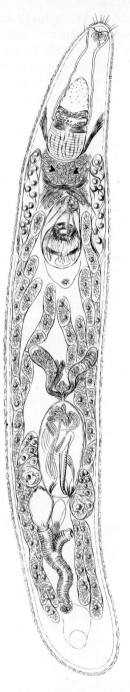

Figure 18 *Utsurus camarguensis* Brunet, 1965, general organization (from life).

Sinoë. The delta may be divided into two regions: the river delta and the river-marine delta; a further area is called the predeltaic region.

The latter region is very important for its biocenoses as well as for recurrence of the processes which led to the building of the delta. Its main features are: shallow (hence its name "melea"), brackish water with a fluctuating salinity, and good aeration. The biocenoses have mixed structures where the Ponto-Caspian elements are prevailing, the Mediterranean and freshwater species occurring at a lesser percentage. This applies to the "melea" of the Sulina arm (Musura or Northern Bay and the Southern Bay) and the "melea" Sahalin (St. George's arm).

The "melea" of the Sulina arm have a depth up to 5 m. In Musura the salinity depends on the wind force and the flow coming from the Danube. In the Southern Bay the salinity is higher.

In Musura, where a marked mixture of Ponto-Caspian, Mediterranean, and freshwater elements occurs, I established two sampling points (Fig. 20): (1) Close to the fog signal, where the bottom is alluvial and has a rich vegeta-

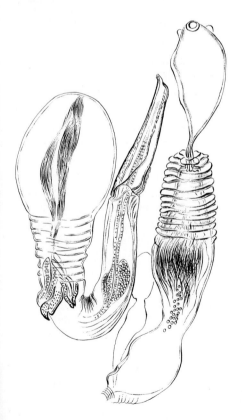

Figure 19 *Utsurus camarguensis* Burnet, 1965, male copulatory organ with stylet devaginated (from life).

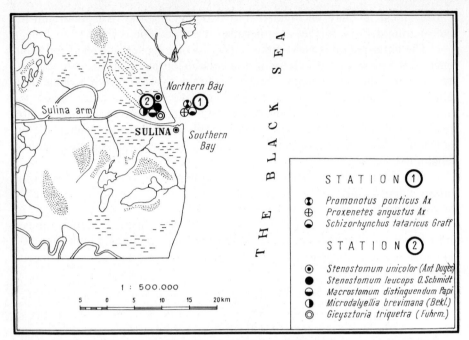

Figure 20 Distribution of the Turbellaria (excepting Tricladida) in the Northern Bay.

tion, the water being fresh and shallow (a few centimeters only), I collected *Macrostomum distinguendum*, *Microdalyellia brevimana*, *Gieysztoria triquetra*, *Strongylostoma radiatum*, *S. cirratum*, *Mesostoma lingua*, and *Gyratrix hermaphroditus*. (2) At the depth of 3 to 5 m, I found *Promonotus ponticus*, *Proxenetes angustus*, and *Schizorhynchus tataricus*, the latter being somehow questionable, since only a not well-preserved specimen was available. In this area, *Promonotus ponticus* appears to be the prevalent species; it was captured also inside the channel up to the fog signal (1 km away from the mouth of the Sulina arm).

In the St. George's "melea" (Island of Sahalin), where no human activity ever took place, the depth is small (0.5 to 1 m) and the salinity varies in general between 1.5 and 3‰, except for a sweetened area facing the mouth of the main arm. I set up four sampling stations (Fig. 21) there. At stations 1, 2, and 4, within the freshwater area (the bottom at station 4 is of alluvial nature, while at the others it is sandy) the following species live: *Macrostomum distinguendum*, *Gieysztoria triquetra*, *Typhloplana viridata*, *Strongylostoma radiatum*, *S. cirratum*, *Papiella otophthalma*, *Mesostoma lingua*, *Gyratrix hermaphroditus*, *Oligochoerus* sp., *Stenostomum leucops*, and *S. unicolor*.

At the southern end of the Sahalin Island (station 3) I collected amidst the vegetation on sandy bottom a huge number of *Macrostomum hystricinum*. This species was also found close to Zăton (3.3‰ salinity), at 10 m or more away from the lake mouth.

3. Littoral Lakes (Fig. 22)

The Lagoon Complex Razelm-Sinoë   This is the largest lagoon complex of Romania and, excepting the Azov Sea, the most important annex of the Black Sea, covering an area of 88,000 hectares. At the beginning it was a part of the early marine gulf, a Danube lagoon stretching south to the Cape of Midia.

It comprises two lake systems: (1) the Northern System (the Razelmul Mare and its annex (the Babadag Lake), Golovitza and Smeica Lakes), and (2) the Southern System (Sinoë, Caranasuf, and Tuzla-Duingi Lakes). These

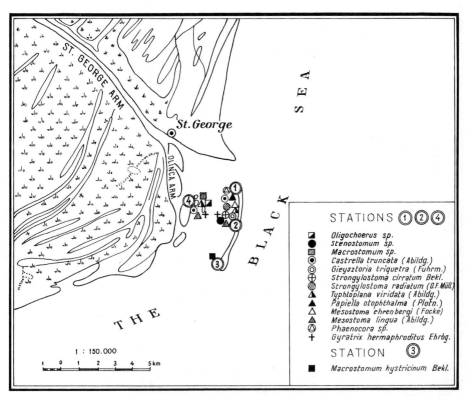

Figure 21 Distribution of the Turbellaria (excepting Tricladida) in the "melea" St. George.

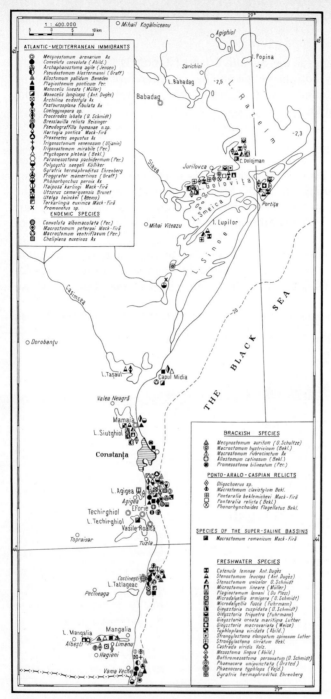

Figure 22 The turbellarian fauna (excepting Polycladida) of the Romanian littoral waters of the Black Sea and its annexes.

systems are linked with the sea at Portitza ("the little door"), the width of which varies annually between 200 m and a complete closure, depending on the Danube flow.

In the Northern System, supplied via several channels with Danube water, the average salinity is 2.5 to 4.5‰, but in the Southern System, which has a direct link with the sea at Periteaşca, the salinity reaches 12 to 20‰. Therefore, the Northern System is mixooligohaline, the Southern System being mixomesohaline.

I investigated the Turbellaria of the Razelmul Mare at the station of Doloşman, Golovitza, Portitza, Sinoë, and Tuzla-Duingi. In the Northern System the freshwater species prevail: *Stenostomum leucops*, *S. unicolor*, *Catenula lemnae*, *Microstomum lineare*, *Microdalyellia fusca*, *Gieysztoria cuspidata*, *G. triquetra*, *G. macrovariata*, *Mesostoma lingua*, *Gyratrix hermaphroditus*, and *Olisthanella* sp. In addition to the species captured also in the other parts of the system, I found in the Golovitza Lake: *Strongylostoma elongatum spinosum*, *S. cirratum*, *Plagiostomum lemani*, *Phaenocora* sp. and the relict species *Oligochoerus* sp., *Pontaralia beklemichevi*, *P. relicta*, *Phonorhynchoides flagellatus*, and the typical brackish species *Macrostomum hystricinum*, which is widely distributed in Europe.

In the Tuzla-Duingi Lakes, which have the same salinity as the sea and sometimes more, I found *Phonorhynchus pernix* and *Promonotus* sp., but no freshwater species.

The System of Lakes Taşaul-Gargalîc-Siut-Ghiol-Tăbăcăria   This system consisted originally of river mouths, which the Black Sea overflowed in the Quaternary, when the Mediterranean waters broke through the Bosporus, and were later submitted to a progressive closure and partial (external complex constituents: Gargalîc and Tăbăcăria) or complete (central constituents: Siut-Ghiol and Taşaul) isolation caused by the formation of offshore bars. These lakes have a meanderlike shape, recalling the early river beds.

The Siut-Ghiol Lake is the main component of this complex. It covers an area of 2,500 hectares and has an average depth of 7 m. It is separated from the sea by an offshore bar. Its water supply is secured by rainfall and the phreatic layer of the Dobrudga. The lake bottom is made up of *Gyttya* mud, underlaid by a sandy layer with subfossil remnants of mollusks and *Balanus* shells. An area of 35 hectares is stony. The prevailing genus among algae is *Chara*, which covers 80 percent of the bottom.

I collected *Stenostomum leucops*, *S. unicolor*, *Microstomum lineare*, *Gieysztoria cuspidata*, *Microdalyellia fusca*, *Typhloplana viridata*, *Plagiostomum lemani*, and *Gyratrix hermaphroditus*.

In the Taşaul Lake, considered to be the ancient mouth of the Casimcea river, salinity 1.5‰, I found *Stenostomum leucops* and *Macrostomum clavistylum*.

Brackish Lakes   Agigea Lake: An early lagoon recently isolated from the sea by human impact. It covers an area of 70 hectares and has a maximum depth of 1.3 m. Large portions of the shore are exposed to drying. The salinity has progressively decreased to 1.3‰, and 1‰ in the region of the freshwater sources. The bottom is covered with a black organic mud and Characeae.

Among the turbellarians *Gieysztoria ornata maritima* and *Mesostoma lingua* occur, the first being one of the prevailing elements of the microfauna in this basin.

Mangalia Lake: This lake is south of the town of Mangalia and has a meanderlike shape, with high banks and an area of 289 hectares and maximum depth of 16 m. The stony bottom is covered with mud or sand. The water supply is a mixed one: rainfall, strong freshwater sources proceeding from the phreatic layer of the Dobrudga, and seawater.

The salinity reaches 12.75‰ at the end of the channel linking it to the sea, 12‰ in the middle of the lake (village Limanu), 2.34‰ at the lake end, and only 0.98‰ in its swamp prolongation. In the three sampling stations I found the following Turbellaria species:

1   In the area of 12‰ salinity (fishing point Limanu): *Macrostomum rubrocinctum*, *Mecynostomum auritum*, *Ptychopera plebeia*, and *Monocelis lineata*, the first and last being prevalent

2   In the swamp at the lake end (0.98‰ salinity): *Microstomum lineare*, *Macrostomum* sp., and *Castrada viridis*, which are typical freshwater species

3   In the freshwater sources which supply the lake: *Microdalyellia armigera*, *Phaenocora unipunctata*, *Gyratrix hermaphroditus*, and among the vegetation developed in the basin of a captured source. *Bothromesostoma personatum*

Supersaline Lakes   Tekirghiol Lake: A former Black Sea gulf, now isolated by a sandy bar with a maximum width of 150 m and a height of less than 5 m, having an area of 1270 hectares and 2500 m^2 and a maximum depth of 10 m. It is situated in a Sarmatic depression where the evaporation caused a concentration in Cl^- and Na^+ ions and a decrease in those of K^+ and Ca^{2+}. H_2S production is particularly high in the border area, with a maximum concentration of 0.161 cm^3/l (Tuculescu, 1965). The lake is supplied by rainfall and the freshwater sources surrounding it. The salinity varies between 55 and 100‰ and is lower in the neighborhood of the sources. The temperature and the density also fluctuate, and the water pH has a range of 7.5 to 8.5.

Because of the high salinity, the fauna and the flora are poor. *Cladophora* is abundant, but it disappears toward the bottom (3 to 4 m). The only turbellarian detected in this basin was *Macrostomum romanicum*.

ECOLOGICAL CONSIDERATIONS

From the short description of the biotopes and the species found in them, it is obvious that the most populated Black Sea area is the phytal zone. Here most of the turbellarians occur in the muddy sand at the vegetation bases or among the vegetation.

Mecynostomum arenarium, Monocelis lineata, M. longiceps, Archilina endostyla, Macrostomum ventriflavum, M. peteraxi, Torkarlingia euxinica, Cheliplana euxeinos, Coelogynopora sp., *Utelga heinckei, Utsurus camarguensis*, and *Itaipusa karlingi* are specific for the sand interstices.

As Ax (1959) pointed out and as my observations demonstrated (Mack-Fira, 1968), *Polycystis naegeli* is an inhabitant of the algal vegetation, whereas *Mecynostomum arenarium* could be found only in the sand. Not less specific for the phytal are *Allostomum pallidum, Trigonostomum venenosum, T. mirabile, Paramesostoma pachidermum*, and *Progyrator mamertinus*, which we always collected among the vegetation. Although the Proseriata are specific for the sand interstices, we found *Monocelis lineata* in great numbers at Constantza among the filaments of *Enteromorpha* and even inside them where it seems to take shelter.

The maximum density occurs in *Convoluta convoluta* and *Polycystis naegeli*, the latter prevailing in the stony region of the Romanian littoral, followed by *Pseudostomum klostermanni, Plagiostomum ponticum, Monocelis lineata, M. longiceps, Archilina endostyla*, and *Itaipusa karlingi*. Without being common species, *Proxenetes angustus* and *Torkarlingia euxinica* are well represented. I encountered rather rarely *Bresslauilla relicta, Hartogia pontica, Macrostomum peteraxi*, and *Gyratrix hermaphroditus*. Compared with *Convoluta convoluta, C. albomaculata* is a rather poorly represented species. I found *Postbursoplana fibulata* only in the sand area.

The greatest density and variety in the turbellarians of the stony area occurs at a low depth (0.5 to 1 or 2 m), where the water is well aerated by wave action.

Promesostoma bilineatum, Allostomum pallidum, and *Pseudostomum klostermanni* are abundant mainly in *Mytilus* agglomerations at a low depth (1 m), where they take shelter between the byssus filaments and in shells of dead individuals after having consumed their content.

The "littoral basins" seem to be particularly favorable for the development of some turbellarian species.These tiny sea basins are very quiet so long as waves do not reach them, but their salinity may fluctuate by evaporation and their temperature is subjected to considerable variations under the influence of the sun's radiation. I feel, therefore, that the Turbellaria which are thriving therein (*Pseudostomum klostermanni, Allostomum pallidum, Promesostoma bilineatum, Archilina endostyla, Monocelis lineata*) are forms showing some inclination to euryhalinity and chiefly to eurythermy. I have also noticed that these species accommodate themselves to laboratory captivity. Moreover, *Monocelis lineata*, a eurytope species, can live at 49‰ salinity in the saline of La Louvelle (France), and together with *Macrostomum rubrocinctum* it is the representative element of Turbellaria in a 12‰ salt concentration in the Mangalia Lake.

I could find *Trigonostomum venenosum* and *Progyrator mamertinus* only at Vama Veche, a locality at the most southern end of the Romanian littoral, not being reached by the Danube waters. I assume, therefore, that these species prefer seawater of fairly constant salinity.

The brackish-water species are confined in the predeltaic region, the lagoon complex Razelm-Sinoë, and the coastal brackish lakes. The salt concentration of these basins varies from one to another and even in the same basin, depending on rainfall and the stream direction.

A review of the turbellarian fauna of the brackish basins of the Romanian Black Sea shore led us to the conclusion that it differs, in both the composing species and the dominant species, not only between the various basins, but even in the same basin, depending on the salinity and the bottom structure.

In this respect a clear-cut delimitation can be found in the predeltaic sector, in both the Northern Bay (Sulina arm) and the "melea" of Sahalin (St. George's arm). The marine or brackish-water species are living in biotopes with sandy bottom and oligo- or mesohaline brackish water, while in the freshwater areas with alluvial bottom and typical swamp vegetation only freshwater forms occur. In Musura, where the maximum salinity is 9.9‰, occur the marine species *Promonotus ponticus* and *Proxenetes angustus*, which show throughout their entire range the ability to tolerate rather wide limits of salinity: the former lives in the Marmara Sea at 20.5‰ salinity (highest limit) and on the Bosporus' Asiatic shore at 4.5‰; the latter in Marmara at 34.2‰ and in the Bosporus (Baltaliman) at 14.6‰ salinity (Ax, 1959).

Macrostomum hystricinum, a typical brackish-water species, was collected in oligohaline waters: 1 to 2‰ salinity in Golovitza Lake, 3‰ at the southern point of the Sahalin Island (station 3, Fig. 21) and 3.4‰ at Ciotic (Zaton), but never in the freshwater area. In mesohaline water I found it in great numbers at the end of Sinoë Lake.

Pontaralia beklemichevi, *P. relicta*, and *Phonorhynchoides flagellatus*, typical brackish-water species, are lacking in the Black Sea and are confined to the lagoon complex Razelm-Sinoë (Golovitza, 1 to 2‰ salinity), but *Pontaralia beklemichevi* also occurs in the freshwater lake of Snagov (Romanian Plain). The brackish *Oligochoerus* sp. is present in the oligohaline water of the Razelm complex and enters far upstream in the Danube.

In the basins having a salinity close to that of the Black Sea (Mangalia Lake 12‰, Tuzla-Duingi 17‰) there are the following brackish-water species of marine origin and marine species: *Mecynostomum auritum, Macrostomum rubrocinctum, Monocelis lineata, Ptychopera plebeia* in Mangalia Lake; *Phonorhynchus pernix, Promonotus* sp. in Tuzla-Duingi.

Macrostomum romanicum occurs in very great numbers among the filaments of *Cladophora* in the supersaline basins (Tekirghiol Lake). I did not find it in other water types, although Ax and Borkott (1968, 1970) report its presence in La Camargue and assume it to be a eurhyaline species, with a lower salinity limit of 5‰. In our country it occurs also in the supersaline lakes of the Romanian Plain (Lake Sărat-Brăila with 100 to 200‰ salinity) (Mack-Fira, 1968).

Macrostomum clavistylum is considered to be a constant oligohaline species, living in Ai-dai Lake (Beklemischev, 1951), and in Romania in the sand interstices in Taşaul Lake, where it appears to be the representative species of the Turbellaria.

Some of the freshwater species in the littoral lakes of the Black Sea are eurytopic: *Stenostomum leucops, S. unicolor, Catenula lemnae, Microdalyellia fusca, Gieysztoria cuspidata, Castrella truncata, Mesostoma lingua, Microstomum lineare,* and *Gyratrix hermaphroditus.* This is also one of the reasons for their presence in the Danube swamps (the flooded area and the delta) and other inner freshwater basins of our country. Their occurrence (except *Castrella truncata*) which develops on piles and submersed stones in the microcoenosis with *Enteromorpha* at the fishing station at Golovitza (fishing point Jurilofca) at a salinity of 1 to 2‰, may be an indication that they have some inclination toward euryhalinity. In the Finnish Gulf *Microdalyellia fusca* can endure a salt concentration of 5 to 6.5‰ (Luther, 1955).

Mesotoma lingua, Gieysztoria triquetra, and *Strongylostoma cirratum* are very abundant in the freshwater portion of the "melea" of Sahalin among the root filaments of *Salvinia natans,* producing a phenomenon of homochromy.

It should be noted that various populations of *Gyratrix hermaphroditus,* a ubiquitous and euryhaline species, have stylet constituents of different length relations, but these relations are constant in individuals of the same population.

THE ORIGIN OF THE TURBELLARIAN FAUNA IN THE BLACK SEA AND ITS ANNEXES

The Black Sea and its related basins, the Caspian and the Aral Seas, originated from the Tertiary Sarmatic Sea, which in turn was separated from the Tethys Sea, this latter surrounding the globe like a girdle during the Eocene and Oligocene. A northern portion of this sea, which in the lower and middle Miocene bordered the southern part of the U.S.S.R. and was still largely connected with the ocean, having a typical Mediterranean salinity and fauna, was

separated toward the end of the middle Miocene, forming the so-called Para-
tethys Sea. During the upper Miocene its Mediterranean link was definitively
lost and it became the brackish Sarmatic Sea. This sea became, gradually,
less saline and the composition of its fauna began to change, owing to the disappearance of the stenohaline Mediterranean forms, which were not able to adapt
to the new living conditions, and the sea became an evolutionary center for an
endemic brackish-water fauna.

 According to the investigations by Ax (1959) in the pre-Bosporus area
of the Pontic basin, and to my own research in the Romanian Black Sea littoral, the turbellarian fauna is composed of Mediterranean eurhyaline immigrants, which adapted themselves to the new living conditions, remaining
as such or producing endemic forms, brackish-water species of marine origin
with a wider dissemination, relics of the ancient Pontic lake, Arctic and Baltic
immigrants, and species of a limnic origin (Mack-Fira, 1970).

1. Atlantic-Mediterranean Immigrants

This category is the largest one, consisting of 85 species [75 species according
to Ax (1959)], 31 of which I have found on the Romanian coast. The following
Mediterranean immigrants were identified on the Romanian seashore:

Convoluta convoluta (Abildgaard, 1806)
Mecynostomum auritum (Schultze, 1851)
Mecynostomum arenarium Ax, 1959
Archaphanostoma agile (Jensen, 1876)
Plagiostomum ponticum Pereyaslawzewa, 1892
Pseudostomum klostermanni (Graff, 1874)
Allostomum pallidum Beneden, 1861
Monocelis lineata (Müller, 1774)
Monocelis longiceps (Ant. Dugès, 1830)
Archilina endostyla Ax, 1959
Promonotus sp.
Postbursoplana fibulata Ax, 1955
Coelogynopora sp.
Procerodes lobata (O. Schmidt, 1862)
Bresslauilla relicta Reisinger, 1929
Pseudograffilla hymanae sp. n.
Hartogia pontica Mack-Fira, 1968
Proxenetes angustus Ax, 1951
Trigonostomum mirabile (Pereyaslawzewa, 1892)
Trigonostomum venenosum (Uljanin, 1870)
Ptychopera plebeia (Beklemischev, 1927)
Promesostoma bilineatum (Pereyaslawzewa, 1892)
Paramesostoma pachidermum (Pereyaslawzewa, 1892)

Polycystis naegeli Kölliker, 1845
Gyratrix hermaphroditus Ehrenberg, 1831
Progyrator mamertinus (Graff, 1874)
Phonorhynchus pernix Ax, 1959
Utelga heinckei Attems, 1897
Itaipusa karlingi Mack-Fira, 1968
Utsurus camarguensis Brunet, 1965
Torkarlingia euxinica Mack-Fira, 1971

Ax (1959) listed 49 species known only in the Black Sea. This includes the following species, which I found on the Romanian seashore but which have a geographical origin difficult to explain:

Convoluta albomaculata (Pereyaslawzewa, 1892)
Macrostomum ventriflavum (Pereyaslawzewa, 1892)
Macrostomum peteraxi Mack-Fira, 1971
Promonotus ponticus Ax, 1959
Schizorhynchus tataricus Graff, 1905
Cheliplana euxeinos Ax, 1959

2. Brackish-water Species of Marine Origin

According to Ax (1959), the brackish-water species of marine origin in the Ponto-Aralo-Caspian basin are to be put in two categories: those widely distributed in Europe but with an origin difficult to establish, and the others confined to the Ponto-Aralo-Caspian area, representing the relics of the great Pliocene lake.

A. Brackish Species of Marine Origin Widely Distributed in Europe Ax (1959) reports from the Black Sea nine species belonging to this category: *Macrostomum hystricinum, Enterostomula graffi, E. catinosum, Pseudosyrtis subterranea, Pseudomonocelis agile, Vejdovskya pellucida, V. helictos, Tvaerminnea karlingi, Promesostoma bilineatum.*

I am able to add *Mecynostomum auritum, Macrostomum hystricinum, Macrostomum rubrocinctun, Allostomum catinosum* (Beklemischev, 1927), and *Promesostoma bilineatum* from the Romanian littoral.

B. Ponto-Aralo-Caspian Relics I found species of this category confined to the lagoons and the predeltaic sector of the Romanian Black Sea shore: *Oligochoerus* sp., *Macrostomum clavistylum, Pontaralia beklemichevi, Pontaralia relicta, Phonorhynchoides flagellatus.*

The only relict species known till now as being common to the three Sarmatic basins was *Thalassoplanina geniculata,* discovered by Ax (1959) on the Anatolian shore (Sile) of the Black Sea, at the mouth of a freshwater tributary. My investigations demonstrate, however, that the number of common relicts living in the three basins is greater as the turbellarian fauna in these basins becomes better known. *Pontaralia relicta* of the Aral Lake and Caspian Sea occurs in some lagoons of the Black Sea with low salinity (Golovitza). On the other hand, the presence of *Phonorhynchoides flagellatus*, from the Aral Lake in the lagoon complex Razelm-Sinoë induces us to assume that more careful investigations in the Caspian would lead to the same conclusion concerning this Sarmatic remnant.

The genus *Oligochoerus* from the Caspian Sea, found by An der Lan (1964) in the Danube, and by Ax and Dörjes (1966) in other Central European rivers, and recently reported by the author from the predeltaic area and from the oligohaline brackish water at Golovitza (Mack-Fira, 1970), but not yet known from the Aral Lake, also may be considered to be a relict in the Ponto-Caspian basin, having produced limnic species which were able to reach the Central European rivers.

Among the relicts of Ponto-Caspian origin I may include also *Macrostomum clavistylum*, described from Ai-Dai (Beklemischev, 1951), a brackish lake in the eastern Ural Mountains. It was found in our country in Taşaul Lake (1 to 1.5‰ salinity). Although its present distribution and its occurrence in basins with a very low salinity would induce us to consider this species as a Sarmatic relict of limnic origin, I completely agree with Ax (1959) that the brackish-water *Macrostomum* species are originally marine forms and I classify *M. clavistylum* among the Ponto-Caspian relicts which along with the relicts of limnic origin formed the populations of the outlying brackish-basin areas.

3. Northern Immigrants

The annexes on the Romanian Black Sea littoral shelter several turbellarian species which according to their geographical distribution should be northern Baltic or Siberian forms. These may be divided into two groups: (1) the brackish-water elements of marine origin and (2) the forms of a limnic origin which can live in oligohaline brackish water.

In the first group is included *Macrostomum rubrocinctum* from the Bay of Kiel (Ax, 1951) and the Swedish shore (Westblad, 1953), both being found in our country in Mangalia Lake (12‰ salinity).

The second group occurs in our oligohaline littoral basins and in the Danube swamps. *Gieysztoria ornata maritima* occurs in the Finnish Gulf, and in Romania in Agigea Lake (1.3‰ salinity). It is surprising, on the one hand, that this species occurs under the same living conditions in two so geographically remote basins, and on the other hand, that the environmental conditions are so similar in Agigea Lake and the lagoon adjacent to Rosore Lake near Pisa (Mediterranean coast of Italy), where it is represented by its vicarial species, *Gieysztoria subsalsa* Luther, 1955. *Strongylostoma elongatum spinosum* from the Finnish Gulf occurs in Romania in Golovitza Lake (Razelm) (1 to 2‰ salinity) and in the Danube flooded area (Mack-Fira, 1968, 1970).

There are other species of Siberian origin. *Strongylostoma cirratum* described by Beklemischev (1922) from Tomsk (Siberia) was found by the author in the Ponto-Aralo-Caspian area in the complex Razelm-Sinoë, the freshwater portion of the "melea" Sahalin, and the Danube swamps (the delta and the flooded area) (Mack-Fira, 1968, 1970).

Strongylostoma cirratum, for instance, is known so far from the Tomsk region, which is a part of the area of the large glacial accumulation lake in the Tobolsk region (de Lattin, 1967) and from the Ponto-Aralo-Caspian area in the Razelm system and the Danube.

4. Species of Limnic Origin

In addition, there are species of limnic origin, some of them being known from Europe and others having a Palaearctic or cosmopolitan repartition, present in the freshwater or brackish annexes of the Romanian Black Sea littoral. Most of them are common in the inner freshwater basins of Romania.

Among these, *Gieysztoria triquetra*, a Central European species, occurring in Germany, Switzerland, Italy, and Yugoslavia, is living in our country in the lagoon complex Razelm-Sinoë, the predeltaic area, and the Danube swamps (delta and the flooded area). *Gieysztoria macrovariata*, also of Central Europe (Italy, Germany) was found by the author in a swamp near Portitza (Razelm-Sinoë complex). *Microdalyellia brevimana* is a North European form, but it was found also in the freshwater portion of the Northern Bay (Sulina arm) and in the Cătuşa swamp (near Galatz), the latter being an early estuary of the Siret River, a Danube tributary.

CONCLUSIONS

The fauna of Turbellaria in the Pontic basin, including its annexes, has a mixed origin.

1. The prevailing element is composed of the Atlantic-Mediterranean immigrants, which entered the Pontic region at several stages after the breaking down of the Aegeais and the cutting through of the Bosporus at the same time as the salt Mediterranean waters. Most of these animals disappeared in the period when the basin became brackish after the raising of the Black Sea level and the interruption of its Mediterranean link (New Euxinic phase).

The Mediterranean stock of immigrants was renewed after the link was re-stored (the second and still existing Mediterranean phase of the Black Sea). Others of these animals adapted themselves to the new conditions and became endemic, remaining confined in the Black Sea or reentering the Marmara.

2. In the brackish Black Sea annexes there are brackish-water species of marine origin. These represent in part species widely distributed in Europe outside the Ponto-Aralo-Caspian region, the geographical origin and age of which are difficult to specify, and in part relicts of the brackish Pliocene sea, which are remainders of the first Mediterranean phase of the Black Sea. They entered there after the overflowing of the river mouths by the Mediterranean water coming across the Bosporus and were able to survive by avoiding the noxious salinity and the competition of the Mediterranean immigrants. These species are apparently lacking in the proper Black Sea.

3. The number of the Ponto-Caspian relicts in the western area of the Pontic basin and of those which are common to the three existing Sarmatic basins proves to be larger than estimated and will probably increase during future investigations. This suggests that they occurred in the area before the Pliocene breaking up of the inner Pontic sea. To the species *Thalassoplanina geniculata*, existing in front of some river mouths in the Aral, Caspian, and Black Seas, may be certainly added *Pontaralia relicta*, recently found in the complex Razelm-Sinoë (Northern System) and probably *Phonorhynchoides flagellatus*, living in the Aral and Black Seas and which may be discovered in the near future in the Caspian Sea also. The limnic elements of Baltic (*Gieysztoria ornata maritima*, *Strongylostoma elongatum spinosum*) or Siberian (*Strongylostoma cirratum*, *Microdalyellia brevimana*) origin are living in the brackish oligohaline or freshwater annexes.

4. The Mediterranean immigrants entered only into the annexes of equal or higher salinity as compared to the Black Sea (Southern System of the lagoon complex Razelm-Sinoë; *Phonorhynchus pernix*).

5. In addition to the Ponto-Caspian elements and the northern immigrants, there are in the oligohaline and freshwater annexes (which were primarily river mouths), Central European, Palaearctic, Holarctic, and cosmopolitan freshwater species, all being remnants of the native limnic fauna able to bear slight salinity fluctuations.

BIBLIOGRAPHY

An der Lan, H. 1964. Zwei neue tiergeographisch bedeutsame Turbellarien-Funde in der Donau. *Arch. Hydrobiol., Suppl. Donauforsch.*, **27**(4):477−480.

Ax, P. 1959. Zur Systematik, Ökologie und Tiergeographie der Turbellarienfauna in den ponto-kaspischen Brackwassermeeren. *Zool. Jahrb. (Syst.)*, **87**(1/2): 43−184.

————. 1970. Organisation und Fortpflanzung von *Macrostomum salinum* (Turbellaria-Macrostomida). *Inst. Wiss. Film. Wiss. Film* C 947/1968: 12 pp.

———— and J. Dörjes. 1966. *Oligochoerus limophilus* nov. spec., ein kaspisches Faunenelement als erster Süsswasservertreter der Turbellaria Acoela in Flüssen Mitteleuropas, *Intern. Rev. Ges. Hydrobiol.*, 51(1):15–44.

———— and H. Borkott. 1968. Organisation und Fortpflanzung von *Macrostomum romanicum* (Turbellaria Macrostomida). *Verhandl. Deut. Zool. Ges. Innsbruck.* Akad. Verlagsges. Geest & Portig: 344–347.

Beklemischev, W. 1922. Les Turbellariés des steppes disposées à l'Quest des Monts Urals. *Trav. Soc. étude Kriguise*, 2:17–42. (In Russian, with French summary.)

————. 1927. Über die Turbellarienfauna des Aralsees. *Zool. Jahrb. (Syst)*, 54:87–138.

————. 1951. The Species Belonging to the Genus *Macrostomum* (Turbellaria Rhabdocoela) of the Soviet Union. *Bull. Soc. Nat. Moscow (Biol.)*, 56(4):31–40. (In Russian.)

Graff, L. von. 1904. Marine Turbellarien Orotavas und der Küsten Europas. I. Einleitung und Acoela. *Z. Wiss. Zool.*, 78:190–244.

————. 1905. Marine Turbellarien Orotavas und der Küsten Europas. II. Rhabdocoela. *Z. Wiss. Zool.*, 83:68–150.

————. 1911. Acoela, Rhabdocoela und Alloeocoela des Osten der Vereinigten Staaten von Amerika. *Z. Wiss. Zool.*, 99:321–428.

Jacubova, L. 1909. Les polyclades de la Baie de Sébastopol. *Mem. Acad. Sci. St. Petersburg.* s.8, T.24. (In Russian.)

Lattin, G. de. 1967. *Grundriss der Zoogeographie.* Fisher Verlag. Jena. 602 pp.

Luther, A. 1955. Die Dalyelliden (Turbellaria Neorhabdocoela). Eine Monographie. *Acta Zool. Fenn.*, 87: xi + 337 pp.

Mack-Fira, V. 1967. Cîţiva reprezentanti din subordinul Dalyellioida (Turbellaria Rhabdocoela) din România. *Anal. Univ. Buc. (Biol.)*, 16:19–26. (Romanian, with French summary.)

————. 1968a. Rhabdocoeliden aus dem Überschwemmungsgebiet der Donau. *Limnol. Berichte X. Jubiläumstagung Donauforsch, Bulgarien, 10–20 Okt., 1966*: 251–258.

————. 1968b. Dalyelliide (Turbellaria, Rhabdocoela) din România. *Stud. Cercet. Biol. (Zool.)*, 20(6):527–534. (Romanian, with English summary.)

————. 1968c. Macrostomide (Turbellaria Macrostomida) din apele interioare ale României. *Stud. Cercet. Biol. (Zool.)*, 20(2):131–136. (Romanian, with English summary.)

————. 1968d. *Macrostomum romanicum* sp.n., un Turbellarié des lacs salés de la Roumanie. *Lucr. ses. şt. staţ. cercet. marine "Prof. I. Borcea," vol. fest.*: 197–202.

————. 1968e. Turbellariés de la Mer Noire. *Rapp. Comm. Int. Mer Médit.*, 19(2): 179–182.

————. 1968f. *Pontaralia beklemichevi* n.g.n.sp., un Kalyptorhynque relique du bassin Ponto-aralo-caspien. *Trav. Mus. Hist. Nat. "Gr. Antipa"* 8. "Le Centennaire Gr. Antipa" 1867–1967: 333–341.

————. 1968g. Sur un nouveau Turbellarié, *Hartogia pontica* n.g.n.sp. (Rhabdocoela

Typhloplanoida) de la Mer Noire. *Rev. Roum. Biol. (Zool.)*, **13**(6):411–415.

――――. 1970a. Sur la position systématique de l'espèce *Castradella otophthalma* (Plotnicov 1906). *Rev. Roum. Biol. (Zool.)*, **15**(1):3–10.

――――. 1970b. "Turbellariate din România (Archoophora Prolecithophora, Proseriata, Rhabdocoela, Lecithoepitheliata), Studiu sistematic, ecologic si zoogeografic." Rezumatul tezei de doctorat. Centrul de multiplicare Univ. Bucuresti. 70 pp.

――――. 1971. Deux Turbellariés nouveaux de la Mer Noire. *Rev. Roum. Biol. (Zool.)*, **16**(4):233–240.

―――― and M. Cristea-Nastasesco. 1970. Sur la faune littorale des Turbellariés, côte roumaine de la Mer Noire. *Rapp. Comm. Int. Mer Médit.*, **20**(3):56–58.

Paradi, K. 1881. Die in der Umgebung von Klaussenburg gefundenen rhabdocoelen Turbellarien. *Med-naturwiss. Anz. (Orvos-Term. Ertesitö)* **6**(2), *Klaussenburg*: 161–174.

――――. 1882. Bericht über die Resultate der die Turbellarien der Siebenbürger Gewässer betreffenden Forschungen. *Math. Naturwiss. Mitt. (Math. és természet. közlemények) Ungar. Akad. Wiss.*, **18**:99–116.

Pereyaslawzewa, S. 1892. Monographie des Turbellariés de la Mer Noire. *Pubbl. Staz. Zool. Napoli*, **22**:291–311.

Tuculescu, I. 1965. *Biodinamica Lacului Techirghiol*. Ed. Acad. R.S.R. Bucureşti.

Uljanin, W. N. 1870. *Turbellarians from the Bay of Sevastopol. Trudy vtor. sesd. Russk. estestv.* Moskva 1869. 96 pp. (In Russian.)

Valkanov, A. 1936. Notizen über die Brackwässer Bulgariens. II. *Ann. Univ. Sofia*, **32**.

――――. 1954. Beiträge zur Kenntnis unserer Schwarzmeerfauna. *Arb. Biol. Meeresstat. Varna*, **18**:49–53.

――――. 1957. Katalog unserer Schwarzmeerfauna. *Arb. Biol. Meeresstat. Varna*, **20**:55–57.

Westblad, E. 1953. Marine Macrostomida (Turbellaria) from Scandinavia and England. *Arkiv Zool.*, **B4**:391–408.

Zenkevich, L. A. 1963. *Biology of the Seas of the U.S.S.R.* London: Allen and Unwin. 955 pp.

Further Studies on the Vertical Distribution of Freshwater Planarians in the Japanese Islands

Masaharu Kawakatsu
Biological Laboratory, Fuji Women's College,
Sapporo (Hokkaidô), Japan

Kawakatsu (1965a, 1967) published a description of the distributional ecology of freshwater planarians in the Japanese Islands based on analysis of the factors controlling the vertical distribution of this animal group from ecological and chorological standpoints. The present paper is a general revision of my 1965 article. The objective is to define the vertical distribution of the stream-dwelling planarians in the Japanese Islands as well as in the Far Eastern countries adjacent to Japan. Further considerations about the factors controlling the vertical distribution are reported below.

For the completion of this study, I wish to express my thanks to many friends of my team—their names are not mentioned here—for their continuing assistance in my studies. Especially, I am grateful to Dr. Tatsuya Yamada, who has been most helpful as my coworker for several productive years and also has given his permission to use the data on field surveys as well as laboratory observation obtained jointly.

JAPANESE SPECIES AND THEIR VERTICAL DISTRIBUTION

Nineteen species of freshwater planarians are now known in the territory of Japan (thirteen in Planariidae, two in Kenkiidae, and four in Dendrocoelidae) (Table 1). In addition there are also a considerable number of undescribed or uncertain species found in Japan. The Japanese freshwater planarian fauna consists of seven genera, including one that is uncertain, namely, *Dugesia*, *Phagocata*, *Polycelis*, *Sphalloplana*, *Bdellocephala*, *Dendrocoelopsis*, and *Monocotylus*? (Kawakatsu, 1965a, 1966b, 1967, 1968b, 1969b; Kawakatsu, Yamada, and Iwaki, 1967). Among them, the rather widely distributed species are *Dugesia japonica*, *Phagocata vivida*, *Polycelis auriculata*, *Polycelis sapporo*, and *Polycelis schmidti*. *D. japonica*, *P. vivida*, *P. sapporo*, and *P. schmidti* also occur in the countries adjacent to Japan, i.e., the northeastern part of the Far East (Kawakatsu, 1965a, 1966b, 1967, 1969b).[1] The other species show rather limited distribution in the Japanese Islands (Kawakatsu, 1966b, 1969b). It is clear that not all the geographical areas have been well investigated with regard to the occurrence of planarians (Fig. 1).

Figure $2a-e$ shows the external appearance of the above-mentioned five species. They show characteristic vertical distribution. Their geographical distribution revised to the latest data is shown in the sketch maps of Figs. 3 to 7. A more detailed geographical distribution of three *Polycelis* species is shown in the sketch maps of Figs. 8 and 9.

Taking into account the geographical distribution of the species and their vertical distribution, Kawakatsu (1965a) thought that five areas could be distinguished on the Japanese Islands. Now, however, I consider my previous conclusion must be subject to revision based on the new data—not described in detail here—of the distributional ecology of this animal group (see also Kawakatsu, 1967, p. 131, note 2).

The Far East, including the Japanese Islands, can be divided into nine areas in reference to the geographical distribution of *D. japonica*, *P. vivida*, *P. auriculata*, *P. sapporo*, and *P. schmidti*, namely, (1) the *japonica* area, (2) the *japonica-vivida* area, (3) the *japonica-vivida-auriculata* area, (4) the *japonica-vivida-sapporo-auriculata* area, (5) the *japonica-vivida-sapporo-schmidti* area, (6) the *japonica-sapporo-auriculata* area, (7) the *japonica-*

---

[1]The occurrence of *Phagocata vivida* in the mountainous region of Korea is confirmed. The taxonomic study of this species will be published elsewhere.

Table 1   Species of Japanese Freshwater Planarians

Class Turbellaria
  Order Tricladida
    Suborder Paludicola or Probursalia
      Family Planariidae (Kenk, 1930 emend.)
        Genus *Dugesia* Girard, 1850
1. *Dugesia japonica* Ichikawa and Kawakatsu, 1964
2. *Dugesia izuensis* Katô, 1943
        Genus *Phagocata* Leidy, 1847
3. *Phagocata vivida* (Ijima and Kaburaki, 1916)
4. *Phagocata kawakatsui* Okugawa, 1956
5. *Phagocata teshirogii* Ichikawa and Kawakatsu, 1962
6. *Phagocata iwamai* Ichikawa and Kawakatsu, 1962
7. **Phagocata papillifera* (Ijima and Kaburaki, 1916)
8. **Phagocata albata* Ichikawa and Kawakatsu, 1962
9. **Phagocata tenella* Ichikawa and Kawakatsu, 1963
        Genus *Polycelis* Ehrenberg, 1831
10. *Polycelis sapporo* (Ijima and Kaburaki, 1916)
11. *Polycelis auriculata* Ijima and Kaburaki, 1916
12. *Polycelis schmidti* (Zabusov, 1916)
13. *Polycelis akkeshi* Ichikawa and Kawakatsu, 1963
      Family Kenkiidae Hyman, 1937
        Genus *Sphalloplana* de Beauchamp, 1931
14. **Sphalloplana* sp. of Mts. Yatsu-gadake (Ichikawa and Kawakatsu, 1967)
15. **Sphalloplana* sp. of Himeji (Ichikawa and Kawakatsu, 1967)
      Family Dendrocoelidae (Kenk, 1930 emend.)
        Genus *Bdellocephala* de Man, 1874
16. †*Bdellocephala annandalei* Ijima and Kaburaki, 1916
17. *Bdellocephala brunnea* Ijima and Kaburaki, 1916
        Genus *Dendrocoelopsis* Kenk, 1930
18. *Dendrocoelopsis ezensis* Ichikawa and Okugawa, 1958
19. ‡*Dendrocoelopsis lactea* Ichikawa and Okugawa, 1958

*Species inhabiting in subterranean water.
†True lake-dwelling species (Lake Biwa-ko).
‡The specific name of a Japanese freshwater planarian referred to in the literature as *Dendrocoelopsis lacteus* is the incorrect gender (cf. Kawakatsu and Ichikawa, 1971).
Source: (Reproduced from Kawakatsu, 1969; partly modified)

sapporo-schmidti area, (8) the *sapporo-schmidti* area, and (9) the *schmidti* area. The Tsugaru Strait in North Japan subdivides the *japonica-vivida-sapporo-auriculata* area into two subareas. In each area, there are considerable differences in the vertical extension of the habitats of these five species. There is a close relationship between the height of their habitats and the latitudes. The types of the vertical distribution in each area are (1) J, (2) J-JV-V, (3) J-JV-JAV-AV-V and J-JV-JVA-VA-A, (4) JSV-JSVA-SVA-VA-A and JSV-SVA-VA-A, (5) JSV-SVC-VC-C, (6) JSA-SA-A, (7) JSC-SC-C, (8) SC-C, and (9) C. Figures 10 and 11 illustrate the nine areas. The results are given in Table 2.

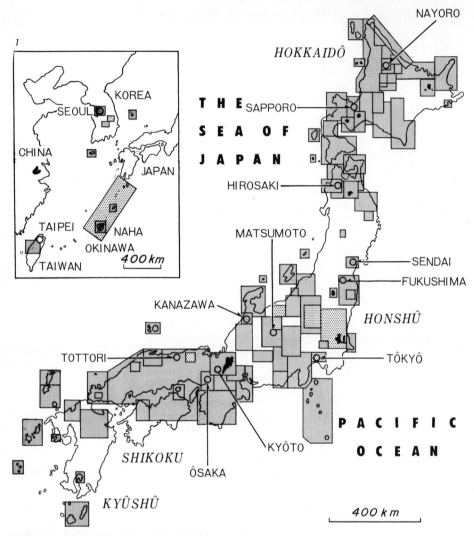

Figure 1 Map of the Japanese Islands and adjacent countries in the Far East, showing the areas in which the survey of distributional ecology of freshwater planarians has been done. Shaded areas are those on which articles have been provided by the members of Kawakatsu's team. Dotted areas are those on which articles have been provided by other researchers. *(Reproduced from Kawakatsu, 1969; partly modified.)*

ANALYSIS OF THE FACTORS CONTROLLING THE VERTICAL DISTRIBUTION

Concerning the vertical distribution of freshwater planarians, the five species *Dugesia japonica*, *Phagocata vivida*, *Polycelis sapporo*, *Polycelis auriculata*, and *Polycelis schmidti* occupy the high position of prominence among the Japanese probursalian fauna (Kawakatsu, 1965a, 1967). Why do only these species achieve primacy as widely distributed planarians?

Most of the Japanese freshwater planarian species are epigean forms. The subterranean species are *Phagocata papillifera*, *Phagocata albata*, *Phagocata tenella*, *Sphalloplana* sp. of Mts. Yatsu-gadake, *Sphalloplana* sp. of Himeji, and a number of undescribed or uncertain planariid species. *D. japonica*, *P. vivida*, *Phagocata kawakatsui*, and *Dendrocoelopsis lactea* are known as the troglophilic species. They are sometimes found in caves and/or in shallow wells. Only *Bdellocephala annandalei* and an uncertain species, *Dendrocoelopsis?* sp. are known as true lake-dwelling species and they are endemic in Lake Biwa-ko (cf. Table 1). Kawakatsu (1965a) expressed the opinion that the Japanese freshwater planarians may be divided into seven groups according to their ecological characters (pp. 395–396; 1967), namely:

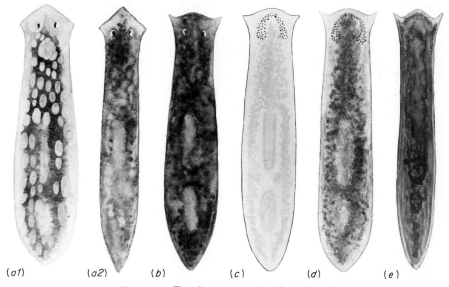

(*a1*)   (*a2*)   (*b*)   (*c*)   (*d*)   (*e*)

Figure 2 The five stream-dwelling planarians *(reproduced from Kawakatsu, 1969; partly modified). (a*1) *Dugesia japonica* Ichikawa and Kawakatsu (sexual race); *(a*2) *Dugesia japonica* Ichikawa and Kawakatsu (asexual race); *(b) Phagocata vivida* (Ijima and Kaburaki); *(c) Polycelis sapporo* (Ijima and Kaburaki; *(d) Polycelis auriculata* Ijima and Kaburaki; *(e) Polycelis schmidti* (Zabusov).

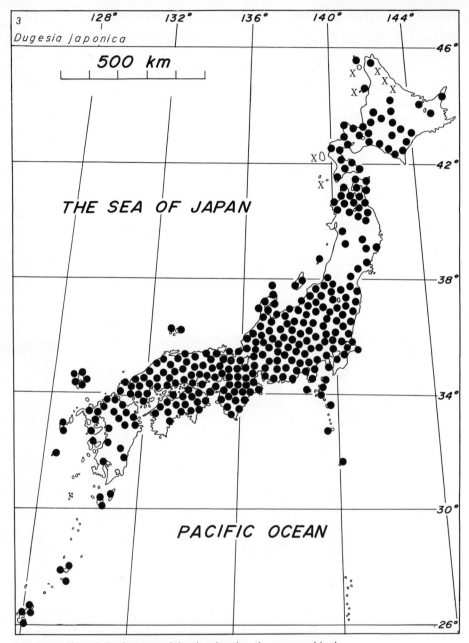

Figure 3 Map of the Japanese Islands, showing the geographical distribution of *Dugesia japonica (reproduced from Kawakatsu, 1969; partly modified)*. Numerous localities lying close together in one district are represented by a single symbol. X: districts where the species do not occur.

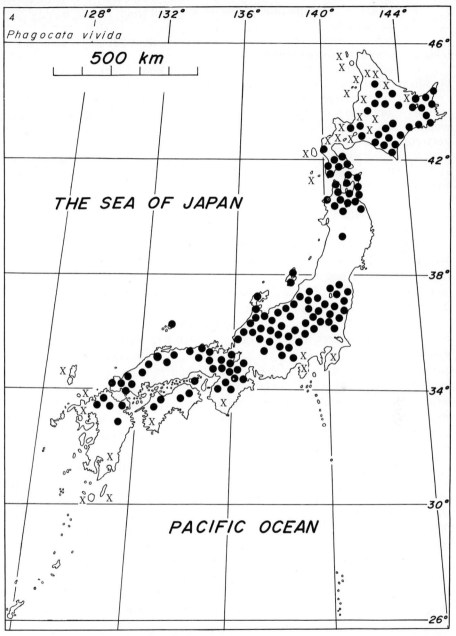

Figure 4 Map of the Japanese Islands, showing the geographical distribution of *Phagocata vivida*. *(Reproduced from Kawakatsu, 1969; partly modified.)*

    a Especially stenothermal—type 1; stagnant cold water; muddy bottom of Lake Biwa-ko

    b Stenothermal—type 1; slowly running or almost stagnant cold water; lakes or springs, or spring-fed brooks

    c Stenothermal—type 2; chiefly running cold water; sometimes occurs in spring-fed brooklets

    d Comparatively stenothermal; slowly running water

    e Comparatively eurythermal; swiftly or slowly running, or standing, waters, such as brooklets, brooks, creeks, rivers, and springs; also found in very cold water

    f Especially stenothermal—type 2; running and standing waters, such as brooklets, brooks, creeks, rivers, pools, or ponds; also found in comparatively cold waters

    g Probably stenothermal; chiefly found in small springs or wells; subterranean inhabitants

    Table 3 shows the known species of Japanese freshwater planarians arranged according to their ecology, geographical distribution, the modes of reproduction, the number of cocoons laid by one worm in one breeding season, and the number of juveniles released from one cocoon. From this table it will be seen that *D. japonica*, *P. vivida*, *P. auriculata*, *P. sapporo*, and *P. schmidti* constitute the so-called rheophilic forms or the inhabitants of running waters. Moreover, these species can also live well in standing waters. Among them, *D. japonica* is especially eurythermal; the other four species are comparatively eurythermal forms (*P. vivida* and *P. sapporo*) or stenothermal forms (*P. auriculata* and *P. schmidti*) (c, e, and f groups). Except *P. sapporo*, four of them are members of the fissioning planarians and can also propagate sexually by egg laying (Fig. 12). In general, the number of juveniles released from each cocoon of the above-mentioned five species is greater than that of the other species. The general conclusion to be drawn from Table 3 and Fig. 12 is that the five species have high ecological adaptability to their surroundings in both the warm- and cold-water regions. They could therefore gradually spread their domains in the Far East including the Japanese Islands throughout the past geological ages. This supposition will be evidenced by the fact that these five widely distributed forms are more or less "polymorphic" species.

    In previous articles (Kawakatsu, 1965a, 1967), I discussed the problem of the vertical distribution of freshwater planarians from two main points, i.e., the analysis based on the zoogeographical and ecological standpoints. In the present paper, I shall discuss the problem in these same steps.

Zoogeographical Points

It is a well-known fact that the Japanese Islands are the continental islands of the Asiatic continent. The repeated emergence of the several land bridges be-

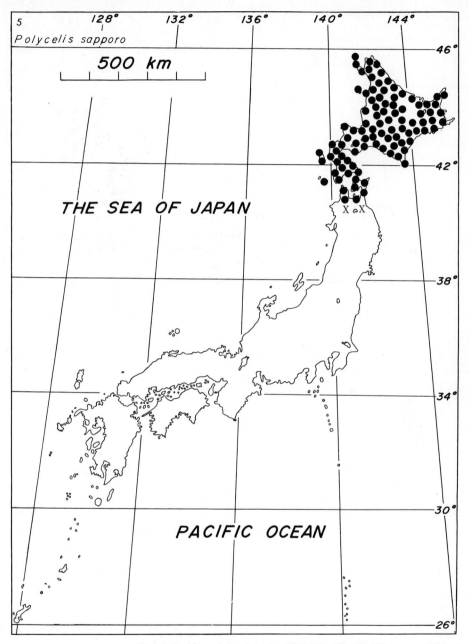

Figure 5   Map of the Japanese Islands, showing the geographical distribution of *Polycelis sapporo. (Reproduced from Kawakatsu, 1969; partly modified.)*

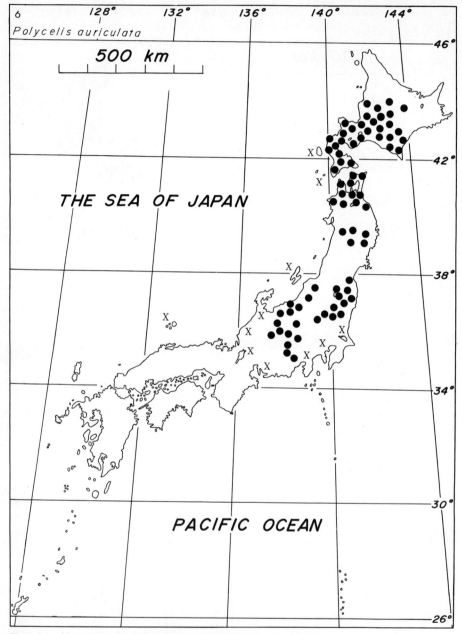

Figure 6 Map of the Japanese Islands, showing the geographical distribution of *Polycelis auriculata*. (Reproduced from Kawakatsu, 1969; partly modified.)

tween the Old Japanese Islands and the continent in past geological ages has been proved by geologists, paleontologists, and archaeologists.[2] The Japanese freshwater planarian fauna is very similar to that of the areas around the North Pacific (the Loochoo Islands, Taiwan, China, Manchuria, Korea, North and East Siberia, Kamchatka, Sakhalin, and the Kuril Islands in the Far East, and Alaska). In the Far East, three undoubtedly Eurasian genera, i.e., *Phagocata*, *Polycelis*, and *Bdellocephala* (including *Rectocephala*), dominate over other genera. Because of the geological history and zoogeography of the Japanese Islands, there is no room for doubt that the Japanese planarians owe their origin to the Asiatic continent (Kawakatsu, 1965a, 1967). In the Far East, the localities of the known species of *Dugesia* are limited to the southern part of the Asiatic continent, Taiwan, the Loochoo Islands, and the Japanese Islands; *D. japonica* may have migrated south to Japan (Kawakatsu, 1965a, 1967).

Concerning the transmigrating routes of the planarian genera in the Far East, a working hypothesis was proposed in my previous articles (Kawakatsu, 1965a, pp. 372–374; 1967). The routes are as follows (cf. Kawakatsu, 1965a, p. 373, fig. 10; 1967, p. 140, fig. 10):

1 The first route: From Siberia to Alaska (*Phagocata*, *Polycelis*, *Dendrocoelopsis*); *Polycelis*, *Bdellocephala*, and *Rectocephala* branch away to Khabarovsk (Okhotsk seaboard area) and Kamchatka.
2 The second route: From Siberia to Sakhalin and Hokkaidô in North Japan (*Phagocata*, *Polycelis*, *Bdellocephala*, *Dendrocoelopsis*, *Monocotylus*?); *Rectocephala* is found in Sakhalin.
3 The third route: From Siberia to Manchuria and China (*Phagocata*, *Polycelis*); *Phagocata* branches away to Primorsk (the areas on the seaboard) and Middle Japan, Sakhalin, and Hokkaidô.
4 The fourth route: From South China to Taiwan (Formosa) and the Loochoo Islands and Kyûshû in South Japan; and from Middle China to Korea and Kyûshû (*Dugesia*).

In recent years, the freshwater planarian faunas of the countries in the North Pacific area have become better known. Two kenkiid species were recorded from Japanese subterranean waters (Ichikawa and Kawakatsu, 1967b). In the southeastern part of the Far East, the freshwater planarian fauna of the Loochoo Islands and Taiwan consists of only one species, *D.*

---

[2]A brief description of the geological history of the Japanese Islands is available in my previous papers: Kawakatsu, 1965a, pp. 372–374, and 1967. The following description about the geological history of the Japanese Islands seems to be a logical conclusion from the biogeographic viewpoint (Tokuda, 1969). (1) The formation of the old land of the Old Okinawa Islands (Miocene or the early age of the Pliocene); (2) the formation of the old land of the Old Honshû Islands including Shikoku and Kyûshû (the early age of the Pleistocene); (3) the formation of the old land of the Old Hokkaidô Island including the South Kuril Islands (the middle age of the Pleistocene); (4) the formation of Taiwan, the Okinawa Islands, and the Tsushima Islands (the later age of the Pleistocene); (5) the formation of Sakhalin and the Kuril Islands; the separation of Honshû, Shikoku, and Kyûshû; the formation of many small islands of Japan (Holocene).

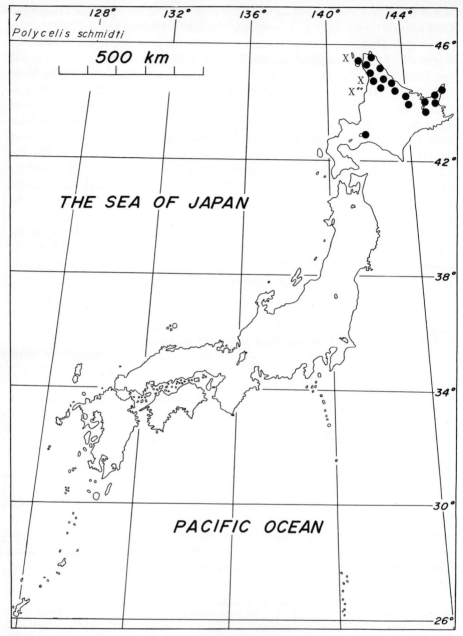

Figure 7 Map of the Japanese Islands, showing the geographical distribution of *Polycelis schmidti. (Reproduced from Kawakatsu, 1969; partly modified.)*

japonica (Ichikawa and Kawakatsu, 1967a; Kawakatsu and Iwaki, 1967b, 1968; Kawakatsu and Tanaka, 1971). The Korean freshwater planarian fauna consists of three species, *D. japonica*, *P. vivida*, and *Sphalloplana coreana* (Kawakatsu, Iwaki, and Kim, 1967; Kawakatsu and Kim, 1966, 1967). It is very similar to that of the Japanese Islands, but differs from the latter in the absence of the genera *Polycelis*, *Dendrocoelopsis*, and *Bdellocephala*. The knowledge about the freshwater planarian fauna of China is scant. However, *D. japonica* may be common in South, Middle, and North China (Ichikawa and Kawakatsu, 1967a). This widely distributed species shows some mor-

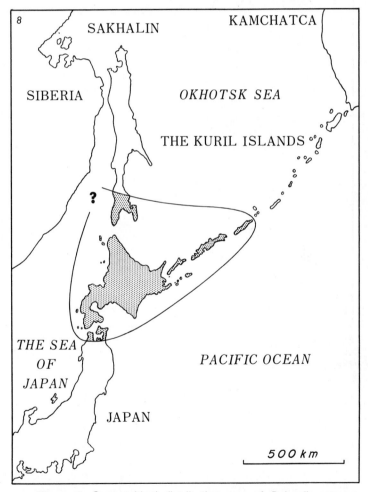

Figure 8 Geographical distribution area of *Polycelis sapporo*. *(Reproduced from Kawakatsu, 1969.)*

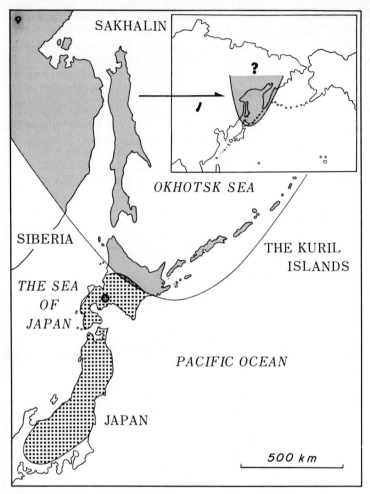

Figure 9 Geographical distribution areas of *Polycelis auriculata* (dotted area) and *Polycelis schmidti* (shaded area). *(Reproduced from Kawakatsu, 1969.)*

phological and physiological variations (i.e., polymorphic species; Kawakatsu, 1971). There is a record of only one *Polycelis* species from the Sanshi district in North China: *Polycelis* sp. Katô (Kawakatsu, 1965a, 1967). In the northeastern part of the Far East including Middle and North Japan, the species of the genera *Phagocata*, *Polycelis*, *Dendrocoelopsis*, and *Bdellocephala* are differentiated (Kawakatsu, 1968c). The northern distribution limit of the *Dugesia* species (*D. japonica* and *D. izuensis*) in the northeastern part of the Far East is the Japanese Islands (Kawakatsu, 1965a, 1966b, 1967, 1968c, 1969b, and others), except for one uncertain record by Zabusova-Zudanova

(1960). The Lake Baikal planarian fauna abounds with species of the dendrocoelid genera (Porfirjeva, 1968, 1969, 1970a, b, c, 1971; Zabusova-
Zudanova, 1970).

Around the South China Sea, the occurrence of a number of *Dugesia*
species from Malaya, the Philippine Islands, and Borneo was confirmed (Ball,
1970; Kawakatsu, 1972a, b). Two *Dugesia* species were reported from India
(Kawakatsu, 1969a; Kawakatsu and Basil, 1971) and Ceylon (Ball, 1970).

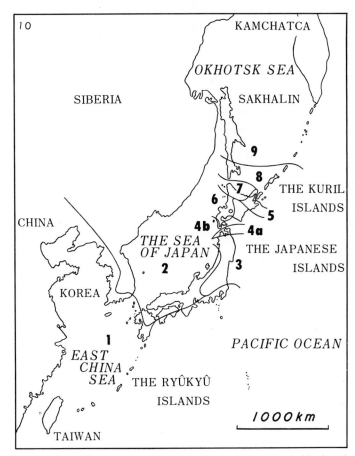

Figure 10 Map showing the nine areas of the geographical and
vertical distributions of the five species of stream-dwelling planarians in the Far East. 1, the *japonica* area; 2, the *japonica-vivida*
area; 3, the *japonica-vivida-auriculata* area; 4a, the *japonica-vivida-
sapporo-auriculata* subarea a; 4b, the *japonica-vivida-sapporo-
auriculata* subarea b; 5, the *japonica-vivida-sapporo-schmidti*
area; 6, the *japonica-sapporo-auriculata* area; 7, the *japonica-
sapporo-schmidti* area; 8, the *sapporo-schmidti* area; 9, the
schmidti area.

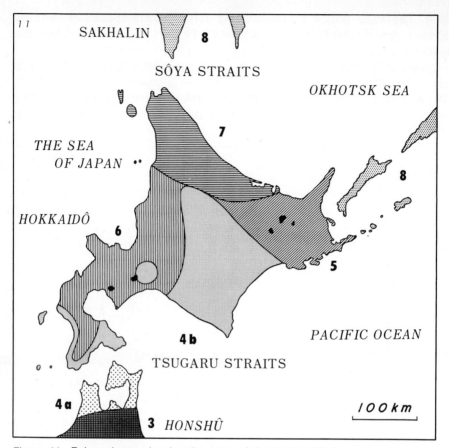

Figure 11 Enlarged map showing the areas of the geographical and vertical distributions of the five species of stream-dwelling planarians in North Japan. For explanation, see Fig. 10.

The freshwater planarian faunas of Alaska and the western part of the United States and Canada were recently studied by Ball (1969a, b), Ball and Fernando (1968), Carpenter (1969), Holmquist (1967a, b), Hyman (1963), Kawakatsu (1965a, b, 1968a), Kenk (1953, 1969, 1970a, b, c), and Mitchell (1968). The distribution ranges of the Eurasian genera *Phagocata*, *Polycelis*, and *Dendrocoelopsis* and of the family Kenkiidae (*Sphalloplana*) have become rather clear.

The Mexican and Central American planarian faunas were studied by Ball (1971), Benazzi and Giannini (1971), and Mitchell and Kawakatsu (in press). In Mexico four *Dugesia* and one *Cura* species were found. They seem to have a southern origin (for detailed discussion, see Mitchell and Kawakatsu, in press).

It should be said that my foregoing consideration of the supposed migration routes of freshwater planarians in the Far East (including the Japanese Islands) based on zoogeographical data is supported by the results of recent taxonomic and chorological studies of this animal group in the North Pacific area.

Concerning the migration routes and the speciation of freshwater planarians in the Japanese Islands, a working hypothesis is proposed by the author (Kawakatsu, 1965a, p. 378, fig. 11; 1967). The main points are as follows:

Dugesia japonica may have extended its distribution during the Pliocene or the Pleistocene and proceeded northeast across the southern land bridge. *Phagocata vivida* came from northwestern routes to Old Honshû Island along the seaside of the Old Sea of Japan (Primorsk–Sadogashima Island and the Okinoshima Islands–Honshû). Afterward, it may have extended its distribution range during the Pleistocene and proceeded southwest and northeast in Old Honshû Island. However, the other species of the genus *Phagocata* and those of the other genera may have had their sources in East Siberia and have come from the northern route to Old Honshû Island (Kawakatsu, 1965a, pp. 375–377).

The author also has a definite point of view about the migration and speciation of the widely distributed five species in the Japanese Islands (Kawakatsu, 1965a, p. 379; 1967):

If we follow the march of migration of planarians in the Japanese Islands in past geological time—then the first immigrant in South and Middle Japan was *Dugesia japonica*; next to it, *Phagocata vivida* migrated to Middle Japan from North Manchuria and Primorsk by the north-western routes. In more recent times, several stenothermal species marched towards Middle Japan from Siberia, Sakhalin and North Japan. These migrations, of course, had been repeated time and time again within the Neozoic Era. Further, there is no doubt that such migrations of freshwater planarians formed the historic background of the present local types of the vertical distribution. As will be seen from the types of the vertical distribution, the population (or the number of habitats) of *Phagocata vivida* dominates over *Polycelis auriculata* in Middle Japan, but the population has brought about a reversal of the order in the Shimokita Peninsula in North Honshû. The fact cannot be explained by the difference of the inhabitable water temperature range in each species, but it seems to the author that Middle Japan would have been occupied by *Dugesia japonica* and *Phagocata vivida* prior to the arrival of *Polycelis auriculata*.

Since my 1965 paper was published, the geographical distribution of the above-mentioned five species in the Japanese Islands, especially in North Japan, has become rather clear. *P. vivida* is distributed in Hokkaidô only in the seaside district of the Oshima Peninsula, the northeastern part of the Lake

Table 2 The Species and the Types of the Vertical Distribution of Freshwater Planarians in the Far East

Inhabitants	Areas	Type of vertical distribution
1 *Dugesia japonica*	JAPAN: South Kyûshû; Amami-Ôshima, Yakushima, Tanegashima, and Tsushima Islands; and other small islands near Kyûshû; southern parts of Shikoku and Honshû; the Izu Islands OTHER COUNTRIES: Taiwan; the Okinawa (Ryûkyû) Islands; the western half of Korea; South Manchuria (Kwantung); East and South China	J†
2 *Dugesia japonica* *Phagocata vivida*	JAPAN: North Kyûshû and Shikoku; Chûgoku, Kinki, Hokuriku, and Tôkai Regions in Honshû; Okinoshima and Sadogashima Islands on the Sea of Japan; Shôdoshima Island on the Inland Sea OTHER COUNTRIES: The eastern half of Korea (including the Sobaek Mountains and the Taebaek Mountains); Ullungdo Island on the Sea of Japan	J-JV-V
3 *Dugesia japonica* *Phagocata vivida* *Polycelis auriculata*	JAPAN: Mountain districts of Chûbu, Kantô, and Tôhoku Regions in Honshû (except for the Shimokita Peninsula and the Tsugaru Peninsula)	J-JV-JAV-AV-V (Chûbu and Kan▮ J-JV-JVA-VA-A (Kantô and Tôh▮
4 *Dugesia japonica* *Phagocata vivida* *Polycelis sapporo* *Polycelis auriculata*	a JAPAN: The Shimokita Peninsula and the Tsugaru Peninsula b JAPAN: South and Central Hokkaidô (seaboard and the Lake Ônuma districts in the Oshima Peninsula, northeastern part of the Lake Shikotsu-ko district, the Erimo Cape, the Hidaka Mountains, the Mts. Daisetsu district and the southern part of the Kitami Mountains); Oshima-kojima and Okushiri Islands on the Sea of Japan	JSV-JSVA-SVA-▮ JSV-SVA-VA-A
5 *Dugesia japonica* *Phagocata vivida* *Polycelis sapporo* *Polycelis schmidti*	JAPAN: East Hokkaidô (the Akan National Park and the Shiretoko Peninsula)	JSV-SVC-VC-C
6 *Dugesia japonica* *Polycelis sapporo* *Polycelis auriculata*	JAPAN: South and Central Hokkaidô (mountain district of the Oshima Peninsula, Niseko, the Syakotan Peninsula, the Shikotsu-Dôya National Park, the Yûbari Mountains, and the Mashike mountainous districts); Teure and Yangeshiri Islands on the Sea of Japan	JSA-SA-A

Table 2 (continued)

	Inhabitants	Areas	Type of verti-cal distribution*
7	Dugesia japonica Polycelis sapporo Polycelis schmidti	JAPAN: North Hokkaidô including the Okhotsk seaboard district; Rishiri and Rebun Islands on the Sea of Japan	JSC-SC-C
8	Polycelis sapporo Polycelis schmidti	OTHER COUNTRIES: South Sakhalin, the South and Middle Kuril Islands (probably including Kunashiri and Itrup Islands)	SC-C
9	Polycelis schmidti	OTHER COUNTRIES: Middle and North Sakhalin; Northeast Siberia and Kamchatka; the North Kuril Islands	C

*In each type of vertical distribution, the capital letters indicate higher levels as they go rightward; i.e., the left side shows the districts of the lower elevation, and the right side, the upper elevation. In transitional zones two or more allied species are found together.

†J, Dugesia japonica; V, Phagocata vivida; A, Polycelis auriculata; S, Polycelis sapporo; C, Polycelis schmidti

Shikotsu-ko district, and on the southern side of the demarcation line drawn from the Hidaka Mountains to Mts. Daisetsu district and Mt. Teshio-dake in the Kitami Mountains to the northern side of the Shiretoko Peninsula (Kawakatsu and Yamada, 1967). One of the possible speculations about the speciations of *P. auriculata* and *P. schmidti* is that they may have had a common source or a protospecies in East Siberia in the past geological age. The present *P. auriculata* may be the old immigrant to North Japan; the species is supposed to adapt itself more to a non-cold-water habitat than that of the present *P. schmidti*, which may be a newcomer in North Japan (Kawakatsu and Yamada, 1966). *P. auriculata* may have passed the Tsugaru Strait in the Pleistocene and reached Middle Japan (Kawakatsu, 1965a). From the results of recent studies on the vertical distribution of freshwater planarians, a reversal of the order of the vertical extensions of *P. vivida* and *P. auriculata* is seen in the Kantô Region in Honshû. *P. sapporo* is distributed in the entire Hokkaidô Region including many isolated islands of Hokkaidô as well as in the northernmost part of Honshû in North Japan (Kawakatsu, 1969b; Kawakatsu, Teshirogi, and Fujiwara, 1970). It is likely that *P. sapporo* migrated from Sakhalin to Hokkaidô comparatively late in the Pleistocene. Afterward, it spread all over Hokkaidô and proceeded south to the Shimokita Peninsula and the Tsugaru Peninsula in Honshû, and also proceeded east to the South Kuril Islands (Kawakatsu, 1969b). It may have expanded its domain into the habitats not suitable for many stenothermic planarian species in Hokkaidô during their migrating period. *P. sapporo* may have passed the Tsugaru Strait immediately before the Würm glacial period (W1–W2) (Kawakatsu, 1965a). A more extensive discussion of this problem is given in my previous papers (Kawakatsu, 1965a, 1967).

Table 3 The Known Species of Japanese Freshwater Planarians Arranged by Types of Habitat, the Geographical Distribution, and the Modes of Reproduction

Species		Habitat;* geographical distribution and the variation	Reproduction			Cocoons		No. of cocoons laid by one worm in one breeding season	No. of juveniles released from one cocoon
			Sexual only	Both	Asexual only	Laying	Hatching		
Dugesia japonica	f	Especially wide; highly polytypic; sexual and asexual	+	+	+	Late winter	Spring	1 to 2 (4)†	1 to 15 or more
Dugesia izuensis	d	Especially narrow	−	+	−	?	?	?	?
Phagocata vivida	e	Wide; polytypic fragmentation	−	+	−	Spring	Spring	1 to 2	2 to 5
Phagocata kawakatsui	d	Comparatively wide	+	−	−	Spring	Spring	1 to 2	10 to 20
Phagocata teshirogii	d	Comparatively wide	+	−	−	Spring	?	?	?
Phagocata iwamai	d	Comparatively wide	+	−	−	Spring to early summer	Summer	1 to 2	2 to 6
Phagocata papillifera	g	Narrow	+	−	−	May‡	?	?	?
Phagocata albata	g	Narrow	+	−	−	Spring	Spring	1	2 to 5
Phagocata tenella	g	Narrow	+	−	−	?	?	?	?
Polycelis sapporo	e	Wide; polytypic	+	−	−	Autumn to early winter	Spring	1	8 to 19
Polycelis auriculata	c	Wide; more or less polytypic	−	+	−	Late autumn to winter	Spring	1	8 to 21
Polycelis schmidti	c	Wide; more or less polytypic	−	+	−	Autumn to early winter	Spring	1	10 to 16

Polycelis akkeshi	d	Comparatively wide	+	—	—	Autumn to early winter	Spring	1	11 to 20
Sphalloplana sp. of Mts. Yatsu-gadake	d	Especially narrow	+?	—	—	?	?	?	?
Sphalloplana sp. of Himeji	g	Especially narrow	+?	—	—	?	?	?	?
Bdellocephala annandalei (Lake Biwa-ko)	a	Especially narrow	+	—	—	?	?	?	?
Bdellocephala brunnea	b	Wide; monotypic	+	—	—	Late winter to spring	Spring	1 to 2	2 to 12
Dendrocoelopsis ezensis	d	Comparatively wide; monotypic	+	—	—	Early winter to late spring (autumn)	Winter to early summer	1 to 2	1 to 15
Dendrocoelopsis lactea	d	Comparatively wide; monotypic	+	—	—	Early winter to spring	Spring to early summer	1 to 2	2 to 6

*The seven groups classified by the ecological characters (Kawakatsu, 1965a, p. 396; see in the text).
†Including data of the Chinese specimens observed by Hisiao (1935).
‡According to the descriptions of Ijima and Kaburaki (1916, p. 163) and Kaburaki (1922, p. 17).
Source: Reproduced from Kawakatsu, Yamada, and Iwaki, 1967.

Ecological Points

An outstanding paper on the distribution and abundance of lake-dwelling triclads in the British Isles was published by Reynoldson (1966). Its content may
be briefly summed up as follows:

The four main lake species of triclad, *Polycelis nigra* (Müller), *Polycelis tenuis*
Ijima, *Dugesia lugubris* (Schmidt) and *Dendrocoelum lacteum* (Müller) all

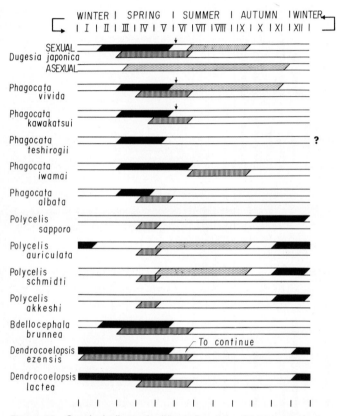

Figure 12 Graphs indicate the life cycles of the thirteen species
of Japanese freshwater planarians *(reproduced from Kawakatsu, Yamada, and Iwaki, 1967).* In each graph, the upper row
indicates the adult worms; the lower row, the newly hatched
young worms. Black areas (breeding season); open areas
(nonbreeding season or the season when planarians are without
sexual organs in almost all specimens); dotted area (season of
asexual reproduction by fission or fragmentation); lined area
(season when the newly hatched young worms occur). The
arrows indicate a period when high mortality is observed.

occupy the shallow littoral zone, etc. . . . The triclad populations of lakes vary in size from complete absence to very high numbers and there is a general, positive correlation with both the calcium content and the total dissolved matter of the water. The pattern of species distribution is also related to these factors. The relation is indirect and triclad abundance is determined primarily by the amount of food available in the littoral zone, itself ultimately dependent on the water-chemistry. . . . The distribution and abundance of the individual species of triclad are determined primarily by intra-specific competition for food. . . . It is suggested that because these important food organisms become both less numerous and less varied in the littoral zone as lake productivity declines, the triclad species are forced to compete with increasing severity. As a result, lakes of low productivity (minerally poor) support smaller numbers of triclads and fewer species than lakes of high productivity (minerally rich). [Reynoldson, 1966, pp. 62–63]

The food supply seems to be one of the chief factors controlling the density of lake-dwelling planarians both in Lake Biwa-ko in Japan and in Lake Tahoe in the Sierra Nevada Mountains in the United States (Kawakatsu, 1965b, 1968a). In the natural habitats it is also one of the effective causes controlling the population size of the stream-dwelling planarians in Japan. However, we can hardly avoid the conclusion that the environmental physico-chemical and the biological factors or the position of planarians in the food-chain relation in nature cannot be explained as decisive factors controlling the geographical distribution of planarians, even if these factors can strongly control the population size and the seasonal migration of planarians in a certain limited area. The difference of the factors controlling the distribution of planarians in the Japanese Islands and in the British Isles may be due to the differences of the ecological niche of each species and of the geological history of both countries.

According to my field observations, many species of stream-dwelling planarians usually crowd to the dead bodies of various terrestrial and aquatic animals. In some cases, planarians attack living aquatic insect nymphs and small crustaceans. When the squash method was used, fragments of crustaceans, setae of aquatic earthworms, and shells of diatoms could be detected in the intestine of several species of planarians. The diatoms found in the intestine of planarians may be derived from the aquatic insect nymphs and *Tubifex*, which seem to be their main food. Table 4 shows the preference for the various kinds of foods of planarians observed in the laboratory. It will be seen from this table that most of the Japanese species do not show any preference in regard to their foods but *P. vivida* and most of the dendrocoelids prefer insect nymphs and crustaceans.

According to the literature, some aquatic animals such as newts, fishes, and certain carnivorous insect larvae and nymphs eat planarians (Young and

Table 4   The Preference for the Various Kinds of Diets in the Different Species of Freshwater Planarians Observed in the Laboratory

Species	Pork liver	Chicken liver and spleen	Frog's liver	Plecoptera (nymph)	Trichoptera (nymph)	*Asellus* and *Gammarus*	Aqu earthw
Dugesia japonica	+++	+++	+++	+++	+++	+++	++
Phagocata vivida	+	++	++	+++	+++	+++	?
Phagocata iwamai	+++	+++	+++	+++	+++	?	++
Phagocata albata	+++	+++	+++	+++	+++	?	?
Polycelis sapporo	+++	+++	+++	+++	+++	+++	++
Polycelis auriculata	+++	+++	+++	+++	+++	+++	?
Polycelis schmidti	+++	+++	+++	+++	+++	+++	++
Polycelis akkeshi	+++	+++	+++	+++	+++	+++	++
Bdellocephala sp. of Rishiri Island	+	+	+	++	+++	?	++
Dendrocoelopsis ezensis	+	+	+	+++	+++	?	++
Dendrocoelopsis lactea	+++	+++	+++	+++	+++	?	?
Dendrocoelopsis sp. of Otoineppu	+	+	+	++	+++	?	++
Monocotylus? sp. of Rishiri Island	−	−	−	+++	++	?	−

Note: −: do not eat; +: scarcely eat; ++: eat moderately; +++: eat well.

Reynoldson, 1965; Reynoldson, 1966). I observed intra- and interspecific cannibalism of planarians in laboratory cultures (Kawakatsu, 1965a) but I have no other firsthand data about predators of planarians. As to the endoparasites of Japanese freshwater planarians, several species are known (Kawakatsu, 1970). A natural infection of a gregarinid species found in the *Dendrocoelopsis lactea* population in Nayoro City was investigated (Kawakatsu, 1970). It was observed in this locality that the number of the infected specimens of planarians has largely increased according to the increment of the pollution of water where planarians live (Table 5). In laboratory culture the infected planarians eventually disintegrated.

The environmental factors affecting the local distribution of four North American species of stream-dwelling planarians were studied by Chandler

(1966).[3] According to this exhaustive study made in two streams in Indiana, *Cura foremanii* (Girard) was abundant at the headwaters of the first stream but was replaced downstream by *Dugesia tigrina* (Girard). In the second stream, *Phagocata gracilis* (Haldeman) occupied only the headwaters and up-stream areas and was replaced downstream largely by *Dugesia dorotocephala* (Woodworth). He demonstrated from this study that temperature, water level, substrate and changes in bottom character, calcium, and dissolved organic matter seemed directly related to the distribution and abundance of these triclads. The abundance of *P. gracilis* and *D. dorotocephala* in an apparent transition zone alternates reciprocally throughout the year, suggesting inter-specific competition and migration behavior in *P. gracilis*. The population size of *P. gracilis* is changed by the season of breeding and hatching in this species. Dr. Chandler's deduction mentioned above is essentially very similar to that of my study on the population ecology of the Japanese stream-dwelling planarians.

In previous papers (Kawakatsu, 1965a, 1967), the seasonal migration of *Dugesia japonica* and *Phagocata vivida* inhabiting the Hachiga-sawa River system in the Chûbu Region in Honshû, Middle Japan, was described. In the main stream of this river system, the lower distribution limit of *P. vivida* is about 930 m in April, but about 1000 m in July. On the other hand, the upper distribution limit of *D. japonica* is about 1070 m in April and about 1150 m in July. Thus, the zone where these two species are found together shows a tendency to decline in winter and spring. In summer, it shows the opposite tendency. The seasonal change of the altitude of the vertical dis-tribution of planarians appears to be chiefly due to the temperature of the water of the stream in different seasons.

Natural populations of *Dugesia japonica* were examined both in a moun-

[3]The North American polypharyngeal *Phagocata* form currently designated as *P. gracilis gracilis* is recognized: *P. gracilis* (Haldeman, 1840) (cf. Kenk, 1970c).

Table 5   The Incidence of a Natural Infection of the Gregarinid Species Found in the *Dendrocoelopsis lactea* Population in Nayoro City

	November 1960	November 1962	November 1963
Number of specimens of *Dendrocoelopsis lactea* examined	95 {sexual, 90 / asexual, 5	83 {sexual, 71 / asexual, 12	62 {sexual, 40 / asexual, 22
Number of specimens of *Dendrocoelopsis lactea* parasitized	0	2 sexual	20 {sexual, 12 / asexual, 8
Parasitic rate	0%	2.4%	32%
Circumstance of the habitats	Very clean water	Not clean water, with some organic matter	Not clean water, with much organic matter

Source: Reproduced from Kawakatsu, 1969.

tain stream and in a stream in the plain near Tottori City, the Chûgoku Region in South Honshû, throughout the year at monthly intervals (Ondô and Inoue, 1965). The seasonal range of water temperature is from about 8.5°C (December) to 23.5°C (July and August) in the mountain stream. The plain stream has a maximum water temperature (July) of 26.0°C in the stream center and of 29.0°C in the shallows of the river side. According to these authors, in general, the population density sharply increases from spring to early summer, but it decreases in midsummer. And the population density is always large in the lower stream (in the case of a mountain stream) or the stations in the stream center (in the case of a plain stream), where the riverbed is composed of loose stones or pebbles. Seasonal change in the microdistribution in the lower stream of a plain was also observed by them. From spring to early summer, animals occur on both sides of the stream, but in midsummer almost all animals occur on the north side of the stream and there are none present on the south side where the water is shallow. But in early autumn, they are found again on the south side of the stream. These authors concluded that the water temperature of the natural habitats is the most effective factor regulating the microdistribution of animals in the stream—especially, during the summer season (the season of diminution of water in streams).

According to Zabusova-Zudanova (1956), who studied the populations of two Kamchatkan *Polycelis* species (*Seidlia schmidti* and *Sorocelides elongata*) inhabiting the Karimaiski Spring (or a stream near the salmon hatchery), planarians were collected from the two stations at 200 m distance from each other, one in a depth of the slow current (25 cm in depth and water velocity of 31.5 cm/sec at the mid-surface) and the other in a shallow of the swift current (15 cm in depth and water velocity of 75.0 cm/sec at the mid-surface). The bottoms of both stations are covered with pebbles (Fig. 13). The water temperature of the stream is below 15°C even in midsummer; the dissolved oxygen content of water nearly reaches saturation even at its bottom. During the period of October 1946 to March 1947, 2862 specimens of *P.* (= *Seidlia*) *schmidti* and 501 specimens of *P.* (= *Sorocelides*) *elongata* were collected from these stations. The population of *P. schmidti* in a shallow of the swift current was very dense. However, no difference in the population density of *P. elongata* in both stations was observed. The natural feature of the Kamchatkan locality of *P. schmidti* is almost coincident with that of the Japanese localities of this species.

As a demonstration of migration of planarians according to the different degrees of water temperature, the following experiment was performed. The apparatus used is shown diagramatically in Fig. 14. An artificial current of water is produced in the aquarium (200 cm long, 30 cm wide, and 25 cm deep). By using several heaters with thermostats for tropical fish culture, water pump, and cooler, the water temperature of the artificial stream in the aquarium is

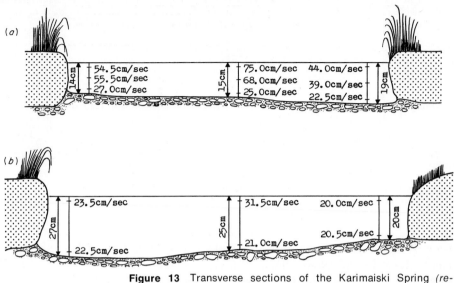

(a)

54.5cm/sec	75.0cm/sec	44.0cm/sec
55.5cm/sec	68.0cm/sec	39.0cm/sec
27.0cm/sec	25.0cm/sec	22.5cm/sec

(b)

23.5cm/sec	31.5cm/sec	20.0cm/sec
		20.5cm/sec
22.5cm/sec	21.0cm/sec	

Figure 13 Transverse sections of the Karimaiski Spring *(reproduced from Zabusova-Zudanova, 1956; modified).* (a) A shallow place of the swift current; (b) a deep place of the slow current. For more detailed explanations, see the text.

controlled about 4°C for station A, 6°C for station B, 12°C for station C, and 15°C for station D. Pebbles are placed in the aquarium as a resting place for animals. The stream velocity is controlled by filters set in the cooler. The apparatus is illuminated with fluorescent lamps (40 W). When the heaters of this apparatus are off, the water temperature of the stream keeps within a range of 4 to 4.5°C. The dissolved oxygen content of the water nearly reaches saturation even at the bottom of the stream. *Polycelis sapporo* and *P. schmidti* collected from the On'nenai locality were used as materials. Three experiments were carried out: (1) both the heaters and the water pump were on, (2) only the water pump was on, and (3) both the heaters and the water pump were on, but several pieces of chicken liver were placed at the bottom of station A. Each species, represented by 500 large-sized worms, was discharged into the water of the apparatus from station D. The animals were kept under each condition for a period of 24 hours, and then the number of worms crowded in each station was counted. The result of this experiment is shown in Table 6.

It is clear from the result of this experiment that planarians are quite sensitive to the difference in water temperature of the habitats, even when the range is very narrow. *P. schmidti* crowds on the lower temperature places than *P. sapporo*. The former species is the most stenothermic stream-dwelling

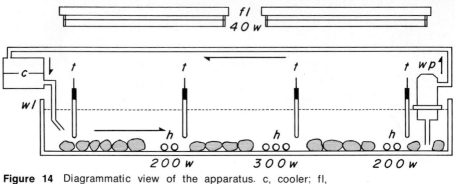

Figure 14 Diagrammatic view of the apparatus. c, cooler; fl, fluorescent lamps (40 W); h, heater; s, switch; t, thermostat; wl, water level; wp, water pump.

form and usually occurs in very cold and clean water in East and North Hokkaidô (Kawakatsu, 1964, 1965a, 1967). *P. sapporo* is also a stenothermic form, but the species can live in warmer waters than that at localities of *P. schmidti*. A crowding of planarians can be affected by the presence of food.

Phagocata kawakatsui is an inhabitant of small springs and spring-fed streams. This species reproduces only sexually with cocoon deposition. The number of sexually mature worms reaches its maximum in winter to early spring, and the cocoons are laid in the spring (March to May). After the breeding season, a high rate of mortality is observed in the populations. From summer to early autumn large worms are rarely found in the habitats, and only small-size, nonsexual worms can be seen. In field observations it is noted that the population size grows larger in late autumn and that many worms of this species hide in the soft mud of the habitat during winter and early spring (Kawwakatsu, 1965a, p. 397; Kawakatsu and Iwaki, 1967a).

The principal cause for the seasonal change of the population size of *P. kawakatsui* can be the intraspecific mode of its life cycle, which is controlled strongly by the seasonal change of water temperature of the habitats. Several other environmental factors may be of only secondary importance in the seasonal change of the population size.

The seasonal migration of several Japanese species of freshwater planarians distributed in North Hokkaidô is discussed below.

1. A spring-fed Stream in Nayoro City An investigation of the seasonal migration of three species inhabiting a spring-fed stream in Nayoro City, North Hokkaidô, was carried out by Yamada (1965a, b). As shown in the sketch map of Fig. 15, there is a large spring-fed stream near the center of the swampy place which has developed on an alluvion of the Nayoro River, a large tributary of the Teshio River (altitude, 100 m). Topographically, the stream is divided into three parts: the wide southern part with several springs along the east side of the stream, the rather narrow middle part with a number of springs along the east side of the stream, and the narrow northern part. The water of the first and second parts of the stream is very clean and plentiful. The third part of the stream is extremely filthy, and the water spreads into the ground. The swampy place is grown with *Fraxinus mandshurica* and several kinds of deciduous plants. It is also thickly grown with *Equisetum hyemale, Lysichiton camtschatense, Oenanthe stolonifera, Polygonum thunbergi, Chrysosplenium macrostemon*, etc. In the stream several common species of aquatic animals are found in addition to the above-mentioned planarian species. They are *Hydra* sp., a number of species of Hirudinea, *Chironomus* sp., *Gammarus* sp., *Pseudocrangonix yezonis, Pungitium tymensis, Misgurnus anguillicaudatus*, etc.

The natural features of the eleven stations where field observations were made are as follows:

Station A. A large spring at the source of the stream: very clean water 40 cm in depth; bottom sandy with stones.

Station B. A small outcrop of spring water located in the stream 6 m from the source: clean water 50 to 70 cm in depth; bottom muddy with few stones.

Station C. A small spring located near the east side of the stream 15 m from the source: clean water 20 to 30 cm in depth; bottom sandy with stones.

**Table 6   Migrations of *Polycelis sapporo* and *Polycelis schmidti*
Observed under Artificial Conditions**

Station and the condition of water temperature	Number of specimens calculated after 24 hours	
	Polycelis sapporo	*Polycelis schmidti*
Switch the A (4.0°C)	111	355
heaters on: B (6.0°C)	242	140
C (12.0°C)	137	5
D (15.0°C)	10	0
Total	500	500
Switch the A (4.0°C)	93	241
heaters off: B (4.2°C)	191	117
C (4.5°C)	104	31
D (4.5°C)	112	111
Total	500	500
Switch the A (4.0°C)	396	414
heaters on: B (6.0°C)	101	79
C (12.0°C)	2	7
D (15.0°C)	1	0
Total	500	500

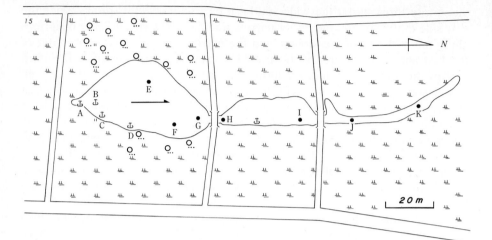

Figure 15 Sketch map showing a spring-fed stream in Nayoro City
(reproduced from Yamada, 1965; partly modified). Arrow indicates
the direction of the stream.

Station D. A small spring located near the east side of the stream 25 m
from the source: clean water 45 cm in depth; bottom muddy.

Station E. A small pool with much vegetation located near the west side
of the stream 35 m from the source: almost all stagnant shallow water; bottom
muddy.

Station F. A station located near the east side of the stream 40 m from
the source: running water 60 cm in depth; bottom muddy with much trash.

Station G. A station in the stream center 50 m from the source: running
water 60 cm in depth; bottom muddy.

Station H. Center of stream 70 m from the source: running water 50 to
60 cm in depth; bottom muddy with sand and organic matters.

Station I. A station with much vegetation located near the east side of
the stream 110 m from the source: running water 20 to 30 cm in depth; bottom
muddy.

Station J. A station located in the center of the stream 130 m from the
source: slowly running water 20 to 30 cm in depth; bottom muddy with much
organic matter.

Station K. A station located near the end of the stream 160 m from the
source. Its natural features are similar to those of station J.

Seasonal migration was studied with the Carpenter-Kawakatsu method
(Kawakatsu, 1965a, 1966a, 1967). Once a month, as many planarians as
possible were collected for 10 min on each visit to the stations, and the number
of worms of each species was noted; the results were recorded, together with
the variations of water temperature of the stations. Notes were also taken of

the degree of sexual maturity of worms, and of the number of cocoons obtained After the observations were made, most of the worms were set free in each original locality. Most of the survey was made from January to May 1962 and from June to December 1963. Figure 16 shows the seasonal change of water temperature and that of air temperature at station H. The seasonal

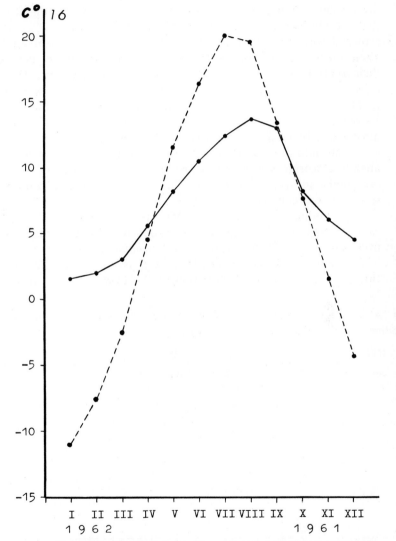

Figure 16 Graph showing the seasonal variation of water temperature (solid line) and air temperature (broken line) in a spring-fed stream in Nayoro City (station H). *(Reproduced from Yamada, 1965.)*

change of the number of worms of the three species in each station is shown in Table 7A to C.

Phagocata iwamai is the dominant species. It scarcely migrates in this stream according to the seasonal change of water temperature. From March to June it develops reproductive organs and from May to June it lays cocoons. Dendrocoelopsis lactea, a densely inhabiting species, is the most sensitive to the seasonal change of water temperature. In colder months, it was observed that the stations near the spring of rather warm water (C, D, and I) were crowded with this species. In the warmer seasons, they spread their domain. From December to April of the next year, almost all the specimens develop their reproductive organs and sporadically lay cocoons. Dendrocoelopsis ezensis is also sensitive to the seasonal change of water temperature. Stations, C, D, F, H, and I are densely inhabited by this species in the warmer seasons. Its reproductive organs develop from December to May of the next year. The most active breeding is observed in April and May.

The data are arranged in two groups, i.e., sexually mature worms and immature worms; the results, expressed in percentages of the total number, are graphically recorded in Fig. 17a–c. The seasonal migration of the three species is summarized in Fig. 18a–c.

According to Matsuda and Matsuyama (1968), in the Motomachi locality in Sapporo City, the maximum density of the populations is observed from January to March and May in P. iwamai, and from November to March of the next year in D. lactea. D. ezensis is not so common in this locality. Interpreting the seasonal migration of P. iwamai, D. lactea, and D. ezensis

Table 7A   Seasonal Variations of the Number of Specimens of Phagocata iwamai Inhabiting a Spring-fed Stream in Nayoro City

Sta-tion	1962					1963						
	Jan.	Feb.	Mar.	Apr.	May	June	July	Aug.	Sep.	Oct.	Nov.	Dec.
A			−	−	−	−	−	−	−	−	−	−
B			−	−	+	+	−	−	+	+	+	+
C			+	+	+	+	+	+	+	+	+	+
D			++	++	++	++	++	+++	++	++	++	++
E			−	−	−	−	−	−	−	−	−	−
F			++	++	++	++	++	+++	+++	+++	+++	++
G			+	+	++	++	++	++	++	++	++	++
H	+++	+++	++	+++	+++	+++	+++	+++	+++	+++	+++	+++
I			++	++	+++	+++	+++	+++	+++	+++	+++	+++
J			−	−	−	−	−	−	−	−	−	−
K			−	−	−	−	−	−	−	−	−	−

Note: The number of specimens collected within 10 min in each station.
−: absence; +: 1 to 10 specimens; ++: 11 to 50 specimens; +++: over 50 specimens.
Source: Reproduced from Yamada, 1965.

Table 7B Seasonal Variations of the Number of Specimens of *Dendrocoelopsis lactea* Inhabiting a Spring-fed Stream in Nayoro City*

Sta- tion	1962					1963						
	Jan.	Feb.	Mar.	Apr.	May	June	July	Aug.	Sep.	Oct.	Nov.	Dec.
A			–	–	–	–	–	–	–	–	–	–
B			–	+	++	++	++	+++	+++	+++	+++	–
C			++	+++	+++	+++	++	++	++	++	++	–
D			+++	+++	+	++	++	+++	+	+	+	+++
E			–	–	–	–	–	–	–	–	–	–
F			+	++	++	++	++	++	++	+++	+++	+
G			+	+	+	+	+	+	+	+	+	
H	+	+	+	++	++	+++	+++	+++	+++	+++	+++	+
I			–	–		+	+	+	+	+	+	+++
J			–	–	–	–	–	–	–	–	–	–
K			–	–	–	–	–	–	–	–	–	–

*See Table 7A.
Source: Reproduced from Yamada, 1965.

Table 7C Seasonal Variations of the Number of Specimens of *Dendrocoelopsis ezensis* Inhabiting a Spring-fed Stream in Nayoro City*

Sta- tion	1962					1963						
	Jan.	Feb.	Mar.	Apr.	May	June	July	Aug.	Sep.	Oct.	Nov.	Dec.
A			–	–	–	–	–	–	–	–	–	–
B			–	+	+	+	+	+	+	+	+	+
C			++	++	++	++	++	+	++	+	+	+
D			+	+	+	+	+	+++	+	+	+	+
E			–	–	–	–	–	–	–	–	–	–
F			+	+	+	+	++	+	+	+	+	+
G			+	+	+	+	+	+	+	+	+	+
H	+	+	+	++	++	++	++	++	++	++	++	++
I				+	+	+	+	++	+	+	+	+
J			–	–	–	–	–	–	–	–	–	–
K			–	–	–	–	–	–	–	–	–	–

*See Table 7A.
Source: Reproduced from Yamada, 1965.

in both localities, we may attribute it to the seasonal change of water temperature of the habitat and the growth in population by the newly hatched juveniles.

2. A spring-fed Stream at Kaminayoro This is a large spring-fed stream running through a hillside at Kaminayoro in North Hokkaidô (the Kaminayoro-

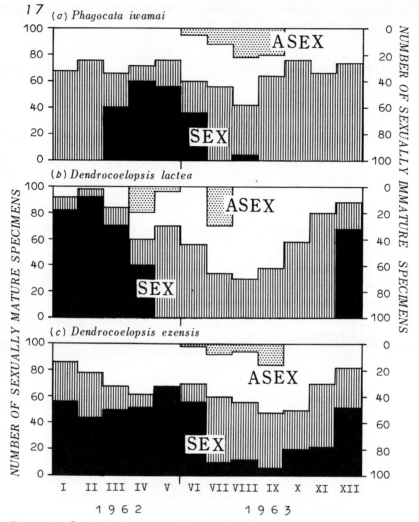

Figure 17 Graphs indicate the seasonal change of number of worms of three species of freshwater planarians inhabiting a spring-fed stream in Nayoro City (station H) *(reproduced from Yamada, 1965; partly modified). (a) Phagocata iwamai; (b) Dendrocoelopsis lactea; (c) Dendrocoelopsis ezensis.* SEX, sexually mature worms (black area: fully mature worms); ASEX, sexually immature worms (dotted area: newly hatched young worms).

issen-zawa valley) and is the type locality of *Polycelis akkeshi* (Ishikawa and Kawakatsu, 1963; Yamada, 1962, p. 49, fig. 1, station 92). As shown in Fig. 19, it is a slowly running narrow stream fed by very cold springwater from station A (altitude, 200 m). The bottom of the source of the stream is muddy, with numerous pebbles (30 cm in depth) and aquatic mosses. The stream extends about 450 m (station F; altitude, 180 m). Swampy places are found around the downstream portion. The field survey was made in the six stations (A to F) in August and December 1965 and June 1966.

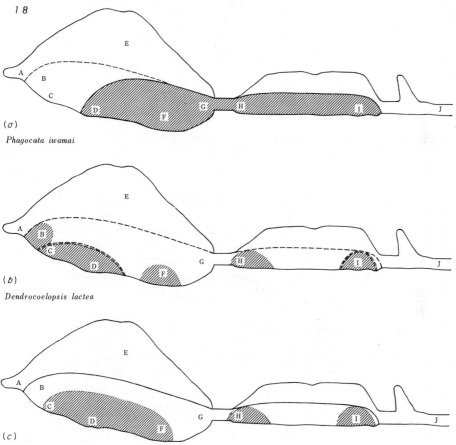

Figure 18 Figures showing the seasonal migration of three species of freshwater planarians inhabiting a spring-fed stream in Nayoro City *(reproduced from Yamada, 1965). (a) Phagocata iwamai; (b) Dendrocoelopsis lactea; (c) Dendrocoelopsis ezensis.* Areas surrounded by the solid line inhabited throughout the year; areas surrounded by the broken line inhabited from spring to autumn; areas surrounded by the double broken line inhabited only in winter. Lined areas indicate the densely populated parts, and the open areas, the sparsely populated parts.

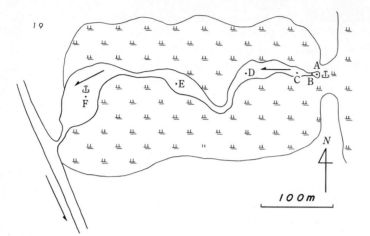

Figure 19 Sketch map showing a spring-fed stream at Kamina-yoro. Arrows indicate the direction of the stream.

A large population of *P. akkeshi* is found in the upper part of the stream. A large number of *Gammarus nipponensis* also occur, and it is frequently observed that these planarians attack living specimens of *Gammarus*. In station F, a small population of *Dendrocoelopsis lactea* occurs. Table 8A shows the seasonal change of the physicochemical data in each station. The seasonal variation of the size of the planarian population, together with that of the *Gammarus* population, is shown in Table 8B.

3. A spring-fed Stream at Naka'aibetsu   This is a narrow and flat mountain stream which is fed by springwaters and flows into a wide swampy place near the upper part of the Ishikari River in Central Hokkaidô (Yamada and Tanji, 1964, fig. 2; the opposite river side of stations 87 to 108). As shown in Fig. 20, it is a slow-running, rather shallow stream and extends about 300 m (altitude, 120 m; 20 to 30 cm in depth). The bottom of station A is muddy, and aquatic algae (*Spirogyra*) grow in summer. The stream is fed by two springs located in stations B and D. The stream bed of the upper part is rocky with scattered stones and the water is clean (stations B to E). Downstream, the bottom becomes muddy and almost all stagnant water is not clean (stations F to I). Larvae of *Chironomus* are common downstream. The field survey was made in August and December 1965 and June 1966.

Phagocata vivida, Polycelis sapporo, and Polycelis auriculata are common in this stream. The seasonal change of the physicochemical data and the seasonal variation of the size of the three planarian populations are shown in Table 9A and B.

4. A spring-fed Stream at On'nenai   This is a small stream fed by a rich spring at On'nenai in North Hokkaidô (Yamada, 1966, p. 131, fig. 1, station 63). As shown in Fig. 21, it is a slow-running narrow stream fed by a cold-water spring (station A; altitude, 120 m). The bottom of the source of the stream is muddy, with many pebbles, covered with fallen leaves (20 cm in depth). Downstream, and in swampy places, several kinds of aquatic plants such as *Lysichiton camtschatense* grow. *Gammarus* sp. and nymphs of *Scopula longa* are common in the source of the stream. The field survey was made in August, September, and November 1965 and May 1966.

Polycelis sapporo and *Polycelis schmidti* are common in this locality (Kawakatsu, 1964). A small population of *Phagocata vivida* also occurs (the northernmost locality of this species; Kawakatsu and Yamada, 1967). The seasonal change of the physicochemical data and the seasonal variation

Table 8A   Seasonal Variations of the Physicochemical Data in a Spring-fed Stream at Kaminayoro

Sta-tion	August 1965 (23.0° C)		December 1965 (−5.8° C)		June 1966 (19.0° C)		
	Water temperature, ° C	pH	Water temperature, ° C	pH	Water temperature, ° C	pH	O_2, mg/l
A	8.6	7.4	6.5	7.2	7.9	7.3	13.9
B	9.0	7.2	5.2	7.3	8.2	7.2	13.5
C	14.1	7.3	2.8	7.4	7.9	7.4	11.2
D	16.2	7.2	2.5	7.5	8.4	7.2	10.1
E	16.4	7.3	2.3	7.2	8.5	7.2	10.3
F	15.8	7.2	4.2	7.2	9.2	7.2	10.7

Table 8B   Seasonal Variations of the Number of Specimens of *Polycelis akkeshi* in a Spring-fed Stream at Kaminayoro

Sta-tion	August 1965		December 1965		June 1966	
	Polycelis akkeshi	*Gammarus* sp.	*Polycelis akkeshi*	*Gammarus* sp.	*Polycelis akkeshi*	*Gammarus* sp.
A	150	+++	340	+++	435	+++
B	98	+++	29	+++	397	+++
C	15	+++	2	+++	45	+++
D	0	+	0	+++	0	+
E	0	−	0	−	0	−
F*	0	−	0	+	0	+
Total	263		371		877	

Note: The number of specimens collected within 10 min in each station. The number of specimens of *Gammarus* species is also shown.
   −: absence; +: 1 to 10 specimens; ++: 11 to 50 specimens; +++: over 50 specimens.
   *A number of specimens of *Dendrocoelopsis lactea* occur at this station.

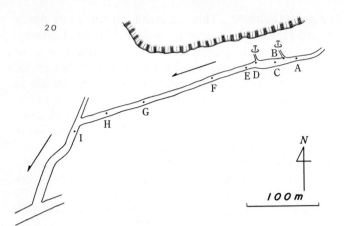

Figure 20   Sketch map showing a spring-fed stream at Naka'aibetsu. Arrows indicate the direction of the stream.

of the size of the three planarian populations are shown in Table 10A and B.

According to the data from the three different localities (Kaminayoro, Naka'aibetsu, and On'nenai) the size of the populations of the five planariid species varies with the seasons of the year. In the Kaminayoro locality, the stations near the source of the stream are crowded with *P. akkeshi* in any season. Their population size is comparatively small in midsummer, but it inclines to the maximum from winter to early summer. The same tendency was observed in the variation of the *Gammarus* population. The range of the seasonal change of water temperature in this stream is narrow at the stations near the source although it increases from the source downstream (Table 8B). The dissolved oxygen content of the water is nearly at its saturated condition in any season, but a slight decrease was observed from the source downstream (Table 8A).

In the Naka'aibetsu locality the seasonal variation of water temperature is narrow at the station near the source. The dissolved oxygen content is near saturation in both the upper and the middle parts of the stream. As will be seen in Table 9B, in general, it may be said that *P. auriculata* inhabits the spring and the cold-water part of the upper stream, and at lower stations it is replaced gradually by *P. vivida*, which, still lower down, gives place to *P. sapporo*. The population size of *P. auriculata* is small in midsummer, but becomes larger in winter. In midsummer they crowd at the stations near the springs. During winter and in early summer, they spread over the stations in the source and lower stream. The population size of *P. vivida* is small from winter to early summer, but it increases in midsummer. The population size of *P. sapporo* is small in winter, but it inclines to the maximum in summer.

The low-temperature stations are crowded with specimens in midsummer; but in the colder seasons they spread their domain over a wide area in the stream.

In the On'nenai stream where *P. sapporo* and *P. schmidti* densely inhabit, a narrow range of the seasonal changes of water temperature and the dissolved oxygen content of water are observed in most of the stations. As will be seen in Table 10B, *P. schmidti* inhabits the stations near the source of the stream, and the lower stations are more commonly inhabited by *P. sapporo*. The population size of *P. schmidti* is large from summer to winter, but it declines in spring. The cold-water stations are crowded with specimens

Table 9A   Seasonal Variations of the Physicochemical Data in a Spring-fed Stream at Naka'aibetsu

Sta-tion	August 1965 (26.8° C) Water temperature, ° C	pH	O_2, mg/l	December 1965 (3.5° C) Water temperature, ° C	pH	O_2, mg/l	June 1966 (18.2° C) Water temperature, ° C	pH	O_2, mg/l
A	15.1	7.2	12.1	6.4	7.1	12.5	11.6	7.3	12.4
B	8.6	7.2	12.3	7.9	7.2	12.4	7.2	7.2	12.7
C	10.1	7.2	11.9	6.5	7.3	12.3	8.9	7.1	12.0
D	10.4	7.2	10.4	6.5	7.1	12.1	8.4	7.1	11.9
E	10.6	7.3	9.2	6.4	7.0	11.0	8.4	7.1	9.8
F	9.8	7.3	12.1	4.5	7.3	12.4	9.2	7.1	12.2
G	14.3	7.3	10.9	4.5	7.2	11.8	9.9	7.1	11.9
H	18.1	7.3	10.2	4.5	7.0	11.5	10.0	7.0	11.7
I	20.1	7.0	8.3	4.6	7.1	10.9	14.4	6.9	10.2

Table 9B   Seasonal Variations of the Number of Specimens of *Phagocata vivida*, *Polycelis sapporo*, and *Polycelis auriculata* in a Spring-fed Stream at Naka'aibetsu

Sta-tion	*Phagocata vivida* 1965 Aug.	1965 Dec.	1966 June	*Polycelis sapporo* 1965 Aug.	1965 Dec.	1966 June	*Polycelis auriculata* 1965 Aug.	1965 Dec.	1966 June
A	2	3	4	1	15	0	1	10	3
B	191	98	48	54	101	39	131	261	189
C	223	187	121	88	64	72	243	211	159
D	96	62	94	96	23	8	18	39	68
E	1	19	43	0	15	2	13	92	89
F	1	36	50	193	5	174	1	13	5
G	10	5	20	41	3	100	9	21	38
H	0	0	0	51	1	61	0	0	0
I	0	0	0	0	0	0	0	0	0
Total	524	410	380	524	227	456	416	647	551

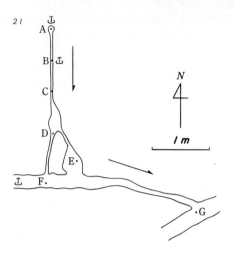

Figure 21 Sketch map showing a spring-fed stream at On'nenai. Arrows indicate the direction of the stream.

in midsummer. In the case of *P. sapporo*, the seasonal change of population size in this stream shows nearly the same tendency as in the Naka'aibetsu locality.

From the foregoing descriptions, it is seen that many stream-dwelling planarian species show seasonal migrations according to the change of water temperature of their habitats. Some of them, such as *P. vivida*, *P. auriculata*, and *P. schmidti*, are more sensitive to the seasonal ups and downs of water temperature than *P. sapporo*. *P. akkeshi* seems not to be so sensitive. Moreover, the differences of the breeding season and/or the mode of reproduction in each species may be said to play an important part. Egg laying and hatching of the cocoons of *P. vivida* occur in spring, and soon afterwards, the high rate of mortality is usually observed in this species. In some populations asexual reproduction by fragmentation occurs in summer (Fig. 12). The decrease in spring and the increase in summer of the size of the *P. vivida* population in the Naka'aibetsu stream can be explained by the seasonal fluctuation of number of both adult and young worms in their life cycle.

The egg-laying season of *P. auriculata* is from late autumn to midwinter; its cocoons hatch out in spring. This species reproduces asexually by active fission in summer (Fig. 12). According to my experience, populations of this species in laboratory cultures die gradually within 2 to 3 years; the rate of mortality is usually somewhat higher in summer than in the colder season (Fig. 12). The midsummer decrease of the *P. auriculata* population in the Naka'aibetsu stream may be due to the death of sexual worms. The winter-and-spring increase is explained by the hatching of their cocoons and the growth of juveniles.

The seasons of egg laying and hatching of cocoons of *P. schmidti* are nearly the same as those of *P. auriculata* (Fig. 12). The midsummer increase in size of the *P. schmidti* population in the On'nenai stream may be related to their life cycle. The egg-laying season of *P. sapporo* is from autumn to early winter; its cocoons hatch in early spring (Fig. 12). The winter decrease and the spring increase of the size of the *P. sapporo* population in the Naka'aibetsu stream and its life cycle are relative to each other. In the cold-water habitats, however, the duration of its breeding season is prolonged extensively. The egg-laying season of *P. akkeshi* is autumn and early winter, and hatching of its cocoons occurs in spring. After the breeding season, a low rate of mortality is seen in this species (Fig. 12). The teeming population of *P. akkeshi* of the Kaminayoro stream in spring is the result of the increase of newly hatched young worms.

In the Nayoro locality where *Phagocata iwamai, Dendrocoelopsis*

Table 10A Seasonal Variations of the Physicochemical Data in a Spring-fed Stream at On'nenai

Sta-tion	August 1965 Water temperature, °C	pH	September 1965 Water temperature, °C	pH	November 1965 Water temperature, °C	pH	May 1966 Water temperature, °C	pH	O_2, mg/l
A	5.9	7.2	5.8	7.2	4.9	7.3	2.9	7.4	14.1
B	5.9	7.2	5.8	7.2	4.9	7.3	2.9	7.4	14.2
C	6.2	7.2	5.8	7.2	3.9	7.3	3.1	7.4	14.2
D	8.8	7.2	5.8	7.2	3.9	7.3	3.2	7.5	14.3
E	10.4	7.2	5.8	7.2	3.5	7.3	3.8	7.3	12.2
F	11.5	7.0	5.8	7.1	3.5	7.0	3.9	7.3	14.5
G	23.4	7.0	7.1	7.1	4.2	7.1	4.5	7.3	11.2

Table 10B Seasonal Variations of the Number of Specimens of *Phagocata vivida*, *Polycelis sapporo*, and *Polycelis schmidti* in a Spring-fed Stream at On'nenai

Sta-tion	Phagocata vivida 1965 Aug.	1965 Sept.	1965 Nov.	1966 May	Polycelis sapporo 1965 Aug.	1965 Sept.	1965 Nov.	1966 May	Polycelis schmidti 1965 Aug.	1965 Sept.	1965 Nov.	1966 May
A	−	−	−	−	+	+	+	+	+++	++	+++	+
B	−	−	−	−	+	+	+	+	+++	++	+	+
C	−	−	−	−	+	+	+	+	++	++	+	++
D	−	−	−	−	+	+	+	+	++	++	+	++
E	−	−	−	−	+	+	++	+++	+	++	++	+
F	−	+	++	++	++	++	++	+	++	+++	++	+++
G	−	−	−	−	−	−	−	−	−	−	−	−

Note: The number of specimens collected within 10 min in each station.
−: absence; +: 1 to 10 specimens; ++: 11 to 50 specimens; +++: over 50 specimens.

lactea, and *Dendrocoelopsis ezensis* inhabit, the increase of the number of young or small-size asexual worms takes place coincidentally with the seasons of hatching cocoons in each species (Yamada, 1965a, b). Moreover, in the habitats of fissioning planarians which seem to be newly regenerated, worms are usually obtained in the warm season. These findings favor the supposition and make the above consideration almost a matter of certainty.

The main points discussed above may be summarized as follows: The principal causes of the seasonal migration and the seasonal fluctuation of the population size are to be sought both in the migration of adult worms according to the seasonal change of water temperature of the habitats and in the influence of newly hatched young worms. The death of adult worms which occurs after the breeding season is also a contributory cause of the seasonal change of the population size.

There are several factors affecting the abundance of Japanese stream-dwelling planarians. The condition of the stream bed is not an important factor. Many species can be found in stony places as well as in muddy places. In general, the velocity of water has more effect on the occurrence of planarians. Several species such as *Dugesia japonica*, *Polycelis sapporo*, and *Polycelis akkeshi* usually occur in slow currents. On the contrary, *Phagocata vivida*, *Polycelis auriculata*, and *Polycelis schmidti* occur in swift currents (Kawakatsu, 1965a, 1967).

Recently, an attempt was made for a qualitative analysis of some water-chemical features (pH, COD value, total hardness, SiO_2, NH_4^+-N, NO_3^--N, PO_4^{3-}, Fe, Cl^-) in the nineteen stations of various kinds in Hokkaidô where several species of planarians inhabit (Kawakatsu, Iwaki, and Yamada, 1969). The species inhabiting these stations are *P. vivida*, *P. sapporo*, *P. auriculata*, *Dendrocoelopsis lactea*, and *Dendrocoelopsis ezensis*. Noticeable pollution of water was not observed in every locality examined. There is a possibility that somewhat higher levels of dissolved ammonium and organic matter in some stations may exclude *P. auriculata*, since the species is a stenothermic stream-dwelling form inhabiting very clean waters. Both *P. vivida* and *P. sapporo* may occur in the stations of more or less high degree of dissolved organic matter. *D. lactea* and *D. ezensis* may not be so sensitive to the putrefactive material in water. However, the water-chemical contents may be of only secondary importance to occurrence and abundance of planarians, except for the cases of high degree of mineral and artificial pollution of water.

SUMMARY

In the present paper, further studies of the vertical distribution of Japanese freshwater planarians, as well as of that of the Far Eastern species, are made, together with some zoogeographical and ecological observations on the possible analysis of the results (cf. Kawakatsu, 1965, pp. 349–408.). In the Japanese Islands, the habitats of the following five stream-dwelling planarians are widely distributed over the geographical areas: *Dugesia japonica* Ichikawa and Kawakatsu, *Phagocata vivida* (Ijima and Kaburaki), *Polycelis auriculata* Ijima and Kaburaki, *Polycelis sapporo* (Ijima and Kaburaki), and *Polycelis schmidti* (Zabusov). *D. japonica*, *P. vivida*, *P. sapporo*, and *P. schmidti* are distributed also in the countries adjacent to Japan. These five species show characteristic vertical distribution.

The Far East, including the Japanese Islands, can be divided into nine areas in reference to the geographical distribution of the above-mentioned species, namely, (1) the *japonica* area, (2) the *japonica-vivida* area, (3) the *japonica-vivida-auriculata* area, (4) the *japonica-vivida-sapporo-auriculata* area, (5) the *japonica-vivida-sapporo-schmidti* area, (6) the *japonica-sapporo-auriculata* area, (7) the *japonica-sapporo-schmidti* area, (8) the *sapporo-schmidti* area, and (9) the *schmidti* area. The Tsugaru Strait subdivides the *japonica-vivida-sapporo-auriculata* area in two subareas. In each area, there are considerable differences in the vertical extension of the habitats of these five species. There is a close relationship between the height of their habitats and the latitudes. The types of the vertical distribution in each area are: (1) J, (2) J-JV-V, (3), J-JV-JAV-AV-V and J-JV-JVA-VA-A, (4) JSV-JSVA-SVA-VA-A and JSV-SVA-VA-A, (5) JSV-SVC-VC-C, (6) JSA-SA-A, (7) JSC-SC-C, (8) SC-C, and (9) C.

From the ecological point of view, they show a high adaptability to their surroundings in warm- or cold-water regions. They could, therefore, have gradually spread their domains in the Far East throughout the past geological age. This supposition will be evidenced by the fact that these widely distributed species are more or less polymorphic. The water temperature seems to be one of the important factors controlling the occurrence and distribution of the stream-dwelling planarians in Japan. The principal causes of the seasonal migration and fluctuation of the population size are to be sought in both the normal seasonal migration of adult worms and in the migration caused by the seasonal change of water temperature in the habitats and its influence on the newly hatched young worms. The death of many specimens of adult worms in a certain population which occurs after the breeding season is a contributory cause of the seasonal change of the population size. The food supply in the natural habitats may be of secondary importance in the occurrence and distribution of the Japanese stream-dwelling planarians.

In my opinion, the fundamental principles of the vertical distribution of freshwater planarians in the Far East including the Japanese Islands have been fixed by their migrations during the past geological age, while the ecological background to the vertical distribution in a certain limited area is probably due to the combination of the physicochemical and biological factors. Among these factors, the water temperature of the natural habitats may be of primary importance in the occurrence and distribution of the stream-dwelling planarians.

BIBLIOGRAPHY

For further references on Japanese planarians, refer to the author's serial articles entitled "A List of Publications on Japanese Turbellarians," etc., published since 1968 (*Bull. Fuji Women's College*, no. 6: 71–127; no. 7, ser II, 23–43; no. 8, ser. II, 107–114; no. 9, ser. II, 29–40 + 1 folder).

Ball, I. R. 1969a. An Annotated Checklist of the Freshwater Tricladida of the Nearctic and Neotropical Regions. *Can. J. Zool.*, **47**:59–64.

——. 1969b. *Dugesia lugubris* (Tricladida: Paludicola), a European Immigrant into North American Fresh Waters. *J. Fisheries Res. Board Can.*, **26**:221–228.

——. 1970. Freshwater Triclads (Turbellaria, Tricladida) from the Oriental Region. *Zool. J. Linnean Soc.*, **49**:271–294 + Pls. 1–2.

——. 1971. Systematic and Biogeographical Relationships of Some *Dugesia* Species (Tricladida, Paludicola) from Central and South America. *Am. Museum Novitates*, no. 2472, 1–25.

—— and C. H. Fernando. 1968. On the Occurrence of *Polycelis* (Turbellaria, Tricladida) in Western Canada. *Can. Field Nat.*, **82**:213–216.

Benazzi, M., and E. Giannini. 1971. *Cura azteca*, nuova specie di planaria del Messico. *Accad. Nazl. Lincei*, ser. VIII, 50 (fasc. 4): 477–481 + Pls. I–II.

Carpenter, J. H. 1969. A New Planarian from Utah, *Phagocata crenophila* n. sp. (Turbellaria, Tricladida). *Trans. Am. Microscop. Soc.*, **88**:274–281.

Chandler, C. M. 1966. Environmental Factors Affecting the Local Distribution and Abundance of Four Species of Stream-dwelling Triclads. *Invest. Indiana Lakes Streams*, **8**(1):1–56.

Hisiao, S. D. 1935. A Preliminary Study of the Seasonal Changes in the Reproductive System of *Planaria gonocephala* Dugès. *Bull. Nat. Hist. Peking*, **9**:161–169.

Holmquist, C. 1967a. Turbellaria of North Alaska and Northwestern Canada. *Intern. Rev. Ges. Hydrobiol.*, **52**:123–139.

——. 1967b. *Dendrocoelopsis piriformis* (Turbellaria Tricladida) and Its Parasites from Northern Alaska. *Arch. Hydrobiol.*, **62**:453–466.

Hyman, L. H. 1963. North America Triclad Turbellaria. 16. Fresh-water Planarians from the Vicinity of Portland, Oregon. *Am. Museum Novitates*, no. 2123, 1–5.

Ichikawa, A., and M. Kawakatsu. 1963. *Polycelis akkeshi*, a New Freshwater Planarian, from Hokkaidô. *Publ. Akkeshi Marine Biol. Sta.*, no. 12, 1–18.

—— and ——. 1967a. Report on Freshwater Planaria from the East China Sea Area. *Nature Life Southeast Asia*, **5**:175–188.

—— and ——. 1967b. Records of Two Planarian Species of the Family Kenkiidae from Japanese Subterranean Waters. *Arch. Hydrobiol.*, **63**:512–519.

Ijima, I., and T. Kaburaki. 1916. Preliminary Descriptions of Some Japanese Triclads. *Annotationes Zool. Japon.*, **9**:153–171.

Kaburaki, T. 1922. On Some Japanese Fresh-water Triclads; with a Note on the

Parallelism in Their Distribution in Europe and Japan. *J. Coll. Sci. Imp. Univ. Tokyo*, **44** (art. 2): 1-71 + Pl. I.

Kawakatsu, M. 1964. Studies on the Taxonomy and Morphology of the Japanese Freshwater Planarian, *Polycelis schmidti* (Zabusov). *Annotationes Zool. Japon.*, **37**:174-184.

———. 1965a. On the Ecology and Distribution of Freshwater Planarians in the Japanese Islands, with Special Reference to Their Vertical Distribution. *Hydrobiologia*, **26**:349-408.

———. 1965b. Some Ecological Notes on the Freshwater Planarians of Lake Tahoe in the Sierra Nevada Mountains in North America and Lake Biwa-ko in Middle Japan. *Japan. J. Limnol.*, **26**:106-112. (In Japanese with English summary.)

———. 1966a. Methods of the Ecological Survey of Freshwater Planarians. *Japan. J. Ecol.*, **16**:123-124. (In Japanese.) The Korean edition of this article in *Korean J. Limnol.*, **3**(1-2):11-14 (1970).

———. 1966b. Synopsis of the Known Species of Freshwater Planarians in Japan. *Bull. Biogeogr. Soc. Japan*, **24**(2):9-28. (In Japanese.)

———. 1967. On the Ecology and Distribution of Freshwater Planarians in the Japanese Islands, with Special Reference to Their Vertical Distribution (rev. ed.). *Bull. Fuji Women's College*, no. 5, 117-177.

———. 1968a. North American Triclad Turbellaria. 17. Freshwater Planarians from Lake Tahoe. *Proc. U.S. Natl. Museum*, **124**(3638):1-21 + Pls. 1-2.

———. 1968b. Illustrated List of Japanese Freshwater Planarians. *Collect. and Breed.* (*Tokyo*), **30**:40-45. (In Japanese.)

———. 1968c. On the Origin and Phylogeny of Turbellarians: Suborder Paludicola. *Japan. Soc. Systematic Zool. Circ.*, nos. 38-41:11-22. (In Japanese with English explanation of figures.)

———. 1969a. Report on Freshwater Planaria from India. *Annotationes Zool. Japon.*, **42**:210-215.

———. 1969b. An Illustrated List of Japanese Freshwater Planarians in Color. *Bull. Fuji Women's College*, no. 7, ser. II: 45-91 + Pls. VII-VIII. (In English and Japanese.)

———. 1970. Parasites of Turbellarians: Records of Several Species of Endoparasites of Freshwater Planarians. *Res. Bull. Meguro Parasitol. Mus.*, no. 3, 37-47 + Pls. I-IV. (In Japanese with English summary.)

———. 1971. Problems on the Morphological Variation and the Physiological Races of a Japanese Freshwater Planarian, *Dugesia japonica* Ichikawa and Kawakatsu. M. Kawakatsu and M. Iba (eds.), Comm. *Compil. Sci. Papers Publ. Occ. of the Retirement of Prof. Hisao Sugino at the Age of Sixty-five, Ser. Turbellarians*: 43-52. Ôsaka. (In Japanese with English summary.)

———. 1972a. Report on Freshwater Planaria from Borneo. *Contrib. Biol. Lab. Kyoto Univ.*, **23**(3-4): 115-122 + Pl. 1.

———. 1972b. The Freshwater Planaria from Batu Caves in Malaya. *Bull. Natl. Sci. Museum Tokyo*, **15**:339-346 + Pls. 1-3.

——— and J. A. Basil, 1971. Records of Freshwater and Land Planarians from India. *Bull. Fuji Women's College*, no. 9, ser. II, 41-50 + Pls. I-II.

——— and A. Ichikawa. 1971. *Dendrocoelopsis lactea* an Emendation of the Specific

Name of a Freshwater Planarian Referred to in the Literature as *Dendrocoelopsis lacteus* Ichikawa et Okugawa, with Remarks on Some Nomina Nuda of Japanese Freshwater Planarians. *Proc. Jap. Soc. Syst. Zool.*, no. 7, 5–12.

——— and S. Iwaki. 1967a. Studies on the Morphology, Taxonomy and Ecology of Freshwater Planarian, *Phagocata kawakatsui* Okugawa, with Remarks on Distribution. *Japan. J. Ecol.*, **17**:214–224.

——— and ———. 1967b. Report on Freshwater Planaria from the Satsunan Islands and Kagoshima (Kyûshû) in South Japan. *Bull. Fuji Women's College*, no. 5, 179–185.

——— and ———. 1968. Report on Freshwater Planaria from Taiwan (Formosa). *Bull. Fuji Women's College*, no. 6, 129–137.

———, ———, and W.-J. Kim. 1967. Report on Freshwater Planaria from Querpart (Cheju) Island, Korea. *Zool. Mag. Tokyo*, **76**:187–189.

———, ———, and T. Yamada. 1969. Report on the Ecological Survey of Freshwater Planarians in the Niseko Quasi National Park, Mt. Muine, the Uryû-numa District, the Southern Part of the Daisetsuzan National Park and the Seaboard District of North Hokkaidô. *Bull. Fuji Women's College*, no. 7, ser. II, 93–107. (In Japanese with English summary.)

——— and W.-J. Kim. 1966. Morphological Studies on the Freshwater Planarian, *Dugesia japonica* Ichikawa et Kawakatsu, from Korea. *Zool. Mag. Tokyo*, **75**:103–107. (In Japanese with English summary.)

——— and ———. 1967. Results of the Speleological Survey in South Korea 1966. VI. Freshwater Planarians from Limestone Caves in South Korea. *Bull. Natl. Sci. Museum Tokyo*, **10**:247–258 + Pls. 1–3.

——— and I. Tanaka. 1971. Additional Report on Freshwater Planaria from the South-west Islands of Japan. *Biol. Mag. Okinawa (Naha)*, **8**:46–52 + Pls. I–II. (In Japanese with English summary.)

———, W. Teshirogi, and H. Fujiwara. 1970. Report on the Ecological Survey of Freshwater Planarians in the Eastern Part of Aomori Prefecture, Honshû, with a Note on the Southern Distribution Limit of *Polycelis sapporo* and *Dendrocoelopsis lacteus*. *Bull. Fuji Women's College*, no. 8, ser. II, 127–141. (In Japanese with English summary.)

——— and T. Yamada. 1966. Additional Studies on the Morphology of *Polycelis schmidti* (Zabusov), with Remarks on Distribution. *Bull. Biogeogr. Soc. Japan*, **24**(1):1–8.

——— and ———. 1967. Report on the Ecological Survey of Freshwater Planarians in the South-eastern Part of the Shikotsu-Dôya National Park (Lake Shikotsu-ko District), Hokkaidô, with a Note on the Distribution of *Phagocata vivida* in North Japan. *Zool. Mag. Tokyo*, **76**:50–56. (In Japanese with English summary.)

———, ———, and S. Iwaki. 1967. Environment and Reproduction in Japanese Freshwater Planarians. *Japan. J. Ecol.*, **17**:263–266.

Kenk, R. 1953. The Fresh-water Triclads (Turbellaria) of Alaska. *Proc. U.S. Natl. Museum*, **103**(3322):163–186 + Pls. 6–8.

———. 1969. Freshwater Triclads (Turbellaria) of North America. I. The Genus *Planaria*. *Proc. Biol. Soc. Wash.*, **82**:539–558.

———. 1970a. Freshwater Triclads (Turbellaria) of North America. II. New or Little Known Species of *Phagocata*. *Proc. Biol. Soc. Wash.*, **83**:13–34.

————. 1970b. Freshwater Triclads (Turbellaria) of North America. III. *Sphalloplana weingartneri* new species, from a Cave in Indiana. *Proc. Biol. Soc. Wash.*, 83:313–320.

————. 1970c. Freshwater Triclads (Turbellaria) of North America. IV. The Polypharyngeal Species of *Phagocata*. *Smithsonian Contrib. Zool.*, no. 80, 1–17.

Matsuda, K., and Y. Matsuyama. 1968. Seasonal Variation of the Population Size of Three Freshwater Planarian Species Inhabiting in a Spring-fed Stream in Sapporo City, Hokkaidô. *Japan. J. Ecol.*, 18:134–136.

Mitchell, R. W. 1968. New Species of *Sphalloplana* (Turbellaria; Paludicola) from the Caves of Texas and a Reexamination of the Genus *Speophila* and the Family Kenkiidae. *Ann. Speleol.*, 23:597–620.

———— and M. Kawakatsu. The First Cavernicole Planarians from Mexico: New Troglobitic and Troglophilic *Dugesia* from Caves of the Sierra de Guatemala. *Ann. Speleol.*, In Press.

Ondô, Y., and N. Inoue. 1965. Ecological Studies on the Natural Population of Japanese Fresh-water Planaria, *Dugesia japonica* (Dugès). *Liberal Arts J. Tottori Univ. Nat. Sci.*, 16(1):1–10. (In Japanese with English summary.)

Porfirjeva, A. N. 1968. Planarii celenginskogo melkovod'ia i prilezhashchikh k nemu uchastkov Ozera Baikal. *Sbornik Kratkikh Soobshchenii, Zool.* II:3–9. Izdatel'stvo Kazanskogo Univ., Kazan. (In Russian.)

————. 1969. Endemichnaia Baikal'skaia Planaria *Rimacephalus arecepta* sp. n. (Tricladida, Paludicola). *Zool. Zh.*, 48(9):1303–1308. (In Russian with English summary.)

————. 1970a. K khracteristike fauny Planarii (Tricladida Paludicola) Baikala. *Boprosy Evoliutsionni Morfologii i Biogeografii*: 77–91. Izdatel'stvo Kazanskogo Univ., Kazan. (In Russian.)

————. 1970b. Ob endemichnom bidoobrazovanii u Baikal'skikh Dendrotselid (Tricladida, Paludicola). *Zool. Zh.*, 49(10):1456–1464. (In Russian with English summary.)

————. 1970c. Ocherki planarii Baikala, 11. *Protocotylus fungiformis* (H. Sab.). *Boprosy Evoliutsionnoi Morforogii i Biotsenologii*: 164–172. Izdatel'stvo Kazanskogo Univ., Kazan. (In Russian.)

————. 1971. Nekotorykh Evoliutsii Baikal'skikh Planarii. *Boprosy Morfologii i Ekologii Bespozvonochnykh*: 81–93. Izdatel'stvo Kazanskogo Univ., Kazan. (In Russian.)

Reynoldson, T. B. 1966. The Distribution and Abundance of Lake-dwelling Triclads —towards a Hypothesis. *Advances Ecol. Res.*, 3:1–71 + pl. I.

Tokuda, M. 1969. *The Biogeography*. Tôkyô: Tsukiji-Shoten. 220 pp. (In Japanese.)

Yamada, T. 1962. Notes on the Freshwater Planarians Found in the Vicinity of Mt. Piyashiri, Kitami Mountains, Hokkaidô. *Seibutsukyôzaino-Kaitaku*, no. 2, 36–49. (In Japanese with English summary.)

————. 1965a. Ecological Studies on Freshwater Planarians Found in a Spring-fed Stream in Nayoro City, Hokkaidô. I. Seasonal Migration of the Three Species. *Zool. Mag. Tokyo*, 74:156–164. (In Japanese with English summary.)

————. 1965b. Ecological Studies on Freshwater Planarians Found in a Spring-fed Stream in Nayoro City, Hokkaidô. II. Life Cycle of the Three Species. *Zool. Mag. Tokyo*, 74:226–237. (In Japanese with English summary.)

————. 1966. Report on the Ecological Survey of Freshwater Planarians in the Teshio Mountains and Its Adjacent Seaboard District on the Sea of Japan. *Japan. J. Ecol.*, **16**:129–133. (In Japanese with English summary.)

———— and H. Tanji. 1964. On the Distribution of Freshwater Planarians Found in the Ushubetsu and Antaroma River Systems in the Mts. Daisetsu District, Hokkaidô. *Rept. Taisetsuzan Inst. Sci. Hokkaidô Gakugei Univ. Asahigawa*, no. 3, 24–42. (In Japanese with English summary.)

Young, J. O., and T. B. Reynoldson. 1965. A Laboratory Study of Predation on Lake-dwelling Triclads. *Hydrobiologia*, **26**:307–313.

Zabusova-Zudanova, Z. I. 1956. Planarii Nyeryestilishch Lososyevuix ruib zapadnogo poberezjya Kamchatki. *Tr. Obshch. Estestvoisp., Uchenye Zapiski Kazan Gosud. Univ.* V. I. Ul'ianova-Lenina, **116**(14):25–39. (In Russian.)

————. 1960. Planarii Daljinyego Vostoka. *Tr. Obshch. Estestvoisp. 63, Kazan Gosud. Univ.*, **120**(6):112–121. (In Russian.)

————. 1970. Rasselenie planarii po Cobetskomu soiuzu. *Boprosy Evoliutsionnoi Morforogii i Biotsenologii*: 164–172. Izdatel'stvo Kazanskogo Univ., Kazan. (In Russian.)

Chapter 15

A Contribution to the Phylogeny and Biogeography of the Freshwater Triclads (Platyhelminthes: Turbellaria)

Ian R. Ball
Department of Entomology and Invertebrate Zoology
Royal Ontario Museum, Toronto, Canada

Hitherto the supraspecific classification of the freshwater triclads has had little to say concerning the history of the group. The following paper, which concentrates on a single family, is intended partly to alleviate this situation. That biological classification consists of the assembling of organisms into groups that are similar as a result of their common descent (Mayr, 1969, p. 121) would be denied by few, if any, taxonomists. Common descent, however, is not enough. It follows from the theory of evolution that all organisms are phylogenetically related; it is the degree of this relationship, the recency of common ancestry, which is important. The diversity that taxonomists attempt to classify is a product of organic evolution, and any classificatory system must take cognizance of this fact. This implies that the system must be based on phylogeny, and not vice versa. Therefore, I have attempted to tackle the numerous problems of the classification of freshwater triclads, as I see them, by taking the phylogenetic approach so ably advocated by Hennig (1966), Brundin (1966), and Crowson (1970). This does not imply that existing classifications are of no value, but the traditional phenetic taxonomists frequently do produce classifications which are at variance with their phylogenetic trees for the very reason that their systems are based primarily upon phenetic distance and not recency of common ancestry. The phylogeny and classification of the Paludicola proposed by Kawakatsu (1968) provides an example of this. I also believe that a true phylogenetic system is a necessary prerequisite for biogeographical analysis. This is a view which has been ably demonstrated by Nelson (1969), who also lays to rest the erroneous idea that biogeographical inquiry is meaningless in the absence of a fossil record.

Since phylogenetic relationship, or cladistic affinity, is best determined on the basis of synapomorphies, or the sharing of derived characters, it is always necessary to make distinctions between the primitive (plesiomorph) and derived (apomorph) states of given characters or character correlations. Frequently the trends are clear, but the direction is equivocal. In the absence of other evidence the principle of parsimony must decide the issue.

There is rarely, if ever, empirical proof for a given phylogenetic system, and certainly not for that developed for the Paludicola in the present paper. The probability of its correctness will be a product of the quality of the data available, and the skill of the taxonomist, and it is its heuristic value over a long period of time that will determine its worth. That there remains a great deal of work to be done before we shall fully understand evolutionary and biogeographical relationships within the Paludicola will be apparent from that which follows.

THE HIGHER CLASSIFICATION OF FRESHWATER TRICLADS

Historical

The classification of the Turbellaria has been in a state of flux for many years, and the categorical ranking of its subdivisions has been altered frequently. For present purposes I follow the arrangement by Ax (1956) in which the order Seriata Westblad, 1935 is divided into two suborders, viz., Proseriata Meixner, 1938 and Tricladida Lang, 1884 (= Euseriata Westblad, 1952). Within the latter the Maricola, Paludicola, and Terricola (Hallez, 1894) are given the status of infraorders, which is a well-used category in vertebrate zoology, and which will avoid the ambiguities of the categories "section" and "series," which have diverse meanings in the zoological and botanical literature.

Steinböck (1925) divided the Tricladida into two groups based primarily on the nervous system. These were the Diploneura, containing the terrestrial forms, and the Haploneura. The latter were further subdivided into the Vaginalia (Uteriporidae, Bdellouridae), Retrobursalia (Procerodidae, Cercyridae, Micropharyngidae), and the Probursalia (= Paludicola). Both Kenk (1930a) and Hyman (1931a) in their revisions opted for this nomenclature.

The exact relationships of the Terricola to the other triclads are not clear, but it does seem that the most primitive family is the Rhynchodemidae, which may be derived from the Maricola (Meixner, 1928; Marcus, 1953). Also, the aptness of Steinböck's name for the Paludicola was lost with the discovery of marine triclads with an anterior bursa (Probursidae Hyman, 1944) and of a freshwater triclad with a posterior bursa (*Rhodax* Marcus, 1946). For these reasons I agree with Marcus (1963) that Hallez's division of the triclads into three groups with equal rank is the most useful on present knowledge.

Hallez (1894, p. 187) recognized nine genera of Tricladida Paludicola and divided them into two families. These were the Planariidae, without anterior adhesive organs, and the Dendrocoelidae, with such organs. Later, von Graff (1912–1917) increased the number of families to five, again basing their definitions on the adhesive and creeping organs, a procedure which has been considered unreliable (Hyman, 1931a; Mitchell, 1968). Three of von Graff's families were erected for the peculiar Lake Baikal triclads, which show a great diversity of size and form, and which possess many gradations of adhesive organs and suckers. It is now accepted that these forms belong to the Dendrocoelidae (Livanov, 1962; Kozhov, 1963).

A precise definition of the families of Paludicola was not attained until the revision of Kenk (1930a). He defined two families by the arrangement of the inner muscle layers of the pharynx. In the Planariidae the circular and longitudinal muscles of the inner muscle zone of the pharynx form two separate layers, whereas in the Dendrocoelidae the circular and longitudinal muscle fibers are intermingled. Kenk's scheme, based on a definite morphological character, confirmed the earlier arrangement of Hallez (1894), which was based on external features. The distribution of the genera was the same in both schemes, although many more genera had been described by the time Kenk proposed his revision.

The correlation between external form and internal morphology in the two families of freshwater triclads was upset by the detailed study and description of an unusual cave triclad from North America (de Beauchamp, 1931). Packard (1879) placed this species in the genus *Dendrocoelum* on the basis of its external appearance. De Beauchamp (1931) was able to show, however, that the musculature of the pharynx was that of a planariid and not of a dendrocoelid, despite the presence of an anterior adhesive organ. For this and other reasons, he erected the new genus *Sphalloplana* for this species. Hyman (1931a) had also noticed the incorrect assignment of the species by Packard, but she felt that it fell naturally into the planariid genus *Fonticola*. Later, however, she described a number of new cave planarians, and in so doing not only recognized the genus *Sphalloplana* but also erected two new ones, *Kenkia* and *Speophila*, together with a new family, the Kenkiidae, to contain them (Hyman, 1937).

As more species of Kenkiidae became known, it became increasingly apparent that Hyman's original definition of the family could not be maintained. On morphological grounds de Beauchamp (1961) had not accepted it, but for purely pragmatic reasons I had retained the family in a checklist of the fresh-water triclads of Nearctis and Neotropica (Ball, 1969a). Mitchell (1968), in a detailed review, synonymized the genera *Sphalloplana* and *Speophila* and presented evidence suggesting that the family could no longer be satisfactorily delimited. He regarded the genera *Sphalloplana* and *Kenkia* as merely planariids with an anterior glandulomuscular adhesive organ. He therefore proposed the elimination of the family Kenkiidae, a proposition with which Kawakatsu (1969a) disagreed.

The thirty or so genera of Tricladida Paludicola which are recognized today are distributed generally among two families as follows:

Family Planariidae: *Planaria* Müller, 1776; *Fonticola* Komarek, 1926 (inc. *Penecurva*); *Atrioplanaria* de Beauchamp, 1932; *Phagocata* Leidy, 1847; *Polycelis* Ehrenberg, 1831 (inc. *Seidlia*, *Polycelidia*, *Ijimia*); *Plagnolia* de Beauchamp and Gourbault, 1964; *Crenobia* Kenk, 1930; *Sphalloplana* de Beauchamp, 1932; *Kenkia* Hyman, 1937; *Hymanella* Castle, 1941; *Rhodax* Marcus, 1946; *Bopsula* Marcus, 1946; *Cura* Strand, 1942; *Dugesia* Girard, 1850 (inc. *Spathula*).

Family Dendrocoelidae: *Dendrocoelum* Oersted, 1844; *Bdellocephala* de Man, 1874; *Rectocephala* Hyman, 1953; *Dendrocoelopsis* Kenk, 1930 (inc. *Amyadenium*); *Acromyadenium* de Beauchamp, 1931; *Thysanoplana* von Graff, 1916; *Hyperbulbina* Livanov and Porfirjeva, 1962; *Procotyla* Leidy, 1857; *Macrocotyla* Hyman, 1956; *Rimacephalus* Korotneff, 1901; *Protocotylus* Korotneff, 1912; *Polycotylus* Korotneff, 1912; *Monocotylus* Korotneff, 1912; *Armilla* Livanov, 1961; *Sorocelis* Grube, 1872; *Baikalobia* Livanov, 1962. *Caspioplana* Zabusova, 1951 is unique in its possession of pharyngeal muscles converse to those of the Planariidae. Its reproductive apparatus is closest to that of the Dendrocoelidae.

The unusual genus *Bdellasimilis*, ectoconsortic on Australian Chelonia, was originally, but tentatively, assigned to the Paludicola (Richardson, 1968,

1970). Since its occurrence in freshwater is paralleled by other Maricola (Ball, 1974), and since an anterior bursa is known in the marine Probursidae, I find the numerous morphological similarities with the Maricola more convincing, and have no hesitation in classifying it with this group. Richardson (1971, pers. comm.) agrees that it is probably maricolous but finds the criteria separating the Paludicola and the Maricola somewhat inadequate. Recent work is confirming his opinion.

Evolutionary Trends

The most important contribution to the study of evolutionary trends within the Tricladida is undoubtedly that of Meixner (1928). Recognizing the taxonomic importance of the reproductive organs, he divided the aquatic triclads into a number of types on the basis of the female reproductive system. In summary, these types were:

I   Atrium undivided. Oviducts and shell glands open into the bursal stalk (Maricola and Paludicola).
II Atrium undivided. Oviducts unite to form a common oviduct which opens into the posterior part of the bursal stalk or the atrium.
   A   Common oviduct and shell glands open into the bursal stalk (Maricola and Paludicola).
   B   Common oviduct opens into the bursal stalk, the shell glands into the common oviduct (Maricola only).
   C   Common oviduct opens into the mouth of the bursal stalk or below it into the atrium; the shell glands open chiefly into the common oviduct (Paludicola only).
III Atrium divided. Common oviduct opens independently of, and anterior to, the bursal stalk into the atrium. Shell glands open into the oviduct (Paludicola only).

All the genera and subgenera of aquatic triclads recognized by Meixner were assigned to one or the other of these categories, together with notes on their distribution. Meixner forbore to erect any new systematic categories but went on to compare other characters and organ systems with his scheme and found many correlations in their categorical distribution. He considered that the Maricola and Paludicola showed independent evolutionary lines, that the Maricola were more primitive than the Paludicola, and that the Paludicola of type III were the most advanced. Many of the details of Meixner's arrangement are now questionable, but the broad principles seem to be eminently sound.

The genera *Cura* and *Dugesia* share many features in common, such as their triangular head shape and the production of stalked cocoons, which are not found in other genera (Kenk, 1930a); they are the only Paludicola widely distributed in the Southern Hemisphere. Meixner's types I and II contain all the known species of these genera, and some species with variable oviducts could be assigned to either type. On the other hand the Dendrocoelidae, which are unique in their possession of intermingled inner pharyngeal muscles, are wholly contained in type III, as are the remainder of the Planariidae.

The principal dichotomy in Meixner's scheme seems to be between type I–II, in which the oviducts are closely associated with the bursal stalk, and type III, in which they are associated with the atrium. In their morphology and ontogeny the Proseriata form the phylogenetic precursors of the triclads (Ax, 1963), and in this group the oviducts are usually associated with the bursal stalk or the equivalent female genital canal (Meixner, 1938; Ax, 1956), as they are in the Paludicola of type I–II and all Maricola with the exception of the aberrant commensal genus Nexilis, in which they enter the penis bulb (Holleman and Hand, 1962). In this respect type I–II is undoubtedly more primitive than type III, and the first evolutionary advance made by the Paludicola over the Maricola appears to be the shifting of the course and position of the oviducts, and the associated shell glands.

Kenk (1930a) is probably correct in saying that the Planariidae are more primitive than the Dendrocoelidae, but it follows from this that the Planariidae of type III must occupy an intermediate position between type I–II and the Dendrocoelidae. The latter could be separated by the adhesive organs (Hallez, 1894) or by their pharyngeal muscles (Kenk, 1930a).

Anterior glandulomuscular organs are not found in the Proseriata or the free-living Maricola, but they are usual in the Dendrocoelidae and thus represent an apomorph (derived) character state when considering the Tricladida as a whole. Similarly, the organization of the pharyngeal muscles of the Proseriata, Maricola, and Planariidae is basically the same. Consequently the arrangement in the Dendrocoelidae must be derived.

It seems to me that there are three principal grades of organization within the Paludicola which need to be distinguished. The most advanced of these are the Dendrocoelidae, followed by the Planariidae of type III, and then finally the Planariidae of type I–II, and I propose that the latter should be separated as a distinct family, with the name Dugesiidae. This follows logically if on the basis of the characters discussed above we attempt to reconstruct the most probable evolutionary sequence.

Although the functional significances of the arrangement of the pharyngeal muscles and the courses of the oviducts are not fully understood, it is very unlikely that these two characters are functionally related. It is difficult to conceive of either of these characters being necessary correlates of the presence or absence of an adhesive organ. The congruence of the apomorph grades of all these characters in the Dendrocoelidae, and of the plesiomorph grades in the Dugesiidae is, therefore, surely of phylogenetic significance.

If it is accepted that the most primitive freshwater triclad possessed oviducts which emptied into the bursal stalk, possessed a pharynx in which the inner muscles formed a distinct inner circular and outer longitudinal layer, and lacked an adhesive organ, as is the case only in the Dugesiidae, then the most probable sequence leading to the other groups is that presented as Fig. 1, if a monophyletic origin for the Paludicola is assumed. The four combinations of the three characters shown are the only ones known in living Paludicola, and it follows from the scheme presented that the Dugesiidae are phylogenetically equivalent to the Planariidae + Dendrocoelidae. This could be recognized by using the superfamily category.

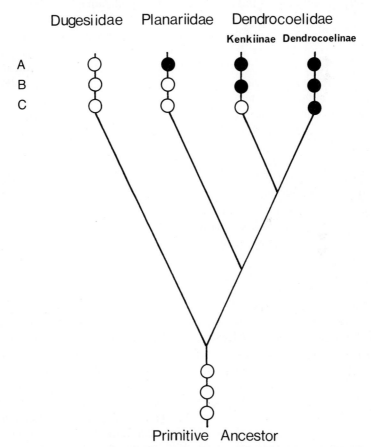

Figure 1 Phylogenetic relationships of the families of Tricladida Paludicola. Open circles are the plesiomorph (p), and solid circles the apomorph (a), grades of given characters. A, oviducts open to bursal stalk (p) or atrium (a). B, adhesive organs absent (p) or present (a). C, inner pharyngeal muscles layered (p) or inter-mingled (a). For further discussion, see text.

Kawakatsu's (1968) valuable discussion of the phylogeny of the Paludicola is summarized in his figure IV-1, and although I disagree with him in many details, the principal dichotomies in his phylogenetic tree are similar to those given here. It is encouraging that to this degree we have reached the same conclusions. However, he maintains the old threefold classification of Planariidae, Kenkiidae, and Dendrocoelidae. This is totally at variance with his phylogenetic tree since his Planariidae become a paraphyletic group, as they would if the Dugesiidae were united with them in Fig. 1. This is the more surprising inasmuch as Kawakatsu has placed his categories on a time scale which clearly indicates that at the familial level there are three major phyletic lines equivalent to those of Fig. 1, the significance of which he appears to have overlooked.

The principal difficulty of the scheme presented in Fig. 1 is that only the Dendrocoelidae are defined by true synapomorphy. It is true that the Dugesiidae are symplesiomorphic by analysis rather than by synthesis, which somewhat lessens the danger of their being a paraphyletic or polyphyletic group, but it is desirable that they should also be defined by further apomorph characters. The only such character I have been able to recognize is the triangular head so characteristic of *Dugesia* and *Cura*, which is an apomorph character within the Paludicola (p. 355). Even this, however, is absent in a few Southern Hemisphere forms which, on other grounds, must be classified with these genera. Nearly all Dugesiidae produce stalked cocoons, in contrast to the other families, but whether this is a primitive retention or a secondary (apomorph) acquisition is impossible to say.

The problem is compounded by current difficulties in adequately separating the marine and freshwater triclads, and thus, by implication, in deciding on the exact origins of the Dugesiidae. Precise definitions of the Maricola and Paludicola based on ecological and morphological (reproductive) criteria can no longer be put forward. In this connection the Probursidae (Maricola) and *Rhodax* (Paludicola) have already been discussed. One *Dinizia* species (Maricola) lacks a bursa copulatrix but has an anteriorly directed bursal stalk, and is found in freshwaters (Ball, 1974), as is *Bdellasimilis* (Maricola), which has an anterior bursa copulatrix. It has long been known that certain Paludicola may invade brackish habitats (Wilhelmi, 1909; Meixner, 1928), and that some Maricola can adapt to almost completely fresh water (Meixner, 1928; Pantin, 1931). It was this ecological overlap which led Steinböck (1925) to drop the ecological names. The genus *Caspioplana* (Paludicola) is endemic to the saline waters of the Caspian Sea, and although it shows many unusual morphological features, it is undoubtedly derived from the Dendrocoelidae (Zabusova, 1951), even though it coexists with the maricole *Pentacoelum* (de Beauchamp, 1961). More important, however, is the fact that a number of Procerodidae have been recorded from freshwaters, most frequently on Southern Hemisphere oceanic islands where, it is assumed, the historical absence of freshwater triclads has enabled the Maricola to invade the freshwater habitat in the face of reduced competition (Ball, 1974). New types of "freshwater triclads" may be expected to arise from these forms, and it would be difficult to distinguish them from the Dugesiidae.

The fact that the Paludicola (Dugesiidae) cannot be separated from the Maricola except by a constellation of characters which, taken individually, could pertain to either group, is a fine example of the Non-Congruence Principle discussed by Crowson (1970). As characters do evolve both concurrently and successively, this may indicate in the case of the aquatic triclads that a great many primitive types, intermediate between the major phyletic lines, have persisted to the present day, possibly with the added complications of convergence. If so, their elucidation remains a formidable task. It could also mean that the assumption of a monophyletic origin of the Paludicola from marine ancestors is erroneous. Possibly the Dugesiidae and the Dendrocoelidae + Planariidae have arisen independently from marine ancestors in the Southern and Northern Hemispheres respectively. There can be little doubt that the evolutionary relationships between the Maricola and the Paludicola are far more complex than is generally realized, and I suspect that a hypothesis of a polyphyletic origin of the higher aquatic triclads from marine ancestors may lead to a better understanding of their systematic and biogeographical relationships. I am unable to finalize such a hypothesis at the present time, for a great deal of fundamental taxonomic research is necessary before this could be attempted. It is only right, however, that one of the main difficulties derived from the present approach should be emphasized, and this is the major reason why the present study is restricted to the systematic and biogeographical relationships of the Dugesiidae.

The status of the newly restricted Planariidae requires further study which is outside the scope of the present paper. As regards the Kenkiidae, Kawakatsu clearly considers them to be a monophyletic group, and upon this we are agreed. I concur with Mitchell (1968), however, in that they cannot be awarded family rank, though I am dubious about merging them with the Planariidae. For the present they may be considered a distinct subfamily, the Kenkiinae, but their precise position in my proposed phylogenetic scheme is problematical.

If the Kenkiinae are derived from the Planariidae, it is necessary to postulate that they have acquired their adhesive organs independently of the acquisition of apparently homologous organs in the Dendrocoelidae. Mitchell (1968) does indeed argue that there is no reason to suppose that adhesive organs have arisen only once within the Paludicola and states that the wide occurrence of glandulomuscular organs in the Turbellaria suggests a multiple origin for these structures. This may well be true if the character occurs sporadically, and in related groups, as is the case with adenodactyls, but the suggestion is unreasonable when the organ appears within a subtaxon which shows other evidence of uniformity or monophyly, which is the case with the Kenkiinae. Hyman (1937) derived the Kenkiinae from the genus *Phagocata* (= *Fonticola*), with which both Mitchell (1968) and Kawakatsu (1968) seem to concur, and there appears to be little disagreement concerning the "naturalness" of the group.

Their association with this genus has been based largely upon the planariid pharynx and on the possession of a copulatory complex of the *Fonticola* type. In fact, the copulatory complex is typical of Meixner's type III and is found in other Planariidae and Dendrocoelidae. It has also been inferred (Mitchell, 1968) that support for the *Fonticola* origin of the Kenkiinae is forthcoming from the fact that *F. albata* is described as possessing an incipient adhesive organ (Ichikawa and Kawakatsu, 1962), although this lacks any muscular differentiation. I am of the opinion, however, that this feature has been overemphasized since adhesive depressions of this type are known in other planariid species such as *Planaria torva* (Ball *et al.*, 1969, p. 111). and modifications of the epithelium of this ventral anterior region, which presumably are of sensory function, are found in *Planaria occulta* and *P. dactyligera* (Kenk, 1970), and probably other species.

There are further difficulties in classifying the Kenkiinae with the Planariidae. Hyman (1937) and Carpenter (MS1970) have shown that fundamentally there are two types of adhesive organ: that typical of the Dendrocoelidae and one typical of the Kenkiinae. Hyman (1956) considered that of the dendrocoelid genus *Macrocotyla* to be structurally closer to that of the Kenkiinae than to the other Dendrocoelidae. The supposed derivation of the Kenkiinae from the Planariidae as a group quite independent of the Dendrocoelidae thus requires two assumptions: the independent acquisition of an anterior adhesive organ and convergence in the same organ in *Macrocotyla*.

If, however, the phylogenetic arrangement presented in Fig. 1 is accepted, neither of these assumptions is necessary, and the classificatory system is more parsimonious. The Kenkiinae are considered to be the most primitive subfamily of the Dendrocoelidae, whose sporadic occurrence in caves, subterranean waters, and deep lakes in central Asia, the Far East, and North America represents the remnants of a formerly much wider distribution, and which results in part from the evolutionary success of the more recent Dendrocoelinae (= Dendrocoelidae *sensu* Kenk, 1930a). The relict nature of cave faunas in general is well known (Vandel, 1965). It may also be noted that Carpenter (MS1970) has been unable to find any significant function related to the cave habit for the adhesive organ of the Kenkiinae, which is surprising if these are to be regarded as a specialized group of planariids

(Hyman, 1937, 1960). However, its occurrence is perfectly consistent with the hypothesis proposed here that they are plesiomorph dendrocoelids.

There are, of course, difficulties in proposing a major classificatory division on the basis of the presence or absence of adhesive organs. It could certainly be considered a retrograde step when compared with Kenk's (1930a) redefinition of the families used by Hallez (1894). But as we have seen, Kenk's reclassification on the basis of pharyngeal musculature did not alter Hallez's disposition of the genera into families on the basis of adhesive organs; it merely provided a more "acceptable" morphological criterion. Hyman (1931a) criticized the use of adhesive organs in the higher taxonomy of the group, but later she recognized that they must be of some taxonomic importance since they are absent in the Planariidae and usually present in the Dendrocoelidae (Hyman, 1937, p. 472).

The fact that adhesive organs exhibit different grades of development in the Dendrocoelinae (Kenk, 1930a) is irrelevant to their use in phylogenetic classification, though it might make diagnosis difficult. However, the only known genera of Kenkiinae are clearly defined (Mitchell, 1968; Carpenter, MS1970), and the dendrocoelids most likely to show secondary reduction of the anterior adhesive organ, as, for example, some species of *Sorocelis*, are all members of the subfamily Dendrocoelinae, which is clearly identifiable by its pharyngeal musculature.

It must be assumed that polypharyngeal Kenkiinae have acquired this character independently of its occurrence in the Planariidae. Polypharyngy is found in the eastern North American genus *Phagocata* and in the European *Crenobia*, two planariid genera which do not show particularly close relationships. The feature is undoubtedly independently acquired in each, as far as we can tell from present evidence. An alternative hypothesis would be that the Kenkiinae had a polypharyngeal ancestor, presumably close to *Phagocata*, and that polypharyngy has been secondarily lost in many species.

The Kenkiinae are also uniform in that the testes are prepharyngeal. Restriction of the testes to the anterior part of the body is frequent in both the Dugesiidae and the Planariidae, but until comparatively recently was not known in the Dendrocoelinae. However, *Macrocotyla*, which possesses a kenkiinid-like adhesive organ, also has testes which are restricted to the prepharyngeal region (Hyman, 1956). It is also interesting to note that the Kenkiinae are restricted to Asia and North America, achieving their greatest diversity in the latter, and *Macrocotyla* is endemic to North America. It is quite possible that the unpigmented, eyeless *Macrocotyla* forms a link between the Kenkiinae and the Dendrocoelinae.

From the knowledge we possess at present it is unnecessary to postulate independent acquisition of an adhesive organ in the Kenkiinae, and convergence in *Macrocotyla*, in order to fit them into a satisfactory phylogenetic scheme of the Paludicola, since they can be classified satisfactorily with the phyletic lineage leading to the Dendrocoelinae.

A Revised Classification

I propose, therefore, a revision of the higher classification of the Paludicola, as follows:

Family Dugesiidae fam. n.

Paludicola in which the oviducts, separately or combined, empty into the bursal stalk, or rarely into the atrium very close to, and posterior to, the entrance of the bursal stalk. Type genus: *Dugesia* Girard, 1850. Usually pigmented. With eyes, and typically with a triangular head. Adhesive organs absent. Adenodactyls rarely present. Inner pharyngeal muscles in two distinct layers. Shell glands usually open into the bursal stalk. Cocoons spherical and stalked (exceptions *Rhodax, D. montana, D. fontinalis*). Genera: *Dugesia, Cura, Bopsula*, and possibly *Rhodax*. Distribution: Worldwide.

Family Planariidae Stimpson, 1857, emend.

Paludicola in which the oviducts unite to a common oviduct which empties into the roof of the atrium, independently of the bursal stalk. Anterior glandulomuscular organ absent. Type genus: *Planaria* Müller, 1776. Pigmented or white. Typically with eyes, never with a triangular head. Inner pharyngeal muscles in two distinct layers. Adenodactyls sometimes present. Shell glands open into the common oviduct. Cocoons spherical or ovoid, without a stalk. Genera: *Planaria, Fonticola, Phagocata, Hymanella, Atrioplanaria, Polycelis, Plagnolia*, and *Crenobia*. Distribution: Holarctis and parts of Orientalis.

Family Dendrocoelidae Hallez, 1894

Paludicola in which the oviducts unite to a common oviduct which empties into the roof of the atrium, independently of the bursal stalk. Shell glands open into the common oviduct. Anterior glandulomuscular organ present (but secondarily lost in a few species). Type genus: *Dendrocoelum* Oersted, 1844.

Subfamily Kenkiinae nom. n.

Inner pharyngeal muscles in two distinct layers. Usually blind, unpigmented cave dwellers. Adenodactyls absent, testes prepharyngeal. Genera: *Kenkia* and *Sphalloplana*. Distribution: Central and east Asia, North America.

Subfamily Dendrocoelinae nom. n.

Inner pharyngeal muscles intermingled. Usually with eyes and usually unpigmented. Adenodactyls frequently present. Testes usually distributed throughout the body length. Genera: *Dendrocoelum, Bdellocephala, Recto-cephala, Dendrocoelopsis, Acromyadenium, Thysanoplana, Hyperbulbina, Procotyla, Macrocotyla, Rimacephalus, Protocotylus, Polycotylus, Mono-cotylus, Armilla, Sorocelis, Baikalobia,* (and *Caspioplana?*). Distribution: Holarctis.

A word on the nomenclature of the new family is necessary at this point. Of the five families recognized by von Graff (1912–1917) one, the Curtisi-idae, was founded for the American species *Cura foremanii* (olim *Curtisia foremanii*), the only species of the genus known at that time. Since von Graff's classification of this triclad was based on a misinterpretation of the description by Curtis (1900), the family was unnecessary and unjustified (Meixner, 1928; Hyman, 1931a), although its retention was favored by Poche (1926). However, it would be misleading to use this name when Poche's concept of the family is so vastly different from that presented here. In addition, the name Curtisiidae was based on the generic name *Curtisia* von Graff, 1916, which has been shown to be a junior homonym, and for which the name *Cura* has been sub-stituted (Strand, 1942). Finally, the genus *Cura* is unusual in a number of respects and the genus *Dugesia* is more typical of the family as it is at present conceived. Consequently it has been selected as the type genus of the new family.

The systematic position of the unusual Brazilian genus *Rhodax* is prob-lematical. According to the original definition it would fall into the family Planariidae, as the common oviduct opens into the genital atrium, more par-ticularly into the part called the common atrium (Marcus, 1946, pp. 133–134). Both the Planariidae and the Dendrocoelidae are restricted to the Northern Hemisphere (with one exception), and it is my contention that the Southern Hemisphere triclads all belong to the Dugesiidae. A careful reexamination of *Rhodax* therefore is necessary.

Marcus describes the genus as being a "collective morphological type" showing affinities with both the Proseriata and with *Phagocata* (= *Fonticola*) of the Planariidae. I hope to show later that the relationships with *Fonticola*, which also are accepted by Kawakatsu (1968), are not very strong and that the genus is indeed primitive. Its assignment in my phylogenetic scheme de-pends, however, solely on the course of the oviducts and associated shell glands.

In the only known species, *R. evelinae*, there is a bursa–intestinal duct posterior to the penis. The posteriorly directed bursal canal is histologically well differentiated from the atrium, and into its dorsal wall, very near to the junction with the atrium, opens what Marcus terms a common oviduct (Marcus, 1946, fig. 149). The shell glands open into this "common oviduct." According to Marcus's figure 152 the bursal stalk first assumes its typical histological structure at the level of the gonopore, and according to his figures 149 and 150 the "common oviduct" opens into the bursal stalk posterior to the level of the gonopore. From his figures I would infer that that part of the duct anterior to the gonopore is atrium, and that posterior must be considered to be bursal stalk. It may also be noted that the oviducts do not unite to form the usual slender common oviduct, but rather open separately from the sides into a broad saclike structure (Marcus, 1946, fig. 149).

For these reasons I consider the "common oviduct" of *Rhodax evelinae* to be a diverticulum of the bursal stalk into which open the shell glands. It is not homologous with the common oviduct of the Planariidae and the Dendrocoelidae. These views are supported by examination of a number of specimens of *Rhodax* kindly sent to me by Mrs. Eveline Marcus. Regrettably, only two specimens showed traces of copulatory organs, but these are consistent with the above interpretation. The similarity of this diverticulum of the bursal stalk of *Rhodax* to the "glandular duct" of the Procerodidae (Maricola) of Meixner's type IIB is very striking, and may be taken as evidence of the close relationship of *Rhodax* with the Maricola. Moreover, this genus possesses vitellaria which are anterior to the germaria, which is usual in the Maricola, but not in the Paludicola. A good case could be made for *Rhodax* having evolved from marine ancestors independently of the other Paludicola, especially as there are other features linking this genus with the Proseriata (Marcus, 1946). In this case a distinct family for the genus would be necessitated, but I decline to take this step until I have had the opportunity to study more material. Consequently, I provisionally assign *Rhodax* to the Dugesiidae and express doubt concerning any close relationships with *Fonticola*.

THE FAMILY DUGESIIDAE fam. n.

It is my purpose here to provide an overall synopsis and preliminary phylogenetic analysis of the family Dugesiidae. It is well known that the Paludicola are a difficult group to study taxonomically. Their lack of definite measurable characters, together with the difficulty of proper preservation, is probably responsible for this. Consequently, attempts to determine their relationships by cytological (Dahm, 1958, 1963) and cytogenetic (Benazzi, 1960, 1963, 1966) studies must be applauded, even if the full fruits of such studies have yet to be reaped. That studies of the Paludicola have for a long time remained at the α-taxonomy level is a fact that cannot be avoided, and any attempt at revision or synthesis must still be based largely upon comparative morphological data.

Our lack of knowledge concerning the functional significances of many of the characters which appear to be useful, such as the musculature of the pharynx or of the bursal stalk, is a considerable drawback. As regards phylogenetic analysis a further complicating factor is the extreme conservatism of the group. As Brundin (1966) has pointed out, morphologically uniform groups are likely to be difficult subjects for phylogenetic analysis. Such groups yield few adequate characters for analysis, and parallelism and convergence occur frequently because there are a limited number of evolutionary pathways open. Thus, strict phylogenetic analysis involving the rule of dichotomy as proposed by Brundin (1966) is difficult, if not impossible. Darlington (1970) too has drawn attention to the difficulties caused by the oversimplifications of the rule of dichotomy. The analysis of characters not considered here, and not available in most species descriptions, possibly would alleviate this situation. The areas presenting the greatest diversity, and the greatest difficulty, all lie in the Southern Hemisphere, particularly in South Africa and Australasia.

In the analysis that follows I have elected to study, by the methods of Camin and Sokal (1965), several sets of characters from the Dugesiidae so that the conclusions may be compared between sets. Where adjustments of the resulting cladograms have been necessary, for reasons to be given, I have followed in general the principle that groupings are best based on synapomorphies. I have relied heavily on the detailed descriptions of many species given by such authors as Ijima (1884), Böhmig (1902), Weiss (1910), Lang (1913), Hyman (1925), Marcus (1946, 1948, 1953, 1954, 1955), de Beauchamp (1939, 1959), Kenk (1930b, 1935, 1944), Ichikawa and Kawakatsu (1964), and Ball (1970, 1971). Owing to the kindness of many colleagues I have also been able to examine directly many specimens of *Dugesia* and *Cura* from most parts of the world.

Taxonomic Characters and Their Correlation

Taxonomic characters may function as diagnostic characters for a particular species or as indicators of relationship (Mayr, 1969). Failure to recognize these two principal functions of a taxonomic character has led to much sterile discussion by taxonomists on the relative values of given characters. It follows that a species description which provides only the minimum of diagnostic characters is of little value to the evolutionist and the biogeographer, even though the description is adequate for the recognition of the species.

In attempting to reconstruct the probable phylogenies of characters and character correlations, the level of plesiomorphy or apomorphy of these must be decided. At present genetic and fossil data are insufficient to be of real value. The criteria I have relied upon are outlined below; my indebtedness to Marx and Rabb (1970) will be apparent.

   1   Uniqueness: A decision concerning the phylogenetic relationships of the families of Paludicola having been made, a character state unique to a derived family is thereby inferred to be apomorph.
   2   Relative abundance: A character state which is widely distributed in divergent taxa of the group under study is likely to be plesiomorph.
   3   Morphological specialization: If a character state is predominant in some particular adaptive specialization, it is likely to be apomorph.
   4   Ecological specialization: A character state is likely to be apomorph if it is relatively much more abundant in taxa with a particular mode of life.
   5   Geographical restriction: Limitation of a character state to most taxa of a particular geographical area suggests that it is apomorph.
   6   Related taxa: A character state which occurs in forms closely related to, but outside of, and not directly descended from, the group under consideration is likely to be plesiomorph.

   Of course it is not possible to apply all these criteria to each of the characters discussed below, but the probability that a decision on the apomorphy or plesiomorphy of a character state is correct will be increased in proportion to the number of criteria used. A further aid to determining the direction of change is Dollo's law, in that a complex character once lost in the course of evolution is unlikely to be reacquired in exactly the same form (Crowson, 1970). This criterion must be used cautiously, however, since exceptions are known (Mayr, 1969). The correlation of given character states in a single taxon is not always useful in the absence of other data. They could be necessary correlates, and there is certainly no reason why plesiomorph and apomorph characters should not sometimes be statistically correlated. Nonetheless, the statistical correlation of apomorph characters which cannot be shown to be necessary correlates is likely to result from common genetic history of the taxa.

   Habitus   A general increase in body size often accompanies evolutionary advancement (Meixner, 1928; Bonner, 1965). The largest Paludicola known belong to the most advanced group, the Dendrocoelinae, the Lake Baikal representatives of which attain lengths of 40 cm and more (Kozhov, 1963). The other Palaearctic and Nearctic Dendrocoelinae rarely exceed 40 mm. It is slightly paradoxical, from the points of view put forward in this paper, that the Dugesiidae in general are larger than the Planariidae, but there are many exceptions and an explanation may be sought in their ecology and reproductive biology.
   Meixner (1928) notes that the more primitive Paludicola are generally

uniformly colored or mottled, and longitudinally striped and white forms occur only in the Planariidae and Dendrocoelidae. We know now of an unpigmented *Dugesia* species, *D. batuensis*, but the absence of pigment is almost certainly related to its cave-dwelling habit (Ball, 1970).

All the Dugesiidae possess two eyes, but supernumerary eyes occasionally occur. Multiocularity occurs in both the Planariidae and Dendrocoelidae, as do eyeless forms. The usual two-eyed condition is almost certainly the ancestral condition in the Paludicola.

Meixner (1928) and Marcus (1946) infer that the sagittate head is the primitive condition, but more recent data lead me to question this opinion. Most, but not all, Dugesiidae possess a triangular head with projecting auricles, as do a few Maricola (Meixner, 1928). In the higher Paludicola truncate or rounded heads are usual, sometimes with the development of tentacles, as in *Polycelis felina*, *Fonticola bursaperforata*, *Crenobia*, and some Dendrocoelidae. Head shape in the Dugesiidae is illustrated in Fig. 2, and the geographical distribution of this character is shown in Fig. 3. Most Proseriata have a rounded or spathulate anterior end (Meixner, 1938; Ax, 1956; Luther, 1960), and there can be little doubt that this condition is primitive within the Turbellaria as a whole. A rounded or spathulate head is rare in the Dugesiidae but is found in some forms from Australasia, the Crozet Archipelago, and South America. The high triangular form characteristic of *D. tigrina* occurs in all the New World species of the genus, and elsewhere in only one species, *D. montana* from New Zealand. The subtriangular shape is typical of the genus *Cura* and of all *Dugesia* species from Palaearctis and Orientalis, and some from Australasia.

I conclude that the ancestral Dugesiidae were pigmented, with two eyes,

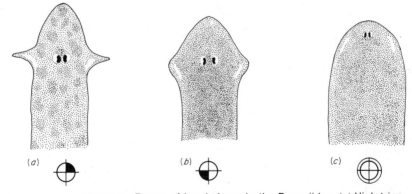

(a)   (b)   (c)

Figure 2   Range of head shape in the Dugesiidae. *(a)* High triangular, *Dugesia tigrina;* *(b)* subtriangular, *D. gonocephala (after de Beauchamp, 1961); (c)* rounded, *Cura pinguis (after Nurse, 1950).* The symbols beneath the sketches form the key to Fig. 3.

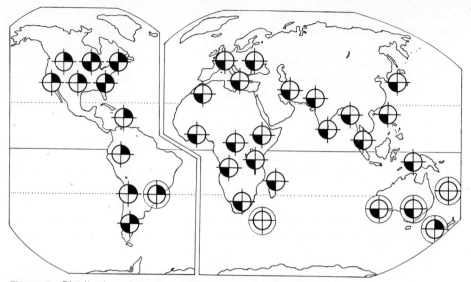

Figure 3 Distribution of the Dugesiidae possessing the different types of head shape. See Fig. 2 for key. Symbols on this, and subsequent maps, are composites of those given in the key. For further discussion see text.

and with a rounded or spathulate anterior end. Such forms are represented today by *Rhodax* (South America), *D. seclusa* (Crozet Islands), *C. pinguis*, *D. schauinslandi*, *D. fontinalis*, and probably most other Australian *Dugesia* species (Weiss, 1910).

Pharynx The relative thicknesses of the various pharyngeal muscles may well have phylogenetic significance. In many Maricola, for example, the circular layers are exceptionally well developed. In certain cases the thickness of the layers may be of diagnostic significance (Ball, 1971).

The phylogenetic significance of the inner pharyngeal muscles has been discussed. The outer pharyngeal musculature of most Dugesiidae consists of two layers, a subepithelial layer of longitudinal fibers underlain by a circular layer. This is the condition found in the Maricola and Proseriata. A few Dugesiidae have developed a third layer, consisting of longitudinal fibers beneath the outer circular layer. This extra layer is found in *Dugesia gonocephala*, *D. congolensis*, *D. ectophysa*, *D. colapha*, *Cura paeta*, and *C. jeanneli*. All these species are from west, central, and south Africa. *D. gonocephala* is also widespread in Europe.

The spatial and taxonomic distribution of this character leaves little doubt that it is an apomorph condition.

Male Reproductive System—Penis   De Beauchamp (1939) divided the genus *Dugesia* into a number of types on the basis of penis morphology. The New World species belonged to his simplest type in their possession of a short penis and a nonmuscular bifid seminal vesicle. The Old World forms were divided into two further groups. The *D. gonocephala* group was characterized by a large folded seminal vesicle and by the diaphragm (often accompanied by extensive eosinophil glands), which separates the vesicle from the ejaculatory duct. The last type was that of *D. lugubris* with a large penis and two very muscular seminal vesicles, one behind the other. Marcus (1953) added two further types, the simple one of *D. neumanni*, *D. glandulosa*, and *D. seclusa*, and the more complicated one represented by the other Australian species described by Weiss (1910).

Hyman (1931a) considered the male sexual apparatus unsatisfactory in general for taxonomic purposes. She did consider, however, the form and presence or absence of the seminal vesicles (bulbar cavities) to be of importance. Certainly such characters as the size and shape of the penis must be disqualified since they vary greatly with maturity and method of fixation. Nevertheless there are a number of other definite morphological features which are of significance.

The form of the seminal vesicle is related to the course of the vasa deferentia, and within the Dugesiidae there are a number of possibilities. The most primitive form would seem to be in those cases where the vasa deferentia unite to a common duct which enters the penis bulb. This is found in *Rhodax*, *Bopsula*, and *Cura pinguis*, and it is also characteristic of the Bothrioplanidae and some Cercyrinae, as well as the Probursidae. The next stage may be represented by the separate entrance of the vasa deferentia into the penis bulb without enlargement to form vesicles. Such is the case in *Dugesia seclusa*, and some other South American and Australian species of *Dugesia*. Enlargement of the ducts within the penis bulb gives the characteristic bifid seminal vesicle of most New World species, and the most advanced stage is represented by the enormous and very muscular vesicle of *D. gonocephala* and its allies, from which, no doubt, the double vesicle of *D. lugubris* is derived. The single enlarged vesicle of some New World *Dugesia* species is undoubtedly a secondary acquisition (Ball, 1971), and in some of these species either condition occurs.

The ejaculatory duct may be straight, and open from the tip of the penis papilla. This is the usual state in the Proseriata and Maricola, and also in the Paludicola, and thus probably is the plesiomorph condition. In a few Dugesiidae it is highly convoluted, and occasionally it may open dorsally or ventrally from the tip of the papilla. These represent different apomorph grades.

The glandular diaphragm separating the seminal vesicle and the ejaculatory duct in *D. gonocephala* and its allies is a new structure in the triclads and is undoubtedly apomorph. So, too, the adenodactyl of *D. cretica*, *D. iranica*, and others is a new and apomorph character in the Paludicola. This type of adenodactyl, which projects from the penis bulb alongside the penis papilla, is quite different from the type found in *D. bactriana* and some Planariidae and Dendrocoelidae, as emphasized by de Beauchamp (1959). In the latter the adenodactyl is a discrete muscular-gland organ projecting into the atrium quite independently of the penis. The atrial adenodactyls of some Australasian forms (Weiss, 1910; Meixner, 1928) appear to present a third type in that they lack a true papilla. Only the first-mentioned type of adenodactyl is considered here.

These various character states may be coded as follows, lowercase letters representing the plesiomorph, and capitals the apomorph, grade:

a   Common vas deferens.
Al   Separate vasa deferentia without enlargement to form a vesicle.
A2   Vasa deferentia enlarge to form a bifid vesicle that is not very
        muscular.
A3   Single large, usually round vesicle, which is very muscular.
A4   Two large, muscular vesicles, one behind the other.
b   Ejaculatory duct straight and narrow.
B   Ejaculatory duct dilated and convoluted.
c   Ejaculatory duct opens terminally.
C1   Ejaculatory duct opens subterminally.
C-1   Ejaculatory duct opens supraterminally.
d   No diaphragm in the ejaculatory duct.
D   Diaphragm present.
e   Adenodactyls associated with the penis bulb not present.
E1   One adenodactyl present.
E2   Two adenodactyls present.

Of the possible character combinations the known Dugesiidae have utilized thirteen, as shown in Table 1. Using the methods of Camin and Sokal (1965), I have constructed a cladogram treating these thirteen types as Operational Taxonomic Units (OTU) irrespective of their present taxonomic status. The result is shown in Fig. 4.

On the basis of this cladogram the Dugesiidae may be divided into six groups as indicated, and it may be noted that Groups II to VI correspond closely to the divisions of de Beauchamp (1939) and Marcus (1953). However, this division does create some difficulties. In particular the position of *Cura paeta* (type 10) is equivocal since in terms of the characters presented by the ejaculatory duct (B and C1) it shows relationships with both Group IV and Group V. However, classifying it with Group V would involve not only convergence in character B but also secondary loss of character D. Similarly, the

Table 1 The Thirteen Types of Male Copulatory Organ Found in the Dugesiidae Defined on the Basis of Five Characters Discussed in the Text

Type	Characters	Examples
1	abcde	*Rhodax, Bopsula, C. pinguis*
2	A1bcde	*C. patagonica, D. arimana, seclusa, rincona, hoernesi, glandulosa, graffi*
3	A2bcde	*D. tigrina*
4	A3bcde	*C. foremanii, tinga, D. mertoni, fontinalis, schauinslandi*
5	A3bcDe	*C. evelinae, D. gonocephala*
6	A3bcDE1	*D. cretica, iranica*
7	A3bcDE2	*D. ilvana, transcaucasica*
8	A3bC1De	*D. japonica, nannophallus*
9	A3Bcde	*D. boehmigi*
10	A3BC1de	*C. paeta*
11	A3BC-1de	*D. montana*
12	A4bcde	*D. lugubris, polychroa*
13	A4bcDe	*D. ectophysa*

status of *D. ectophysa* (type 13) remains uncertain by virtue of its possessing a diaphragm (D) indicating relationship with Group V. Marcus (1953) has drawn attention to the similarity in the male copulatory organ between *D. ectophysa* and *D. lugubris* (type 12), but he also pointed out a number of organizational differences. There are also some important differences in the female apparatus, and it is thus possible that type 13 is best derived directly from type 5, and type 12 independently as given in the cladogram. This would also fit with their disjunct distribution (Fig. 5).

Of the groups delimited, Group I is confined to South America (Brazil) and Australasia. The geographical distributions of the remaining Groups II to VI are shown in Fig. 5. The following broad patterns may be discerned. The most primitive Groups I to III are confined to the Southern Hemisphere and North America. The most advanced types of Group V are spread throughout Africa and the Old World only, and Group IV shows possible relationships between Africa, Australasia, and possibly also the New World. However, the American representative of this Group, *C. foremanii* and *C. michaelseni*, are probably examples of convergence or parallelism since they show so many other close relationships with *C. pinguis* and *C. patagonica* of Group II. Certainly the crenate vesicle of *C. michaelseni* is quite different from the seminal vesicle of the *D. gonocephala* group (Böhmig, 1902, plate II, fig. 37). It may also be noted that there are no close connections between Europe and North America, and Asia and North America.

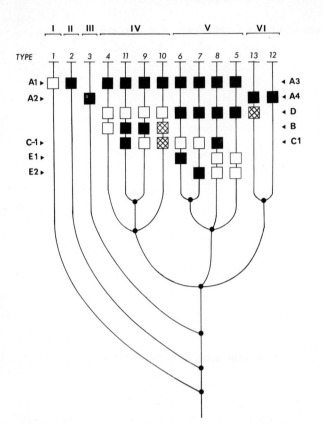

Figure 4 Cladogram representing the evolutionary relationships of the thirteen types of male copulatory organ found in the Dugesiidae, and listed in Table 1. Open squares represent the plesiomorph, solid squares the apomorph, character state. Cross-hatching indicates noncongruent apomorph character states.

Male Reproductive System—Testes   The male gonads present a number of character states for consideration, viz., their form, position, number, and distribution. Unfortunately, it is not easy to decide which of these various character states represents the plesiomorph, and which the apomorph, condition.

In most Paludicola, and all Maricola and Proseriata, the testes are basically round or oval follicles. Partially fused testes have been described in some Planariidae, as *Crenobia alpina* (Chichkoff, 1892, as *Planaria montana*), *Fonticola opisthogona* (Kenk, 1936), and *Fonticola velata* (Ball, 1972). As these species show few other close relationships, it is likely that the condition is secondarily acquired in each.

Fully fused testes are known only in *Hymanella retenuova* (Kenk, 1944, as *Phagocata vernalis*?; see Ball, 1973) of the Planariidae, and in *Rhodax evelinae*. The latter species exhibits a number of other primitive features (Marcus, 1946) whereas *Hymanella* is clearly a specialized form closely related to *Fonticola*, and especially *F. velata*, with which it is easily confused (Ball, 1973). It is my opinion that the fusion of the testes is a secondarily acquired character, attained independently in each genus, follicular testes representing the primitive, and still general, condition.

The position of the testes, whether dorsal or ventral, is variable. Within the Turbellaria as a whole the original condition is probably represented by the dorsal follicles of the Acoela (Ax, 1963; Beklemishev, 1969). At the proseriate level of organization, however, they may be ventral, lateral, or dorsal (Meixner, 1928, p. 586), such variation occurring even within a single family. In *Bothrioplana* they are dorsal (Luther, 1960) or dorsolateral (Marcus, 1946), and in the Otomesostomidae they may be ventral or ventrolateral, as in the Monocelidae with the exception of *Monocelis anta* Marcus, 1954. In the Otoplanidae they are lateral or dorsolateral according to Meixner (1928), or lateral or ventrolateral according to Ax (1956, p. 601), who considers the latter condition to be primitive.

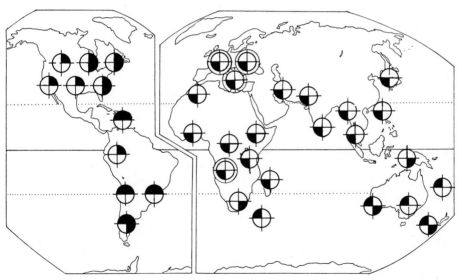

Figure 5 The distribution of the Dugesiidae grouped according to the male copulatory organ, as shown in Fig. 4.

   ⊕ Group II   ⊕ Group III   ⊕ Group IV

       ⊕ Group V   ⊕ Group VI

Group I occurs in Brazil and Australasia.

Meixner (1928) considered that within the Maricola there is a relatively constant arrangement of the testes. Thus, in the primitive Stummerinae, Micropharynginae, and Uteriporidae, as in the Procerodidae of type IIA, they were described as ventral, in the Bdellouridae as lateral, and dorsal or dorsolateral in the Cercyrinae (type I) and the Procerodidae of type IIB. More recently, however, *Leucolesma corderoi*, a procerodid of type IIB, has been described with ventral testes (Marcus, 1948), which condition is also found in *Miava evelinae* Marcus, 1954, a member of the Bdellouridae. *Vatapa gabriellae* Marcus, 1948 of the Cercyrinae is also described with ventral testes. The recently defined Nesionidae, close to both the Bdellouridae and the Uteriporidae, possess ventral testes (Marcus, 1963), and in the Probursidae they are dorsal (Corrêa, 1960). Clearly the variation is greater than was suspected by Meixner.

So, too, with the Paludicola species with dorsal, and others with ventral, testes are found distributed within the three principal families. Even species which are otherwise very closely related show divergence in this character, as, for example, in the Australasian forms *Dugesia fontinalis* (dorsal) and *D. schauinslandi* (dorsal and ventral).

The systematic importance of the position of the testes has been denied by de Beauchamp (1939), but there can be little doubt that it has some value since within the Dugesiidae forms with ventral testes show a well-defined distribution pattern in that they are confined to the Southern Hemisphere and Nearctis, and they also possess other morphological features in common (Ball, 1971). The position of the testes thus may be useful for delimiting groups, but it would be difficult to use it alone for inferring phylogeny.

The number and distribution of the testes also show considerable variation. In most Paludicola they are very numerous and distributed over the whole length of the body. In the genus *Cura*, as defined by Marcus (1955), they are restricted to the prepharyngeal region. *C. schubarti* was an exception in that they extended to the bursa copulatrix, but this species belongs more properly in the genus *Dugesia* (Ball, 1971). In some species the number of testes is very small, perhaps two or three follicles on each side of the midline. Among the proseriates, very few, prepharyngeal testes are characteristic of the Bothrioplanidae and Otomesostomidae. In the Monocelidae the testes tend to be more numerous, although they are still prepharyngeal, and this is also the case in the Otoplanidae. The Maricola, too, show great variation in the number of testes (Marcus and Marcus, 1951).

Reference to the freshwater Proseriata would thus suggest that the primitive condition is represented by the few prepharyngeal follicles present in the species of *Cura* from Australasia and North and South America. On the other hand this condition is rare in the Dugesiidae and there is very great variation in the Maricola. This arrangement of the testes is not found in the

Planariidae or in the Dendrocoelidae, although there are species in these families in which the numerous testes are restricted to the prepharyngeal region.

On the basis of the three characters afforded by the male gonads, the Dugesiidae may be divided into six types, as in Table 2. The geographical distribution of the various types is shown in Fig. 6. The restriction of type 2 to the New World and New Zealand is especially noteworthy, as is the predominance of type 1 in Palaearctis and Aethiopia. Type 5 shows a curious disjunction in the New World which, possibly, will be dispelled by further collecting in northern South America.

As there is little evidence upon which to base judgments concerning the plesiomorphy or apomorphy of these types, their evolutionary relationships can be inferred only by reference to the cladogram (Fig. 4).

Superimposition of the various testis types on the penis morphology cladogram can be achieved with a fair degree of success. The principal difficulties lie with the African species of *Cura*. I have suggested elsewhere that the genus is not a uniform assemblage (Ball, 1971), and it is no longer accepted by de Beauchamp (1968). Under Marcus's (1955) definition of the genus the principal character is the restriction of the testes to the prepharyngeal region. However, he included *C. schubarti*, in which they extend to the bursa, as already mentioned. It may also be noted that in *D. montana* (Fig. 4, type 11) the testes do not extend beyond the copulatory apparatus, and this species shows closer relationships with the African *Cura* species than with the New World and Australasian *Cura* species. The latter forms are additionally characterized by the fact that the number of testes is reduced to a few discrete

Table 2   The Six Types of Testis Arrangement Found in the Dugesiidae

Type	Characters	Examples
1	Testes numerous, dorsal, pre- and postpharyngeal	*D. gonocephala* group, some New World and Australasian forms
2	Testes numerous, ventral, pre- and postpharyngeal	*D. tigrina* group, *D. schauinslandi*
3	Testes numerous, dorsal, prepharyngeal	*C. paeta, evelinae*
4	Testes numerous, ventral, prepharyngeal	*C. tinga*
5	Testes very few, dorsal, prepharyngeal	*C. foremanii, patagonica*
6	Testes very few, ventral, prepharyngeal	*C. pinguis*

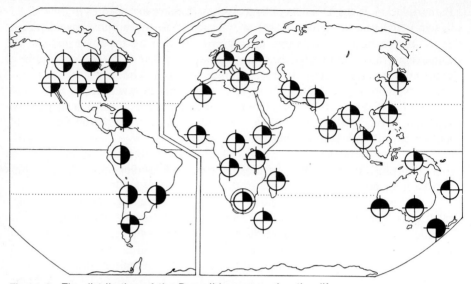

Figure 6 The distribution of the Dugesiidae possessing the different types of testis arrangement listed in Table 2.

◑ Type 1   ◔ Type 2   ◕ Type 5

◔ Type 6   ⊕ Types 3 and 4

follicles, which lie anterior to the pharynx. As will become apparent later, these forms also show some remarkable similarities in the female reproductive system. On the other hand, Dugesiidae which possess numerous testes have them distributed to the pharynx, copulatory apparatus, or the tail.

Rejecting the distribution of the testes as a phylogenetic character, we may recognize the following four groups of triclads:

Group I   Testes numerous and dorsal
Group II   Testes numerous and ventral
Group III Testes very few and dorsal
Group IV   Testes very few and ventral

Of course the distribution of the testes in African forms remains a useful discriminatory character.

These groups are superimposed on the penis morphology cladogram in Fig. 7. The phyletic lines leading to types 2 and 4 clearly require further resolution. The numerous ventral testes of line 4 are exemplified only in *Cura tinga*, one of the most problematical species in many respects, and *D. schauin-*

slandi. As the latter species has been described as having either dorsal or ventral testes (Neppi, 1904; Nurse, 1950), its taxonomic position with respect to this character can be resolved without invoking parallelism or convergence. Considering *Cura tinga*, we may note that it is the only African species reported with ventral testes, and this condition seems likely to have been secondarily acquired. The analysis of the female system supports this conclusion. I have suggested earlier (p. 359) that *C. foremanii* and *C. michaelseni* belong more properly in type 2 than in type 4, which also will be confirmed by the female system, and thus the phyletic line leading to type 4 becomes homogenous.

It is not clear whether the dorsal testes of the New World *Dugesia*

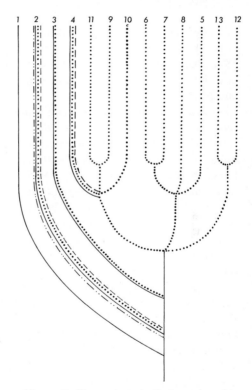

Figure 7 The various testis types found in the Dugesiidae superimposed on the cladogram of Fig. 4.

———————————Testes numerous and ventral
·······················Testes numerous and dorsal
- - - - - - - - - - -Testes very few and dorsal
— ·· — ·· — ··Testes very few and ventral

species (types 2 and 3) have been derived independently of those of the Old World species. Certainly the testes of New World forms differ in that they are usually discrete, and large dorsoventrally, whereas those of the *D. gonocephala*-like forms (types 5 to 8) are usually small and aggregated into clusters. As the integrity of these two large subgroups is also supported by their external features (p. 355) and by the female system (p. 370), we may suspect this possibility.

These data suggest that the primitive condition is represented by ventral testes, but whether numerous (*Rhodax, Bopsula*) or few (*Cura pinguis*) cannot be stated with certainty. The most apomorph conditions appears to be represented by the very numerous and small dorsal follicular groups of *D. gonocephala* and its allies.

Female Reproductive System   Although the female reproductive system is generally acknowledged as being of great taxonomic importance (Meixner, 1928; Kenk, 1930a; Hyman, 1931a), it has been little used other than in generic definitions. Yet, it possesses many characters which are of diagnostic and phylogenetic significance.

A bursa copulatrix is usually present in the Paludicola, but it is greatly reduced in *Dugesia microbursalis* and absent in *Cura foremanii*. In my material of *C. pinguis* it is absent; Kawakatsu (1969b) found it present but reduced in size. Both Weiss (1910) and Nurse (1950) found a relatively large bursa in their material. Kenk (1935) considers the absence of a bursa in *C. foremanii* to be a primitive feature, and he is supported by my findings for *C. pinguis*. On the other hand it has been shown that the former species self-fertilizes, and does not cross-copulate, so that a well-developed bursa is unnecessary (Anderson, 1952; Anderson and Johann, 1958). Consequently it is unwise initially to place too much phylogenetic significance on this character.

Likewise, the presence of a bursa–intestinal duct is of little phylogenetic significance. Such a duct is found in divergent species in the Dugesiidae, Planariidae, and Dendrocoelidae (Steinböck, 1924; Marcus, 1946) and in a few Maricola (Corrêa, 1960), and possibly represents a recurrent ancestral character. Frequently its occurrence is constant within a species, and is thus a diagnostic character, but in other species such as *Polycelis tenuis*, it may or may not occur.

Of greater phylogenetic significance is the position of the bursa copulatrix. The Paludicola differ from the Terricola and most of the Maricola with a single genital pore in that the bursa is usually anterior to the penis. The posterior position of this organ in *Rhodax* is, therefore, yet another primitive feature retained by this genus.

The possible taxonomic importance of the musculature of the bursal stalk has not always been recognized, with the result that in many species descriptions there are no data concerning this point. In most Paludicola, as in nearly all Maricola for which data are available, the musculature of the bursal stalk consists of two layers. In most Paludicola the inner layer is of circular fibers and the outer layer is of longitudinal fibers, and thus the sequence of muscle layers found in the body wall and atrium continues to the bursal stalk. However, in some Dugesiidae the sequence of muscles in the bursal stalk is reversed so that the circular layer is on the outside. That this situation is restricted to some species from Australasia and most from Africa, Orientalis, and Palaearctis suggests that is is an apomorph condition. It also appears that most Maricola have the longitudinal layer on the outside, an exception being *Procerodes ohlini*, in which the muscle layers of the bursal stalk are intermingled (Böhmig, 1906), a condition which may exist also in *Probursa veneris* Hyman, 1944, although I have found the available slides of the latter difficult to interpret in this respect. Regrettably, more recent descriptions of the Maricola do not include the bursal stalk musculature, but I have studied as yet undetermined species of Procerodidae from eastern Canada, and from Saint Helena, together with original slides of *Procerodes pacifica* Hyman, 1954 and *Nesion arcticum* Hyman, 1956 and in all but one of these the outer of the two muscle layers is longitudinal. The exception, from Saint Helena, has lost the outer layer of longitudinal musculature. It may be noted, however, that in *Nesion* the circular layer is extraordinarily well developed, and less so in *Probursa veneris*, but in most *Procerodes* species the bursal musculature is very weak. These data are consistent with the suggestion that a bursal stalk musculature of two layers, the outer being longitudinal, is plesiomorph, and it is relevant to note that the musculature of the atrium femininum of the proseriate *Otomesostoma auditivum* is similar in this respect (Hofsten, 1907).

Most variation in the bursal stalk musculature occurs in those species in which the inner layer is longitudinal. Some species, viz., *D. astrocheta* and *D. burmaensis*, have developed an extra layer of longitudinal fibers outside the circular layer, and in many others there is an outer layer of longitudinal fibers reinforcing the ectal region of the bursal stalk. These are undoubtedly apomorph conditions, as is the unusual and very strong sphincter composed of circular fibers which is found in the ectal region of the bursal stalk of *D. fontinalis*.

In a few Dugesiidae there is an extraordinary thickening of the outer circular layer, so that this equals or exceeds the diameter of the rest of the bursal stalk as in *Cura paeta* and *tinga* from South Africa, and in *Dugesia montana* from New Zealand. A tendency in this direction is also noticeable in a number of other African and Australian *Dugesia* species. This excessive thickening is unknown in the other Paludicola, and is almost certainly apomorph, but is paralleled by the marine form *Nesion*, although here it is bounded by the outer longitudinal fibers.

The interpretation of the bursal stalk musculature of some species has proved to be difficult. Kawakatsu's (1969b) description of *Cura pinguis*, for example, conflicts with that of de Beauchamp (1968) and Weiss (1910), who indicate clearly that the outer layer is circular. Since drawing attention to this (Ball, 1970), I have had the opportunity to examine this species and find the bursal stalk to be very similar to that of *Cura foremanii* in that the circular muscle fibers are quite prominent whereas the longitudinal fibers are very difficult to discern. However, I am of the opinion that in both species there are scattered, not continuous, longitudinal fibers outside the circular muscles. The entire musculature, however, is very weak, and the histology of the bursal stalk is very reminiscent of the Procerodidae of Meixner's type IIA which I have examined.

A further difficulty is presented by *Dugesia schauinslandi*. Neppi (1904) indicates that the outer muscle layer of the bursal stalk is longitudinal, whereas my own observations on this species indicate the opposite. The probability of an error by Neppi is supported by the fact that in the same paper she also described *D. neumanni* as having an outer layer of longitudinal muscles, which conflicts with the more recent description of Marcus (1955), who had better-preserved material.

The bursal stalk is usually a smoothly curved duct of relatively uniform diameter, which appears to be the primitive condition, and in some cases the thickness of the duct may be of diagnostic significance (Ball, 1970, 1971) at the species level. In two species, however, *Dugesia montana* and *Cura tinga*, the bursal stalk is exceptionally dilated and its walls are thrown into folds or creases. It is of more than passing interest to note that these are two of the three species which are characterized by excessive thickening of the circular muscle layer (Fig. 9).

The course of the oviducts and shell glands is an important taxonomic character which has been discussed previously. It was concluded that the primitive condition was that in which the oviducts and shell glands open into the bursal stalk. In Meixner's type IIC, comprising the Australian species *D. hoernesi* and *D. boehmigi* of Weiss (1910), a condition apparently intermediate between the Dugesiidae and the Planariidae occurs. A long common oviduct, receiving the shell glands, opens into the atrium at the base of the bursal stalk. A common oviduct occurs in other Dugesiidae, but it opens into the bursal stalk, as do the shell glands. However, in many species there is variation in this character, and in some individuals there is a common oviduct, and in others not (see Ball, 1971), and so it is of little phylogenetic significance. Similarly, I attach little phylogenetic importance to the position of the openings of the oviducts into the bursal stalk, whether proximal or distal to the atrium, although it is a useful diagnostic character.

The caudal dichotomy in the oviducts of some species has been a controversial character. Neppi (1904) was the first to observe this peculiarity when she described *Dugesia schauinslandi* from New Zealand, and noted that the oviducts dispatched a caudal dichotomy to the vitellaria of the posterior part of the body. Meixner (1928, p. 576, note 5), however, intimated that Neppi had mistaken a simple doubling back of the oviducts, or recurvature, for a true dichotomy, and consequently Marcus (1946) considered that *Rhodax* was unique in its possession of branched oviducts. In her study of the New Zealand triclads Nurse (1950) redescribed *D. schauinslandi* under the name *Spathula limicola* (see de Beauchamp, 1951a) and described a new and closely related species as *Spathula fontinalis*. She found that both species possessed branched oviducts, thus confirming the findings of Neppi (1904). Unfortunately, de Beauchamp (1951a) misunderstood Nurse's descriptions and believed that she was erecting her new genus *Spathula* on the presence of a common oviduct, rather than on the branching of each oviduct, an error which was carried over into later papers (de Beauchamp 1959, 1961). However, I have examined specimens of both *D. schauinslandi* and *D. fontinalis* and can confirm the accuracy of Nurse's and Neppi's descriptions in this respect.

The question may now be raised as to whether the branched oviducts of *Rhodax* and the two Australian species represent a plesiomorph or apomorph condition. The only other Seriata with branched oviducts are *Otomesostoma* and *Bothrioplana* (Marcus, 1946), and this link with the freshwater Proseriata indicates that the condition is most likely a primitive retention. It is also relevant to note that *Rhodax* is one of the few Paludicola known which reproduces paratomically; whether or not *D. schauinslandi* and *D. fontinalis* possess this capability is not known. In some Dugesiidae it has proved possible to induce the formation of supernumerary reproductive organs artificially, and in such cases branching of the oviducts has been observed (Okugawa, 1955, p. 7). Further, I have slides of an undescribed species of *Dugesia* from South Africa which show evidence of both paratomy and branching of the oviducts. It is possible, therefore, that branched oviducts are related to the phenomenon of paratomy, but this does not invalidate the conclusion concerning the plesiomorphy of the character since paratomy is primitive within the Turbellaria as a whole, it being the normal method of reproduction in many Macrostomida and Catenulida (Beklemishev, 1969).

The nine characters afforded by the female system, as discussed above, may be coded as follows. As previously, lowercase letters indicate the plesiomorph condition.

a   Bursa copulatrix posterior to penis.
A   Bursa copulatrix anterior to penis.
b   Bursal stalk with two muscle layers.
B   Bursal stalk with three muscle layers.

c   Inner muscle layer of the bursal stalk circular.
C   Inner layer longitudinal.
d   Ectal reinforcement absent.
D   Ectal reinforcement present.
e   Bursal stalk without strong sphincter.
E   Bursal stalk with strong sphincter.
f   Circular muscles of bursal stalk normal.
F   Circular muscle layer greatly thickened.
g   Bursal stalk relatively uniform.
G   Bursal stalk dilated and convoluted.
h   Oviducts with caudal dichotomy.
H   Oviducts without caudal dichotomy.
i   Oviducts and shell glands enter bursal stalk.
I   Oviduct enters atrium, shell glands enter common oviduct.

Of the numerous possible character combinations the known Dugesi-idae have utilized thirteen, as shown in Table 3, and a cladogram of these types has been constructed (Fig. 8). In this scheme all characters were coded as two state characters, and consequently *D. polychroa* is not included since it could not be coded for characters B, C, and F. In this species the muscle layers of the bursal stalk are intermingled and it is impossible to say whether this was derived from Group II or from Groups III and IV (Fig. 8, char-acter C).

Table 3   The Thirteen Types of Female Reproductive Apparatus Found in the Dugesiidae, Defined on the Basis of Nine Characters Discussed in the Text

Type	Characters	Examples
1	a(b) (c)defghi	*Rhodax*
2	AbcdefgHi	*Bopsula, C. foremanii, pinguis, patagonica, D. tigrina, annandalei, seclusa, glandulosa, graffi*
3	AbCdefghi	*D. schauinslandi*
4	AbCdEfghi	*D. fontinalis*
5	AbCdefgHi	*D. indica, nannophallus, izuensis, batuensis, gonocephala, monomyoda, ectophysa*
6	AbcdefgHI	*D. hoernesi*
7	ABCdefgHi	*D. burmaensis, astrocheta*
8	AbCDefgHi	*D. japonica, gonocephala, lindbergi, iranica bactriana, neumanni*
9	AbCdefgHI	*D. boehmigi*
10	AbCdeFgHi	*C. paeta*
11	AbCdeFGHi	*D. montana*
12	AbCDefgHi	*C. evelinae*
13	AbCDeFGHi	*C. tinga*

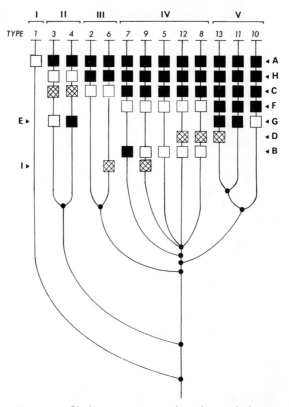

Figure 8 Cladogram representing the evolutionary relationships of thirteen types of female reproductive apparatus found in the Dugesiidae, and listed in Table 3. Open squares represent the plesiomorph, solid squares the apomorph, character state. Cross-hatching indicates noncongruent apomorph character states.

Further consideration of characters D (ectal reinforcement of the bursal stalk) and I (common oviduct enters atrium) is necessary. Character D is clearly incompatible with character F (exceptionally thick circular muscles of the bursal stalk), and I have had to reach a decision as to which is the most useful character. I have selected F on the basis that it brings together *Cura paeta*, *C. tinga*, and *D. montana*, three species which show close relationships in the male copulatory organ and the distribution of the testes. Further, the stability of character D is open to question. *Dugesia gonocephala* has been described by de Beauchamp (1961, p. 82, fig. 39) as possessing a third muscle layer reinforcing the ectal region of the bursal stalk, whereas Marcus (1953) does not indicate this. Therefore I have accepted character F as being the most reliable.

Meixner's (1928) reproductive apparatus type IIC comprises those Australian species of *Dugesia* which possess character I—a common oviduct into which the shell glands open, and which itself enters the atrium at the base of, and posterior to, the bursal stalk. This would seem to be a good apomorph character uniting *D. hoernesi* (Fig. 8, type 6) and *D. boehmigi* (type 9), but there are good reasons for not placing too much weight on it. In the first place the apparent position of the oviduct may depend upon the degree of contraction of the preserved specimen. It is quite obvious, for example, that Weiss based her description of *D. hoernesi* on a contracted animal (Weiss, 1910, fig. 21). There are other species in which the oviducts open into the bursal stalk extremely close to its junction with the atrium (e.g., *D. nannophallus*), and it is easy to see how misinterpretation could occur.

A second, more important, factor results from my examination of paratypical material of *D. montana* from New Zealand. Nurse's (1950) description of this species is perfectly adequate for its recognition, and there can be no doubt of the identity of the material examined by me. Nevertheless, there are some minor points which need clarification. My own interpretation of the reproductive apparatus of *D. montana* is shown in Fig. 9; it may be compared with that of Nurse (1950, plate 46, fig. 1). The blind posterior diverticulum of the atrium, not described by Nurse, receives many eosinophil glands, presumably the shell glands, and could easily be mistaken for a common oviduct. In some respects it is comparable with the nonglandular diverticulum of *D. tigrina* (Ball, 1971). The problem is compounded by the fact that the oviducts do not extend beyond the bursal stalk and recurve to it in the usual way. On the contrary they run directly to the stalk and enter it on its frontal, rather than lateral, face. This course of the oviducts is difficult to trace, but it shows on both sets of sections I have examined, and it is curiously reminiscent of the condition in *Probursa* (Corrêa, 1960). A third Australasian species, *D. mertoni* of the Kai Islands, is figured in the original description as having a common oviduct which enters the atrium at the base of the bursal stalk (Steinmann, 1914, p. 116, fig. 3: the labels for "Ovidukt" and "Drüsensack" have been transposed); the shell glands are not described. Meixner (1928) does not place this species in his type IIC, but allies it with *D. tigrina, D. glandulosa*, and others in type IIA.

A reexamination of the course of the oviducts in these Australian species is certainly warranted. From the data outlined above I am unable, for the present, to accept Meixner's type IIC as representing a natural grouping, and prefer to base decisions concerning the relationships of *D. hoernesi* and *D. boehmigi* on less equivocal characters. On the basis of the male apparatus and of the musculature of the bursal stalk I consider *D. hoernesi* to be much closer to the *D. tigrina* group than *D. boehmigi* is.

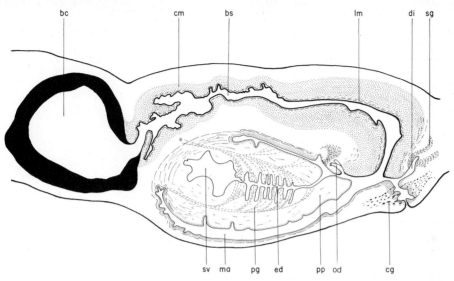

Figure 9 *Dugesia montana.* Paratype. Reconstructed sagittal section of the copulatory organs, viewed from the left side. Diagrammatic. bc, bursa copulatrix; bs, bursal stalk; cg, cement glands; cm, circular muscle; di, diverticulum; ed, ejaculatory duct; lm, longitudinal muscle; ma, male atrium; od, left oviduct; pg, penis glands; pp, penis papilla; sg, shell glands; sv, seminal vesicle.

Returning to the cladogram of the female reproductive system (Fig. 8), I have divided the various types into six Groups, as shown. The sixth Group is formed by *D. lugubris*. The geographical distribution of these Groups is shown in Fig. 10, which may be compared with Fig. 5. Again it is noteworthy that the most primitive Groups, I to III, are confined to the Southern Hemisphere and North America. The most advanced types of Groups IV and VI are spread throughout Africa and Palaearctis and Orientalis, and possibly Australia. Group V shows possible relationships between South Africa and Australasia. These data compare well with the broad patterns obtained from the analysis of the male system (p. 359), and once more the absence of Europe–North America and Asia–North America relationships is prominent.

In comparing the female reproductive system (F Groups) with the male system shown in Fig. 4 (M Groups), a number of similarities and differences become apparent. M Group VI and F Group VI are identical, with the qualification concerning *D. ectophysa* discussed earlier (p. 359). M Group V is equivalent to F Group IV, with the exception of *D. boehmigi*, whose position appears equivocal. M Groups II and III are equivalent to F Group III with the exceptions of *Cura pinguis* and *Bopsula*, which appear in M Group I. The association of these forms with other members of M Group II is, however,

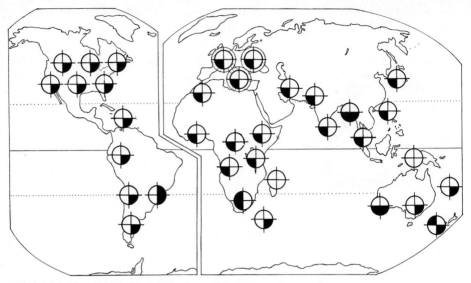

Figure 10 The distribution of the Dugesiidae grouped according to the female reproductive apparatus, as shown in Fig. 8.

◐ Group I ◑ Group III ◐ Group IV

◑ Group V ⊕ Group VI *(D. polychroa)*

Group II is confined to New Zealand.

justifiable. *Bopsula* is known from only one specimen, and Marcus (1946) considers it to be derivable from a *Dugesia* of the *D. tigrina* type, which it resembles in outward appearance. However, *Bopsula* is so aberrant in a number of respects that its exact placing in my phylogenetic scheme is best deferred until further specimens are found. The unusual duct between the bursal stalk and the penis bulb is paralleled by the unusual oviducts of the marine form *Nexilis*. The close relationships of *C. pinguis* and *C. foremanii* are confirmed by other characters relating to the testes and the female reproductive apparatus, and these carry more classificatory weight than the primitive retention of a common vas deferens in *C. pinguis*. M Group IV is equivalent, with the exception of *D. boehmigi*, to F Groups II and IV, and since the differences between the latter two Groups are so marked, their separation seems justified.

With these limitations the correlations between the three cladograms of Figs. 4, 7, and 8 and between the character maps of Figs. 3, 5, 6, and 10 are remarkably good, so much so that it appears both useful and justifiable to propose a classification of the Dugesiidae based on the major groupings demonstrated. *Rhodax* would be representative of the most plesiomorph form, and *D. gonocephala* of the most apomorph.

The interrelationships of the more plesiomorph forms are difficult to determine; this is often the case for primitive groups. If *Rhodax* is considered to be closest to the ancestral freshwater triclad, then its phylogenetically closest relatives would appear to be within either the *D. tigrina-* or *C. foremanii*-like forms. On the basis of head shape, the male copulatory organ, and the female reproductive apparatus either group could qualify as closest relative by virtue of some of their members. What does seem clear is that the *D. tigrina* group has not given rise to any of the higher forms and has diversified in isolation in the New World, and to a lesser extent across the Southern Hemisphere. The independent derivation of the *D. gonocephala*-like (and subsequently *D. lugubris*-like) forms and the *D. montana*-like forms from the *Cura pinguis* group not only fits the biogeographical data rather well (see below), but also agrees with the suggestion made earlier (p. 365) that the dorsal testes of these higher groups are derived independently of the dorsal testes of some members of the *D. tigrina* group.

It remains to be seen, from future work, whether other characters are confirmatory of the groupings proposed here. Some preliminary comments may be made on the basis of limited data from the literature, but to clarify discussion, it seems advisable to define these groupings formally.

Revision and Synopsis

In view of the difficulties resulting from parallelism and convergence of producing a phylogenetically based classification of the Dugesiidae, I have adopted a conservative approach to the nomenclature of the group. Recognizing that this preliminary attempt is open to considerable refinement in the future, I have kept nomenclatural changes to a minimum and given the new categories the rank of subgenera. This permits discussion and argument without affecting the established binominal nomenclature. One change, however, has been unavoidable. The genus *Cura* clearly cannot be allowed to stand as recognized by Marcus (1955) and Kawakatsu (1969b). According to the analyses performed here *C. paeta*, *C. tinga*, *C. wimbimba*, and *C. evelinae* are far removed from *C. pinguis*, *C. patagonica*, and *C. foremanii*, and their association in the same genus, distinct from *Dugesia*, does not reflect their phylogenetic relationships. Thus, I have restricted *Cura*, as a subgenus, to the American and Australasian forms. The genus *Cura* has previously been abandoned by de Beauchamp (1951a, 1968), although other workers have declined to follow him. It will be noted that the principal subgenera proposed below are similar to the divisions of the genus *Dugesia* proposed by de Beauchamp (1939) on the basis of the male copulatory organ.

A Synopsis of the Dugesiidae   In the list which follows, species marked
with an asterisk have been studied by me. A series of notes dealing with problematical taxa follows the list.

Genus *Rhodax* Marcus, 1946
   *Rhodax evelinae Marcus, 1946

Genus *Bopsula* Marcus, 1946
   Bopsula evelinae Marcus, 1946

Genus *Dugesia* Girard, 1850

Subgenus *Dugesia* Girard, 1850

   Head of low triangular form. Seminal vesicle an enlarged muscular cavity.
Diaphragm present. Bursal stalk musculature of inner longitudinal fibers surrounded by circular muscles. Testes numerous, forming dorsal clusters scattered throughout the body length (Note 1).
   D. (D.) gonocephala (Dugès, 1830); **D. (D.) cretica* (Meixner, 1928;
Kenk, 1930); **D. (D.) iranica* Livanov, 1951 (= *Euplanaria cretica*: de Beauchamp, 1936); *D. (D.) transcaucasica* Livanov, 1951; *D. (D.) taurocaucasica*
Livanov, 1951; *D. (D.) praecaucasica* Livanov, 1951; *D. (D.) etrusca* Benazzi, 1946; *D. (D.) ilvana* Lepori, 1948; *D. (D.) sicula* Lepori, 1948; *D. (D.)*
benazzii Lepori, 1951; *D. (D.) neumanni* (Neppi, 1904) (Note 2); *D. (D.)*
congolensis de Beauchamp, 1951; *D. (D.) milloti* de Beauchamp, 1953;
D. (D.) machadoi de Beauchamp, 1953; *D. (D.) lamottei* de Beauchamp,
1953; **D. (D.) monomyoda* Marcus, 1953; *D. (D.) astrocheta* Marcus,
1953; *D. (D.) colapha* Dahm, 1967; *D. (D.) ectophysa* Marcus, 1953 (Note
3); **D. (D.) japonica* Ichikawa and Kawakatsu, 1964; *D. (D.) indica* Kawakatsu, 1969; **D. (D.) nannophallus* Ball, 1970; **D. (D.) batuensis* Ball, 1970;
?*D. (D.) annandalei* (Kaburaki, 1918) (Note 4); ?*D. (D.) izuensis* Kato, 1950;
?*D. (D.) mertoni* (Steinmann, 1914); ?*D. (D.) jeanneli* (de Beauchamp, 1913);
*?*D. (D.) evelinae* (Marcus, 1953); *?*D. (D.) lindbergi* de Beauchamp, 1959;
*?*D. (D.) bactriana* de Beauchamp, 1959.

Subgenus *Girardia* nom. n.

   Head typically high triangular, but may be truncate. Seminal vesicle
absent or of the bifid nonmuscular type. Diaphragm absent. Bursal stalk
musculature of inner circular muscles surrounded by longitudinal fibers.
Testes numerous, distributed throughout the body length (exception *schu-
barti*), and usually ventral.
   **D. (G.) tigrina* (Girard, 1850); *D. (G.) microbursalis* (Hyman, 1931b);
**D. (G.) dorotocephala* (Woodworth, 1897); *D. (G.) aurita* (Kennel, 1888);
D. (G.) festai (Borelli, 1898); *D. (G.) polyorchis* (Fuhrmann, 1914); **D. (G.)*
anceps (Kenk, 1930) (= *dubia* Borelli, 1895); *D. (G.) nonatoi* Marcus, 1946;
D. (G.) arndti Marcus, 1946; *D. (G.) rincona* Marcus, 1954; *D. (G.) sanchezi*
Hyman, 1959; *D. (G.) jimi* Martins, 1970 (doubtfully distinct from *D. tigrina*);

D. (G.) dimorpha (Böhmig, 1902); **D. (G.) antillana* Kenk, 1941; *D. (G.) schubarti* (Marcus, 1946); *D. (G.) hypoglauca* Marcus, 1948; **D. (G.) arimana* Hyman, 1957; *D. (G.) paramensis* (Fuhrmann, 1914); *D. (G.) andina* (Borelli, 1895); **D. (G.) chilla* Marcus, 1954; *D. (G.) veneranda* Martins, 1970 (doubtfully distinct from *D. chilla*); *D. (G.) seclusa* (de Beauchamp, 1940); *D. (G.) graffi* (Weiss, 1910); *D. (G.) glandulosa* (Kenk, 1930) (= *striata* Weiss, 1910); *D. (G.) hoernesi* (Weiss, 1910) (Note 5).

Subgenus *Cura* Strand, 1942

Head truncate, or of the low triangular form. Seminal vesicle primitively absent, but enlarged form present in two species (*foremanii, azteca*). Bursal stalk musculature of inner circular muscles surrounded by a fine layer of longitudinal fibers. Testes very few, prepharyngeal. Bursal stalk frequently expanded at entrance into roof of male atrium to form a female atrium, which receives the shell glands.

*D. (C.) foremanii (Girard, 1852); *D. (C.) patagonica* (Borelli, 1901); *D. (C.) michaelseni* (Böhmig, 1902); **D. (C.) pinguis* (Weiss, 1910); *D. (C.) azteca* (Benazzi and Giannini, 1971); *?D. (C.) falklandica* (Westblad, 1952) (Note 6).

Subgenus *Neppia* nom. n.

Head typically of low triangular form. Seminal vesicle a single muscular cavity, diaphragm absent. Bursal stalk musculature of inner longitudinal fibers surrounded by an exceptionally thick layer of circular fibers. Testes numerous, dorsal (exception *tinga*), not extending beyond the copulatory apparatus. Ejaculatory duct typically convoluted.

*D. (N.) montana Nurse, 1950; **D. (N.) paeta* (Marcus, 1955); **D. (N.) tinga* (Marcus, 1955); *D. (N.) wimbimba* (Marcus, 1970); *?D. (N.) boehmigi* (Weiss, 1910) (Note 7).

Subgenus *Spathula* Nurse, 1950

Head rounded or spathulate, or of the low triangular form. Seminal vesicle a single cavity. Diaphragm absent. Bursal stalk musculature of inner longitudinal muscles surrounded by circular fibers. Testes numerous, dorsal or ventral, and extending throughout the body length. Oviducts branched caudally (Note 8).

*D. (S.) schauinslandi (Neppi, 1904); **D. (S.) fontinalis* (Nurse, 1950).

Subgenus *Schmidtea* nom. n.

Head of the low triangular form. Seminal vesicle consisting of an intrabulbar muscular cavity and an extrabulbar muscular cavity. Diaphragm absent. Bursal stalk musculature of intermingled circular and longitudinal fibers. Testes numerous, dorsal, extending throughout the body length.

D. (S.) lugubris (Schmidt, 1862) sensu Reynoldson and Bellamy, 1970;
*D. (S.) polychroa (Schmidt, 1862) sensu Reynoldson and Bellamy, 1970.

The following species, described under the name *Planaria*, but presumably of the genus *Dugesia*, are inadequately described and are here considered to be *taxa dubia*.

P. *aborensis* Whitehouse, 1913a; P. *tiberiensis* Whitehouse, 1913b; P. *salina* Whitehouse, 1913b; P. *barroisi* Whitehouse, 1913b; P. *fissipara* Kennel, 1888 (possibly identical with *Rhodax*); P. *similis* Böhmig, 1902; P. *ambigua* Böhmig, 1902; P. *laurentiana* Borelli, 1897; P. *hymanae* Sivickis, 1928; P. *bilineata* Kaburaki, 1918; P. *andamanensis* Kaburaki, 1925; P. *rava* Weiss, 1910; P. *tanganyikae* Laidlaw, 1906; P. *venusta* Böhmig, 1897; P. *brachycephala* Böhmig, 1897; P. *iheringii* Böhmig, 1887.

Concerning D. *aberana* Neppi, I can find no mention of this species other than in de Beauchamp (1951b). Similarly, a species of *Dugesia* belonging to Meixner's type IIC seems never to have been described (see Meixner, 1928, p. 583). It is characterized by the strengthening of the musculature of the ectal part of the bursal stalk, which is a distinctive feature of D. *fontinalis*.

Notes to the List

1 A number of species which one would expect to fall into the subgenus *Dugesia* appear to be excluded by virtue of certain characters. D. *izuensis, annandalei*, and *lindbergi* all lack a diaphragm in the ejaculatory duct. The first two species are known only from the types, now lost, as described by the original authors, and so cannot be considered further. I have specimens from Malaya of a species which I assign to D. *lindbergi* on the basis of penis morphology and the distribution of the muscles covering the bursal stalk. These latter are weak entally and become stronger ectally (de Beauchamp 1959, p. 35). The histological condition of my material is poor, but a diaphragm is discernible. There can be little doubt of the close relationship of D. *lindbergi* to the other Oriental species of the subgenus (Ball, 1970). It may also be noted that in the specimens of D. *bactriana* available to me I cannot recognize a diaphragm, although this is described and figured by de Beauchamp (1959).

The exact status of D. *jeanneli* is also problematical. Originally described by de Beauchamp (1913) with later corrections (de Beauchamp, 1939), it was placed in the genus *Cura* by Marcus (1955). Of the subgenera recognized here it certainly cannot be assigned to either *Cura* or *Neppia*, and de Beauchamp (1939) considers its affinities to be with the D. *gonocephala*-like forms, in which case the reduced musculature of the bursal stalk, and the restriction of the testes to the prepharyngeal region must be considered secondary. Similarly, the diaphragm and bursal stalk musculature lead me to place D. *evelinae* in this group for the present, though its status needs clarification. In the slides available to me a diaphragm is not apparent.

D. burmaensis and *D. astrocheta* are unusual in that they possess a three-layered musculature of the bursal stalk, but as in all other respects they approach the *D. gonocephala*-like forms, notably in the possession of a distinct diaphragm in the ejaculatory duct, I consider the acquisition of the third, outer longitudinal layer to be secondary.

Dugesia mertoni (Steinmann, 1914) is inadequately described for its proper placement. I assume it belongs in this subgenus.

Dugesia absoloni (Komarek, 1919) regrettably is known only from not fully mature material (Meixner, 1928, p. 577, note 8), and so its exact relationships cannot be determined. It was originally described as a terrestrial triclad, under the name *Geopaludicolia*, which name Kenk (1930) retains with subgeneric rank.

2   Both de Beauchamp (1939) and Marcus (1953) refer to *D. neumanni* as having a simple type of male copulatory organ. It seems to me that the presence of a diaphragm, the musculature of the bursal stalk, the arrangement of the testes, and the head shape, all justify its placement in this subgenus.

3   Placed in this subgenus, and not *Schmidtea* on the basis of the presence of a diaphragm, the bursal stalk musculature, and the comparative remarks of Marcus (1953).

4   *Dugesia annandalei* separates with the *D. tigrina* group according to the analysis of the female system (Fig. 8). I consider such a relationship unlikely. The original description is based on a single specimen, now lost. De Beauchamp (1951a), however, has suggested that this species is identical with *D. glandulosa*, which would be of great biogeographical interest.

5   *Dugesia hoernesi* belongs to Meixner's unusual type IIC, which I do not accept in this revision. The folds and contractions of the penis as figured by Weiss (1910, plate XIX, fig. 21) are very reminiscent of the condition figured in *D. festai* by Hyman (1939) and in *D. tigrina* by Kenk (1935). The muscular gland organ, or adenodactyl, missed by Weiss but recognized in the original sections by Meixner (1928, p. 577, note 10), similar to the one found in *D. boehmigi*, is in my opinion without phylogenetic significance. I disagree with Meixner's view that the two species are closely related; penis morphology and the histology of the bursal stalk argue strongly against this.

6   The description of *Cura falklandica* is inadequate for its proper placement. I assume it belongs in this subgenus.

7   The relatively strong circular musculature of the bursal stalk and the convoluted or dilated ejaculatory duct are the principal reasons for tentatively assigning this problematical species to the subgenus *Neppia* (compare Fig. 9 with Weiss, 1910, plate XXI, fig. 28).

8   Contrary to the philosophy stated earlier in this paper this category is defined largely by plesiomorph characters. However, only two species are involved, and their peculiarities are such that their separation seems to be both justifiable and useful at the present.

The proposed subgenera could, of course, be divided into further subunits, perhaps with the rank of "species groups"; I have already, for example, separated off some Caribbean and South American forms of the subgenus *Girardia* into a *Dugesia antillana* species group (Ball, 1971). Other possible species groups within this subgenus may be exemplified by *Dugesia tigrina*, *D. chilla*, and *D. glandulosa*, which differ among themselves in small ways in the morphology of the penis and atrium, and in the course of the bursal stalk and the position of the testes. Within the subgenus *Dugesia* there are two large subgroups evident, the *D. gonocephala* group with a terminal opening of the ejaculatory duct, and a more apomorph *D. japonica* group with a subterminal opening of the ejaculatory duct. It is possible that *D. burmaensis* and *D. astrocheta* form a third group (Fig. 8, type 7) or even another subgenus. Also, it is mainly within the subgenus *Dugesia* that we find species with a three-layered outer musculature of the pharynx. However, despite the success that this type of approach has had in elucidating systematic and evolutionary relationships in other groups (e.g., Vuilleumier, 1969), I feel that it would be premature to continue this line of inquiry in the Paludicola at present. It seems advisable to test and refine the revision here proposed first, and to leave further detailed analyses of the new taxa to the revisers of regional faunas.

On the basis of the characters used in the subgeneric definitions I have constructed a further cladogram to suggest the phylogenetic relationships of the genera and subgenera of Dugesiidae (Fig. 11). It will be seen that this differs from Meixner's (1928) scheme not only in the number of divisions, but also in the distribution of many of the species. Thus, Meixner type I is composed principally of primitive elements, but it also contains forms such as *D. gonocephala*, *D. lugubris*, and *D. burmaensis*, which are here considered to be the most advanced of the Dugesiidae.

Meixner placed more emphasis on the morphology of the atrium than is done here (see p. 343). His more primitive types I and II are described as having an undivided atrium. In fact, it is characteristic of a number of species of *Girardia*, and of *Cura*, that the bursal stalk enters the roof of the male atrium and does not travel to the gonopore and, especially in *Cura*, enlarges there to form a female atrium. Steinböck (1924) has commented previously on the arrangement of the bursal stalk and oviducts in *D. (Cura) pinguis*; he considered this condition to be primitive. In the higher Dugesiidae, as in the large and diverse subgenus *Dugesia*, a divided atrium is the exception rather than the rule.

The similarities in the atrial muscles and penis morphology of *D. (C.) pinguis* and *D. (D.) evelinae* have been discussed by Marcus (1955) and Ball and Fernando (1969). The present analysis has failed to resolve the problem of their possible close relationship, and the exact systematic position of the latter species remains equivocal. It may be noted that in one important respect, the presence of a diaphragm in the ejaculatory duct of *D. evelinae*, there may

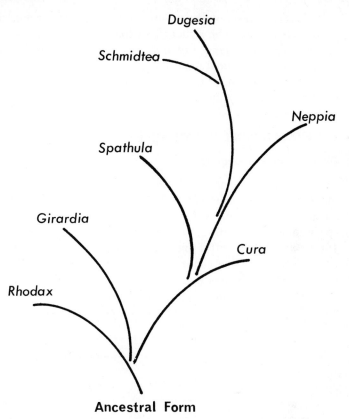

Ancestral Form

Figure 11   Suggested phylogenetic relationships of the genera and subgenera of Dugesiidae. *Bopsula* is omitted.

be good reason to reinvestigate the status of this species, since the diaphragm is not apparent in the original figure (Marcus, 1955, fig. 20) or in the original slides, although it is described. It is also of note that the atrial muscles of *D. montana* are thicker than is usual in members of the Dugesiidae (Fig. 9).

Concerning the cement glands it may be noted that in my experience these are very well developed in *Girardia* and *Cura*, and less so in *Dugesia*. Similarly, the shell glands are most extensive in the lowest forms, and in both *D. pinguis* and *D. foremanii* I have detected extensive cyanophil and eosinophil secretion of the shell glands by Mallory-Heidenhain staining. Usually the shell glands produce only an eosinophil secretion.

Pigmentation of the pharynx is rare in freshwater triclads, and it is of interest to note that it occurs only in species of the subgenus *Girardia*, but whether or not in all of them is not known.

It would certainly be of interest to have further data to compare with the proposed scheme. The structure of the anterior nervous system, for example, in the various subgenera of differing head shape may prove to be of significance. Biogeographical data, too, are of importance, and it will be the purpose of the next section of this paper to review and explain the distribution and interrelationships of the taxa.

BIOGEOGRAPHICAL RELATIONSHIPS

Distribution and Dispersal

All other things being equal, the area of distribution of a taxon is proportional to its age, but as Emerson (1952) has pointed out, all other things are rarely equal. The "Age and Area" theory of Willis (1922), which depends upon such space-time correlations, is now generally discredited (Croizat, 1958; Udvardy, 1969), and it is doubtful that the idea that older taxa have wider distributions is more than a loose generalization. Yet the comparative morphological evidence clearly indicates that the Dugesiidae are older and more primitive than the other families, and they are also the most widely distributed (cf. Figs. 12 and 13). The restriction of the Dendrocoelidae and the Planariidae to areas

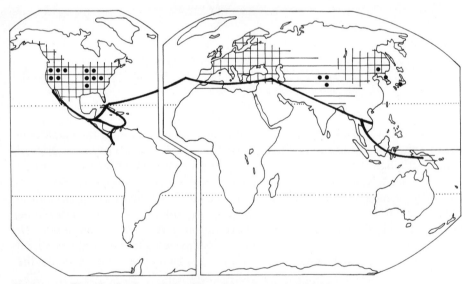

Figure 12 The distribution of the Planariidae (cross-hatching), Dendrocoelinae (vertical hatching), and Kenkiinae (dots) in relation to the Tethys and associated geosynclines (bold lines). *(Basic map after Croizat, 1962.)*

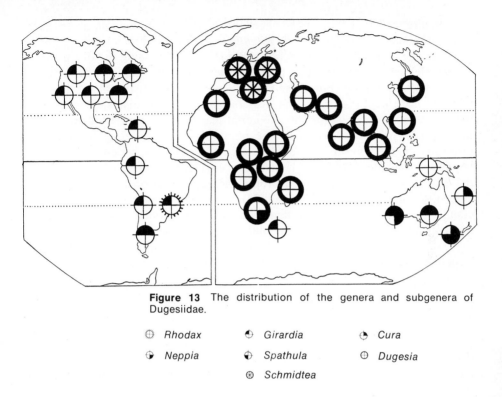

Figure 13 The distribution of the genera and subgenera of Dugesiidae.

⊙ Rhodax	◑ Girardia	◑ Cura
◑ Neppia	◐ Spathula	⊕ Dugesia
	⊛ Schmidtea	

north of the Tethys geosyncline, and the curious Europe–east Asia disjunction in the distribution of the Dendrocoelinae (Porfirjeva, 1960), are interesting facts which may be noted (Fig. 12), but a detailed explanation is as yet not possible, and an attempt is outside the scope of the present paper.

An explanation of the distribution patterns of the Dugesiidae first requires a consideration of their capacity for active and passive dispersal. The efficacy of passive dispersal as a means of extending range in many freshwater organisms cannot be denied, and there are many such organisms which may be considered cosmopolitan (Carpenter, 1928; Macan, 1963). This, however, is not true of the Paludicola; each species appears to be restricted to a particular geographical area, with two notable exceptions to be discussed further below.

Ullyott (1936) is quite adamant in maintaining that Paludicola disperse only by their own activity and that passive dispersal is impossible, and Leloup (1944) is right in saying that most authors have taken this view. It is important to remember that freshwater triclads do not possess any resting stages which are resistant to extremes of temperature, or to desiccation, that the adults are very fragile, and that they spend their entire life cycle in an aquatic environment. An exception is *Hymanella retenuova*, which inhabits temporary ponds and produces a thick-shelled cocoon capable of withstanding the dry periods. Nonetheless, this species has a fairly restricted distribution in eastern North America (Ball, 1969a). For these reasons anemochore dispersal would seem to be of little importance for the dispersal of these organisms, especially as they have not been recorded in the aerial plankton (Gislèn, 1948). Zimmerman (1963), however, has made the point that in considering transoceanic dispersal it is the abnormal conditions, such as hurricanes, which play the most important role. Bousfield (1961) has also considered the role of hurricanes in the dispersal of littoral marine arthropods in eastern North America, but concluded that a hypothesis of chance influx alone would scarcely account for the complexity of the established fauna and for the amphiatlantic distributions of many of the species. Experimental or observational evidence relating to the transport of freshwater triclads in this way is impossible to obtain, but their structure, habits, and present-day distributions argue strongly against their having been so dispersed.

About the only agents for biochore dispersal of freshwater triclads which have been proposed are birds. Although there is some evidence that birds may have aided the dispersal of *Crenobia alpina* and *Polycelis felina* over short distances in northwestern Europe (Dahm, 1958; Reynoldson, 1966), such dispersal seems to have been of little significance on a wider scale. The studies of Maguire (1963) on biochore dispersal do not afford much evidence for the transport of triclads by other animals, and in a recent review Reynoldson (1966) concludes that birds have not been an important factor in long-range dispersal. The biochore dispersal of cocoons is a possibility, but those of the Dugesiidae are attached to the substratum, as are many of those of the Planariidae and Dendrocoelidae. Certain stream-dwelling forms produce free, unattached cocoons, but according to Voigt (1904), they are so placed in the streams as to exclude the probability of transport by birds or other animals. I have found a trichopteran larva which had incorporated a single cocoon of *Phagocata woodworthi* into its case structure, but this is unlikely to be of any significance other than in dispersing the cocoon within the lake.

The occurrence of freshwater triclads in the invertebrate drift of streams is a rare event (Minshall and Winger, 1968), but the hydrochore dispersal of cocoons or adults, either in floodwaters or on floating objects, is well established (Leloup, 1944). Their resistance to salinity, however, is low and transoceanic hydrochore dispersal is improbable, if not impossible. There is no evidence to indicate that hydrochore dispersal has played any important role in the determination of present-day distribution patterns.

Two species of *Dugesia* appear to present exceptions to what has been said above. *D. tigrina*, the most widespread North American species, occurs in scattered localities in Europe, where it appears to be extending its range (Gourbault, 1969). In the other direction the common European form *D. polychroa* has established itself in the Saint Lawrence River and environs in North America. There can be no doubt, however, that these species have been introduced within the last 60 years or so, *D. tigrina* via the trade in aquarium plants, and *D. polychroa* probably with shipping (Ball, 1969b). Anthropochore dispersal is also probably responsible for the occurrence of triclads on Anticosti Island (Ball and Fernando, 1970) and of *Rectocephala* in Washington, D.C. (Hyman, 1953). That such accidental dispersal is a rare event is demonstrated by studies of the Paludicola of the Canadian Maritime Provinces, which show that the reduced fauna is typical of eastern North America (Ball, 1973) and that no European elements occur here, despite the activities which have been responsible for the accidental introduction of many other types of organisms (Lindroth, 1957) by European traders in this area for two or three hundred years.

Perhaps the best evidence against passive dispersal as a general phenomenon is afforded by the comparisons of the triclad faunas on opposite sides of narrow sea straits which were subject to Pleistocene glaciations. Such comparisons have been made in northwestern Europe (Ullyott, 1936; Reynoldson, 1966) and the Gulf of Saint Lawrence (Ball and Fernando, 1970). The dissimilarities in the faunas argue very strongly against passive dispersal.

The difficulty of disproving a hypothesis of chance dispersal is, of course, great. Schopf (1970, p. 658) has made this point in saying that "the stochastic hypothesis is expressly designed to take advantage of the impossibility of providing a negative and in nullifying systematic methods of explanation by emphasizing the improbable." Strong words, but containing much truth.

It is likely that triclads disperse mainly by their own activities. It is true that the speed or ease of locomotion of organisms is in no direct relation to the speed or ability of dispersal (Udvardy, 1969), but nevertheless the Paludicola are generally slow to colonize new areas. Reynoldson (1966), for example, notes that in the north of Great Britain they are generally underdispersed, and presumably are still extending their range northwards following the last glaciation. A recent study of the distribution of triclads in eastern Canada indicates that this may also be true here (Ball, 1973). The freshwater triclads of Anticosti Island have not succeeded in crossing to the northern watershed since their presumed introduction into the Port Menier area 70 years ago (Ball and Fernando, 1970).

If freshwater triclads disperse principally by their own activities, and only through contiguous freshwater bodies, and perhaps the groundwater where soil conditions are suitable, then a causal explanation of their distribution must take careful consideration of historical events. Further, since the history of a taxon in nature is reflected by both its morphology and its distribution, a causal explanation of distribution is intimately concerned with the evolutionary relationships of its members. Using the systematic data elucidated previously, and taking cognizance of the data on vagility just discussed, a biogeographical analysis of the Dugesiidae must provide a reasonable explanation of the distribution patterns exhibited by the various genera and subgenera (Fig. 13).

Historical Biogeography

The problem of the "center of origin" of the Dugesiidae, and thus of the Paludicola as a whole, is a difficult one to solve in the absence of a fossil record. It is a corollary of Brundin's biogeographical methods that "within the total distribution area of a group the species possessing the most primitive characters are found within the earliest, those with the most derivative characters within the latest occupied part of the area" (Brundin, 1966, p. 56), but this concept has been criticized severely by Darlington (1970), who considers it a rule of thumb to expect most primitive forms to be in distant-peripheral areas (1957), an idea which dates back to the work of Matthew (1915). The idea that the center of origin of a taxon is indicated by the point of occurrence of the greatest number of its members is also an old one (Wulff, 1950), and it is one of the important factors in the geographical-morphological method used by many botanical taxonomists (Davis and Heywood, 1963). However, Cain (1944) has justified the view that to equate center of diversity with center of origin, without supporting evidence, is a dangerous policy. Such coincidence is only to be expected in those cases where the area of the taxon has not been subjected to later influences, such as glaciations.

It must be accepted, therefore, that any conclusion concerning the center of origin, or of dispersal, of the freshwater triclads must be considered as hypothesis, open to rejection, acceptance, or modification according to the weight of evidence available. If, however, it can be shown that the criterion of diversity agrees with the conclusions reached from an examination of the distribution of a phylogenetic series, then the likelihood of attaining a correct solution is increased.

Kawakatsu (1968, figs. III-12 and IV-1) unequivocally places the origin of the Dugesiidae in the Balkan Peninsula, from where they have dispersed to most parts of the world (Fig. 14). The merits of this scheme are that it places the center of dispersal of the Dugesiidae in an area well known as an evolutionary center (Stankovic, 1960) and provides an adequate explanation of the broad distribution of the subgenus *Dugesia*. Further, the heterogeneity of the triclad fauna of Australasia is explained in that it results from three distinct immigration sources, viz., south Africa, South America, and southeastern Asia.

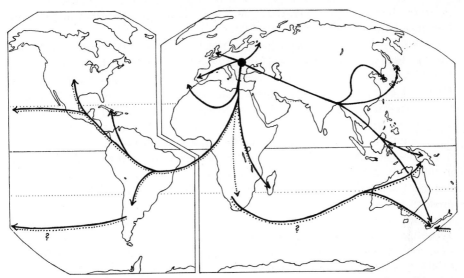

Figure 14 The distribution and dispersal of the Dugesiidae according to Kawakatsu. Solid lines represent the genus *Dugesia* s. l., broken lines the genus *Cura* s. l. *(Redrawn from Kawakatsu, 1968.)*

There are, however, considerable drawbacks to this proposal. In the first place the Dugesiidae are represented by only two or three species in the Balkan area, and these are apomorph species. The diversity and endemism of the Planariidae and Dendrocoelidae in Lake Ohrid (Stankovic, 1960, 1969) clearly indicate a center of diversification, and possibly dispersal, for these forms, but this does not appear to be true for the older and more primitive Dugesiidae. On the other hand, if the apomorph (and therefore specialized?) subgenus *Dugesia* is a more recent arrival in this area, diversification to the degree exhibited by the other families would not be expected.

Kawakatsu notes, correctly, the absence of amphiatlantic relations in the Northern Hemisphere. He shows the main migratory routes of the Dugesiidae as being southwards through Africa, westwards across the Atlantic Ocean to South America and then North America, and across the Pacific to New Zealand, and also eastwards from south Africa across the Indian Ocean to Australasia. The genus *Cura* s. l. arises in central Africa and follows these routes. Again the difficulties here are that the migratory routes run contrary to the phylogenetic series (cf. Figs. 11 and 13). Thus the African species of *Cura* (here referred to a distinct subgenus *Neppia*) are evolutionarily more recent than the American and Australasian species; and the same applies to *Dugesia*. What is more, Kawakatsu's scheme suggests that there should be close relationships between the Dugesiidae of south Africa and South America, relationships which I cannot detect.

A further contradictory feature of this hypothesis is that it fails to account for the lack of amphiatlantic connections. The route from the Balkans to North America is circuitous, to say the least, and it would appear that the Dugesiidae should have had ample time to migrate directly from Europe to North America, or even from Asia to North America. Yet it is apparent, and Kawakatsu seems to agree, that the Dugesiidae of North America have been derived entirely from the south. The fact that most primitive forms are found in the Southern Hemisphere, and that here, too, is found the greatest diversity in numbers of both species and subgenera, suggests to me that the center of dispersal, and probably the center of origin, of the Dugesiidae lies south of the present-day equator.

Before taking this conception further, it is desirable to consider in more detail the transoceanic migratory routes followed by the Paludicola. Since I have rejected hypotheses of chance dispersal, which I agree with Brundin (1966) are an admission of defeat, there appear to be only two alternatives left—former land (freshwater?) bridges, now sunken, and continental drift.

The general question of land bridges has been reviewed in detail by van Steenis (1962), Illies (1965a), Hallam (1967), and Udvardy (1969). It is the proponents of continental stability who continue to make most use of them. Croizat, for example, in his valuable discussion of the origin and dispersal of the angiosperms, proposes several land bridges and considers the world now to be the disconnections of the Jurassic-Cretaceous period (Croizat, 1952, fig. 98). Later, Croizat (1962) makes the point that the biogeographer must look to his own data when considering transoceanic relationships and not trouble himself with detailed problems of disintegrated land bridges versus drifting continents. This is not entirely responsible although the point concerning the validity and independence of biogeographical data is well taken. Van Steenis (1963) considers that for a satisfactory explanation of the major features of spermatophytan plant geography a minimum of five major land bridges is required. He does not find the zoogeographical data to be in conflict with these ideas, and they are largely consistent with the distribution patterns exhibited by the Paludicola.

Hallam (1967) has pointed out that the major difficulty in the transoceanic land bridge hypothesis is the isostatic problem involved in sinking an extensive sialic continent without trace. One solution would be to reduce the size of the required land bridges so that they become isthmian links, and Croizat (1952) has emphasized that a continuous land bridge at any one time is unnecessary. The geological evidence, however, does not favor hypotheses involving sunken bridges or continents (Hallam, 1967). The rejection of such hypotheses does not mean that land bridges have played no part on a smaller scale. It is quite evident, for example, that there has been a bridge across the Bering Strait (Moore, 1961; Stegman, 1963; Hopkins, 1967), and biogeographical evidence suggests the possibility, which is not in conflict with geological data, of such a bridge, or land mass, in the Caribbean area (Ball, 1971).

The alternative to transoceanic land bridges is the hypothesis of continental drift. This is not the place to review all the evidence in favor of this hypothesis; this has already been done for the geological evidence by Runcorn (1962), Blackett *et al.* (1965), Garland (1966), Phinney (1968), Kay (1969), and Dietz and Holden (1970); for the zoological evidence by Jeannel (1961), Brundin (1966), Illies (1965a), and Hallam (1967); and for the botanical evidence by Schuster (1969) and Schopf (1970). It suffices to say that the evidence from paleoclimatology, paleomagnetism, oceanic rift structures, and the excellence of fit of the continents overwhelmingly corroborates the hypothesis.

The most recent evidence favors the idea of a single supercontinent, Pangaea, in Permian times, about 200 to 250 million years ago (Schopf, 1970; Dietz and Holden, 1970). By the end of the Triassic period, 180 million years ago, the northern part of the supercontinent, Laurasia, had split away from the southern part, Gondwanaland, and was drifting northwards, the two parts being separated by the Tethys Sea. There are excellent summaries of the timing of breakup by Heirtzler (1968) and Dietz and Holden (1970), the latter providing a particularly graphic account.

I have suggested that the center of origin of the Dugesiidae is an austral one, and I further suggest that this was in Gondwanaland, probably in what is now Antarctica. I postulate that by the commencement of the Mesozoic, some 220 million years ago, the early diversification of the Dugesiidae was complete, with a main massing of *Girardia* in the west, and of *Neppia* and *Spathula* in the east. The northwards dispersal of these elements coincided with the early stages of Gondwanaland breakup, leading to a concentration of *Girardia* in the Americas, with relatives across the South Atlantic to Australasia, and a main massing of *Neppia* in the east. *Cura* seems to have been particularly widespread, and perhaps it was ancestral to both the above subgenera, though this cannot be demonstrated at present. After separation was well under way, the subgenus *Dugesia* arose in Africa, and after closure of the Tethys Sea, of which the Mediterranean is a remnant, it spread northwards into Palaearctis, and eastwards to India and southeastern Asia. In general, therefore, these migratory routes are the opposite of those proposed by Kawakatsu (1968).

Additional to the fact that the proposed austral center of dispersal is near the present center of frequency and diversity, this scheme has the advantages of explaining most of the disjunctions and continuous distributions. The relatively early separation of Africa from South America fits with the quite different triclad faunas which these continents have, and the relatively late separation of Madagascar from Africa explains why its triclad fauna consists mostly, if not entirely, of the subgenus *Dugesia*. India was probably populated entirely from the north, which again explains why its triclads belong almost exclusively to the *D. japonica* species group of the subgenus *Dugesia*. The possibly close connections between *D. astrocheta* of Africa and *D. burmaensis* (p. 380) are of interest here, but need further elucidation. The present distributions of *Neppia* and *Spathula* are the remnants of a former and probably wider distribution in eastern Gondwanaland. It may also be noted that until the Tethys Sea was closed, and the junction of North and South America accomplished, late in the Cretaceous, there was no possibility of exchange between North America and Europe. It is probable, therefore, that the Dugesiidae were not distributed in the northern continents in time to take advantage of the Laurasian supercontinent as a migratory route to North America. Heirtzler (1968), for example, dates the separation of North America at about 230 million years ago, very much earlier than Dietz and Holden (1970) do. The freshwater triclads of the Cape Verde Islands (Luther, 1956) and the Azores (Marcus and Marcus, 1959) may be expected to throw some light on these problems, but unfortunately their affinities are unknown.

It thus seems possible, even likely, that the Dugesiidae diversified rapidly early in Mesozoic times, or even earlier in the aftermath of the Permocarboniferous glaciations, which both Croizat (1965, p. 609) and Brundin (1966) consider to be of great importance to the history of biotas. During the early stages of breakup of Gondwanaland, they dispersed northwards in the main land masses, without the possibility of interchange between Australasia and southeastern Asia. The final patterns of distribution in the Northern Hemisphere were probably achieved in the Cenozoic, when the continents had attained their present form and position.

Views similar in many ways to those proposed here have been put forward for the Plecoptera (Illies, 1965b), Chironomidae (Brundin, 1966), and Dermaptera (Popham and Manly, 1969). It is encouraging that study of such diverse groups should lead to broadly similar hypotheses. The continental drift hypothesis has also been invoked in biogeographical studies of freshwater Ostracoda (McKenzie and Hussainy, 1968), freshwater fishes (Gery, 1969), and marsupials (Cox, 1970). In addition Besch (1969) has concluded that the distributions and relationships of South American Arachnida support the concept of southern land connections. The recent discovery of Triassic tetrapods in Antarctica provides further confirmatory evidence (Elliot *et al.*, 1970).

The main objections to the proposals will probably concern the origin and history of the insular faunas (e.g., de Beauchamp, 1940b), such as those of the Crozet Islands, New Caledonia, and Hawaii, especially as stratigraphical studies have shown these islands to be relatively young. Croizat (1962, p. 258), however, has made the point that the marriage between geology and biogeography should involve tectonics and not stratigraphy, and he criticizes the distinction between "oceanic" and "continental" islands from a biogeographical standpoint. For the present it may be noted that the triclads of Crozet [*D. (Girardia) seclusa*] and New Caledonia [*D. (Cura) pinguis*] belong to primitive groups; those of Hawaii have not yet been described.

It should be emphasized that I have presented here only a hypothesis. Considerable reservations concerning the idea of continental drift are being expressed by biologists (Croizat, 1952; van Steenis, 1962) and geologists (Meyerhoff, 1970a, 1970b; Meyerhoff and Teichert, 1971), and the data discussed in this paper are not proof for the drift hypothesis. They are consistent with it, however, and indeed are best explained with its aid, unless the land bridge theory is to be resurrected. A hypothesis of Palaearctic origins for the Dugesiidae, coupled with continental stability and permanence of the oceans is incompatible with the known facts of their distribution and phylogenetic relationships. The systematic and biogeographical scheme presented here is not intended, however, as the last word, which clearly it cannot be, but as a stimulus to future thought and research, and is presented in the hope that it soon will be replaced by a scheme based on additional and new data, and of greater refinement.

> Thus, the task is not so much to see what no one has seen yet, but to think what nobody has thought yet, about that which everybody sees. [Schopenhauer]

ACKNOWLEDGMENTS

For the loan or gift of specimens I wish to express my sincere thanks to Dr. F. R. Allison (Christchurch, N.Z.), Dr. D. Barr (Royal Ontario Museum), Professor P. de Beauchamp (Paris), Dr. P. C. G. Benoit (Tervuren, Belgium), Professor M. Benazzi (Pisa), Dr. E. L. Bousfield (Ottawa), Dr. Per Brinck (Lund, Sweden), Mr. H. Feinberg (American Museum of Natural History), Dr. C. H. Fernando (Waterloo, Ontario), Dr. A. D. Harrison (Waterloo, Ontario), Dr. H. B. N. Hynes (Waterloo, Ontario), Mr. P. B. Karunaratne (Colombo, Ceylon), Professor M. Kawakatsu (Sapporo, Japan), Dr. R. Kenk (Smithsonian Institution), Mrs. Eveline du Bois Reymond Marcus (São Paulo, Brazil), and Dr. T. B. Reynoldson (Bangor, U.K.).

Many colleagues have aided me considerably by their critical readings of this manuscript in earlier drafts. I thank especially Drs. Harrison, Hynes, and W. B. Kendrick, and Messrs. B. Coad and D. D. Williams, all of the University of Waterloo, and Dr. E. L. Bousfield of the National Museums of Canada. The editor, Dr. N. W. Riser, made many valuable suggestions for the improvement of the final draft. I have a much older debt to Dr. T. B. Reynoldson. However, these gentlemen must not be held accountable for any of the shortcomings of this paper.

The work was supported jointly by the Department of Biology of the University of Waterloo and the National Museum of Natural Sciences, Ottawa.

BIBLIOGRAPHY

*References not seen in the original.
†English translations of these articles will be deposited in the Library of the National Museum of Natural Sciences, Ottawa, Canada.

Anderson, J. M. 1952. Sexual Reproduction without Cross-copulation in the Freshwater Triclad Turbellarian, *Curtisia foremanii. Biol. Bull.*, **102:**1–8.

———— and J. C. Johann. 1958. Some Aspects of Reproductive Biology in the Freshwater Triclad Turbellarian, *Cura foremanii. Biol. Bull.*, **115:**375–383.

Ax, P. 1956. Monographie der Otoplanidae (Turbellaria). *Akad. Wiss. Lit. Mainz Abhandl. Math. Nat. Kl.*, 1955, **13:**499–796.

————. 1963. "Relationships and Phylogeny of the Turbellaria." In E. C. Dougherty (ed.), *The Lower Metazoa*, pp. 191–224. Berkeley and Los Angeles: University of California Press.

Ball, I. R. 1969a. An Annotated Checklist of the Freshwater Tricladida of the Nearctic and Neotropical Regions. *Can. J. Zool.*, **47:**59–64.

————. 1969b. *Dugesia lugubris* (Tricladida: Paludicola), a European Immigrant into North American Freshwaters. *J. Fisheries Res. Board Can.*, **26:**221–228.

————. 1970. Freshwater Triclads from the Oriental Region. *Zool. J. Linnean Soc.*, **49:**271–294.

————. 1971. The Systematic and Biogeographical Relationships of Some *Dugesia* Species (Turbellaria, Tricladida) from Central and South America. *Am. Museum Novitates*, no. 2472. 1–25.

————. 1973. The Haploneura of Eastern Canada (in preparation).

————. 1974. La Faune Terrestre de l'Ile de Sainte Hélène: Turbellaria Tricladida. *Ann. Mus. Roy. Afr. Centrale*, in 8°, Zool. (in press).

———— and C. H. Fernando. 1969. Freshwater Triclads (Platyhelminthes, Turbellaria) and Continental Drift. *Nature*, **221:**1143–1144.

———— and ————. 1970. Freshwater Triclads (Platyhelminthes, Turbellaria) from Anticosti Island. *Naturaliste Can.*, **97:**331–336.

————, T. B. Reynoldson, and T. Warwick, 1969. The Taxonomy, Habitat and Distribution of the Freshwater Triclad *Planaria torva* (Platyhelminthes: Turbellaria) in Britain. *J. Zool. London*, **157:**99–123.

Beauchamp, P. de. 1913. Turbellariés, Trématodes et Gordiacés. *Voy. Alluaud et Jeannel en Afr. Orient. Res. Scient.* Paris: Libraire Albert Schulz., 22 pp.

————. 1931. Biospéologica 56(2). Campagne spéologique de C. Bolívar et R. Jeannel dans l'Amérique du Nord (1928). Turbellariés triclades. *Arch. Zool. Exptl. Gen.*, **71:**317–331.

————. 1936. A Propos d' *Euplanaria cretica* Meixner. *Bull. Soc. Zool. Fr.*, **66:**433–440.

————. 1939. Results of the Percy Sladen Trust Expedition to Lake Titicaca V. Rotifères et Turbellariés. *Trans. Linnean Soc. London*, 1:51–79.

————. 1940a. Croisière du Bougainville aux Iles australes Françaises XII. Turbellariés et Rotifères. *Mem. Museum Natl. Hist. Nat. Paris*. 14:313–326.

————. 1940b. Sur les triclades paludicoles de l'hémisphere sud. *Compt. Rend. Soc. Biogéog.*, 17:7–9.

————. 1951a. A propos d'une planaire du Congo Belge. *Rev. Zool. Bot. Afr.*, 45: 90–97.

————. 1951b. Turbellariés de l'Angola. *Publ. Cult. Comp. Diamantes de Angola, Museo do Dundo, Lisbão*, no. 11, 79–84.

————. 1953. Sur les *Dugesia* (Turbellariés Triclades) d'Afrique tropicale et de Madagascar. *Bull. Soc. Zool. Fr.*, 77:362–370.

————. 1959. Triclades paludicoles d'Afghanistan. *Kgl. Fysiograf. Sallskap. Lund Forh.*, 29:27–43.

————. 1961. "Classe des Turbellariés." In P. P. Grassé (ed.), *Traité de Zoologie*, vol. IV, pp. 35–212. Paris: Masson et Cie.

————. 1968 Turbellariés d'eau douce de Nouvelle-Caledonié. *Cah. O.R.S.T.O.M.*, sér. *Hydrobiologia*, 11:67–68.

Beklemishev, W. N. 1969. *Principles of Comparative Anatomy of Invertebrates.* Edinburgh: Oliver & Boyd. 2 vols., xxx + 490, vii + 529.

Benazzi, M. 1946. Sopra una nuova planaria di acqua dolce. *Arch. Zool. Ital.*, 31: 93–102.

————. 1960. Evoluzione cromosomica e differenziamento razziale e specifico nei tricladi. *R. C. Accad. Lincei, Quaderno*, 47:273–296.

————. 1963. "Genetics of Reproductive Mechanisms and Chromosome Behaviour in Some Freshwater Triclads." In E. C. Dougherty (ed.), *The Lower Metazoa*, pp. 405–422. Berkeley and Los Angeles: University of California Press.

————. 1966. Cariologia della planaria americana *Dugesia dorotocephala*. *Lincei Rend. Sci. Fis. Mat. Nat.*, 40:999–1005.

———— and E. Giannini. 1971. *Cura azteca*, nuova specie di planaria del Messico. *Lincei Rend. Sci. Fis. Mat. Nat.*, (VIII)50:477–481.

Besch, W. 1969. "South American Arachnida." In Fittkau, Illies, Klinge, Schwabe, and Sioli (eds.), *Biogeography and Ecology in South America*, pp. 723–740. The Hague: W. Junk.

Blackett, P. M., E. Bullard, and S. K. Runcorn. 1965. A Symposium on Continental Drift. *Phil. Trans. Roy. Soc. London*, A258:1–323.

Böhmig, L. 1887. *Planaria Iheringii*, eine neue Triclade aus Brasilien. *Zool. Anz.*, 10:482–484.

————. 1897. Die Turbellarien Ost-Afrikas. K. Mobius, *Die Thierwelt Ost-Afrikas*, 4:1–15. Berlin.

————. 1902. Turbellarien, Rhabdocoeliden und Tricladiden. *Ergeb. Hamburg. Magalh. Sammelreise*, 3:1–30.

————. 1906. Tricladenstudien I. Tricladida Maricola. *Z. Wiss. Zool.*, 81:344–504.

Bonner, J. T. 1965. *Size and Cycle.* Princeton, N.J.: Princeton University Press. vii + 219 pp.

Borelli, A. 1895. Viaggio del Dott. Alfredo Borelli nella Republica Argentina e nel Paraguay. 13. Planarie d'acqua dolce. *Boll. Museo Zool. Anat. Comp. Torino*, 10:1–6.

————. 1897. Viaggio del Dott. Alfredo Borelli nel Chaco Boliviano, etc. V. Planarie d'acqua dolce. *Boll. Museo Zool. Anat. Comp. Torino*, **12**:1–4.

————. 1898. Viaggio del Dr Enrico Festa nel'Ecuador etc. IX. Planarie d'acqua dolce. *Boll. Museo Zool. Anat. Comp. Torino*, **13**:1–6.

————. 1901. Di una nova Planaria d'acqua dolce della Republica Argentina. *Boll. Museo Zool. Anat. Comp. Torino*, **16**:1–5.

Bousfield, E. L. 1961. Studies on Littoral Marine Arthropods from the Bay of Fundy Region. *Bull. Nat. Museum Canada*, **183**:42–62.

Brundin, L. 1966. Transantarctic Relationships and Their Significance, as Evidenced by Chironomid Midges, with a Monograph of the Subfamilies Podonominae and Aphroteniinae and the Austral Heptagyiae. *Kgl. Svenska Vetenskapsakad. Handl.*, **11**:1–472.

Cain, S. A. 1944. *Foundations of Plant Geography*. New York: Harper, xiv + 556 pp.

Camin, J. T., and R. R. Sokal. 1965. A Method for Deducing Branching Sequences in Phylogeny. *Evolution*, **19**:311–326.

Carpenter, J. H. MS1970. "Systematics and Ecology of Cave Planarians of the United States." Ph. D. thesis, University of Kentucky, 212 pp.

Carpenter, K. 1928. *Life in Inland Waters*. London: Sidgwick & Jackson. xviii + 267 pp.

*Chichkoff, G. D. 1892. Recherches sur les dendrocoeles d'eau douce (Triclades). *Arch. Biol.*, **12**:435–568.

Corrêa, D. D. 1960. Two New Marine Turbellaria from Florida. *Bull. Marine Sci. Gulf Caribbean*, **10**:208–216.

Cox, C. B. 1970. Migrating Marsupials and Drifting Continents. *Nature*, **226**: 767–770.

Croizat, L. 1952. *Manual of Phytogeography*. The Hague: W. Junk. viii + 587 pp.

————. 1958. An Essay on the Biogeographic Thinking of J. C. Willis. *Arch. Bot. Biogeogr. Ital.*, **34**:1–27 (4 série, vol. III: 90–116).

————. 1962. *Space, Time, Form: The Biological Synthesis*. Caracas: The Author. xix + 881 pp.

————. 1965. An Introduction to the Subgeneric Classification of "Euphorbia" L., with Stress on the South African and Malagasy Species. I. *Webbia*, **20**:573–706.

Crowson, R. A. 1970. *Classification and Biology*. London: Heinemann. vii + 350 pp.

Curtis, W. C. 1900. On the Reproductive System of *Planaria simplissima*, a New Species. *Zool. Jahrb. (Anat.)*, **13**:447–466.

Dahm, A. G. 1958. *Taxonomy and Ecology of Five Species Groups in the Family Planariidae*. Malmö: Nya Litografen. 241 pp.

————. 1963. The Karyotypes of Some Freshwater Triclads from Europe and Japan. *Arkiv Zool.*, **16**:41–67.

————. 1967. A New *Dugesia* "Microspecies" from Ghana Belonging to the *Dugesia gonocephala* Group. *Arkiv Zool.*, **19**:309–321.

Darlington, P. J. 1957. *Zoogeography: The Geographical Distribution of Animals*. New York: Wiley. xi + 675 pp.

————. 1970. A Practical Criticism of Hennig-Brundin "Phylogenetic Systematics" and Antarctic Biogeography. *Systematic Zool.*, **19**:1–18.

Davis, P. H., and V. H. Heywood. 1963. *Principles of Angiosperm Taxonomy.* Edinburgh: Oliver & Boyd. xvi + 558 pp.

Dietz, R. S., and J. C. Holden. 1970. Reconstruction of Pangaea: Breakup and Dispersion of Continents, Permian to Present. *J. Geophys. Res.,* **75:**4939–4956. (Also in *Sci. Am.,* **223:**30–41.)

*Dugès, A. 1830. Aperçu de quelques observations nouvelles sur les planaires et plusieurs genres voisins. *Ann. Sci. Nat.,* **21:**72–90.

Elliot, D. H., E. H. Colbert, W. J. Breed, J. A. Jensen, and J. S. Powell. 1970. Triassic Tetrapods from Antarctica: Evidence for Continental Drift. *Science,* **169:** 1197–1201.

Emerson, A. E. 1952. The Biogeography of Termites. *Bull. Am. Museum Nat. Hist.,* **99:**217–225.

Fuhrmann, O. 1914. Turbellariés de Colombie. *Mem. Soc. Neuchâtel Sci. Nat.,* **5:**748–804.

Garland, G. D. (ed.). 1966. *Continental Drift.* Toronto: University of Toronto Press. ix + 140 pp.

Gery, J. 1969. "The Freshwater Fishes of South America." In Fittkau, Illies, Klinge, Schwabe, and Sioli (eds.), *Biogeography and Ecology in South America,* pp. 828–848. The Hague: W. Junk.

Girard, C. 1850. A Brief Account of the Freshwater Planariae of the United States. *Proc. Boston Soc. Nat. Hist.,* **3:**264–265.

———. 1852. Descriptions of Two New Genera and Two New Species of Planaria. *Proc. Boston Soc. Nat. Hist.,* **4:**210–212.

Gislèn, T. 1948. Aerial Plankton and Its Conditions of Life. *Biol. Rev.,* **23:**109–126.

Gourbault, N. 1969. Expansion de *Dugesia tigrina* (Girard), planaire américaine introduite en Europe. *Ann. Limnol.,* **5:**3–7.

Graff, L. von. 1912–1917. "Tricladida." In Bronn, H. G. (ed.), *Klassen und Ordnungen des Tierreichs.* Bd. IV, Abt. 1c, III. Leipzig.

Hallam, A. 1967. The Bearing of Certain Palaeozoogeographical Data on Continental Drift. *Palaeogeography, Palaeoclimatol., Palaeoecol.,* **3:**201–241.

Hallez, P. 1894. *Catalogue des Rhabdocoelides, Triclades et Polyclades du Nord de la France,* 2d ed. Lille: L. Danel. 240 pp.

Heirtzler, J. R. 1968. Sea Floor Spreading. *Sci. Am.,* **219:**60–70.

Hennig, W. 1966. *Phylogenetic Systematics.* Urbana: University of Illinois Press. 263 pp.

Hofsten, N. von. 1907. Studien über Turbellarien aus dem Berner Oberland. *Z. Wiss. Zool.,* **85:**391–654.

Holleman, J. T., and C. Hand. 1962. A New Species, Genus, and Family of Marine Flatworms (Turbellaria: Tricladida, Maricola) Commensal with Mollusks. *Veliger,* **5:**20–22.

Hopkins, D. M. (ed.). 1967. *The Bering Land Bridge.* Stanford, Calif.: Stanford University Press. ix + 495 pp.

Hyman, L. H. 1925. The Reproductive System and Other Characters of *Planaria dorotocephala* Woodworth. *Trans. Am. Microscop. Soc.,* **44:**51–89.

———. 1931a. Studies on the Morphology, Taxonomy, and Distribution of North American Triclad Turbellaria. IV. Recent European Revisions of the Triclads, and Their Application to the American Forms, with a Key to the Latter and New Notes on Distribution. *Trans. Am. Microscop. Soc.*, **50**:316–335.

———. 1931b. Ibid. V. Descriptions of Two New Species. *Trans. Am. Microscop. Soc.*, **50**:336–343.

———. 1937. Ibid. VIII. Some Cave Planarians of the United States. *Trans. Am. Microscop. Soc.*, **56**:457–477.

———. 1939. New Species of Flatworms from North, Central and South America. *Proc. U.S. Natl. Museum*, **86**:419–439.

———. 1944. Marine Turbellaria from the Atlantic Coast of North America. *Am. Museum Novitates*, no. 1266, 1–15.

———. 1951. *The Invertebrates*, vol. II. New York: McGraw-Hill. vii + 550 pp.

———. 1953. North American Triclad Turbellaria. 14. A New, Probably Exotic, Dendrocoelid. *Am. Museum Novitates*, no. 1629, 1–5.

———. 1954. A New Marine Triclad from the Coast of California. *Am. Museum Novitates*, no. 1679, 1–5.

———. 1956. North American Triclad Turbellaria. 15. Three New Species. *Am. Museum Novitates*, no. 1808, 1–14.

———. 1957. A Few Turbellarians from Trinidad and the Canal Zone, with Corrective Remarks. *Am. Museum Novitates*, no. 1862, 1–8.

———. 1959. On Two Freshwater Planarians from Chile. *Am. Museum Novitates*, no. 1932, 1–11.

———. 1960. Cave Planarians in the United States. *Am. Midland Naturalist*, **64**: 10–12.

Ichikawa, A., and M. Kawakatsu. 1962. *Phagocata albata*, a New Probably Subterranean Freshwater Planarian, from Hokkaidô. *Annotationes Zool. Japon.*, **35**:29–37.

——— and ———. 1964. A New Freshwater Planarian, *Dugesia japonica*, Commonly but Erroneously Known as *Dugesia gonocephala* (Dugès). *Annotationes Zool. Japon.*, **37**:185–194.

Ijima, I. 1884. Untersuchungen über den Bau und die Entwicklungsgeschichte der Süsswasserdendrocoelen (Tricladen). *Z. Wiss. Zool.*, **40**:359–464.

Illies, J. 1965a. Die Wegenersche Kontinentalverschiebungstheorie im Lichte der modernen Biogeographie. *Naturwissenschaften*, **18**:505–511.

———. 1965b. Phylogeny and Zoogeography of the Plecoptera. *Ann. Rev. Entomol.*, **10**:117–140.

Jeannel, R. 1961. La Gondwanie et le peuplement de l'Afrique. *Ann. Mus. Roy. Afr. Centrale, Tervuren*, **102**:1–161.

Kaburaki, T. 1918. Freshwater Triclads from the Basin of the Inlé Lake. *Rec. Indian Museum*, **14**:187–194.

———. 1925. Planarians from the Andamans. *Rec. Indian Museum*, **27**:29–32.

Kato, K. 1950. A New Freshwater Triclad from Japan. *Annotationes Zool. Japon.*, **24**:45–48.

†Kawakatsu, M. 1968. On the Origin and Phylogeny of Turbellarians: Suborder Paludicola. *Japan. Soc. Systematic Zool. Circ.*, nos. 38–41: 11–22.

———. 1969a. An Illustrated List of Japanese Freshwater Planarians in Color. *Bull. Fuji Women's College*, no. 7, 45–91.

——. 1969b. Report on Freshwater and Land Planarians from New Caledonia. *Bull. Osaka Museum Nat. Hist.*, **22**:1–14.

——. 1969c. Report on Freshwater Planaria from India. *Annotationes Zool. Japon.*, **42**:210–215.

Kay, M. (ed.). 1969. *North Atlantic Geology and Continental Drift*. Memoir 12. The American Association of Petroleum Geologists. ix + 1082 pp.

Kenk, R. 1930a. Beiträge zum System der Probursalier. *Zool. Anz.*, **89**:145–162, 289–302.

——. 1930b. *Euplanaria cretica* Meixner, eine Triklade mit eigentümlichen Drusenorgan. *Zool. Anz.*, **92**:247–253.

——. 1935. Studies on Virginian Triclads. *J. Elisha Mitchell Sci. Soc.*, **51**:79–125.

——. 1936. Eine neue Hohlentriklade, *Fonticola opisthogona* n. sp. *Zool. Anz.*, **113**:305–311.

——. 1941. A Freshwater Triclad from Puerto Rico, *Dugesia antillana* New Species. *Occasional Papers Museum Zool. Univ. Mich.*, **436**:1–9.

——. 1944. The Freshwater Triclads of Michigan. *Misc. Publ. Museum Zool. Univ. Mich.*, **60**:1–44.

——. 1970. Freshwater Triclads (Turbellaria) of North America. I. The Genus *Planaria. Proc. Biol. Soc. Wash.*, **82**:539–558.

Kennel, J. 1888. Untersuchungen an neuen Turbellarien. *Zool. Jahrb. (Anat.)*, **3**:447–486.

Komarek, J. 1920. Uber hohlenbewohnende Tricladen de balkanischen Karste. *Arch. Hydrobiol.*, **20**:822–828. (German translation of the original Czech paper of 1919.)

Kozhov, M. 1963. *Lake Baikal and Its Life*. The Hague: W. Junk. vii + 344 pp.

*Laidlaw, F. F. 1906. Report on the Turbellaria. Zoological Results of the Third Tanganyika Expedition. *Proc. Zool. Soc. London*, 1906, 777–779.

*Lang, A. 1884. "Polycladen." *Fauna und Flora des Golfes von Neapel*, 11: ix + 688 pp. W. Engelmann, Leipzig.

Lang, P. 1913. Beiträge zur Anatomie und Histologie von *Planaria polychroa. Z. Wiss. Zool.*, **105**:136–155.

Leloup, E. 1944. Recherches sur les triclades dulçicoles epigés de la Forêt de Soignes. *Mem. Mus. Roy. Hist. Nat. Belge*, **102**:1–112.

Lepori, N. G. 1948a. Descrizione di *Dugesia sicula* n. sp. di tricladi di acqua dolce dei dintorni di Catania. *Arch. Zool. Ital.*, **33**:461–472.

——. 1948b. Descrizione di *Dugesia ilvana* n. sp. di planaria di acqua dolce dei dintorni di Catania. *Arch. Zool. Ital.*, **33**:183–193.

——. 1951. Sulle caratteristiche morfologiche e sulla posizione sistematica della planaria di Sardegna e Corsica gia ascritta a *Dugesia* (*Euplanaria*) *gonocephala* (Dugès). *Atti Soc. Toscana Sci. Nat.*, **58B**:28–47.

Lindroth, C. H. 1957. *The Faunal Connections between Europe and North America*. Stockholm: Almqvist and Wiksell. 344 pp.

†Livanov, N. A. 1951. Planarians of Kopet-Dag and Related Species of Crimea, Caucasia and Transcaucasia. *Tr. Murgabsk. Gidrobiol. St.*, **1**:103–114. (In Russian.)

——. 1962. "Report on Triclads in Lake Baikal." In G. I. Galazy (ed.), *Oligochaeta and Planarians of Lake Baikal*, pp. 152–188. Moscow: Proc. Limnol. Inst. (In Russian.)

Luther, A. 1956. Turbellaria Tricladida von den Cap Verde-Inseln. *Soc. Sci. Fennica Commentationes Biol.*, **15**:1–8.

———. 1960. Die Turbellarien Ostfennoskandiens. I. Acoela, Catenulida, Macrostomida, Lecithoepitheliata, Prolecithophora, und Proseriata. Soc. F. Fl. Fenn., F. F., **7**:1–155.

Macan, T. T. 1963. *Freshwater Ecology*. London: Longmans. x + 338 pp.

McKenzie, K., and S. U. Hussainy, 1968. Relevance of a Freshwater Cytherid (Crustacea, Ostracoda) to the Continental Drift Hypothesis. *Nature*, **220**: 806–808.

Maguire, B. 1963. The Passive Dispersal of Small Aquatic Organisms and Their Colonization of Isolated Bodies of Water. *Ecol. Monographs*, **33**:161–185.

Marcus, E. 1946. Sôbre Turbellaria brasileiros. *Bol. Fac. Filosof. Ciências Letras Univ. São Paulo, Zool.*, **11**:5–254.

———. 1948. Turbellaria do Brasil. *Bol. Fac. Filosof. Ciências Letras Univ. São Paulo, Zool.*, **13**:111–243.

———. 1953. *Turbellaria Tricladida*. Exploration du Parc National de l'Upemba, Fascicule 21: 1–62. Brussel: Institut des Parcs Nationaux du Congo Belge.

———. 1954. Reports of the Lund University Chile Expedition 1948–49. 11. Turbellaria. *Lunds Univ. Årsskr.*, **49**:3–115.

———. 1955. "Turbellaria." In Hanström, Brinck, and Rudebeck (eds.), *South African Animal Life*, vol. 1, pp. 101–151. Stockholm.

†———. 1963. Eine neue Meerestriklade von São Paulo. *Zool. Beitr.*, **9**:441–446.

———. 1970. "Turbellaria (Addenda)." In Hanström, Brinck, and Rudebeck (eds.), *South African Animal Life*, vol. 14, pp. 9–18. Stockholm.

——— and E. Marcus. 1951. Contributions to the Natural History of Brazilian Turbellaria. *Com. Zool. Museo Hist. Nat. Montevideo*, **3**:1–25.

——— and ———. 1959. Turbellaria from Madeira and the Azores. *Bol. Mus. Mun. Funchâl*, **12**:15–42.

Martins, M. E. Q. P. 1970. Two New Species of *Dugesia* (Tricladida Paludicola) from the State of São Paulo, Brazil. *Anais Acad. Brasil. Ciênc.*, **42**:113–118.

Marx, H., and G. B. Rabb. 1970. Character Analysis: an Empirical Approach Applied to Advanced Snakes. *J. Zool. London*, **161**:525–548.

Matthew, W. D. 1915. Climate and Evolution. *Ann. N.Y. Acad. Sci.*, **24**:171–318. (Reprinted 1939, *Spec. Publ. N.Y. Acad. Sci.*, **1**:1–223.)

Mayr, E. 1969. *Principles of Systematic Zoology*. New York: McGraw-Hill. xi + 428 pp.

Meixner, J. 1928. Der Genitalapparat der Tricladen und seine Beziehungen zu ihrer allgemeinen Morphologie, Phylogenie, Ökologie und Verbreitung. *Z. Morphol. Ökol. Tiere*, **11**:570–612.

———. 1938. "Turbellaria (Strudelwürmer) I." In G. Grimpe and E. Wagler (eds.), *Die Tierwelt der Nord-und Ostsee*, **IVb**:1–146.

Meyerhoff, A. A. 1970a. Continental Drift: Implications of Palaeomagnetic Studies, Meteorology, Physical Oceanography, and Climatology. *J. Geol.*, **78**:1–51.

――――. 1970b. Continental Drift, II: High Latitude Evaporite Deposits and Geologic History of Arctic and North Atlantic Oceans. *J. Geol.*, **78**:406–444.

―――― and C. Teichert. 1971. Continental Drift, III: Late Paleozoic Glacial Centers, and Devonian-Eocene Coal Distribution. *J. Geol.*, **79**:285–321.

Minshall, G. W., and P. V. Winger. 1968. The Effect of Reduction in Stream Flow on Invertebrate Drift. *Ecology*, **49**:580–582.

Mitchell, R. W. 1968. New Species of *Sphalloplana* (Turbellaria: Paludicola) from the Caves of Texas, and a Re-examination of the Genus *Speophila* and the Family Kenkiidae. *Ann. Speleol.*, **23**:597–620.

Moore, J. 1961. The Spread of Existing Diurnal Squirrels across the Bering and Panamanian Land Bridges. *Am. Museum Novitates*, no. 2044, 1–26.

Nelson, G. J. 1969. The Problem of Historical Biogeography. *Systematic Zool.*, **18**:243–246.

Neppi, V. 1904. Über einige exotische Turbellarien. *Zool. Jahrb. (Syst.)*, **21**:303–326.

Nurse, F. R. 1950. Freshwater Triclads New to the Fauna of New Zealand. *Trans. Roy. Soc. New Zealand*, **78**:410–417.

Okugawa, K. I. 1955. On the Supernumerary Sexual Organs of *Dugesia gonocephala* (Dugès) Induced by the Low Temperature. *Bull. Kyoto Gakugei Univ.*, **B6**: 1–14.

Packard, A. S. 1879. *Zoology for Students and General Readers.* New York: Henry Holt. 719 pp.

Pantin, C. F. A. 1931. The Adaptation of *Gunda ulvae* to Salinity. *J. Exptl. Biol.*, **8**:63–94.

Phinney, R. A. (ed.). 1968. *The History of the Earth's Crust.* Princeton N.J.: Princeton University Press. viii + 244 pp.

*Poche, F. 1926. Das System der Platoderia. *Arch. Naturgeschichte*, 91, Abt. A:1–459.

Popham, E. J., and B. F. Manly. 1969. Geographical Distribution of the Dermaptera and the Continental Drift Hypothesis. *Nature*, **222**:981–982.

†Porfirjeva, N. A. 1960. On the Zoogeography of Planarians in the U.S.S.R. *Zool. Zh.*, **40**:454–456. (In Russian, English summary.)

Reynoldson, T. B. 1966. The Distribution and Abundance of Lake-dwelling Triclads— towards a Hypothesis. *Advances Ecol. Res.*, **3**:1–71.

―――― and L. S. Bellamy. 1970. The Status of *Dugesia lugubris* and *D. polychroa* (Turbellaria, Tricladida) in Britain. *J. Zool. London*, **162**:157–177.

Richardson, L. R. 1968. A New Bdellourid-like Triclad Turbellarian Ectoconsortic on Murray River Chelonia. *Proc. Linnean Soc. N.S. Wales*, **93**:90–97.

――――. 1970. A Note on *Bdellasimilis barwicki* and an Indication of a Second Species (Turbellaria: Tricladida). *Australian Zool.*, **15**:400–402.

Runcorn, S. K. (ed.). 1962. *Continental Drift.* Int. Geophysical Series 3: 1–140. New York: Academic Press.

Schopf, T. M. 1970. Relations of Floras of the Southern Hemisphere to Continental Drift. *Taxon*, **19**:657–674.

Schuster, R. M. 1969. Problems of Antipodal Distribution of Lower Land Plants. *Taxon*, **18**:46–91.

Sivickis, P. B. 1928. The Freshwater Planarians of the Philippines. *Trans. Am. Microscop. Soc.*, **47**:356–365.

Stankovic, S. 1960. *The Balkan Lake Ohrid and Its Living World*. The Hague: W. Junk. 357 pp.

———. 1969. Turbellariés triclades endémiques nouveaux du Lac d'Ohrid. *Arch. Hydrobiol.*, **65**:413–435.

Steenis, C. G. G. van. 1962. The Land-bridge Theory in Botany. *Blumea*, **11**:235–372.

———. 1963. "Transpacific Floristic Affinities, Particularly in the Tropical Zone." In J. L. Gressitt (ed.), *Pacific Basin Biogeography*, pp. 219–231. Bishop Museum Press, Hawaii.

Stegman, B. 1963. "The Problem of the Beringian Continental Land Connection in the Light of Ornithogeography." In J. L. Gressitt (ed.), *Pacific Basin Biogeography*, pp. 65–78. Bishop Museum Press, Hawaii.

Steinböck, O. 1924. Untersuchungen über die Geschlechtstrakt-Darmverbindung bei Turbellarien nebst einem Beitrag zur Morphologie des Trikladendarmes. *Z. Morphol. Ökol. Tiere*, **2**:461–504.

———. 1925. Zur Systematik der Turbellaria Metamerata. *Zool. Anz.*, **64**:165–192.

Steinmann, P. 1914. Beschreibung einer neuen Susswasserticlade von den Kei-Inseln. *Abk. Senck. Naturf. Ges.* **35**:111–121.

Strand, E. 1942. Miscellanea nomenclatoria zoologica et palaeontologica. X. *Folia Zool. Hydrobiol.*, **11**:386–402.

Udvardy, M. F. 1969. *Dynamic Zoogeography with Special Reference to Land Animals*. Toronto: Van Nostrand Reinhold. xviii + 445 pp.

Ullyott, P. 1936. A Note on the Zoogeographical History of North-western Europe. *Proc. Prehist. Soc.*, **2**:169–177.

Vandel, A. 1965. *Biospeleology*. Oxford: Pergamon Press. xxiv + 524 pp.

*Voigt, W. 1904. Ueber die Wanderungen der Strudelwürmer in unseren Gebirgsbachen. *Verhandl. Naturh. Ver. Rheinl.*, 61.

Vuilleumier, F. 1969. Systematics and Evolution in *Diglossa* (Aves, Coerebidae). *Am. Museum Novitates*, no. 2381, 1–44.

Weiss, A. 1910. Beiträge zur Kenntniss der australischen Turbellarien I. Tricladen. *Z. Wiss. Zool.*, **94**:541–604.

Westblad, E. 1935. *Pentacoelum fucoideum*, ein neuer Typ der Turbellaria metamerata. *Zool. Anz.*, **111**:65–82.

———. 1952. Turbellaria (excl. Kalyptorhynchia). *Further Zool. Results Swed. Antarctic Expedition, 1901–1903*, **4**:1–55.

Whitehouse, R. H. 1913a. Freshwater Planaria (Zoological Records of the Abor Expedition 1911–1912, Part III, No. 22). *Rec. Indian Museum*, **8**:317–321.

———. 1913b. The Planarians of the Lake of Tiberias. *J. Proc. Asiatic Soc. Bengal*, **9**:459–463.

Wilhelmi, J. 1909. "Tricladen." *Fauna und Flora des Golfes von Neapel* 32. xii + 405 pp. R. Friedlander, Berlin.

Willis, J. C. 1922. *Age and Area*. Cambridge: Cambridge University Press. 259 pp.

Woodworth, W. McM. 1897. Contributions to the Morphology of the Turbellaria II. On Some Turbellaria from Illinois. *Bull. Museum Comp. Zool. Harvard Coll.*, **31**:1–16.

Wulff, E. V. 1950. *An Introduction to Historical Plant Geography*. Waltham, Mass.: Chronica Botanica Co. xi + 223 pp.

†Zabusova, Z. 1951. A New Species of Planarian from the Caspian Sea. *Tr. Murgabsk. Gidrobiol. St.*, **1**:115–126. (In Russian.)

Zimmerman, E. C. 1963. "Summary Discussion." In J. L. Gressitt (ed.), *Pacific Basin Biogeography*, pp. 477–481. Bishop Museum Press, Hawaii.

Chapter 16

The Cave-adapted Flatworms of Texas; Systematics, Natural History, and Responses to Light and Temperature

Robert W. Mitchell
Department of Biology, Texas Tech
University, Lubbock

This paper is the result of observations on the cave-adapted planarians of the genus *Sphalloplana* of Texas and experimental studies of one of them, *S. zeschi*.

Among the triclad turbellaria, only the Paludicola contains troglobite (obligate cavernicole) species, these restricted almost entirely to the Palae-arctic region (Vandel, 1964).

The genus *Sphalloplana* de Beauchamp, 1931, contains 19 species.

Kenk (1970) pointed out that future studies of these planarians may see synonomies proposed. *Sphalloplana* has not been without its taxonomic controversy, both in the constitution of the genus itself and in its assignment to family. The latest revisions are those of Mitchell (1968) in which *Speophila* Hyman, 1937, is synonomized with *Sphalloplana*, and in which it is proposed that the family Kenkiidae Hyman, 1937, be eliminated with removal of its genera (*Kenkia* and *Sphalloplana*) to the family Planariidae.

Most sphalloplanids are North American, but two occur in Japan (Ichikawa and Kawakatsu, 1967), one in southern Korea (Kawakatsu and Kim, 1967), and one in Siberia (Livanow and Zabusova, 1940). Except for *S. percoeca* (Packard, 1879), all the North American species have been described by Hyman (1937, 1939, 1945, 1954) and Mitchell (1968).

The spalloplanid flatworms are one component of the troglobite fauna of Texas (see Mitchell and Reddell, 1971), which is concentrated in the limestone caves of the central part of the state (Fig. 1). The first of these planarians, *S. mohri*, was described by Hyman in 1939 from specimens collected the previous year in Ezell's Cave, Hays County (Fig. 2). No other planarians were found in Texas caves until the 1960s, when four additional species were discovered in the central part of the state. These were described in 1968 by Mitchell as *S. kutscheri* from Spanish Wells Cave, Travis County; *S. sloani*, Harrell's Cave, San Saba County; *S. zeschi*, Zesch Ranch Cave, Mason County; and *S. reddelli*, Cascade Caverns, Kendall County (see Fig. 2).

There are several cavernous areas in Texas (Fig. 1), but it is the central portion of the state which is by far the most extensive and biologically the most important. The primary cave formers here are the limited Ordovician limestones and dolomites of the Llano Uplift with fringing areas of Devonian limestone and the Lower Cretaceous limestones and dolomites of the Edwards and Stockton Plateaus (Figs. 1, 2). So extensive is the Edwards Plateau that frequently this term is used to denote the whole of the central cavernous region. At its eastern and southern boundaries, the Edwards Plateau is sharply delimited by the Balcones Escarpment (Figs. 1, 2), a greatly dissected zone of much faulting.

Uplift of central Texas occurred in the Miocene, but there was no exposure of the present cavernous limestones until the subsequent erosional removal of overlying Upper Cretaceous deposits. The discontinuities at the Balcones Fault Zone unquestionably dictated that the cavernous limestones here were the first exposed, whenever this might have been. It is barely possible that some exposure occurred along the Fault Zone in the late Miocene, but probably this did not occur until the Pliocene or early Pleistocene. Cavernous limestones distant from the Fault Zone were probably not exposed until mid-Pleistocene.

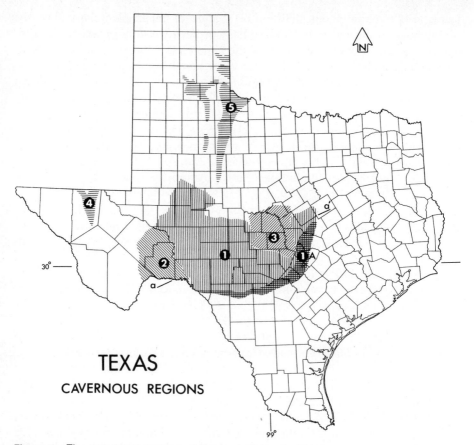

Figure 1 The cavernous regions of Texas. 1, Edwards Plateau; 1A, Balcones Escarpment; 2, Stockton Plateau; 3, Llano Uplift; 4, gypsum plain of Culberson County; 5, gypsum region of northwestern Texas; a–a′, reference points keyed to Fig. 2.

SYSTEMATICS AND EVOLUTION OF THE TEXAS

Sphalloplana

In my earlier paper (1968) describing four of the five Texas sphalloplanids, I pointed out that although all the Texas species were obviously closely related (based upon their morphology), I felt the differences which did exist justified recognition at the specific level. It may be argued in future publications that this separation is too emphatic, but the fact remains that there *are* differences between the worms studied. In this section, I shall attempt,

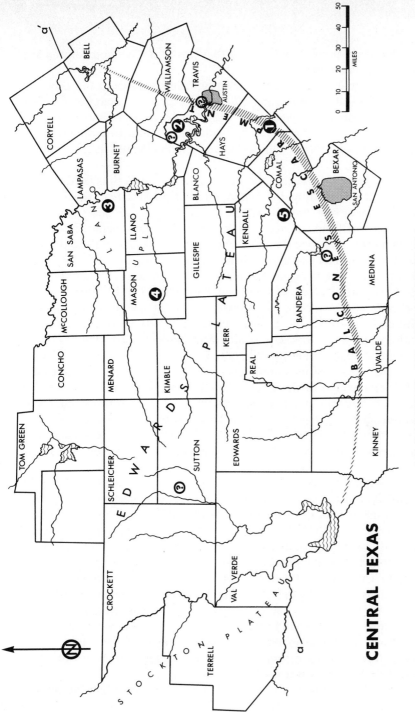

LOCALITIES OF *SPHALLOPLANA*

Figure 2 Central Texas, showing the localities of *Sphalloplana*. 1, Ezell's Cave, *S. mohri*; 2, Spanish Wells Cave, *S. kutscheri*; 3, Harrell's Cave, *S. sloani*; 4, Zesch Ranch Cave, *S. zeschi*; 5, Cascade Caverns, *S. reddelli*; ?, localities where sphalloplanids have been seen or collected but remain unidentified; a–a′, reference points keyed to Figure 1.

405

from an evolutionary point of view to further justify my opinions on the systematics of these planarians.

The Texas sphalloplanids differ from all other species in the genus by their polypharyngy (about 40 proboscides). This pecularity argues for their derivation from a common, epigean, polypharyngeal ancestor rather than for the evolution of this condition as a particular adaptation to a subterranean environment, since the other cave-adapted sphalloplanids remain unipharyngeal.

There is no question but that the Texas sphalloplanids thus far described are each highly isolated from the others. It is a general feature of central Texas that the shallower aquifers (those accessible to collection via caves) tend to be rather small and isolated. This results from stratigraphic discontinuities, intervening areas of noncavernous strata, and dissection of the cavernous strata into segregated blocks through stream erosion, and it is reflected in the extensive speciation occurring within many groups of terrestrial and aquatic invertebrate cavernicoles (see Mitchell and Reddell, 1971). Further, the habitats of the presently described Texas sphalloplanids are all discontinuous. These habitats (Fig. 2) are all in rather close proximity, but in terms of the geomorphology of central Texas, absolute distances mean very little.

The unique fauna of Ezell's Cave, habitat of *S. mohri*, has long evidenced the complete separation of its waters, even from those of other caves located nearby. The composition of the Ezell's fauna, in particular the aquatics, and the degree of cave adaptation of some of its species argue for this being one of the first subterranean systems of central Texas to be colonized, certainly far in advance of those distant from the Balcones Escarpment. Mitchell and Reddell (1971) discuss the fauna of Ezell's Cave and provide a complete faunal listing.

The sphalloplanid cave nearest to Ezell's Cave is Cascade Caverns, habitat of *S. reddelli*, but it shares none of the unique Ezell's species, harboring, for example, a salamander quite distinct from that in Ezell's Cave (see Mitchell and Smith, 1971). Spanish Wells Cave, habitat of *S. kutscheri*, is a very small cave located at the summit of an isolated mesa capped by Edwards limestone. The waters inhabited by the planarians are probably those of a perched water table. Harrell's Cave, the habitat of *S. sloani*, is located in Ordovician limestone of the Llano Uplift. In this cave the planarians occur in an intermittent pool of moderate size whose waters are probably replenished during the rise of normally lower-lying groundwaters. Zesch Ranch Cave, habitat of *S. zeschi*, is formed in Devonian limestone on the fringes of the Llano Uplift. Planarians in this cave also inhabit intermittent waters which are probably replenished with a rise in lower groundwaters.

Holsinger's (1967) excellent study of the systematics, distribution, and evolution of the subterranean amphipods of the genus *Stygonectes* provides invaluable assistance, I believe, in clarifying the relationships of the Texas sphalloplanids. All the caves inhabited by described species of flatworms also harbor stygonectid amphipods. Ezell's Cave contains *Stygonectes flagellatus*; Cascade Caverns, *S. dejectus*; and Harrell's Cave, *S. bifurcatus*. Such speciation as this clearly argues for the isolation of the respective cave systems. On the other hand, distribution of the amphipod *Stygonectes russelli* seems, at first, to offer contradictive evidence for such cave separation and further seems to suggest that the sphalloplanids might all be of the same species. *Stygonectes russelli* occurs in four of the five flatworm caves in question (not present in Ezell's Cave), as well as several other caves. Thus in Cascade Caverns it is sympatric with *Stygonectes dejectus* and in Harrell's Cave with *S. bifurcatus*. Apparently it is the only amphipod species in Spanish Wells and Zesch Ranch Caves. Does the widespread distribution of *Stygonectes russelli* suggest gene flow between all these caves with the implications in interpreting the systematics of the sphalloplanids being obvious? The answer is no, as Holsinger's study suggests.

Holsinger states that "*Stygonectes russelli* is the most highly variable species in the genus." While he was not altogether successful in correlating morphological variation with geographical distribution, he did demonstrate that differences exist between some of the populations. Populations from caves of the Llano Uplift (including Harrell's Cave) are homogeneous and differ consistently from populations in caves in northern Travis County. Not included in the latter, however, is Spanish Wells Cave, habitat of *Sphalloplana kutscheri*, which contains an amphipod showing minor differences even from other caves of this area (based, unfortunately, upon but two specimens, however). As already stated, the waters of Spanish Wells Cave are probably of a highly isolated perched water table. Amphipods from Zesch Ranch Cave seem to have characteristics overlapping those of the amphipods of the Llano Uplift and northern Travis County. A single female amphipod from Cascade Caverns, habitat of *Sphalloplana reddelli*, seems to show closest affinities with populations from caves in southern Travis County. Even though Holsinger applied the name "*Stygonectes russelli*" to this entire complex, morphological differences do exist between many populations. This indicates their geographical isolation and that speciation is occurring, regardless of the level at which it should be recognized. To this, Holsinger states, "*S. russelli* is almost certainly a polytypic form, and when completely studied and fully understood, it may turn out to be a cluster of several sibling species."

As Holsinger discusses, sympatry in these amphipods indicates multiple invasions of the subterranean habitats, and, further, the fact that *Stygonectes russelli* is the most widespread species suggests that it is a recent invader.

As discussed by Mitchell and Reddell (1971), much of the complexity of the troglobite fauna of central Texas is due to the alternating glacials and interglacials of the Pleistocene which brought into the area multiple waves of cave colonizers.

The implications of the foregoing to the *Sphalloplana* problem are obvious. I would suggest that the common epigean ancestor of the Texas sphalloplanids invaded central Texas only once, probably in late Pleistocene. Arguments for such a single, late invasion are as follows. The planarians are morphologically similar, widely distributed, and geographically isolated, indicating their derivation from a single invasion wave. Were the present planarian species derived from more than one wave of invading colonizer (even the exact same species), then there would be every reason to expect greater differences between the cave species because of the disparate times of cave colonization. Such are the differences between the Texas species of stygonectid amphipods that Holsinger (1967) recognizes two major evolutionary lines within them, one derived from an ancient invader, the other from a more recent invader or invaders.

Had this single invasion of the ancestral sphalloplanid come early in the exposure of cavernous central Texas limestones (early Pleistocene), then its present distribution would be confined to caves at or near the Balcones Escarpment. The parallel to this suggestion in the stygonectid amphipods may be found in the old *flagellatus* group (Holsinger, 1967), where the four included species occur only in a cave (Ezell's) and two springs on the Escarpment itself and in a cave not far from the Escarpment (Cascade Caverns). Since the morphologically similar sphalloplanids inhabit a cave on the Escarpment, two near it, one fairly distant, and one far from the Escarpment, it seems apparent that the caves were not colonized by the ancestral species until such time as all the caves were simultaneously available to the colonizer, and this was not especially long ago geologically, as detailed earlier. In this paper I shall present evidence of a more physiological nature which supports the contention that these planarians are rather recent troglobites.

I have implied a parallel between the Texas sphalloplanids and *Stygonectes russelli*, and that they have had essentially the same sort of evolutionary history. I apply five specific names to the sphalloplanid complex while Holsinger applies but one to the "*russelli*" complex; however, Holsinger suggests that several closely related species may be represented in his *Stygonectes russelli*. I simply choose to recognize the sphalloplanid differences at the specific level. There is a rather limited number of features whereby planarians may differ one from another (opposed, for example, to an arthropod such as an amphipod with its complexity of parts), and so any differences should, perhaps, be regarded of considerable taxonomic consequence.

STUDIES OF *Sphalloplana zeschi*

The Habitat

In most of the Texas sphalloplanid caves, it is never possible, on any one visit, to see more than a very few planarians, and sometimes none at all are seen. But Zesch Ranch Cave is truly remarkable for its flatworm population. This cave (Figs. 3, 4, 5) consists of but a single, small room about 12 by 18 m, accessible through a small entrance in the roof. In the southeastern end of the cave, there is an intermittent pool whose level may fluctuate drastically. Usually, this pool exposes no more than 2 to 3 m² of surface area, but at time of high water it may fill the cave to that level portion of floor just beneath the entrance (Fig. 5), and at time of prolonged drought it may disappear completely (pers. comm., J. R. Reddell). Since the pool has no direct vertical or horizontal connections to larger bodies of water, it appears that it is filled with the rise of lower-lying groundwaters. Measured on various occasions, water temperature has varied only slightly, from 21 to 21.5 °C.

The entrance of the cave is so constructed that in late afternoon a beam of direct sunlight streams through the entrance, striking the northeastern wall and illuminating the cave sufficiently well that one may easily move about without the aid of artificial light. Illumination at the planarian pool (Figs. 4, 5) (at low water) is scant, however, and not measurable with a light meter of a 0.2-fc sensitivity.

Along with the planarians, the pool is inhabited by the troglobite amphipod *Stygonectes russelli*, an undescribed troglobite ostracod of the genus *Candona*, and an unidentified copepod. The terrestrial cavernicole fauna of the cave (which may be gleaned from Reddell's checklists of the cave fauna of Texas, 1965, 1966, 1970a, 1970b) is neither striking nor varied. The only terrestrial species whose occurrence must be noted are the cave crickets of the genus *Ceuthophilus*.

Even when quite small, the pool in Zesch Ranch Cave may contain hun-

Figure 3 Entrance to Zesch Ranch Cave, locality of *Sphalloplana zeschi.*

Figure 4 Interior of Zesch Ranch Cave. Arrow points out pool inhabited by the planarians.

dreds of planarians (Fig. 6). This population thus provides unparalleled opportunities for study since everywhere troglobite planarians are usually uncommon at best. Although the findings to be detailed are based upon studies of *Sphalloplana zeschi* (Fig. 7), they are probably generally applicable to all Texas sphalloplanids in view of their close relationship.

Studies of the biology of troglobite planarians, aside from isolated comments in various papers, are essentially limited to those of Buchanan (1936) on *Sphalloplana percoeca*. Although Buchanan's material was limited and his methods not especially refined, his data do offer useful information about this unipharyngeal sphalloplanid, and it is of interest to compare, when possible, this eastern species with the multipharyngeal Texas species.

Movement

Observations in the field and especially in the laboratory have shown that *S. zeschi* passes most of its time in a quiescent state. There seems to be no preference for type of substrate on which the worms may come to rest: bottom mud, rocks, surface film, sides of aquaria, etc. When quiescent, the bodies of the worms are somewhat contracted, the head and especially the auricles partially retracted, and the body margins rather scalloped. The worms may be activated by a variety of disturbances: light, agitation of the water, a tap on the glass of the aquarium, etc.

When active, the worms exhibit two types of movement: typical gliding and a leechlike movement employing the anterior adhesive organ and posterior adhesive secretions. Data on gliding rate are summarized in Fig. 8. These data were obtained by measuring the time (to the nearest $\frac{1}{10}$ sec with a stopwatch) required for a worm to travel a straight-line distance of 2 cm on the side of an aquarium while moving near, and parallel to, the edge of a scale. Fifty planarians were so timed, yielding a mean of 6.8 sec to travel 2 cm, or a rate of about 3 mm/sec. The fastest was 5.5 sec, the slowest, 8.9 sec. Buchanan

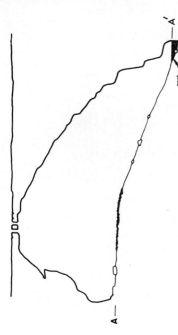

ZESCH RANCH CAVE

MASON COUNTY, TEXAS

BRUNTON TRANSIT AND TAPE SURVEY BY
WILLIAM ELLIOTT AND ROBERT MITCHELL
8 NOVEMBER, 1970
DRAFTED BY WILLIAM ELLIOTT, 1970

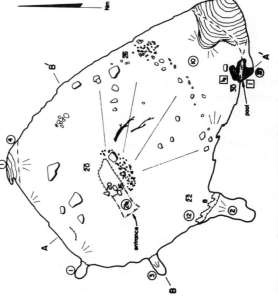

Figure 5 Plan and profiles of Zesch Ranch Cave.

Figure 6 Pool in Zesch Ranch Cave showing many planarians resting or gliding on the surface film.

(1936) did not measure the gliding rate of *S. percoeca*, but he described it as "comparatively rapid," also an appropriate description for *S. zeschi*. Buchanan also mentioned that when gliding, the body margins of *S. percoeca* undulated; this does not occur in *S. zeschi*. Buchanan reported that *S. percoeca* would leave the water (in flasks in the laboratory) and creep about above water level. *S. zeschi* does not do this.

One of the most prominent features of the sphalloplanids is their anterior glandulomuscular adhesive organ, discussed and sketched by Mitchell (1968). It is conical in shape and may be extended and retracted through an anterior opening. Except when in use, this organ is retracted, though usually its tip protrudes through the opening (Fig. 9*a*). My 1968 paper contains observations of the function of this structure in the leechlike movements which these planarians may perform. I should like to elaborate, chiefly through photographs, this method of locomotion, especially the action of the adhesive organ. Figure 9*b* to *i* shows the adhesive organ sequentially through one complete movement. The photographs of the worms were made from the ventral aspect as they moved on the sides of a glass aquarium. The series does not represent a sequence of one worm, but is rather a composite selected from many photographs of many worms. The worms were stimulated to employ this type of locomotion by touching their posterior ends with a small camel's hair brush. In Fig. 9a, the head is seen during normal gliding with the small auricles expanded and the conical adhesive organ slightly protruded, its typical position. Figure 9b and c shows early stages in the movement with the body greatly extending forward and the adhesive organ extruded. I believe that the adhesive organ reaches a maximum extension even greater than that shown in Fig. 9c, although I was never able to capture the view on film because of the great

Figure 7 *Sphalloplana zeschi*, ventral view of animal gliding on surface film. Ingested food outlines gut. Note the small but obvious auricles and the tip of the slightly protruded adhesive organ.

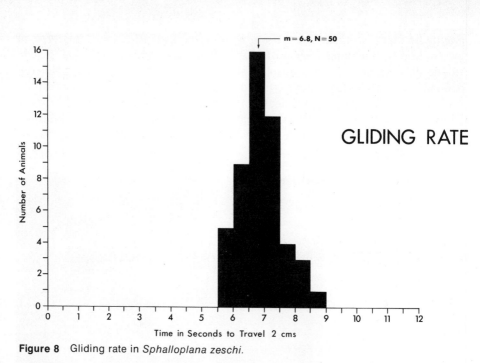

Figure 8   Gliding rate in *Sphalloplana zeschi.*

rapidity with which the organ is extended as it lunges forward. Figure 9*d* shows the organ just at the moment of attachment to the wall of the aquarium. The organ bends ventrally after maximum extension, whereupon it contacts the substrate. Figure 9*e* shows an early view in the pulling of the body forward after attachment. The remaining views (Fig. 9*f* to *i*) show progressive forward movement of the body which is so pronounced that the adhesive organ and auricles come to be surrounded by the anterolateral body margins. Figure 9*i* shows both the anterior and posterior ends of the body now attached just prior to detachment of the adhesive organ and, perhaps, another anterior extension. The number of successive movements of this type which may be made by a worm is variable. Usually a stimulus will elicit a series of two or three movements, occasionally only one, but sometimes several ensuing movements will move the worm a distance of a few centimeters. In my 1968 paper, I suggested that this type of movement was probably most often used as a means of predator escape. Perhaps this is true in those caves containing potential predators (salamanders), but the majority of the Texas sphalloplanid caves are without likely planarian predators. I view the use of the adhesive organ in movement as a secondary function, the primary function being associated with food capture, as will be discussed later.

   S. zeschi is efficient at righting itself from a completely inverted position.

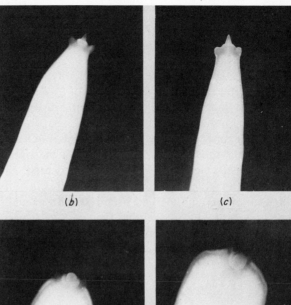
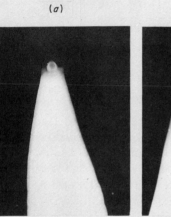
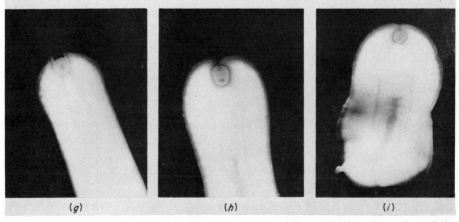

Figure 9 Sequence of illustrations showing the action of the adhesive organ of *Sphalloplana zeschi*. See discussion in text.

In its natural habitat, this planarian frequently detaches itself from the surface film to sink to the substrate, always landing on its dorsum as a result of its body contours. It would seem logical for efficient righting to be associated with this behavior.

The righting times of 20 individuals were measured. Each animal was inverted in a glass dish with the aid of a camel's hair brush from which point they were timed until the entire venter was once again in contact with the substrate. A mean righting time of 12 sec was obtained with the fastest time being 5 sec and the longest 24 sec. This is much more rapid than in *S. percoeca*, for which Buchanan (1936) determined the righting time to be about 40 sec (nine measurements). As would be expected, *S. zeschi* first turns the head to attach the adhesive organ and then rolls the body over in an anteroposterior direction. The animals had usually begun to glide away before the posterior was fully righted. The righting time of *S. zeschi* compares favorably with that of *Phagocata gracilis*, an epigean polypharyngeal species, which Buchanan (1936) reported as being "completed within 10 sec."

Feeding

In my 1968 paper, I briefly mentioned having seen *S. zeschi* feed on the bodies of the troglobite amphipod *Stygonectes russelli* and cave crickets of the genus *Ceuthophilus*. The literature otherwise contains no data on feeding habits in this group of cave planarians. Buchanan (1936) was never successful in his attempts to elicit feeding in his captive *S. percoeca* although he offered them many kinds of food.

Now that I have more extensively observed *S. zeschi* both in its habitat and in the laboratory, I am able to detail several aspects of its feeding. The animals are carnivores and in nature probably feed almost altogether upon the bodies of larger arthropods. I have offered them a variety of such animals (amphipods, crickets, flies, roaches, and even a butterfly), and they have fed readily on all. However, they will take only injured or moribund individuals. For example, they will not attack healthy, active individuals of the amphipods which share the cave waters, but an injured amphipod is quickly approached and fed upon (Fig. 10e). Perhaps the issue of body fluids through a ruptured integument signals the presence of subduable prey. Because of their abundance, amphipods and cave crickets probabily constitute the chief food sources of *S. zeschi*. In some of the other Texas sphalloplanid caves, other arthropods probably assume importance as prey; for example, cirolanid isopods in Ezell's Cave.

Hyman (1937) speculated that the adhesive organ was used for food capture, and this proves correct. If the prey is dead or is not struggling, a worm will simply glide onto it and begin to feed. If the prey is thrashing about, however, the worm will glide up to the prey, and its head will dart forward

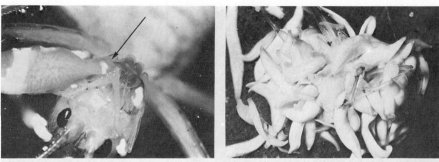

Figure 10 *(a) Sphalloplana zeschi* attaching by its adhesive organ (at arrow) to the head of a cave cricket. *(b)* Many *Sphalloplana zeschi* feeding on the body of a cave cricket. *(c) Sphalloplana zeschi* showing invagination of dorsum of body during maximum extension of the proboscides. *(d)* Cave cricket fed upon by *Sphalloplana zeschi.* Near-complete removal of soft parts indicated by visibility of tracheal tubes at left arrow. One proboscis of a worm visible at right arrow extending midway into femur. *(e)* Troglobite amphipod, *Stygonectes russelli,* fed upon by *Sphalloplana zeschi.* Proboscides visible at arrows. *(f) Sphalloplana zeschi* showing several extended proboscides probing an area from which food had just been removed.

attaching the adhesive organ (Fig. 10a). This done, the body contracts strongly, and the worm then anchors itself more securely by lateral and posterior secretions. Because the planarian population is so large in Zesch Ranch Cave, the body of a prey is soon covered by dozens of worms (Fig. 10b).

The planarians quickly insert their proboscides into the prey, removing almost all soft parts. The proboscides are highly extensible (Fig. 10f) and may reach a length equal to that of the worm itself. The proboscides ramify throughout the prey's body, even reaching into smaller appendages (Fig. 10d, e). The dorsum of a worm will frequently invaginate (Fig. 10c), forcing maximum exsertion of the proboscides. So thorough is the removal of the prey's internal parts that little more than a transparent exoskeleton usually remains.

Both moving and motionless prey are found rather quickly. It seems that the former are located more rapidly, and this suggests the importance of mechanoreception. Perfectly motionless prey (pieces of dead crickets) are located by means other than random wanderings, indicating that chemoreception is also important. Figure 11 illustrates a striking alignment of planarians along what is obviously a chemical track of fluids issuing from the body of a cave cricket. During photography of this phenomenon, the light would cause dispersion of the planarians, but they would quickly realign themselves. Other worms crossing the track would usually turn into it.

In all save those caves inhabited by salamanders (Ezell's and Cascade), the planarians are probably the top carnivores. In Zesch Ranch Cave, the flatworms always feed before the amphipods do (at least on larger arthropod bodies). During planarian feeding, the amphipods usually hover about on the fringes of the food or the feeding area (Fig. 12a), often in appreciable numbers and in a very active state. After the planarians have departed from the food, the amphipods move in and begin their feeding (Fig. 12b). Shortly, the prey is covered by a seething mass of amphipods (Fig. 12c).

Rheotactic Responses

Buchanan (1936) reported that *S. percoeca* showed no oriented movements in flowing water (1 ft/sec). *S. zeschi*, on the contrary, is negatively rheotactic. The animals were tested in a plexiglass chamber 38 by 4 by 4 cm, through which water was moved by a recirculating pump to produce a current of 2 cm/sec. Water depth was 2.5 cm. In the first of two experiments, introduction of the worms was into moving water. The worms were tested individually, and the response of each was entered appropriately into Cole's Closed Sequential Test Design (Cole, 1962). A significant ($P = 0.05$) negative response for the population was indicated by this test after a run of seven individual negative responses (Fig. 13). The times (Fig. 13) to full response (i.e., time from which the worm touched the substrate after introduction to time of

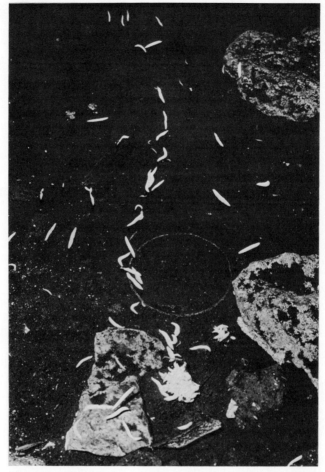

Figure 11 *Sphalloplana zeschi* aligned along a chemical track issuing from the body of a cave cricket covered by worms at center bottom. (Circle is an old can lid.)

straight-line movement away from direction of water flow) were measured, and the mean time for the seven worms was 85 sec.

In a second experiment, the worms were introduced initially into still water. They were allowed to move about until, when they were gliding "upstream," the water flow was begun. Here, again, a significant response was indicated by a run of seven individual negative responses (Fig. 13). Times (Fig. 13) to full response (beginning when current was started) were measured, and the mean time for the seven worms was 36 sec.

Figure 12 *(a)* Body of fresh, injured cave cricket being fed upon by many *Sphalloplana zeschi*. Amphipod, *Stygonectes russelli* (at arrow), hanging from cricket's leg awaiting opportunity to feed. *(b)* Body of cave cricket being abandoned by *Sphalloplana zeschi*. *Stygonectes russelli* moving in to feed. *(c)* Body of a cave cricket covered by mass of feeding *Stygonectes russelli*.

In spite of the small sample sizes, the mean time to full orientation was much greater when the animals were introduced directly into moving water. Even considering that some time was spent in righting (not all animals inverted because of shallowness of the water), this time was still inordinately long. It was obvious that these worms were greatly disoriented. The worms would not begin to glide immediately, but usually there would be considerable head swinging followed perhaps by the worm's turning two or three tight circles. Finally, however, the animal would begin a straight-line movement opposite to the current. When the current was not begun until the worm was gliding "upstream," the animal would respond immediately by pausing or by diverting, and, often after some slight erratic movement, would begin to move downstream.

Light Responses

Buchanan (1936) reported that *S. percoeca* shows no "orientation to light in any way, except . . . writhing when exposed to direct sunlight." *S. zeschi*, however, shows a pronounced negative response to light, even of very low intensity.

When light is shined on a gliding worm, it makes a diversion as seen in Fig. 14. With a point source of light, this response results only when light is shined on the brain. There is no response whatever when light is shined on any other body part, including the nerve cords.

The worms' responses to light were caused by a small flashlight to the end of which had been fitted a small tube so as to produce a circle of light 3 mm in diameter at the level of the worm. In the various tests, this light was held against the side of an aquarium so that the light was shined at right angles to the venter of the worm and directly upon the brain. The light was not turned on until properly positioned. Accessory lighting necessary to conduct the tests was of such low intensity that it could not be read with a light meter of 0.2-fc sensitivity.

Direction of the diversion was found to be random (of 50 worms, 29 turned to the right, 21 to the left) with the light positioned as above. In fact, the worms make no directed response regardless of the position of the light. The diversion angle proved to be rather constant (measured to an accuracy of about 5° as that angle formed by the paths of straight-line movement before and after diversion) as shown in Fig. 15. Diversion results from a fairly strong unilateral contraction of the body musculature. That there is some consistency in the angle of diversion indicates that there exists something akin to a "standard" contraction of this musculature.

Figure 16 summarizes data on times to response by actively gliding and quiescent animals. Timing (with a stopwatch to nearest $1/10$ sec) was begun

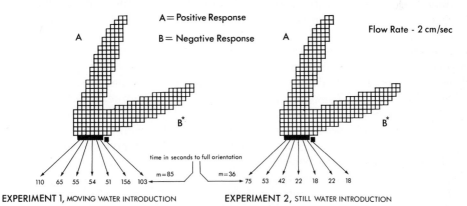

RHEOTACTIC RESPONSES

Figure 13   Rheotactic responses of *Sphalloplana zeschi*. See also text.

Figure 14 *Sphalloplana zeschi* diverting from light.

when the light was turned on and ceased when the head began a lateral movement indicating an ensuing diversion. In the first experiment (Fig. 16), light
intensity was 5 to 6 fc, causing rapid (mean of about 2 sec) and fairly consistent
times to diversion (minimum of 1.0 sec, maximum of 3.1 sec). Although not
quantified, light of greater intensity did not seem to cause more rapid diversion
time. Light of quite low intensity elicits maximum response. A second experiment (Fig. 16) was run identically with the first except that light of only 0.2
fc was used. Time to diversion was slower (mean of 3.7 sec) and less consistent
(minimum of 1.3 sec, maximum of 7.9 sec) than with the stronger illumination. A third experiment (Fig. 16) was identical to the first except that quiescent animals were tested. The results were quite interesting in that the mean
time to response (about 6 sec) was much longer than that for active animals
and was far more variable (minimum of 2.7 sec, maximum of 15.8 sec). Neither
was the manner of diversion typical of that of gliding worms. Upon activation,
the worms would usually exhibit weak and slow contraction of the body,
sometimes drawing the anterior end more or less directly posteriorly rather
than to the side. These data indicate that when the planarians are quiescent,
there is a slowing down of many more physiological processes than those
directly involved in movement.

If a light is held constantly on one of these planarians, the worm will
usually make a few diversions and then become totally unresponsive to the
light. Further response can then be elicited only after an intervening dark
period. To test for the length of this refractory period, a light (24 to 28 fc)
was shined continually on the brain of an animal for a duration of 30 sec, well
beyond the time during which the initial diversions occurred. Dark periods

(of varying length in a series of experiments) were then allowed, followed by another illumination which was maintained until diversion or for a maximum of 30 sec. Here, too, the Closed Sequential Test Design of Cole (1962) was used. The response of each worm after the second illumination (either diversion or no diversion) was entered appropriately into the design. Results are shown in Fig. 17. Dark periods of 5 and 10 sec were insufficient to permit subsequent response in any of the worms tested. Dark periods of 15 and 20 sec were sufficient for a minority of the animals tested, but too short to cite either of these times as sufficient for the population as a whole. A dark period of 25 sec yielded results which, as interpreted by the design, indicated that this is a sufficient interval to permit response in 95 percent of the population. Therefore, a refractory period of about 25 sec may be regarded as characteristic for this planarian. It is noted in Fig. 17 (4, 5, 6) that mean times to diversion (based on small samples and estimations by counting) decreased with increasing length of dark interval. Even after 30 sec of darkness, mean time to diversion was still twice that obtained from experiment 1 in Fig. 16 on times to diversion. Apparently a refractory period considerably longer than those tested here is necessary before maximum response can be elicited.

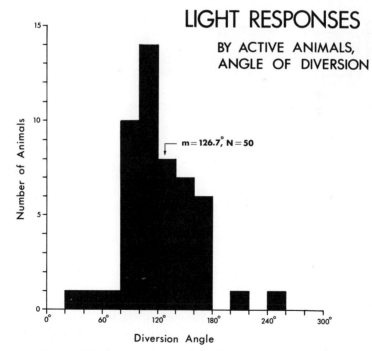

Figure 15 Diversion angle made by *Sphalloplana zeschi* when exposed to light.

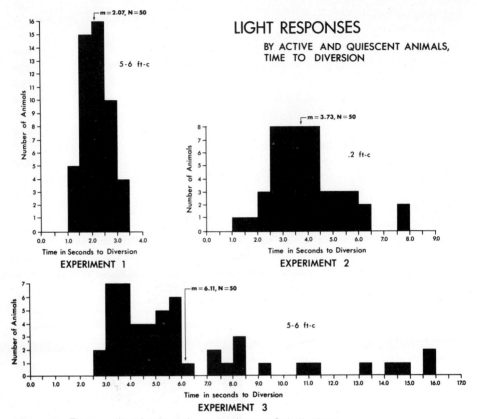

Figure 16   Times to diversion by active and quiescent *Sphalloplana zeschi* when exposed to light.

It is generally held that light is quickly lethal to cave planarians. This is based not so much upon experimental studies as it is upon casual observations made incidental to the collection, transport, photography, etc., of the worms. Vandel (1964) states, "Les Planaires cavernicoles y sont particulière-ment sensibles. Elles se dissolvent si rapidement sous l'action de la lumière qu'il est extrêmement difficile de les photographier ou de les filmer." Although he tested but very few specimens and his experiment was not tempera-ture-controlled, Buchanan (1936) found that *S. percoeca* disintegrated "within two minutes" when exposed to an illumination of 4000 fc (sunlight).

To test the light tolerance of *S. zeschi*, I constructed an apparatus in which the planarians could be exposed to a maximum illumination (artificial) of 2000 fc while holding the temperature of their water at that of the cave (21° ± 0.5 °C). This apparatus consisted of three major components: a light bank of nine high-intensity lights, a heat interceptor through which water flowed, and a base consisting essentially of a temperature-controlled water bath surrounding nine dull black glass specimen chambers, each of 3 cm³ volume. One such chamber was a blank housing the temperature controller's thermistorized sensing probe. In each run, only one animal was placed in each chamber, and these were divided into four experimentals exposed to the light and four controls whose chambers were covered by opaque lids. Selection of a chamber as experimental or control was random.

Two experiments were run, differing only in time of exposure, 1 hour in the first, 3 hours in the second. Light intensity was 2000 fc. After exposure, the planarians were removed to covered glass dishes containing a substrate from the cave pool. Ninety days after exposure, both experimentals and controls from the first experiment were normal. In the second experiment, one experimental worm was largely disintegrated slightly less than one day fol-

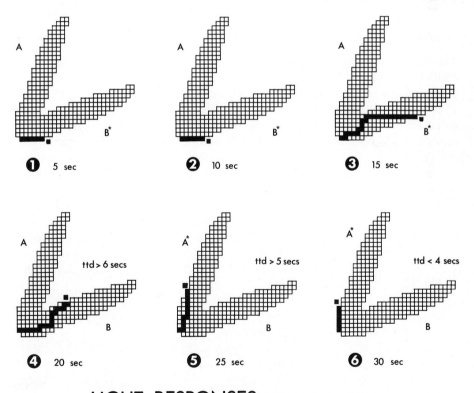

LIGHT RESPONSES, REFRACTORY PERIODS

Figure 17 Refractory period data for *Sphalloplana zeschi*. See text for explanation.

lowing termination of the experiment, and another experimental was dead and disintegrating four days later. The first worm to die was already showing some posterior disintegration before the end of exposure. Ninety days after exposure, the two remaining experimentals and the four controls were normal. During the course of these experiments, especially during the latter portion of the second, some of the worms showed adverse responses to the light even though they were not killed. These consisted of arching the body, curling the posterior end upward, and extending the proboscides. For the most part, though, the worms glided normally, sometimes resting, during exposure. Three days later, two of the worms from the second experiment responded abnormally to the light from a small flashlight, writhing when exposed. One of these worms died the following day; the other returned to normal. Perhaps exposure to sunlight with its infrared radiation would be more severe to these planarians. However, these experiments suggest that *S. zeschi* has some degree of tolerance to visible light.

Temperature Preference

The preference responses of *S. zeschi* were determined in a linear temperature gradient chamber described by Bull and Mitchell (1971). Basically, this apparatus consisted of an experimental chamber 120 cm in length and 12 cm in width containing water of a depth of 1.5 cm in which a temperature gradient could be established. Temperatures were sensed with thermistorized probes at each end of the chamber and at 10-cm intervals between. These probes also served to delineate the chamber into 12 areas for purposes of data recording. Control data (no gradient; water temperature of 21 °C) and two sets of experimental data (gradient of 15 to 30 °C) were gathered. In each, sample size was 48 with individual runs of 12 worms each being made. Ten minutes after introduction, the areas of occurrence of the worms were recorded, and this was repeated at each 10-min interval (sufficient time for a worm to travel the length of the chamber) for a period of 4 hours. Thus for each experiment, a total of 1152 position records was obtained. These data are summarized graphically in Fig. 18.

The graph of control data is typical of the responses of animals in a rectangular chamber, the U shape reflecting the fact that captive animals tend to move to the limits of their confinement.

Two sets of experimental data were gathered. In the first experiment, animals were introduced randomly (with use of a random numbers table) into chamber areas 1 through 12. While in this experiment there was something of a peak in occurrences near cave temperature, the graph of the data (Fig. 18) also shows that the worms, as an entire sample, responded rather erratically. There were many records of occurrence in the two coldest areas and an

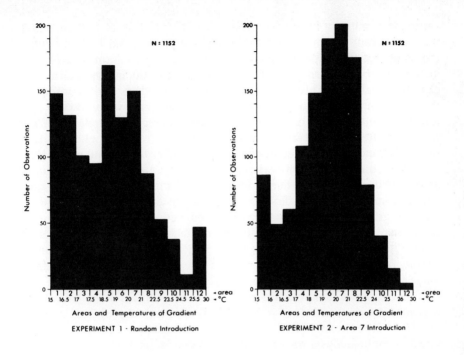

EXPERIMENT 1 - Random Introduction

EXPERIMENT 2 - Area 7 Introduction

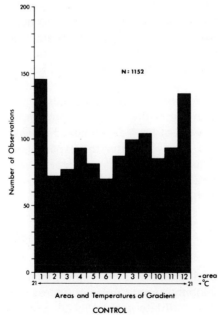

CONTROL

Figure 18 Temperature preference data for *Sphalloplana zeschi*. See text for additional explanation.

appreciable number in the warmest area. Observation of the animals during the experiment indicated that these data reflect not end bias but rather the effects of heat and cold trapping. In fact, one worm died during one run, trapped in the chamber's warm end. The manner of introducing the worms into the chamber might have been a source of the erratic data. As a consequence of their random introduction, the worms were taken from water at cave temperature and immediately dropped into a designated area, the temperature of which was quite often different from that at which they were being held. This temperature shock might have affected the animals' ability to respond properly.

A second experiment was run in which all worms were introduced into area 7, that nearest cave temperature. The worms' responses were quite different from those in the first experiment, the data plotting an almost symmetrical graph (Fig. 18). Most occurrences were in the area at cave temperature. There was some cold trapping but no heat trapping. A temperature range from 17 to 24°C may be regarded as the temperature preferendum for *S. zeschi*; over 75 percent of the records of occurrence were included in this range.

Each set of temperature preference data differs from the other two: experiment 1 vs. control, $\chi^2 = 224$, $P < 0.001$; experiment 2 vs. control, $\chi^2 = 362$, $P < 0.001$; experiment 1 vs. experiment 2, $\chi^2 = 154$, $P < 0.001$. Contingency table analysis was used in the comparisons.

All worms used in gathering these data were removed from the gradient chamber back to water at cave temperature to determine whether any lethal effects would ensue from the worms' movements into hostile temperatures. At the end of 90 days, 45 of the 48 control worms were still alive and normal. Unfortunately, different laboratory accidents befell the two groups of experimental worms, resulting in data less than the best but not without value. One week after completion of the first experimental runs, 34 of these 48 worms had died. Ninety days after the second experimental runs, at least 12 of these 48 worms had died, but at least 28 remained alive and normal. Nine of the 12 deaths came within one week after the experimentation. These data suggest that these planarians are not highly efficient at sensing and removing themselves from temperatures which may be lethal to them. This ability is further disrupted by temperature shock.

EVOLUTIONARY STATUS OF THE TEXAS *Sphalloplana*

Earlier, I suggested that the Texas sphalloplanid planarians are probably rather recent troglobites whose epigean ancestor possibly did not colonize central Texas caves until late in the Pleistocene. This speculation was based upon the wide distribution of the genus and the geographical isolation of the several species which nevertheless remain morphologically similar. The experimental studies of *S. zeschi* have yielded data which support this contention. Because of the stability which characterizes the cave environment, it seems likely that there would often be a diminution or loss in a long-evolving troglobite of those responses adaptive primarily to the troglobite's ancestor in its variable epigean environment. Retention of such response abilities may, then, provide valuable clues to the relative ages of troglobites, especially where related species may be compared. Thus *Sphalloplana percoeca* would seem to be an older troglobite than the Texas sphalloplanids because it has a slow righting time, is intolerant of light, and displays no oriented movements when exposed to moving water or to light. By contrast, *Sphalloplana zeschi* would seem to be a more recent troglobite because it is characterized by a rather rapid righting time, negative rheotaxis and phototaxis, ability to withstand moderate exposure to light, and a well-defined temperature preferendum.

ACKNOWLEDGMENTS

I wish to thank Mr. and Mrs. Kurt Zesch for permitting me to study in the interesting cave located on their ranch near Mason, Texas. From time to time several persons assisted me with various aspects of the fieldwork and the laboratory studies, and I sincerely appreciate their efforts. They are Mrs. Ann Barton, Mr. Jerry Cooke, Mr. William Elliott, Miss Thais Gordon, and Mrs. Suzanne Wiley.

BIBLIOGRAPHY

Beauchamp, P. de. 1931. In C. Bolívar and R. Jeannel, Campagne Spéologique de C. Bolívar et R. Jeannel dans l'Am érique du Nord (1928). Biospeologica 56. 2. Turbellariés Triclades. *Arch. Zool. Exptl. Gen.*, **71**(3):317–331.

Buchanan, J. W. 1936. Notes on an American Cave Flatworm, *Sphalloplana percaeca* (Packard). *Ecology*, **17**:194–211.

Bull, E., and R. W. Mitchell. 1972. Temperature and Relative Humidity Responses of Two Texas Cave Adapted Millipedes, *Cambala speobia* (Cambalida: Cambalidae) and *Speodesmus bicornourus* (Polydesmida: Vanhoeffeniidae). *Int. J. Speleol.*, **4**:365–393.

Cole, L. C. 1962. A Closed Sequential Test Design for Toleration Experiments. *Ecology*, **43**(4):749–753.

Holsinger, J. R. 1967. Systematics, Speciation, and Distribution of the Subterranean Amphipod Genus *Stygonectes* (Gammaridae). *Bull. U.S. Natl. Museum*, 259, 176 pp.

Hyman, L. H. 1937. Studies on the Morphology, Taxonomy, and Distribution of the North American Triclad Turbellaria. VIII. Some Cave Planarians of the United States. *Trans. Am. Microscop. Soc.*, **56**:457–477.

———. 1939. North American Triclad Turbellaria. X. Additional Species of Cave Planarians. *Trans. Am. Microscop. Soc.*, **58**:276–284.

———. 1945. North American Triclad Turbellaria. XI. New, Chiefly Cavernicolous, Planarians. *Am. Midland Naturalist*, **34**(2):475–484.

———. 1954. North American Triclad Turbellaria. XIII. Three New Cave Planarians. *Proc. U.S. Natl. Museum*, **103**(3333):563–573.

Ichikawa, A., and M. Kawakatsu. 1967. Records of Two Planarian Species of the Family Kenkiidae from Japanese Subterranean Waters. *Arch. Hydrobiol.*, **63**(4): 512–519.

Kawakatsu, M., and W.-J. Kim. 1967. Results of the Speleological Survey in South Korea. 1966. VI. Freshwater Planarians from Limestone Caves of South Korea. *Bull. Natl. Sci. Museum Tokyo*, **10**(3):247–262.

Kenk, R. 1970. Freshwater Triclads (Turbellaria) of North America. III. *Sphalloplana weingartneri* New Species, from a Cave in Indiana. *Proc. Biol. Soc. Wash.*, **83**(29):313–320.

Livanow, N. A., and Z. I. Zabusova. 1940. Paludicola of Lake Telezkoe and New Data on Several Siberian Forms. *Tr. Obshch. Estestvoisp. Kazan Gosud. Univ.*, **56**(3–4):83–159. (In Russian)

Mitchell, R. W. 1968. New Species of *Sphalloplana* (Turbellaria; Paludicola) from the Caves of Texas and a Reexamination of the Genus *Speophila* and the Family Kenkiidae. *Ann. Speleol.*, **23**(3):597–620.

——— and J. R. Reddell. 1971. "The Invertebrate Fauna of Texas Caves." In E. L. Lundelius and B. H. Slaughter (eds.), *Natural History of Texas Caves*, pp. 35–90. Dallas, Texas: Gulf Natural History. 174 pp.

——— and R. E. Smith. 1971. Some Aspects of the Osteology and Evolution of the Neotenic Spring and Cave Salamanders (*Eurycea*, Plethondontidae) of Central Texas. *Texas J. Sci.*, **23**(3):341–362.

Packard, A. S. 1879. *Zoology for Students and General Readers*. New York: Henry Holt. 719 pp.

Reddell, J. R. 1965. A Checklist of the Cave Fauna of Texas. I. The Invertebrata (Exclusive of Insecta). *Texas J. Sci.*, **17**(2):143–187.

———. 1966. A Checklist of the Cave Fauna of Texas. II. Insecta. *Texas J. Sci.*, **18**(2):25–56.

———. 1970a. A Checklist of the Cave Fauna of Texas. IV. Additional Records of Invertebrata (Exclusive of Insecta). *Texas J. Sci.*, **21**(4):390–415.

———. 1970b. A Checklist of the Cave Fauna of Texas. V. Additional Records of Insecta. *Texas J. Sci.*, **22**(1):47–65.

Vandel, A. 1964. *Biospéologie, la biologie des animaux cavernicoles*. Paris: Gauthier-Villars. 619 pp.

Chapter 17

Reproductive Ecology of the Polyclad Turbellarian *Notoplana acticola* (Boone, 1929) on the Central California Coast

Alan B. Thum
Oregon State University, Marine Science Center,
Newport, Oregon

Research dealing with the ecology of North American polyclad flatworms has been generally confined to aspects of oyster predation; yet the systematics of the group is fairly well understood. Work on the Pacific coast is limited to studies on the oyster predator *Pseudostylochus ostreophagus* in Puget Sound (Smith, 1955; Woelke, 1954, 1956), comments on polyclads by Bock (1925), and incidental observations by Heath and McGregor (1912) and Freeman (1933). Studies on the east coast of North America have also dealt mostly with polyclad oyster predation (Pearse and Wharton, 1938; Loosanoff, 1956).

The ecology of reproduction in *Notoplana acticola* (Boone, 1929) was studied over a 13-month period and included a definition of the physical and biological environmental conditions in the study area, a description of the habitat, and investigations on density, zonation, behavior, and feeding. Special attention was given to a seasonal analysis of reproduction, a relative estimate of age, and relationships between length and width, size and eyes, substrate, zonation, and reproduction.

AREA AND METHOD OF STUDY

The study was conducted along the rocky shore of Bodega Bay, 67 km north of San Francisco at 38°15′N and 123°58′W. The area includes approximately 2.4 km of coastline and is bordered on the north by the creek Estero de San Antonio and on the south by the sandy area of Dillon Beach, Marin County, California (Fig. 1).

Geologically, the area may be defined as a wave-eroded secondary coast (Shepard, 1963) with prominent irregularities of heterogeneous material, produced largely by wave erosion. Numerous rocky outcrops extending seaward for several hundred feet divide occasional stretches of sandy or rock rubble beach. The rocks in the area are highly metamorphosed undifferentiated sandstones of the Franciscan group (Daetwyler, 1966).

A series of offshore bars protects the shore from the full force of the sea, defining the area as a "protected outer coast" (Ricketts, Calvin, and Hedgpeth, 1968). The shore is depositional from May through December, and erosional from January through April. The average depth of Bodega Bay is 16.8 m and depth contours generally parallel the shoreline. Shoreward, contours are more irregular and when coupled with the west-northwest direction of predominant swell, result in waves washing across the rock outcrops obliquely to the normal of the shoreline, producing a southerly longshore drift and sorting of sediment (Cherry, 1964). The average wave height ranges from 0.9 to 1.5 m. The tides are mixed and occur semidiurnally with a mean tide level of 0.8 m, a mean tide range of 1.1 m, and a mean high high water of 1.6 m, with the zero reference at mean low low water.

The local climate is characterized by wet and dry seasons, with most of the precipitation occurring from November to February. Air temperatures are maximal during summer and autumn (mean daily maximum in October: 18.9°C), and minimal during winter and spring (mean daily minimum in February: 5.3°C). Upwelling, coincident with prevailing west-northwest winds, cools the surface waters from April through July. Monthly mean sea surface temperatures taken at a shore station 11 mi north of Bodega Bay range from 9.2 to 13.5°C.

Ten sites for collection transects were chosen throughout the region from an aerial photograph. The distribution of polyclads throughout the area was considered homogeneous. A transect site exhibiting rock rubble

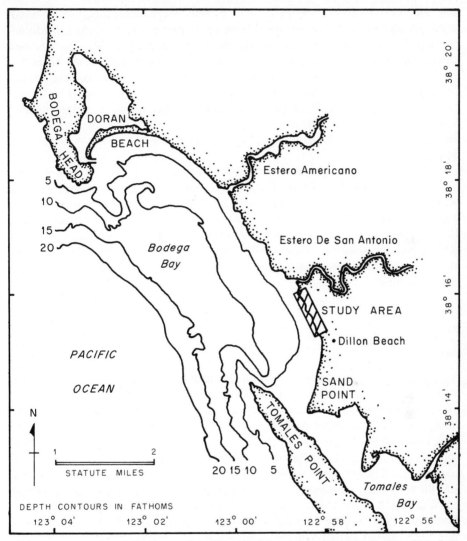

Figure 1   Location of study area (hatched) in Bodega Bay, California. Depth contours are given in fathoms.

throughout the intertidal was selected in the field by inspection of the station area. A "line-of-sight" transect was established and the undersides of all rocks within a meter-wide band were inspected for specimens.

Field data routinely included rock size, substrate character, animal density, zone, etc. In the laboratory live animals were measured. Specimens were fixed in mercuric chloride and stained with Mayer's hemalum and eosin, and the reproductive state and ocelli density were determined from fixed whole mounts or sections. No specimens smaller than 10 mm were considered because it was impossible to identify them with certainty. *N. sciophila* occurred sporadically in the study area and could be distinguished from *N. acticola* only by careful inspection of the copulatory apparatus. The tentacular and cerebral eyespot clusters were distinguished and individual eyes counted on the right half of each organism with the aid of a camera lucida. Specific identity was confirmed by comparison of sectioned specimens with the type material of *N. acticola* obtained from the California Academy of Science.

The sampling program covered a 13-month period, and the sample size ranged from 24 to 180 animals. Lack of data for certain months necessitated a seasonal analysis. Standard statistical methods were used where sufficient material was available and a normal distribution could be assumed (Snedecor, 1956). Nonparametric methods were employed in cases of small or large sample size or when normality could not be assumed (Siegel, 1956). In both methods, the test of significance used was Student's "T" test at the 0.05 level of significance.

RESULTS

Habitat

N. acticola was found commonly on the underside of cobbles and boulders in the rocky intertidal in an area which received a regular tidal exchange. The prevailing composition of the underlying substrate was poorly sorted coarse sand, or coarse sand and pebbles (Table 1). Individuals tend to aggregate or seek rock crevices at low tide, and may cluster about the base of the turban snail *Tegula funebralis* in the upper intertidal.

Eighty percent of the total of 426 specimens collected inhabited boulders and the remainder occupied cobbles, according to the size classification defined by Wentworth (1922). However, during winter more worms were recovered from cobbles than from boulders.

The annual density of individuals under boulders and cobbles did not differ significantly except during spring, 1966, when average boulder densities exceeded cobble densities; 5.78 and 2.67 worms per rock, respectively.

N. acticola lives between 0.0- and 5.0-ft tide levels (Fig. 2). A total of 209 animals were collected from the high intertidal (2.0 to 5.0 ft) and 217 from the low intertidal (0.0 to 2.0 ft), but during spring more animals were found in the high intertidal than the low intertidal.

Worms were found to have a horizontal distribution coincident with the sorting of sediment across rocky outcrops; greater densities were found with poorly sorted sediments.

Table 1   Analysis of Sediment Size from the Horizontal Zones across a Rocky Outcrop at Bodega Bay, California

Sieve size, mm	Percent weight of total fraction			
	Break zone	Rock zone	Rip zone	Beach zone
2.380	40.2	65.3	8.4	5.9
0.969	28.7	13.7	9.2	2.6
0.701	8.5	4.1	9.4	2.1
0.515	5.0	2.7	22.7	8.7
0.208	8.9	6.4	45.4	72.5
0.208	8.7	7.8	4.7	8.1
Sorting coefficient	4.80	2.50	.95	.71

Size

Size, represented by length and width, was measured on live animals. Average values of both dimensions increased from spring through summer and decreased thereafter to winter. The frequency distribution of length appeared to be bimodal in autumn and early spring, 1967 (Fig. 3).

An "average" bilateral worm 23.31 by 4.19 mm possessed an average of 45.37 cerebral and 23.35 tentacular eyes on the right side. The numbers of cerebral and tentacular eyes were significantly correlated ($n = 25$, $p = 0.01$) and their number increased linearly (Fig. 4). They were also significantly correlated with the length measure of size ($n = 25$, $p = 0.01$). The number of ocelli, therefore, increase with animal size, providing means of estimating relative age (Fig. 5).

Worms on low intertidal boulders were significantly larger (mean = 25.73 mm) than those on high intertidal boulders (mean = 23.03 mm), and tended to inhabit boulders rather than cobbles. The high intertidal autumn and winter worms decreased in size rapidly, while the decrease in the low intertidal animals was gradual.

Reproduction

The reproductive state of each individual was classified on the basis of visible evidence of gonads or gametes. Individuals showing identifiable testes or sperm were classified as male; those showing ovaries or eggs as female; and those showing gonads or gametes of both sexes simultaneously were classified as hermaphrodites (Table 2).

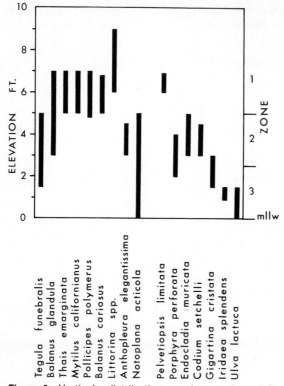

Figure 2 Vertical distribution and zonation *(after Ricketts, Calvin, and Hedgpeth, 1968)* of *Notoplana acticola* and common rocky shore, marine organisms, in the study area of Bodega Bay (feet above mean low low water).

Throughout the year the majority of the individuals were hermaphrodites (68.31 percent), and males (22.77 percent) were more common than females (6.10 percent), while individuals lacking identifiable gonads or gametes (2.82 percent) were rare. Of the males, 13.38 percent possessed testes showing evidence of active spermatogenesis, 8.69 percent had both active testes and sperm, and only 0.70 percent had sperm alone. Females were poorly represented with 1.41 percent bearing ripe ovaries, 4.46 percent with both ripe ovaries and eggs in the oviduct, and 0.23 percent only possessing eggs in the oviduct. The more common reproductive states amongst hermaphrodites were: testes and ovaries both ripe (7.28 percent), ripe testes and ovaries with sperm present (11.27 percent), ripe testes and ovaries with eggs present (8.92 percent), and both gonads active and both kinds of mature gametes present (39.20 percent). (Percentage breakdowns based on total population.)

Males bearing some ripe testes and sperm occurred throughout the year, being least abundant in spring and becoming most abundant in autumn and winter. Females were present only in spring. Hermaphrodites with ripe testes, ovaries, and mature sperm were most frequent in winter, and those with mature eggs most frequent in spring. Individuals with both ripe gonads and mature eggs and sperm were most frequent in late spring and summer.

The percentage of male worms increased from a low of 10 percent in spring to 50 percent in autumn and winter, and then decreased. The hermaphrodites remained more abundant than either males or females throughout the year and increased in frequency from 50 percent in autumn and winter to 75 percent in late spring and summer.

In assessing the amount of eggs and of sperm available in the population throughout the year, male and hermaphrodite conditions and female and hermaphrodite conditions were combined for analysis. The percentage of individuals with both ripe testes and sperm remained high throughout the year, and decreased rapidly in early spring. The percentage of animals with ripe ovaries and eggs increased from early spring through summer and declined thereafter to a low in winter. Egg production seems to precede sperm production in early spring. This is opposite to the sequence found by Gamble (1910) in *Leptoplana tremellaris*.

The presence of sperm in the seminal receptacle or oviduct was taken as evidence of copulation. Throughout the year nearly one-third of the animals

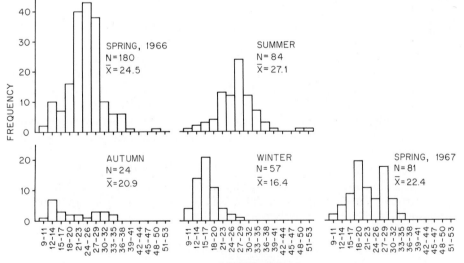

Figure 3 Seasonal size frequency distributions of length of *Notoplana acticola* at Bodega Bay, California.

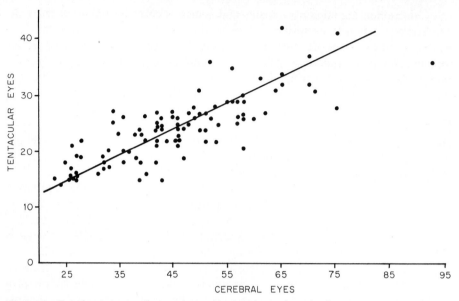

Figure 4 Relation between the numbers of tentacular and cerebral ocelli in *N. acticola* of varying lengths.

showed such evidence of copulation, with the exception of autumn, when the percentage rose to 62 percent. Copulation was more common in the lower intertidal during most of the year, except for early spring, when conditions were reversed.

The reproductive states of animals of different sizes were combined to determine if significant differences existed. Males with ripe testes and sperm were no larger than individuals with ripe testes. However, females with ripe ovaries and eggs were significantly larger than individuals with only ripe ovaries. Females and hermaphrodites were significantly larger than males. Females with active ovaries were larger than hermaphrodites with ripe gonads and sperm present. Females with active gonads and eggs were larger than hermaphrodites with active gonads or with sperm present or absent. No distinction in size could be made between individuals with active gonads and those with sperm in addition; however, animals with eggs present, with sperm present or absent, were always larger than animals without eggs, with sperm present or absent. Animals with active gonads and both gametes were larger than those of the same state, but lacking sperm. It appears then, that there is a significant change in size of animals producing eggs, which is absent in individuals producing sperm.

Males changed little in size from autumn through winter, then increased significantly in size from winter to spring and spring to summer. Hermaphrodites increased significantly in size from winter through summer, then decreased through autumn to winter.

The percentage of male individuals was consistently larger in the high intertidal than in the low intertidal, and hermaphrodites were generally more numerous in the lower intertidal.

DISCUSSION

Habitat

The habitat of *N. acticola* is apparently limited to rock rubble underlain by poorly sorted sediment, as few worms were ever found occupying rocks embedded in medium or well-sorted sediment. Adherence to a direct mode of embryological development may be expected to reduce mortality of young by decreasing dispersal to unfavorable well-sorted habitat conditions. The lack of a free larval stage does not prevent dispersal since adults are capable of swimming.

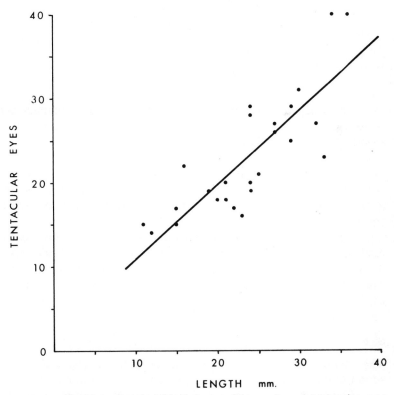

Figure 5 Relationship between the number of tentacular eyes and length (in millimeters) in *N. acticola*.

Table 2   Possible Reproductive States in *Notoplana Acticola*

Not reproductive:
    Lacking copulatory apparatus
    Immature copulatory apparatus
    Mature copulatory apparatus
Reproductive:
  Male:
    Testes ripe
    Testes ripe + sperm
    Sperm
  Female:
    Ovaries ripe
    Ovaries ripe + eggs
    Eggs
  Hermaphrodite:
    Testes ripe + ovaries ripe
    Ovaries ripe + sperm
    Testes ripe + eggs
    Sperm + eggs
    Testes ripe + ovaries ripe + sperm
    Ovaries ripe + testes ripe + eggs
    Testes ripe + sperm + eggs
    Ovaries ripe + eggs + sperm
    Testes ripe + ovaries ripe + sperm + eggs

The finding of a high percentage of worms on boulders and the great improbability of active selection of rock size by individuals suggest that rock size influences survival in the high-energy environment of the rocky intertidal. The higher percentage of worms under cobbles during winter may be accounted for by the interaction of other factors, such as age, density, and reproduction. Apparent bimodality in the seasonal size frequency curves indicates the appearance of juveniles and males in winter, and perhaps occupation of more cobbles.

The occurrence of more individuals in the high intertidal only during early spring is puzzling. However, appearance and redistribution of juveniles, dispersal, selective mortality, or varied local topography may cause this. There is some evidence to suggest that the appearance of juveniles occurs earlier in the higher than in the lower intertidal.

Size

The character of the size frequency distributions in autumn and early spring supports the probability of the coexistence of several generations.

Polyclads have a limited capacity for regeneration (Child, 1904, 1905, 1910; Levetzow, 1939; Olmsted, 1922). In dealing with organisms which possess regenerative powers, measures of size are doubtful indicators of even relative age. However, if such measurements include or correlate with conservative tissues, size measures are justifiable indicators of relative age. Hyman (1951) noted, in reference to freshwater planarians, that nervous tissue may ultimately be affected by starvation, whereas the reproductive system degenerates promptly. Wheeler (1894) observed that the tentacle and brain region of the polyclad *Planocera inquilina* were the last tissues to degenerate. Since the type of ocellus disintegration alluded to by Hyman was never observed, enumeration of eyespots and correlation with size is presumed to be a valid indication of relative age.

The general lack of significance between animal size and rock size, with some exceptions, indicates that it is unlikely that any active, direct selection of substrate size exists.

The significant occurrence of larger and, therefore, older individuals in the lower intertidal suggests either that survival is greater in the lower intertidal or that the requirements for further growth are met by the environmental conditions more effectively. Increased longevity, slower growth rate (Dunbar, 1968), and a colder and more uniform temperature regime (Hyman, 1941) are possible mechanisms influencing reproductive state and activity. Moreover the attainment of a sufficient size may be necessary for a thermal reproductive response and for the onset of hermaphroditism (Hyman, 1941). The more rapid decrease in worm size in the high intertidal and increased abundance of juveniles may mean that young worms appear earlier in the high intertidal (Fig. 6).

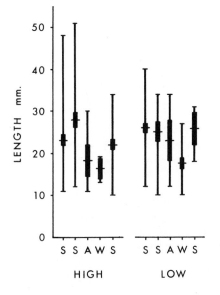

Figure 6 Seasonal analysis of the variation in length of *N. acticola* in the high and low intertidal, indicated by the range mean, and ±2 times the standard error.

Reproduction

The percentages of males and females never exceeded that of hermaphrodites, although the number of males nearly equaled the number of hermaphrodites in autumn and winter, while females were absent. Most of the male and female gametes may be expected to have been contributed by hermaphrodites. Sperm production by hermaphroditic animals increased from late spring through summer and decreased gradually during autumn through early spring. In contrast, sperm production by males rose slowly from late spring, reaching maximum production in autumn and winter, and fell off sharply in early spring. The younger male flatworms apparently function as a sperm reservoir during the slack autumn and winter production period of hermaphrodites (Fig. 7).

The production of eggs by hermaphroditic animals rose markedly from late spring through the summer and dropped suddenly through autumn to a winter low. A definite rise began again in early spring. Egg production by females appears insignificant as they composed only 6.1 percent of the population and were absent in autumn and winter. The female sex condition may in fact merely be produced by depletion and reduction of testes and sperm in older hermaphrodites, rather than being a separate sexual state. Eggs are laid chiefly during the summer. Hermaphrodites apparently produce eggs before sperm. A life-span of 12 months agrees with Gamble's (1910) estimate for

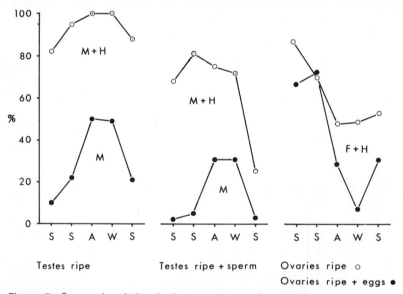

Figure 7 Seasonal variation in the percentage of male (M), female (F), and hermaphroditic (H) flatworms in single or combined reproductive states.

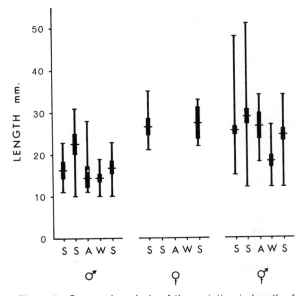

Figure 8 Seasonal analysis of the variation in length of the *N. acticola* represented by male, female, and hermaphrodite sex conditions.

Leptoplana tremellaris. The proposition that protogynous gametogenesis occurs in a marine flatworm appears to be new.

The presence of sperm in the seminal receptacle throughout the year suggests that either copulation occurs throughout the year or sperm is readily stored. The peak period of copulation, autumn, was also the period for maximum sperm production by males and just after that for hermaphrodites. Hermaphrodite production of eggs also preceded autumn. Copulation, therefore, probably takes place chiefly between summer and autumn, with the surmised time of egg laying in late spring, summer, and early autumn, which is similar to the sequence in *L. tremellaris* (Gamble, 1910).

The results of the analysis of the relationships between the size of worms and their reproductive status revealed that significant size variations exist between reproductive states, and these differences facilitated the interpretation of the reproductive cycle. Evidence was found indicating that oogenesis is probably a longer process than spermatogenesis. Significant differences in size of males and hermaphrodites (Fig. 8) suggests the existence of two generations. Autumn and winter males become mature hermaphrodites in spring and summer (Fig. 8). No evidence of degeneration of tissue of the magnitude necessary to account for the observed difference in size between winter and summer animals was ever found.

The presence of two generations might explain the difference in the zonal concentrations of the worms. The fact that males were more abundant in the high intertidal and hermaphrodites in the low intertidal suggests that the environment of the low intertidal may be conducive to the onset of hermaphroditism or to the survival of larger individuals and, therefore, hermaphroditism. More juveniles, potential males, on the other hand, may be produced in or dispersed to the high intertidal. Local topographic differences among the collection stations could also affect the distribution of different-sized animals and animals in various states of reproduction.

ACKNOWLEDGMENTS

I am indebted to Drs. Edmund Smith and Gary Brusca of the Pacific Marine Station of the University of the Pacific, Dillon Beach, California, for providing laboratory facilities, and to Dr. J. Gonor, Oregon State University, Marine Science Center, Newport, Oregon, for reading the manuscript. This research was supported by grant FWPCA-01061-01 during 1967.

BIBLIOGRAPHY

Bock, S. 1925. Papers from Dr. Th. Mortensen's Pacific Expedition 1914–1916, Planarians Parts I–IV. *Vidensk. Medd. Dansk Naturh. Foren.* Bd., **79**:1–84, 97–184.

Boone, E. S. 1929. Five new Polyclads from the California Coast. *Ann. Mag. Nat. Hist.*, **3**:3–46.

Cherry, J. 1964. Sand Movement along a Portion of the Northern California Coast. *Univ. Calif. Hydr. Eng. Lab. Tech. Rept.*, 150 pp.

Child, C. M. 1904a, b, c. Studies on Regulation. IV–VI. *J. Exptl. Zool.*, **1**:95–133.

———. 1905. Studies on Regulation. VII. *J. Exptl. Zool.*, **2**:253–285.

———. 1910. The Central Nervous System as a Factor in the Regeneration of Polyclad Turbellaria. *Biol. Bull.*, **19**:333–338.

Daetwyler, C. C. 1966. *Marine Geology of Tomales Bay, Central California*. University of the Pacific, Pacific Marine Station Research Report No. 6. 169 pp.

Dunbar, M. J. 1968. *Ecological Development in Polar Regions, A Study in Evolution.* Englewood Cliffs, N.J.: Prentice-Hall. 119 pp.

Freeman, D. 1933. The Polyclads of the San Juan Region of Puget Sound. *Trans. Am. Microscop. Soc.*, **52**(2):107–146.

Gamble, F. W. 1910. *Platyhelminthes and Mesozoa.* The Cambridge Natural History, vol. 2, pp. 3–50. London: Macmillan.

Heath, H., and E. A. McGregor. 1912. New Polyclads from Monterey Bay, California. *Proc. Acad. Nat. Sci. Phila.*, **64**:455–488.

Hyman, L. H. 1941. Environmental Control of Sexual Reproduction in a Planarian. *Anat. Record*, **81**:108.

———. 1951. *The Invertebrates.* II. *Platyhelminthes and Rhynchocoela, The Acoelomate Bilateria.* New York: McGraw-Hill. 550 pp.

Levetzow, K. 1939. Regeneration der polycladen Turbellarien. *Arch. Entwicklungsmech. Organ.*, **139:**780–818.

Loosanoff, V. L. 1956. Two obscure Oyster Enemies in New England Waters. *Science*, **123:**1119–1120.

Olmsted, J. M. D. 1922. The Role of the Nervous System in the Regeneration of Polyclad Turbellaria. *J. Exptl. Zool.*, **36:**49–56.

Pearse, A. S., and G. W. Wharton. 1938. The Oyster "Leech" *Stylochus inimicus* Palombi, Associated with Oysters on the Coasts of Florida. *Ecol. Monographs*, **8:**605–655.

Ricketts, E. F., J. Calvin, and J. W. Hedgpeth. 1968. *Between Pacific Tides.* Stanford, Calif.: Stanford University Press. 614 pp.

Shepard, F. P. 1963. *Submarine Geology.* New York: Harper & Row. 557 pp.

Siegel, S. 1956. *Nonparametric Statistics for the Behavioral Sciences.* New York: McGraw-Hill. 312 pp.

Smith, L. S. 1955. "Observations on the Polyclad Flatworm *Pseudostylochus ostreophagus.*" Unpublished master's thesis, University of Washington, Washington.

Snedecor, G. W. 1956. *Statistical Methods.* Ames: Iowa State College Press.

Wentworth, C. K. 1922. A Scale of Grade and Class Terms for Clastic Sediments. *J. Geol.*, **30:**377–392.

Wheeler, W. M. 1894. *Planocera inquilina,* a Polyclad Inhabiting the Branchial Chamber of *Sycotypus canaliculatus* Gill. *J. Morphol.*, **9:**195–201.

Woelke, C. E. 1954. A Newly Identified Oyster Predator. *Wash. Dept. Fish. Fish. Res. Papers,* **1**(2):1–2.

———. 1956. The Flatworm *Pseudostylochus ostreophagus* Hyman, a Predator of Oysters. *Proc. Natl. Shellfisheries Assoc.*, 1956(4):62–67.

7

Chapter 18

Analysis of Substances Inhibiting Regeneration of Freshwater Planarians

Etienne Wolff
Laboratoire D'Embryologie Expérimentale
Collège de France
Nogent-Sur-Marne, France

Planarians are very remarkable animals. Every part of the body can regenerate all the other parts of the organism. Thus the problem posed by regeneration is not the same as in other organisms. One must determine not only how any part is able to reconstitute the whole organism, but also why many organs of the same nature do not regenerate during the reconstruction; for example, why many brains, many eyes, many pharynges do not regenerate simultaneously or successively.

Following the results of several research workers in my laboratory, especially those of Dubois (1949), Lender (1950, 1952, 1956), Sengel (1951, 1953), Ziller-Sengel (1967a, b), and Fedecka-Bruner (1967, 1968), we were led to the conception that regeneration of organs in planarians is controlled on the one hand by inducing factors, on the other hand by inhibiting factors. Inducing factors are responsible for the reconstruction of organs; inhibiting factors prevent the same organ from regenerating several times in the same or in other parts of the body. This eventuality would happen in organisms in which every part is capable of reconstituting all the others, unless there were a mechanism which controls the regeneration process.

Experiments on the brain, on the eyes, and on the pharynx have proved that both kinds of factors play a part in the regeneration of planarians.

INDUCING FACTORS

I shall recall very briefly results concerning induction phenomena in the course of regeneration. They were established by several authors in three cases: eyes, pharynges, and copulatory organs (Fig. 1).

Induction processes consist of the unlatching of organ differentiation by another organ of a different nature, in the same way as in embryonic differentiation.

Induction of Eyes

Wolff and Lender (1950) and Lender (1950) have shown that eye differentiation depends on the brain (Fig. 1b). This differentiation is independent of nerves. It is exerted by the intermediary of substances which diffuse through the parenchyma. These substances are present in homogenates of cephalic regions, even in the lyophilized supernatant of these tissues (Fig. 3).

Induction of Pharynges

Induction of pharynges was studied successively by Santos (1929), Okada and Sugino (1934, 1937), Okada and Kido (1943), and P. Sengel (1951 and 1953), who have defined more and more accurately the sequence of inducing factors responsible for this differentiation. These researches led to the following conclusions: When a planarian is amputated from the region anterior to the genital pore, the head regenerates first. Then it induces the formation of the prepharyngeal region, which in its turn induces the pharyngeal zone, and finally this zone induces the pharynx (Fig. 1c, d, and e). Thus regeneration of this organ is caused indirectly by the head and the prepharyngeal region, and directly by the pharyngeal zone.

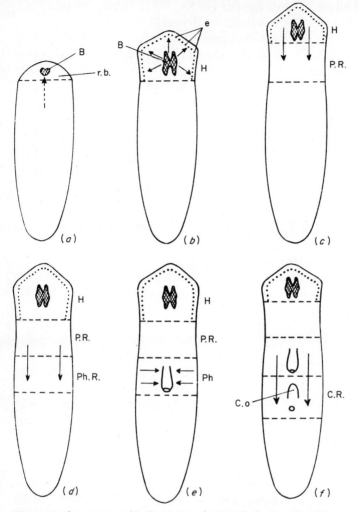

Figure 1 Sequences of induction processes during regeneration of the anterior region of planarians (type *Polycelis nigra*). (*a*) Determination of the head blastema by the remaining part. B, regenerating brain; r.b., regeneration blastema. (*b*) Induction of eyes, e, by the brain, B. H, head. (*c*) Induction of prepharyngeal region, P.R., by the head, H. (*d*) Induction of pharyngeal region, Ph.R., by the prepharyngeal zone, P.R. (*e*) Inducing effect of the pharyngeal region on the reconstruction of the pharynx, Ph. (horizontal arrows). (*f*) Induction of the copulatory organs, C.R., by testes situated in the anterior areas. C.o., genital aperture. (*After E. Wolff and T. Lender, 1962.*)

Induction of the Copulatory Apparatus

Experiments were done by Kenk (1941a, b) and Okugawa (1957) on hetero-genous grafts between sexual and asexual strains of *Dugesia tigrina, D. gonocephala,* or *D. japonica.* They have shown that gonads (especially testes) are probably the inductors of copulatory apparatus (Fig. 1*f*). Fedecka-Bruner in our laboratory (1961 and 1968) has confirmed this result by castrating individuals of *Dugesia lugubris* using the x-ray technique. The technique utilized is based on the fact that all testicular follicles are situated on the ventral side. Thus irradiations limited to the ventral layers of a planarian do not affect the whole body but are able to destroy the testes completely. This result is obtained by irradiating planarians with x-ray apparatus working under low tension (10 kV) and high intensity (10 to 25 mA). Under these conditions, the radiations are only feebly penetrating; they spare almost all tissues except the genital cells. The copulatory apparatus does not regenerate before a number of testicular follicles have regenerated.

INHIBITING FACTORS

On the other hand, there are factors which prevent disorders in the regenerating organism.

Brain

First I shall recall briefly the results of Lender (1960), working on brain regeneration. He demonstrated:

   1   That a head graft inhibits the regeneration of the normal brain, but not that of the eyes (Fig. 2).
   2   In the same way, a decapitated planarian, grown in a dish containing a head homogenate, does not regenerate its brain, whereas the eyes regenerate normally. In several planarians, no brain at all was regenerated after 12 days; in the others, a small brain was regenerated, measuring 50 μm in length on average, whereas the length of normally regenerating planarians was 320 μm on average (Fig. 3).

Pharynx

I should like to give some details concerning inhibitory factors intervening in pharynx regeneration which were recently studied by C. Ziller.

   We know that the regenerating pharynx is induced by the pharyngeal zone, which is itself induced by the cephalic region. When a pharynx is regenerating, it prevents regeneration of other pharynges, as was shown by C.

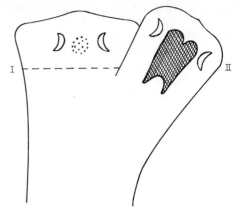

Figure 2 Result of graft of a supernumerary head (II) in the anterior region of the primary head amputated (I) from *Dugesia lugubris.* The regenerating blastema reconstitutes its eyes, but no brain. The dotted area in the regenerating blastema I schematizes a concentration of neoblasts which are not capable of differentiating, as a result of inhibition of the grafted head II. *(After E. Wolff, T. Lender, and C. Ziller-Sengel, 1964.)*

Ziller (1967a, b), by graft experiments, and also using homogenates. We know that, after removal of the pharynx, planarians such as *Dugesia lugubris, D. tigrina,* and *Polycelis nigra* reconstitute their pharynx after 7 to 10 days (Fig. 4). Planarians deprived of their pharynx were placed in water containing homogenates of different regions of the planarian: cephalic, pharyngeal, and caudal zones (Fig. 5). The general results were as follows, the inhibiting effect being evaluated according to the delay of regeneration (Fig. 6).

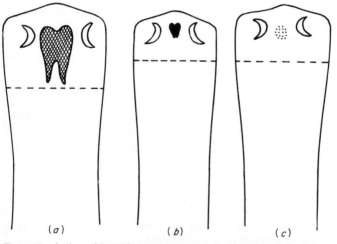

Figure 3 Action of head homogenates on the regeneration of the brain. (a) Controls: normal regeneration of the head. (b, c) Action of head homogenates: regeneration of a small brain (b); no brain regeneration (c). Note that in all cases the eyes are regenerated because head homogenates contain an eye-inducing factor. *(After E. Wolff, T. Lender, and C. Ziller-Sengel, 1964.)*

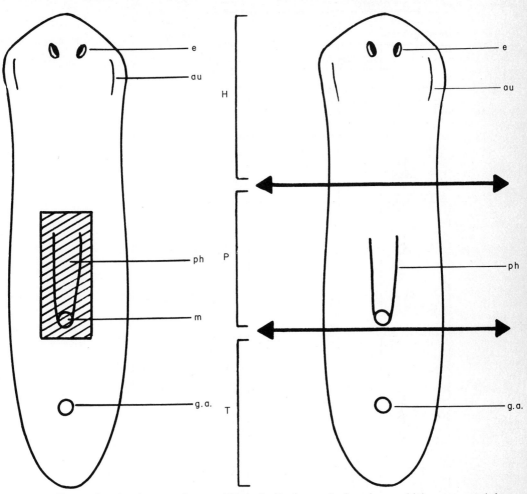

Figure 4 Removal of pharynx from *Dugesia*. au, auricles; e, eyes; g.a., genital aperture; m, mouth; ph, pharynx.

Figure 5 Regions of planarians, which were used for preparation of homogenates: H, head region; P, pharynx region; T, tail region. Other labels as in Fig. 4.

    1   Crude homogenates of the head and tail regions have no effects on regeneration (7 to 10 days).

    2   Crude homogenates of pharyngeal regions have a significant inhibitory effect (9 to 13 days).

    3   Homogenates of pharyngeal regions without pharynx have a greater inhibitory effect (12 to 14 days).

    4   Homogenates of pharyngeal regions prepared from planarians 1 to 3 days after operation have a less inhibitory effect.

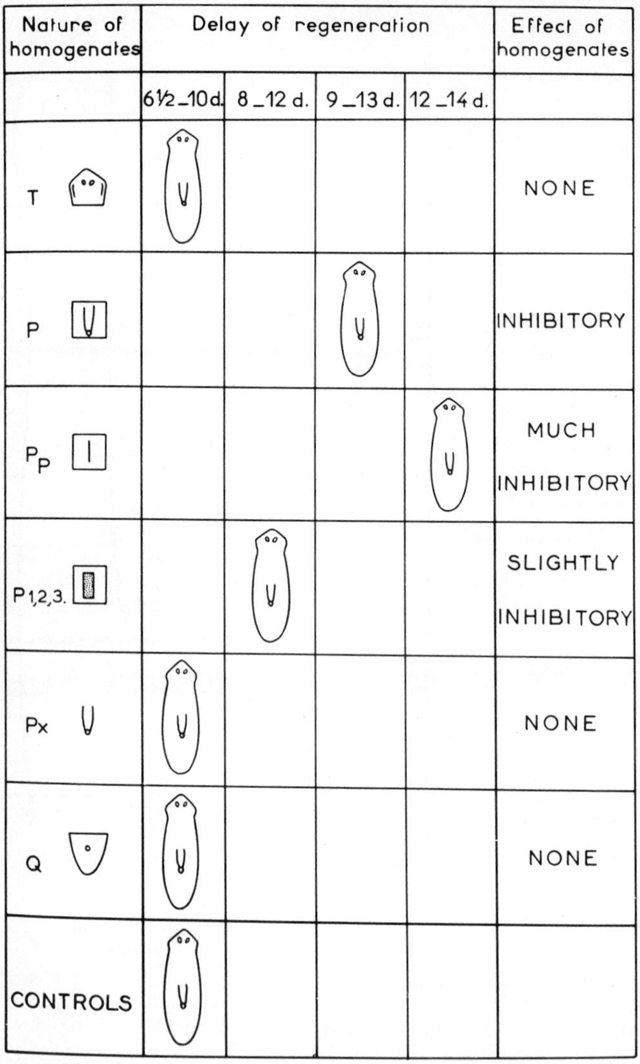

Nature of homogenates	Delay of regeneration				Effect of homogenates
	6½–10d.	8–12 d.	9–13 d.	12–14 d.	
T	✓				NONE
P			✓		INHIBITORY
P_P				✓	MUCH INHIBITORY
P1,2,3.		✓			SLIGHTLY INHIBITORY
Px	✓				NONE
Q	✓				NONE
CONTROLS	✓				

Figure 6 Inhibitory effect of different homogenates on pharynx regeneration: T, heads; P, pharyngeal zones; P_P, pharyngeal zones without pharynx; P1, 2, 3, pharyngeal zones 1, 2, 3 days after amputation; Px, pharynges; Q, tails. (*After C. Ziller-Sengel, 1967b.*)

These results demonstrate that the tissues of the *pharyngeal region* contain specific factors capable of delaying *pharynx* regeneration (Fig. 7).

C. Ziller, with D. Beaupain (unpublished results), undertook an analysis of the chemical nature of these inhibiting substances. The results are still in a preliminary stage, but the following date have been obtained at the present time:

    1   When a homogenate is prepared from two pharyngeal zones minced in 1 ml of water, this preparation contains the inhibitory factor. It is active at a concentration of 1/1 to 1/2. The activity decreases considerably when the homogenate is diluted (at a concentration of 1/20) (Fig. 8).

    2   After boiling for 5 min, the activity of homogenates decreases (Fig. 9).

    3   The inhibitory activity diminishes considerably if the homogenate is left for several days at a temperature of 12°C.

    4   The inhibiting factor is not dialyzable. After dialysis, the activity remains in the nondialyzable fraction, whereas the *dialyzate has no inhibitory activity* (Fig. 10).

    5   The effect of homogenates of pharyngeal zones is compared in each experiment with the effect of control homogenates of tails and heads. Crude extracts of tails have no influence on regeneration; the dialyzate and the dialyzed residue of tails are not inhibiting. On the other hand, *the dialyzed residue of pharynx has an inhibiting activity*. The dialyzate itself *seems to stimulate regeneration* (Fig. 10).

    The general results of these experiments are summarized in Table 1. Thus there are two factors in the pharynx homogenate: one, in the nondialyzable residue, is responsible for inhibition; the other, dialyzable, causes a stimulation of regeneration.

    That is to say, large molecules (with a molecular weight of more than 15,000) have an inhibiting effect, whereas small molecules are stimulatory.

    The analysis of these two fractions is now being continued by means of

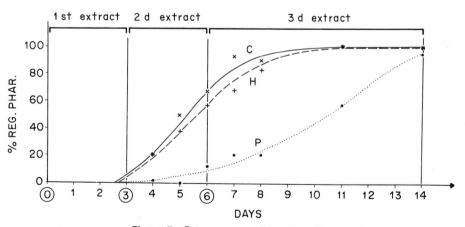

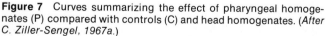

Figure 7 Curves summarizing the effect of pharyngeal homogenates (P) compared with controls (C) and head homogenates. (*After C. Ziller-Sengel, 1967a.*)

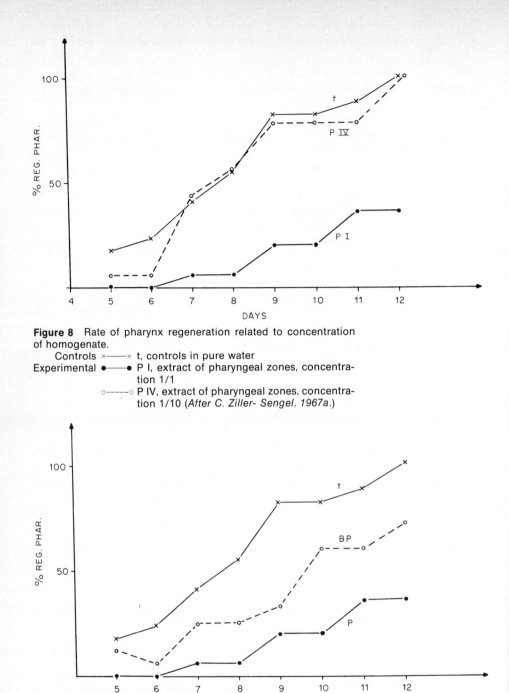

Figure 8 Rate of pharynx regeneration related to concentration
of homogenate.
        Controls ×——× t, controls in pure water
Experimental ●——● P I, extract of pharyngeal zones, concentra-
                       tion 1/1
                 o-----o P IV, extract of pharyngeal zones, concentra-
                       tion 1/10 (*After C. Ziller- Sengel, 1967a.*)

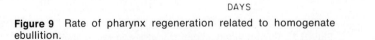

Figure 9 Rate of pharynx regeneration related to homogenate
ebullition.
        Controls ×——× t, controls in pure water
Experimental o-----o BP, boiled extract of pharyngeal zones, con-
                       centration 1/1
                 ●-----● P, unboiled extract of pharyngeal zones, con-
                       centration 1/1 (*After C. Ziller-Sengel, 1967a.*)

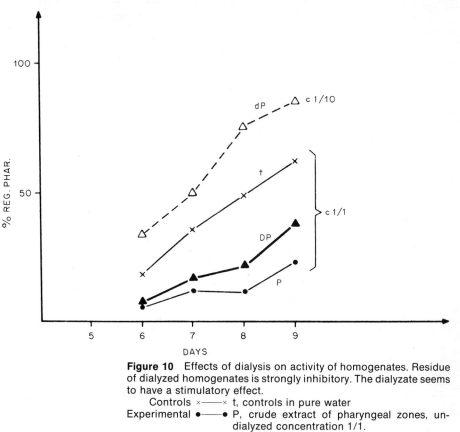

Figure 10   Effects of dialysis on activity of homogenates. Residue of dialyzed homogenates is strongly inhibitory. The dialyzate seems to have a stimulatory effect.
Controls ×———× t, controls in pure water
Experimental ●———● P, crude extract of pharyngeal zones, un-dialyzed concentration 1/1.
△-----△ dP, dialyzed extract of pharyngeal zones, concentration 1/1.
▲———▲ DP, residue of dialyze. (*From C. Ziller-Sengel, 1967a.*)

the usual methods of fractionation. Therefore, we cannot yet draw any conclusion as to the nature of these substances, but it is now possible to start biochemical investigation.

CONCLUSIONS

In conclusion, we can formulate the following hypothesis: Specific inhibitory substances are produced by a regenerating organ or region, and prevent this organ or this region from being regenerated several times. Thus regeneration is restricted to the reconstruction of the missing part.

Table 1 Effect of the Dialyzate and of the Dialyzed Residue of Pharyngeal Homogenates Compared with Controls

	Regeneration time (for 50% of the individuals)*		
	Inhibited regeneration (more than 11–12 days)	Normal regeneration (8–10 days)	Stimulated regeneration (less than 8 days)
Pure water (controls)		+	
Extract of pharyngeal zones			
Untreated extract normal dose 1/1	+		
Untreated extract reduced dose 1/10		+	
Extract aged or boiled		+	
Dialyzed extract	+		
Extract-dialyzate			+
Control extracts			
Tail extract dialyzed residue		+	
Tail extract dialyzate			+

*The mean duration of regeneration is given in terms of the number of days required for 50 percent of the animals in any group to regenerate their pharynx.

The regeneration of the missing parts of a planarian results from a balanced succession of inductions and inhibitions (Wolf and Lender, 1962; Wolf, Lender, and Ziller-Sengel, 1964) (Figs. 11 and 12). In the case of the anterior regeneration, the brain is regenerated by a process of self-differentiation. Then a sequence of inductions proceeding from the head toward the tail (eyes, prepharyngeal region, pharyngeal region, pharynx, copulatory apparatus) (Fig. 11a) is controlled by antagonistic factors which are present in the remaining part and which act in the opposite direction (Figs. 11b and 12b).

Child's theory of axial gradients (1920), with the concepts of dominance and of dominated regions, is confirmed. This theory can be explained by the

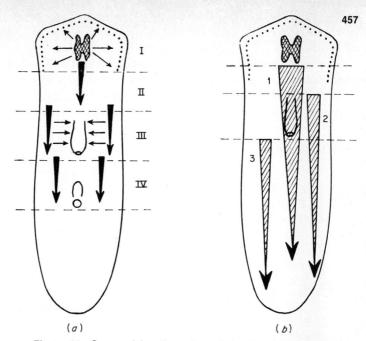

(a)   (b)

Figure 11 Summarizing the antagonistic effects of inducing factors (a) (see Fig. 1) and inhibitory factors (b) originating from different regions (brain, pharynx, copulatory organs).

In Fig. 11b, only inhibitory factors diffusing from cephalic toward posterior regions are represented. (From *E. Wolff and T. Lender, 1962.*)

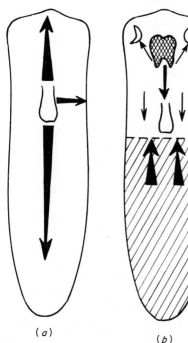

(a)   (b)

Figure 12 (a) Inhibitory substances diffuse from the pharyngeal region in all directions. (b) Antagonistic factors are represented in a planarian regenerating from its posterior part: inducing factors in the regenerating part, inhibitory factors (big arrows) in the tail. (*From E. Wolff, T. Lender, and C. Ziller-Sengel, 1964.*)

presence of specific inhibitory substances which diffuse from a source where their concentration is maximum to distant regions where their concentration is minimum (Fig. 12a). Child's physiological gradients can be interpreted in terms of gradients of diffusion of inhibitory substances and inductors.

BIBLIOGRAPHY

Child, C. M. 1920. Some Considerations concerning the Nature and Origin of Physiological Gradients. *Biol. Bull.*, **39**:147–187.

Dubois, F. 1949. Contribution à l'étude de la migration des cellules de régénération chez les Planaires dulcicoles. *Bull. Biol. France Belg.*, **83**:213–283.

Fedecka-Bruner, B. 1961. La régénération de l'appareil copulateur chez la Planaire *Dugesia lugubris*. *Arch. Anat. Microscop. Morphol. Exptl.*, **50**:221–231.

———. 1967. Etudes sur la régénération des organes génitaux chez la Planaire *Dugesia lugubris*. I. Régénération des testicules aprés destruction. *Bull. Biol. France Belg.*, **101**:255–319.

———. 1968. Etudes sur la régénération des organes génitaux chez la Planaire *Dugesia lugubris*. II. Régénération de l'appareil copulateur. *Bull. Biol. France Belg.*, **102**:4–44.

Kenk, R. 1941a. Induction of Sexuality in the Asexual Form of *Dugesia tigrina*. *J. Exptl. Zool.*, **87**:55–69.

———. 1941b. Induction of Sexuality in *Dugesia tigrina* by Transplantation. *Anat. Record*, **79**: suppl. 12.

Lender, T. 1950. Démonstration du rôle organisateur du cerveau dans la régénération des yeux de *Polycelis nigra* par la méthode des greffes. *Compt. Rend. Soc. Biol.*, **144**:1407–1409.

———. 1952. Le rôle inducteur du cerveau dans la régénération des yeux d'une Planaire d'eau douce. *Bull. Biol. France Belg.*, **86**:140–215.

———. 1956. Analyse des phénomènes d'induction et d'inhibition dans la régénération des planaires. *Ann. Biol.*, **32**:457–471.

———. 1960. L'inhibition spécifique de la différentiation du cerveau des Planaires d'eau douce en régénération. *J. Embryol. Exptl. Morphol.*, **8**:291–301.

Okada, Yô, and T. Kido. 1943. Further Experiments on Transplantation in Planaria. *J. Fac. Sci. Univ. Tokyo Sect. IV*, **6**:1–23.

——— and H. Sugino. 1934. Transplantation Experiments in *Planaria gonocephala*. I and II. *Proc. Imp. Acad. Japan*, **10**:36–40, 107–110.

——— and ———. 1937. Transplantation Experiments in *Planaria gonocephala* Dugès. *Japan. J. Zool.*, **7**:373–439.

Okugawa, K. 1957. An Experimental Study of Sexual Induction in the Asexual Form of Japanese Fresh-water Planarian, *Dugesia gonocephala* Dugès. *Bull. Kyoto Gakugei Univ.*, **11**:8–27.

Santos, V. 1929. Studies on Transplantations in Planaria. *Biol. Bull.*, **57**:188–198.

Sengel, P. 1951. Sur les conditions de la régénération normale de pharynx chez la Planaire d'eau douce *Dugesia lugubris*. *Bull. Biol. France Belg.*, **85**:376–391.

———. 1953. Sur l'induction d'une zone pharyngienne chez la Planaire d'eau douce

Dugesia lugubris O. Schm. *Arch. Anat. Microscop. Morphol. Exptl.,* **42:**57–66.

Wolff, E., and T. Lender. 1950. Sur le rôle organisateur du cerveau dans la régénération des yeux chez une Planaire d'eau douce. *Compt. Rend. Acad. Sci.,* **230:**2238–2239.

—— and ——. 1962. Les néoblastes et les phénomènes d'induction et d'inhibition dans la régénération des Planaires. *Ann. Biol.,* (4)**1:**499–529.

——, ——, and C. Ziller-Sengel. 1964. Le rôle de facteurs auto-inhibiteurs dans la régénération des Planaires (une interprétation nouvelle de la théorie des gradients physiologiques de Child). *Rev. Suisse Zool.,* **71:**75–98.

Ziller-Sengel, C. 1967a. Recherches sur l'inhibition de la régénération du pharynx chez les Planaires. I. Mise en évidence d'un facteur auto-inhibiteur de la régénération du pharynx. *J. Embryol. Exptl. Morphol.,* **18:**91–105.

——. 1967b. Recherches sur l'inhibition de la régénération du pharynx chez les Planaires. II. Variations d'intensité du facteur inhibiteur suivant les espéces et les phases de la régénération. *J. Embryol. Exptl. Morphol.,* **18:**107–119.

Chapter 19

The Role of Neurosecretion in Freshwater Planarians

T. Lender
Université de Paris, Faculté des Sciences
Paris-Sud Orsay, France

The presence of neurosecretory cells was put into evidence by Lender and Klein (1961) in the freshwater planarians *Polycelis nigra*, *Dugesia lugubris*, and *Dendrocoelum lacteum*. These results have been confirmed on other species by other investigators: Vendrix (1963), Ude (1964), Grasso (1965a, c), Bondi (1966), Liotti, Bruschelli, and Rosi (1966), Liotti (1968), and Liotti and Rosi (1968).

In 1950, Wolff and Lender established that the brain emits substances during the regeneration of eyes in *Polycelis nigra*. But at the present time it is impossible to prove that these inductors are real neurosecretory substance.

Once the presence of the neurosecretory cells was established, we tried to analyze its role during the freshwater planarians' life. Three problems have been studied: neurosecretory substance and regeneration; neurosecretory substance and gonad development; and neurosecretory substance and asexual reproduction.

EVIDENCE OF THE NEUROSECRETORY CELLS

The nervous cells of the freshwater planarians form the external layer of the cerebral ganglia. We can find them also along the nerve cords. Some of these cells contain a secretion stained by paraldehyde fuchsin or chrome-hematoxylin phloxine after permanganic oxidation (Fig. 1). There are many of these cells in the posterior part of the brain. In some cases the neurosecretory material drain-off can be noticed along the axons (Fig. 2).

The neurosecretory granules of 1300 Å diameter can be seen with the electron microscope inside the cytoplasm of the nervous cells (Fig. 3) and the nervous fibers (Fig. 4). According to Morita and Best (1965) the granules in *Dugesia dorotocephala* have a diameter of 400 to 1100 Å.

The neurosecretory activity depends on the environment. Ude (1964) shows a daily cycle in *Dendrocoelum lacteum* with a maximum by 6 P.M. and an annual cycle with a maximum in August.

NEUROSECRETORY SUBSTANCE AND REGENERATION

We cut a planarian behind the ovaries. At 18°C the regeneration is finished after 7 days. In the anterior segment the number of neurosecretory cells increases from the first day of regeneration and reaches a maximum by the second or third day (Fig. 5). Afterwards the number of neurosecretory cells decreases to the level of the controls (Lender, 1963, 1965). It appears then, there is a relation between the number of neurosecretory cells and the tail regeneration. This fact was proved by Ude (1964). The neurosecretory substance is most copious when the differentiation starts (Fig. 6).

Grasso (1965b, 1966b), Bondi (1966), Liotti (1968), and Liotti *et al.* (1966, 1968) claim that neurosecretory substance plays a part in regeneration. Neurosecretory material would always exist even during the regeneration of embryos.

Study with the light microscope does not show the neurosecretory substance during head regeneration. But with the electron microscope, Mme Monnot in my laboratory pointed out the neurosecretory substance inside the nervous cells of the nerve cords (Sauzin-Monnot, Lender, and Gabriel, 1970).

After the sixth hour of regeneration (Fig. 7) neurosecretory granules are copious inside the nervous cells behind the blastema. The nervous fibers

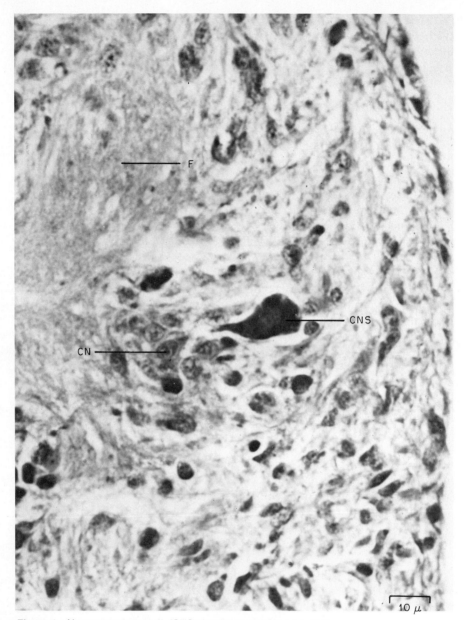

Figure 1 Neurosecretory cell (CNS) in the posterior part of the brain of *Polycelis nigra* stained with paraldehyde fuchsin. F, nervous fiber; CN, nervous cell.

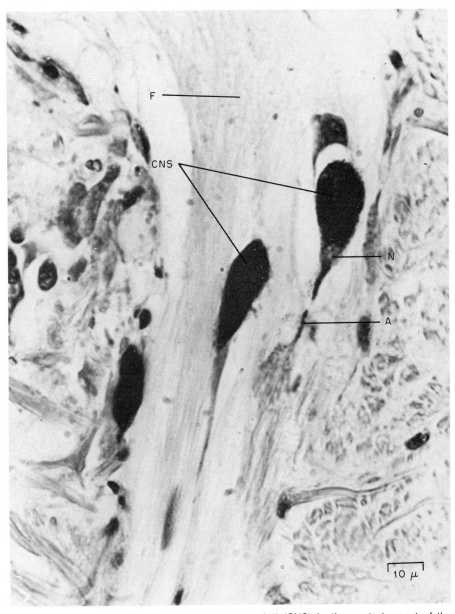

Figure 2 Neurosecretory cell (CNS) in the posterior part of the
brain of *Dendrocoelum lacteum* stained with paraldehyde fuchsin.
A, axon with neurosecretion; F, nervous fiber; N, nucleus.

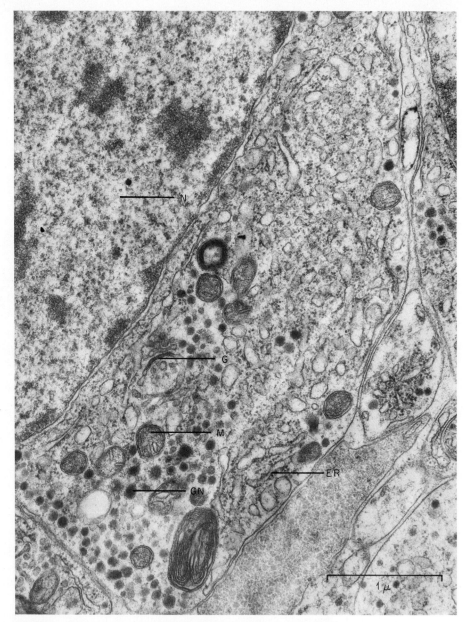

Figure 3 Electron micrograph of portion of a neurosecretory cell of *Dugesia gonocephala* which is not regenerating. ×31,000. ER, ergastoplasm; G, Golgi; GN, neurosecretory granule; M, mitochondria; N, nucleus.

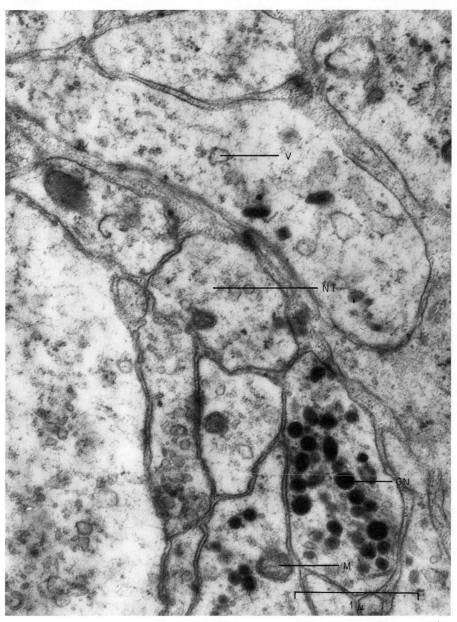

Figure 4 Electron micrograph of the neurosecretory granules (GN) in the nerve fibers of *Dugesia gonocephala*. ×33,000. M, mitochondria; NT, neurotobule; v, vesicle.

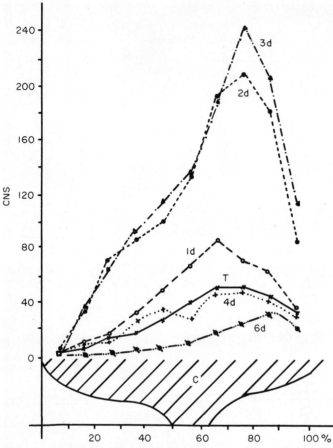

Figure 5 Variation of the number of neurosecretory cells in the
brain of *Polycelis nigra* during the regeneration of a tail. C, profile
of half a nervous ganglion; T, curve of controls; 1d to 6d, curves
after 1 to 6 days of regeneration. Abscissa, percentage of the total
length of the brain; ordinate, number of the neurosecretory cells
(CNS).

which go into the blastema are well filled with them. After 48 hours of regeneration the neurosecretory granules can still be seen.

Neurosecretory substance and regeneration are then certainly bound
together. But we do not know whether neurosecretory substance is the ground
for regeneration or which stage of the regeneration is promoted. The wound
stimulus could instigate the releasing of the neurosecretory substance. This
could promote the migration of regeneration cells (Grasso 1965b).

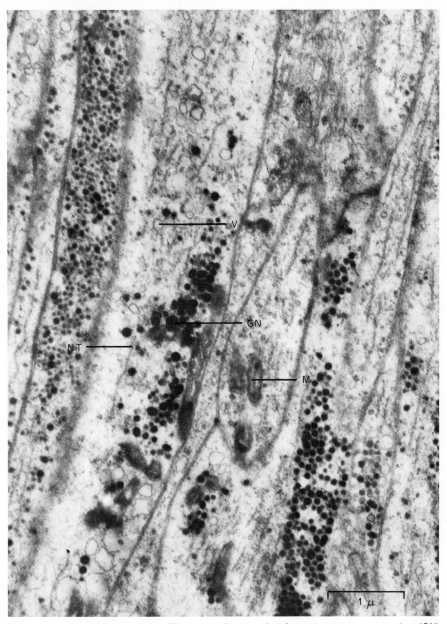

Figure 6 Electron micrograph of neurosecretory granules (GN) in the nerve fibers of *Polycelis nigra* behind a tail blastema 6 hours old. ×21,000. M, mitochondria; NT, neurotubule; v, vesicle.

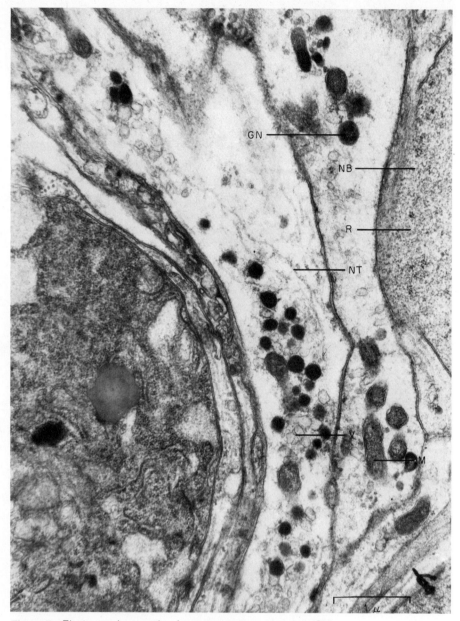

Figure 7 Electron micrograph of neurosecretory granules (GN) in the nerve fibers in connection with a head blastema 24 hours old by *Dugesia gonocephala*. ×21,000. M, mitochondria; NB, neoblast; NT, neurotubule; R, ribosome; v, vesicles.

NEUROSECRETORY SUBSTANCE
AND GONAD DEVELOPMENT

The results are still partial. The number of neurosecretory cells increases during the laying of the eggs by *Dendrocoelum lacteum* (Fig. 8). The neurosecretory cells are always very easy to identify in *Polycelis nigra*, which lays eggs in the aquarium all the year round (Fig. 9). Fedecka-Bruner (1967, 1968) claimed that the gonads and the genital tract maintain themselves in the presence of the brain. Grasso (1965b, 1966a, b) established that the gonads are much innervated, and in his opinion the neurosecretory substance would take part in the gametes' ripening.

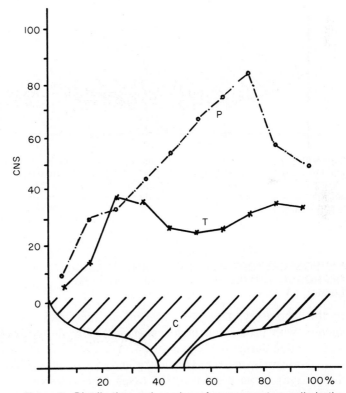

Figure 8 Distribution and number of neurosecretory cells in the brain of *Dendrocoelum lacteum* during the laying (P). C, profile of half a nervous ganglion; T, curve of controls. Abscissa, percentage of the total length of the brain; ordinate, number of neurosecretory cells (CNS).

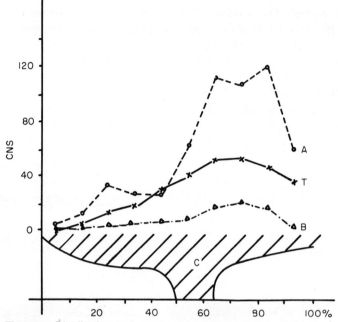

Figure 9 Distribution and number of neurosecretory cells in the brain of *Polycelis nigra* during A and after B the laying. T, curve of controls; C, profile of half a nervous ganglion. Abscissa, percentage of the total length of the brain; ordinate, number of neurosecretory cells (CNS).

NEUROSECRETORY SUBSTANCE AND ASEXUAL REPRODUCTION (Lender and Zghal, 1968, 1969; Lender, 1970).

In the laboratory, *Dugesia gonocephala* multiplies itself by sexual reproduction with production of cocoons especially during spring or by asexual reproduction during the whole year.

During asexual reproduction, planarians divide by a fission level behind the pharynx in 94 percent of the cases. The scissiparity is cyclical, and the cycle duration changes with the seasons. In spring and summer the cycle lasts on an average of 21 days. In autumn the cycle lasts longer, and in January-February planarians are not scissiparous. Temperature and illumination are factors which condition these fluctuations. A temperature increase or maintenance in darkness promotes the scissiparity.

For the study of the relation between scissiparity and neurosecretion, planarians are maintained at 16 to 18°C in semidarkness. The scissiparity cycle lasts then for an average of 21 days.

The comparison of the scissiparity cycles of both segments descended from the same planarian shows that the scissiparity of the posterior segment is delayed an average of 5.5 days compared to the anterior segment. The posterior segment is brainless after the division. The brain regenerates in 5 days; this corresponds to the scissiparity delay. That is why we investigated the role of the brain during asexual reproduction.

The brain of planarians is removed at the beginning of the scissiparity cycle. A part of the tail is removed at the same moment on the controls. The fission of the controls is delayed 2 days; the fission of the brainless planarians is delayed 8 days. If the brain is removed every 4 days, the scissiparity of these planarians disappears during the 2 months of the experiment. If these removals are stopped, the scissiparity reappears. It is then possible to conclude that the brain controls asexual reproduction.

Does the neurosecretion of the brain control the scissiparity? We studied the changes in neurosecretion during the scissiparity cycle (Fig. 10). The number of nervous cells with neurosecretory substance decreases quickly after fission. On the sixth day there are only spoors of neurosecretory substance. Afterwards the number of neurosecretory cells increases very quickly with a maximum 1 or 2 days before the fission and is followed by a drastic decrease. If a brain rich in neurosecretory cells is transplanted on a planarian with a brain poor in neurosecretion, the scissiparity cycle is decreased by 3 to 6 days.

The scissiparity cycle is accompanied by a neurosecretory cycle. The fission takes place after the neurosecretory substance is discharged. There exists then a relation between the two phenomena. The neurosecretion starts the scissiparity.

How is the neurosecretion acting? Scissiparity influences regeneration. After fission each part regenerates the missing parts. Gabriel (1970) showed that the regeneration started by an activation of neoblasts which synthetize RNA and later proteins. The same synthesis can be studied during the scissiparity by incorporation of tritiated uridine and leucine.

The synthesis of RNA starts by the tenth day of the cycle and increases progressively until the fission in the neoblasts of the scissiparity area (Fig. 11). The curve is displaced when compared with the neurosecretion curve, but the increase is parallel. On the contrary, the protein synthesis begins only after the fission and the regeneration of the missing part occurs.

The neoblasts of the planarians are activated during the second part of the scissiparity cycle. The activation prepares the regeneration which starts after the fission. The neurosecretion and the RNA synthesis activities develop in parallel. It is inferred that the neurosecretion acts on the neoblasts and induces RNA synthesis necessary to the next regeneration.

We may conclude that the regeneration, the ripening of the gametes, and the asexual reproduction of the planarians happen in parallel with a neurosecretion increase. As far as scissiparity is concerned, the necessity of the neurosecretory substance is proved by way of experiment. The neurosecretion might promote RNA synthesis in the neoblasts, before the fission during the scissiparity, in the blastema during traumatic regeneration.

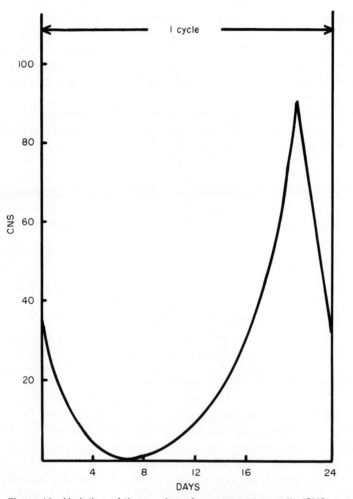

Figure 10 Variation of the number of neurosecretory cells (CNS) during a scissiparity cycle of 21 days' length by *Dugesia gonocephala.*

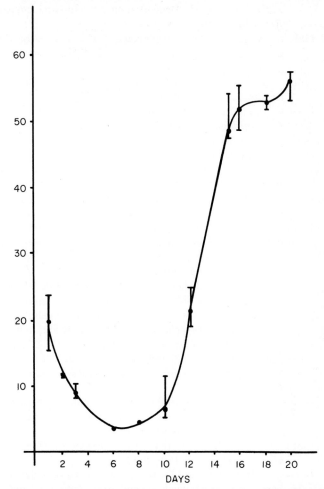

Figure 11 Incorporation curve of tritiated uridine (20 μCi/ml during 15 hours) in the scissiparity area during a cycle by *Dugesia gonocephala*. Abscissa, days of the cycle; ordinate, number of silver grains per cell.

BIBLIOGRAPHY

Bondi, C. 1966. Osservazioni sui rapporti tra neurosecrezione e rigenerazione in *Dugesia lugubris*. *Acta Embryol. Morphol. Exptl.*, **9**:88.

Fedecka-Bruner, B. 1967. Etudes sur la régénération de organes génitaux chez la Planaire *Dugesia lugubris*. I. Régénération des testicules après destruction. *Bull. Biol. France Belg.*, **101**:255–319.

———. 1968. Etudes sur la régénération des organes génitaux chez la Planaire *Du-*

gesia lugubris. II. Régénération de l'appareil copulateur. *Bull. Biol. France Belg.*, **102**:4–44.

Gabriel, A. 1970. Etude morphologique et évolution biochimique des néoblastes au cours des premières phases de la régénération des Planaires d'eau douce. *Ann. Embryol. Morphol. Fr.*, **3**:49–69.

Grasso, M. 1965a. Presenza e distribuzione delle cellule neurosecretrici in *Dugesia lugubris. Monitore Zool. Ital.*, **73**:182–187.

——. 1965b. Prime indagini sul significato funzionale della neurosecrezione in *Dugesia lugubris* e *Dugesia tigrina. Boll. Zool.*, **32**:1037–49.

——. 1965c. Dimostrazione di cellule neurosecretrici in *Dugesia tigrina. Atti Accad. Nazl. Lincei Rend. Classe Sci. Fis. Mat. Nat.*, **38**:712–714.

——. 1966a. Rapporti fra sistema nervoso, gonadi e neurosecrezione in *Polycelis nigra. Riv. Biol. Ital.*, **59**:157–172.

——. 1966b. Sui fenomeni di neurosecrezione durante la rigenerazione di dischetti isolati di *Dugesia lugubris. Arch. Zool. Ital.*, **51**:327–335.

Lender, T. 1963. Mise en evidence et rôles de la neurosécrétion chez les Planaires d'eau douce (Turbellariés, Triclades). *Gen. Comp. Endocrinol.*, **3**:716–717.

——. 1965. Mise en évidence et rôle de la neurosécrétion chez les Planaires d'eau douce (Turbellariés, Triclades). *Ann. Endocrinol. Fr. Suppl.*, **25**:61–65.

——. 1970. Le rôle de la neurosécrétion au cours de la régénération et de la reproduction asexuée des Planaires d'eau douce. *Ann. Endocrinol. Fr.*, **31**:463–466.

——— and N. Klein. 1961. Mise en évidence de cellules sécrétrices dans le cerveau de la Planaire *Polycelis nigra*. Variation de leur nombre au cours de la régénération postérieure. *Compt. Rend. Acad. Sci. Fr.*, **253**:331–333.

——— and F. Zghal. 1968. Influence du cerveau et de la neurosécrétion sur la scissiparité de la Planaire *Dugesia gonocephala. Compt. Rend. Acad. Sci. Fr.*, **267**:2008–2009.

——— and ———. 1969. Influence des conditions d'élevage et de la neurosécrétion sur les rythmes de scissiparité de la race asexuée de *Dugesia gonocephala. Ann. Embryol. Morphol. Fr.*, **2**:379–385.

Liotti, F. S. 1968. Rapporti tra rigenerazione in *Dugesia lugubris. Acta Embryol. Morphol. Exptl.*, **10**:199–200.

——, G. Bruschelli, and G. Rosi. 1966. Variazioni dell'attività neurosecretoria durante la rigenerazione in exemplari di *Dugesia lugubris* trattati o non con simpamina. *Riv. Biol. Ital.*, **59**:353–384.

——— and G. Rosi. 1968. Osservazioni sui diversi stadi del ciclo neurosecretorio in *Dugesia lugubris*. Ricerche su exemplari interi e durante la rigenerazione. *Riv. Biol. Ital.*, **61**:419–432.

Morita, M., and J. B. Best. 1965. Electron Microscopic Studies on Planaria. II. Fine Structure on the Neurosecretory System in the Planarian *Dugesia dorotocephala. J. Ultrastruct. Res.*, **13**:396–408.

Sauzin-Monnot, M.-J., T. Lender, and A. Gabriel. 1970. Etude autoradiographique ultrastructurale des néoblastes dans le blastéme de régénération de 24 h. et dans les tissus en arrière du blastème, chez la Planaire *Dugesia gonocephala* (Turbellarié, Triclade). *Compt. Rend. Acad. Sci. Fr.*, **271**:1892–1895.

Ude, J. 1964. Untersuchungen zur Neurosekretion bei *Dendrocoelum lacteum* Oerst. (Plathelminthes-Turbellaria). *Z. Wiss. Zool.*, **170**:223–255.

Vendrix, J. J. 1963. Existence de cellules neurosécrétrices chez *Polycelis nigra* Ehrenberg et *Dugesia gonocephala* Dugès (Triclades Paludicoles). Caractéristiques cytologiques et histochimiques. *Bull. Soc. Roy. Sci. Liége*, **2**:293–303.

Wolff, E., and T. Lender, 1950. Sur le rôle organisateur du cerveau dans la régénération des yeux chez une Planaire d'eau douce. *Compt. Rend. Acad. Sci. Fr.*, **230**: 2238–2239.

Chapter 20

Fissioning in Planarians from a Genetic Standpoint

Mario Benazzi
Department of Zoology, University of Pisa, Italy

Asexual reproduction by fission is frequent in freshwater triclads of the family Planariidae. In most of the species it consists in an architomic mechanism, i.e., the animal divides transversally into two pieces which then regenerate the missing parts. (General references in Hyman, 1951; de Beauchamp, 1961; Brøndsted, 1969.)

Obviously fissioning may be found only in species which possess a high regenerative power (it is therefore lacking in the species of the family Dendrocoelidae). There are, however, some species which, even though they dis-

476

play remarkable power of regeneration, never show asexual reproduction; moreover in one species there may be races or populations which multiply only sexually or only by fission.

Fissioning is always correlated to an asexual condition, which represents a peculiar characteristic of the fissiparous animals; this had been noticed by Dugès in *Planaria subtentaculata* as long ago as 1828. Divisions in sexual specimens may sometimes occur, but they are cases of accidental autotomy and not real asexual reproduction.

Fissioning as a reproductive mechanism may enter into the biological cycle of the species in various ways. There are races which are only fissiparous and races with both sexual and fissiparous individuals which may coexist or alternate during the course of the year. Curtis (1902) studied populations of *Planaria maculata* (= *Dugesia tigrina*) in the vicinity of Falmouth, Massachusetts, and found that in one locality they reproduced only sexually, in another exclusively by fission, and in two others both by cocoons and by fission at different seasons. The populations which display sexual cycles breed in early spring and lay many cocoons; the copulatory apparatus then disappears and the planarians reproduce by fission.

Races with these different reproductive modalities were found in palaearctic and nearctic species of the genera *Dugesia, Fonticola, Polycelis,* and in the European *Crenobia alpina*. The importance of fissioning in the survival of the populations varies greatly; in some it is the only means of reproduction, in others a temporary phenomenon with little reproductive value (see, for example, Jenkins, 1967, on *Dugesia dorotocephala*).

With regard to the respective importance of the environmental and internal factors, there have been many different opinions. Certainly from the data collected in nature the influence of external factors may appear predominant. However, laboratory cultures have shown that the different characteristics of the various races remain unchanged also in the new condition: Kenk, 1937 for *D. tigrina*; Benazzi, 1936, 1938, 1942 for *D. gonocephala* s. l.; Okugawa and Kawakatsu, 1954a, b, 1956, 1957 for *D. japonica*; this proves that they are genetically controlled.

All living things are the products of both "nature and nurture"; this fundamental concept is fully valid also for asexual reproduction in planarians; therefore the influence of the external factors (above all temperature) on the life cycle cannot be underestimated. I also emphasize the importance of the physiological mechanisms which control fissioning, clearly demonstrated by Child (1911), Lender and Zghal (1968), and Best *et al.* (1969). The aim of my research, however, is to give some data about the genetic basis of this phenomenon, and I believe it is not necessary to discuss the other aspects.

FISSIONING IN *Dugesia gonocephala* s. l.

The research on the genetic control of fissioning that I have been working on for many years has been accomplished on species of the "*Dugesia gonocephala* group." As I have shown in previous papers (cf. Benazzi, 1960), *D. gonocephala* s. l., corresponding to a superspecies or *Artenkreis*, includes many species (or "microspecies") similar in external moropholgy to *D. gonocephala* (Dugès) of central Europe, but differentiated by morphologic characteristics, more or less evident, of the copulatory apparatus, and reproductively isolated. In some of these species, which are spread over the Mediterranean area, fissioning is very frequent. In particular, *D. benazzii* from Corsica and Sardinia is suitable for this research since there are races with both sexual and fissiparous specimens. The sexual specimens are quite fertile, and it is therefore possible to establish whether their offspring become sexual or fissiparous and in what ratio.

The characteristics of all these different races remain unchanged in laboratory cultures; this confirms that they are genetically controlled.

I have already pointed out that the fissiparous individuals are always devoid of the sexual apparatus; the copulatory organ is lacking, and this permits one to recognize them even *in vivo*. The microscopical examination on microtomical sections confirms that the reproductive apparatus is completely undeveloped, although in some individuals traces of gonads and the copulatory organ may be present (Benazzi, 1938). This condition remains also in individuals which, owing to a temporary arrest of fissions, become very large (Fig. 1*a*, *b*, *c*). It is therefore evident that the asexual condition does not depend on a delay in the regeneration of the reproductive apparatus following a previous division, but it is a peculiarity of the fissiparous individuals. As a matter of fact, sexual individuals which are artificially divided always regenerate the reproductive organs and reach sexual maturity.

Vandel (1921) found that in *Polycelis cornuta* (= *Polycelis felina*) the regeneration of the gonads is much slower (4 to 5 months) in comparison with their development in specimens born from cocoons (1 to 2 months). He admitted that the planarians are at the stage in which gonad regeneration is still possible but has become slow and difficult, and thought that this could explain the fact that the species in which fissioning appeared generally become asexual. In 1938 I pointed out that the interpretation of the distinguished French zoologist did not explain the persistence of the asexual state in the fissiparous specimens. I can now add a further and more evident confirmation that the asexual state depends on genetic factors. As I shall show later in this paper, sexual specimens of fissiparous strains produce offspring which may become either sexual or fissiparous. The former show precociously the development of the copulatory organ, while the latter remain asexual even when the

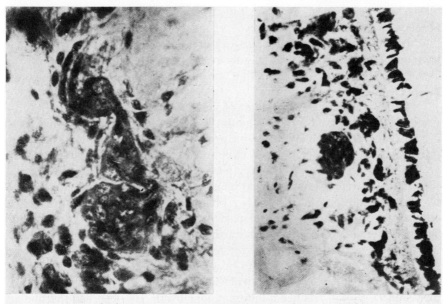

(a)   (b)

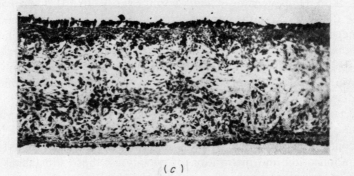

(c)

Figure 1 *Dugesia benazzii*: large asexual specimen showing traces of (a) ovaries, (b) testicles, (c) copulatory organ. (*From Benazzi, 1938.*)

first division takes place quite a long time after birth and when, therefore, they have become very large. Some specimens remain asexual for even one or more years although they never divide.

In my opinion *the asexual state of the fissiparous specimens depends on the same genetic factors which induce the division;* the primary effect exerted by these factors is likely to consist in preventing the transformation of neoblasts into germ cells. The lack of development of the gonads causes the absence of the other reproductive organs. In fact Vandel (1920) suggested that the development of the copulatory organ is determined by an influence of hormonal type exercised by the gonads. Recent research by Fedecka-Bruner (1968) on *D. lugubris* confirms that the copulatory organ regenerates only in the presence of testes.

From a physiogenetical point of view, the problem is, therefore, how the fission-controlling factors may inhibit the transformation of neoblasts into germ cells. I do not believe that our present knowledge permits a satisfactory interpretation. However, Kenk's experiments on *D. tigrina* (1941) and Okugawa's on *D. japonica* (1957) may perhaps give some pertinent suggestions; they united a fragment of a sexual planarian and a fragment of an asexual one by grafting, and in some cases they obtained the development of gonads and copulatory organ in the asexual part of the composite animal. Evidently a sexual stimulus was transmitted by the sexual fragment, but it is not clear whether this sexual induction is due to hormonal substances or to the migration of neoblasts which are able to become germ cells. Okugawa believed the latter improbable, while Vowinckel (1970) on the basis of his own experiments on *D. tigrina* reached a different conclusion. He observed that the germ cells (testicular primordia) are recognizable in newly hatched animals of a sexual strain. Homogenates of adult worms exposed to a 10°C drop in temperature for 24 hours have the capacity to stimulate an increase in the number of aggregated germ cells; this experimental procedure, however, is inefficient on an asexual strain in which germ cell aggregates are missing. Therefore Vowinckel believes it is more probable that the results obtained by Kenk and Okugawa are due to the invasion of cellular elements coming from the transplanted sexual fragment.

I have not accomplished this type of research and, therefore, I cannot formulate an opinion; I think, however, that Vowinckel's data are interesting. They confirm that hormonal substances (probably neurosecretions) can stimulate multiplication and maturation of gonocytes; these substances, however, have no influence on fissiparous specimens in which a peculiar genetical constitution inhibits the formation of the gonocytes. We will see in the following section how in fissiparous strains some individuals succeed in overcoming this inhibition and attain sexual maturity.

THE EX-FISSIPAROUS SPECIMENS

As we have already mentioned, in fissiparous strains some specimens may reach sexual maturity; these specimens, which I call ex-fissiparous, have been very useful in my research. It is not easy to understand why fissiparous specimens become sexual. They appear, in laboratory cultures, among other specimens which continue to divide; therefore the influence of external factors can be excluded, at least as the principal cause. We can suppose that at a certain point the inhibiting action on the development of the sexual organs exercised by fission-controlling factors is overcome.

Most of the ex-fissiparous lose their ability for spontaneous division; moreover, if they are artificially cut, they regenerate the reproductive apparatus; this shows that they have definitely acquired the sexual state. In some populations, however, many ex-fissiparous return to fissioning, at least temporarily.

It is worth noting that the ability to produce ex-fissiparous varies greatly in the different populations. Some of them never breed sexual specimens, and some do so in an extremely low number, e.g., below 0.5 percent of all the cultured animals; moreover, these few ex-fissiparous are completely or almost completely sterile. All these populations can of course survive only by means of asexual reproduction. In other populations, on the contrary, the sexual specimens are much more frequent and lay cocoons which are quite fertile, although their fertility is inferior to the average of the populations which reproduce only sexually (Benazzi, 1942).

There exists, therefore, a relationship between the frequency of ex-fissiparous and their fertility, and it is likely that all these manifestations are due to the action of the fission-controlling factors which are still present in the ex-fissiparous. The differences observed with regard to the frequency and fertility of the ex-fissiparous specimens may be explained by supposing that either the number or penetrance of these factors varies a great deal among different populations.

With the aim of giving a clearer idea of the above I summarize the data referring to some of the populations which I have studied in laboratory cultures.

1   Camogli (Ligurian Riviera). A population which reproduces exclusively by fission; in the laboratory culture no specimen showed signs of sexual evolution. The neoblasts have a set of 24 chromosomes (except for some mitoses with one or two extra elements) which apparently correspond to a triploid condition with respect to the basic number, which is 8.

This planarian had already been found many years ago by Borelli in the same area of the Ligurian Riviera (Rapallo) and referred to as *Planaria subtentaculata* Draparnaud; his student Lobetti Bodoni (1918) illustrated the modality of fission. The specific name *subtentaculata* should not be maintained since it can be applied to all fissiparous strains of the various species (or microspecies) of the "*D. gonocephala* group."

2   Borjas del Campo, Terragona (Spain). These planarians are similar to the previous ones; in the culture no specimen showed signs of gonads. The neoblasts have a set of 24 chromosomes.

3   Tavolara (a small island in the Gulf of Olbia, Sardinia). The specimens collected in two wells were all asexual, and in the laboratory cultures they reproduced exclusively by fission. Many specimens, however, showed an initial development of the ovaries, which were discernible in living individuals; furthermore three or four of them reached sexual maturity (indicated by the development of the copulatory organ), but none laid cocoons. The somatic set is 24 chromosomes.

Presumably this planarian belongs to *D. benazzii*, which is widespread in Sardinia and Corsica. It is interesting to point out that in Molara (another small island near Tavolara) I found a diploid population of *D. benazzii* which is only sexual.   .

4   Island of Ponza (Tyrrhenian sea). Some asexual specimens were found in a little stream. In laboratory culture they reproduced vigorously by fission; some of them, however, showed signs of ovaries and a few reached sexual maturity, laying a small number of sterile cocoons. The somatic set is 24 to 28 chromosomes, and the oocytes have a variable number of bivalents and univalents which are not always easily recognized; this set is approximately triploid or perhaps highly aneuploid.

The histological examination of the copulatory apparatus showed that this planarian belongs to *D. sicula*, a species of the "*D. gonocephala* group."

5   Island of Marettimo (Egadi archipelago, west of Sicily). Two asexual specimens collected in 1968 produced in the laboratory culture numerous fissiparous offspring. In many of them signs of ovaries appeared, but very few became sexual. The only cocoon laid showed oocytes with bivalents and univalents besides some chromatic fragments; this indicates a highly aneuploid set.

The histological examination of the copulatory organ showed that this planarian belongs to *D. sicula*.

6   Algerian populations. Asexual planarians of the "*D. gonocephala* group" were collected in four different places. In the first period of culture they remained asexual, then some specimens (particularly those from "Ruissean des Singes" near Blida) showed a tendency toward sexuality; some of them developed the copulatory organ but did not lay cocoons. Planarians of the *gonocephala* type from Algeria were also reared by de Beauchamp (1961), but he was not able to obtain sexual specimens. Thanks to my sexual specimens, it will be possible to gain a more exact knowledge of the taxonomic rank of the Algerian planarians.

7   Jordan River (Israel). The culture was begun in 1964 with some asexual specimens which divided intensively. Later on some of them became sexual, reaching very large dimensions. Their fertility, however, was almost nil, since the few cocoons laid were sterile, except one which hatched and gave a planarian. This planarian divided and originated a new fissiparous clone; many of these individuals then became sexual but were unable to produce fertile cocoons. The fact that the only specimen born sexually became fissiparous clearly demonstrates the transmission of the fission-controlling

factors through the gametes. However, the percentage of individuals which originated by fission from this specimen and which then became sexual is much higher than the average of the population; this shows that in the sexually born specimens the factors inducing sexuality were more frequent or had greater penetrance.

This planarian from the Jordan River certainly belongs to the "*D. gonocephala* group," but it is probably a separate species or "microspecies" which is now being studied. Its chromosome set is near triploid or more probably aneuploid.

8   Rio Viglietu (Sardinia). This population, belonging to *D. benazzii*, is diploid or aneuploid. Both sexual and fissiparous specimens are present; the former lay a large number of cocoons, but their fertility is rather low owing to a high embryonal lethality (Benazzi, 1942; Benazzi and Benazzi Lentati, 1945). The sexually born specimens may become either sexual or fissiparous; moreover, sexual individuals may originate also from fissiparous lines (ex-fissiparous specimens). This planarian was used for crosses with a sexual race, and we shall refer to this in the next section.

9   Castello Pino (Corsica). This population also belongs to *D. benazzii* and shows the same characteristics as the previous one. On this planarian I am making genetic research about which I shall speak later.

10   Ragusa (Sicily). This planarian, belonging to *D. sicula*, has both sexual and fissiparous specimens; the fertility of the former is rather low because most of their eggs do not become fertilized. However, starting with a few sexually born specimens, I have obtained offspring in which fissioning seems to have disappeared, while their fertility has increased greatly. Evidently a strain has been selected in which the fission-controlling factors have been either completely eliminated or repressed.

The different reproductive behavior of the populations is evident. Populations 1 and 2 give examples of strains in which sexuality seems to be completely lost. Populations 3 to 7 also multiply only by fission; however, they show a certain tendency toward sexuality, since in many individuals traces of gonads appear; the few of them which reach sexual maturity lay either unfertile cocoons or none at all. Finally, populations 8, 9, and 10 show both modalities of reproduction; the fertility of the sexual specimens is lower than that of the average of the populations which reproduced only sexually, but they may give offspring which assure the continuity of sexual reproduction.

Now, it is necessary to examine with more detail the reproductive apparatus of the ex-fissiparous, particularly of those which are completely or almost completely sterile.

The copulatory organ and the vitellaria may fully develop. The testes are numerous although frequently they do not reach complete maturity, and therefore there are few ripe sperms. Ovaries, on the contrary, show great anomalies; they become very large and spread over a wide area of the anterior part of the animal. Histological examination reveals a very large number of oocytes which, however, are blocked at the prophase of the first meiotic division (Figs. 2 to 6).

Obviously there is an intense multiplication of neoblasts and oogonia which then become oocytes, but very few of these oocytes reach the prometaphase I, i.e., the stage in which they leave the ovaries and migrate into the oviducts. A close examination of these oocytes at an ultrastructural level is being conducted by Dr. Gremigni.

The accumulation of oocytes may, in some cases, produce pathological situations, i.e., degeneration of the cephalic region of the animal. In my opinion, this hyperplasia of the female germ cells depends not only on the fact that none or only a few of the oocytes leave the ovary, but above all on a more intense transformation of neoblasts into oogonia. Therefore the situation of the fissiparous specimens, in which the transformation of neoblasts into germ cells is inhibited, seems to be inverted in the ex-fissiparous specimens.

It is necessary to emphasize again that this phenomenon is evident only in ex-fissiparous specimens which are almost completely sterile. Ex-fissiparous specimens which are fertile possess normal ovaries. In these cases, however, other causes of sterility may occur as we have already seen: particularly a high embryo mortality (Benazzi, 1942; Benazzi and Benazzi Lentati, 1945).

The sexual conditions of the ex-fissiparous specimens are therefore very peculiar; certainly they depend on fission-controlling factors, but it is difficult to find a satisfactory interpretation.

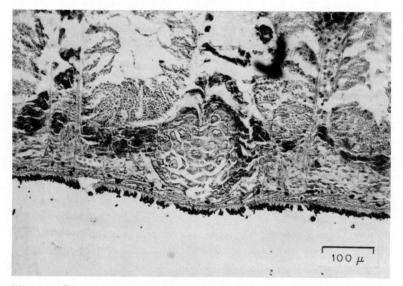

100 μ

Figure 2 *Dugesia benazzii*: ovary of a specimen from a sexual strain.

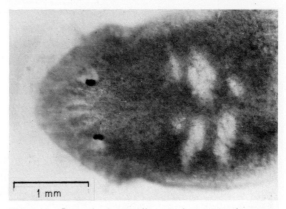

Figure 3 *Dugesia benazzii:* anterior part of an ex-fissiparous specimen showing the ovaries spread over a wide area.

FIRST DATA FOR AN ANALYSIS OF THE FISSION-CONTROLLING FACTORS

As we have seen, the ability to produce sexual individuals varies greatly in different fissiparous populations, probably owing either to a different frequency in the fission-controlling factors or to a difference in their penetrance.

It was therefore interesting to make an analysis of these factors from the point of view of the formal and population genetics, which may be accomplished through two types of crosses: (1) between sexual specimens of fissiparous populations and specimens of populations which reproduce only sexually; the latter may be regarded as lacking fission-controlling factors; (2) between sexual specimens of fissiparous populations.

I had already made an attempt at this 30 years ago, crossing two races of *D. benazzii* from Sardinia, one of which was purely sexual and the other with sexual and fissiparous specimens. The results (Benazzi, 1942) were quite significant since they demonstrated the transmission, although in a limited number of cases, of the fission-controlling factors through the sperm. In fact, some offspring born from specimens of the purely sexual race fertilized by ex-fissiparous specimens became fissiparous. I deduced, therefore, that the fission-controlling factors are located in the chromosomes and presumably they correspond to Mendelian genes.

Recently I have taken up this research again, principally carrying out crosses between sexual specimens of the population of Castello Pino (Corsica) referred to in the previous section as population 9. A first cross, started in the autumn of 1968, was accomplished with 40 sexual specimens, chosen at random from the laboratory culture; of course, I did not know whether they were born from cocoons or whether they were ex-fissiparous. These specimens were

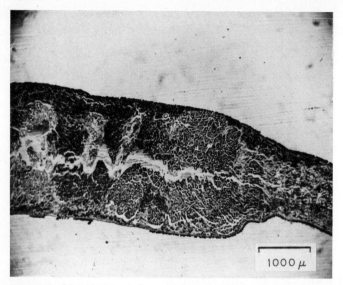

Figure 4 *Dugesia benazzii*: sagittal section of the anterior part of an ex-fissiparous specimen showing the anomalous development of the ovaries.

kept together for the whole laying period and their offspring were kept in similar conditions and frequently controlled in order to establish if they became fissiparous or sexual. In fact, offspring from the same cocoon bred together may divide or become sexual; this is a clear confirmation of the genetic control of fission. In most cases evolution toward sexuality or fissioning takes place precociously, although strong individual differences occur. Some specimens undergo the first fission a few weeks after birth, some others remain asexual for a long time though they do not undergo fission.

However, I wish to emphasize above all the numerical aspect of this cross. As we can see from Table 1, the ratio of fissiparous to sexual individuals (80:67) is 1.2:1, which is close to 1:1.

The chi-square test shows that the difference between the two types of offspring is not significant:

$$\chi^2_{(1)} = 1.14 \qquad P = 0.3$$

This may suggest a simple Mendelian mechanism; however, the data obtained do not permit reliable conclusions to be drawn. Investigations into the progeny of single individuals are required, and with this aim I planned new research employing the same specimens used in the previous experiment. It is necessary to recall that in *D. gonocephala* (as in almost all freshwater triclads) self-fertilization fails to occur and that copulation is mutual, both partners acting as male and female. Keeping in mind these reproductive behaviors, I conducted the following experiments:

1   Ten individuals (numbers 19 to 28) were cut in half, and the specimens obtained after regeneration of the two pieces were mated together (*automates*). The offspring of each couple may be considered as derived from the single original individual. All couples laid a high number of cocoons, but because many of them were sterile, only four produced offspring (Table 2). Although the data obtained are very scarce, it seems that no great differences occur among the individuals.

2   Eight couples were formed with 16 different individuals (*heteromates*); the partners were separated when I had ascertained that cross-fertilization was accomplished. The offspring of the two partners can be considered similar with respect to their genetic constitution; therefore, the comparison between their offspring may give some reliable information about the genetic control of fissioning.

Unfortunately the results obtained till now do not permit this comparison (owing to the great number of cocoons which are sterile) (Table 3). However, I have thought it useful to mention them since, even if they do not permit conclusions about the genetic constitution of the single individuals, they

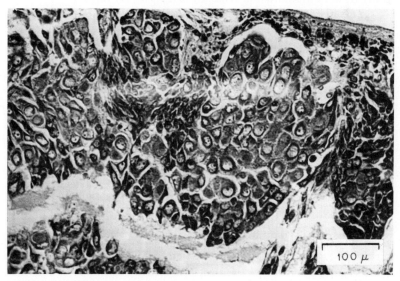

Figure 5   *Dugesia benazzii*: ovary of an ex-fissiparous specimen.

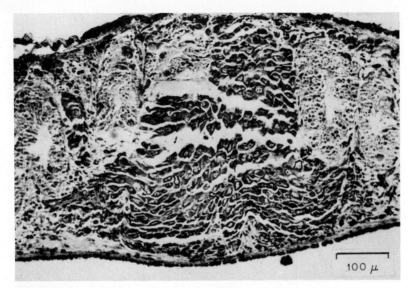

Figure 6   *Dugesia gonocephala* s. l. from the Jordan River: ovary of an ex-fissiparous specimen.

confirm that in the offspring on the whole the ratio between fissiparous and sexual specimens remains near 1:1 (19:21). This ratio is also seen in the off-spring of the automates, in which it is 11:10 (Table 2).

This segregation undoubtedly shows that Mendelian genes are involved in establishing sexual vs. asexual reproduction; however a much greater quantity of data is required for a factorial analysis, and I hope that I shall have time to collect them.

From what I have said, one is induced to think that fissioning is determined by one or more genic mutations which, of course, are not detectable at the chromosomal level. However, the cytogenetic studies that we have been carrying out on triclads for many years have shown some facts which may suggest a relationship between fissioning and numerical variations of the chromosomes.

Dahm (1958) in his valuable monograph on triclad karyology and ecology

Table 1

Total	Fissiparous	Sexual	Undetermined*
152	80	67	5

*These individuals though asexual never divide.

Table 2

	Automates				
Number	21	25	26	28	Total
Fissiparous	2	3	5	1	11
Sexual	2	5	3	. . .	10
Undetermined	. . .	. . .	3	. . .	3

has already called attention to this question; he writes (p. 123): "the existence of the various chromosome numbers found in the planarians treated might also be considered in the light of the common occurrence of asexual reproduction. All viable deviation will have the chance to build up clones as a result of the ability of fission."

Benazzi Lentati (1964) was also interested in this problem. During her research on polysomy in some races of *Dugesia etrusca* (a species of the "*D. gonocephala* group") she found some interesting data: *D. etrusca monoadenodactyla* is always diploid ($2n = 16$) and reproduces only sexually even after a long period of laboratory culture. *D. etrusca biadenodactyla* and *D. etrusca labronica* show, in nature, very few polysomic individuals; however, they became more numerous in the laboratory and reached high chromosome numbers (30 to 40), after fissioning appeared. Many other fissiparous populations of *D. gonocephala* s. l. are aneuploid as demonstrated by the data referred to in the previous pages of this paper. However, it does not seem that the relationship between these two phenomena may be considered constant, and at the present phase of our studies it is not clear if they may assume a causal significance.

SOME CONSIDERATIONS ABOUT THE RELATIONSHIP BETWEEN REGENERATIVE POWER AND FISSIONING

At the beginning of this paper I emphasized that the regenerative power is a necessary but not sufficient condition for the rise of fission. In fact this is lacking in many species of Planariidae, although they display a high regenerative power; moreover in one same species, races or populations either purely sexual or fissiparous may occur.

Vandel (1922) suggested the hypothesis (later accepted by other authors) that fissioning is caused by a mutation, and my genetic research undoubtedly represents an experimental proof. This hypothesis fully explains why fission may be present or absent in different populations of the same species. The fact that fission is always lacking in certain species and frequent in others of the same genus, although all the species possess a high regenerative power,

Table 3

	Heteromates															
Number	1 × 2		3 × 4		5 × 6		7 × 8		9 × 10		11 × 12		13 × 14		15 × 16		Total
Fissiparous	...	2	2	...	1	2	1	4	1	...	...	2	...	2	2	...	19
Sexual	...	2	1	...	...	...	2	2	4	...	5	...	...	1	1	3	21
Undetermined	...	...	...	...	...	1	...	1	...	...	...	...	...	1	...	...	3

appears less easy to explain; we can suppose, however, that in the former the fission-determining mutation cannot manifest itself or that its frequency is so low that the formation of fissiparous strains is highly improbable.

A recent discovery referring to *Dugesia lugubris* s. l. seems to support the second hypothesis. The genus *Dugesia*, which is the most widespread of all freshwater planarians, is clearly divisible in two sections which are well differentiated by many characteristics (cf. Benazzi, 1966). One section comprises the species of the Old World pertaining to the "*D. gonocephala* group" and the American species *D. dorotocephala* and *D. tigrina*; in all these species fissiparous strains are frequent. The other section comprises the species of the "*D. lugubris-polychroa* group" in which fissioning was believed to be wholly absent; this peculiarity had attracted the attention of scientists, more so because these planarians are frequently used for research on regeneration by European scientists. Recently, however, Benazzi *et al.* (1970) have reported a planarian (found near Barcelona, Spain) which in its external morphology is quite similar to *D. lugubris* s. l. but is exclusively asexual; the diagnosis, although the reproductive apparatus is lacking, is confirmed by karyological data (Benazzi, Ballester, and Baguñá, unpublished).

I think this finding may be considered in agreement with the mutational origin of fission and also agrees with the hypothesis of differences in the frequency of the mutation in different species. The asexual form from Barcelona may be regarded as an exceptional event which occurred in the southern part of the geographic range of the species, perhaps favored by the environmental conditions.

Also in the genus *Polycelis* we may distinguish two sections which are clearly differentiated: one comprising the species *felina*, which is frequently fissiparous, the other comprising the two species *nigra* and *tenuis*, in which fissioning has never been found, although even they display a remarkable regenerative power. We may suppose that in *nigra* and *tenuis* the fission-controlling mutation is extremely rare so that it has not thus far been observed; but it may also be supposed that in these planarians either genetic or morphologic conditions which enable the fission to take place are lacking.

BIBLIOGRAPHY

Beauchamp, P. de. 1961. "Classe des Turbellariés." In P. P. Grassé (ed.), *Traité de Zoologie*, vol. 4, pp. 35–212. Paris: Masson et Cie.

Benazzi, M. 1936. Razze fisiologiche di *"Euplanaria gonocephala"* differenziate dalla diversa attitudine alla scissiparità. *Rend. Accad. Nazl. Lincei*, **23**:361–365.

———. 1938. Ricerche sulla riproduzione delle planarie tricladi paludicole con particolare riguardo alla moltiplicazione asessuale. *Mem. Accad. Nazl. Lincei*, **7**:31–89.

———. 1942. Ricerche genetiche sulla scissiparità di *Dugesia* (*Euplanaria*) *gonocephala*. *Arch. Ital. Anat. Embriol.*, **47**:72–94.

———. 1960. Evoluzione cromosomica e differenziamento razziale e specifico nei Tricladi. *Accad. Nazl. Lincei Quaderno N.*, **47**:273–296.

———. 1966. Cariologia della planaria americana *Dugesia dorotocephala*. *Rend. Accad. Nazl. Lincei*, **40**:1001–1005.

———, J. Baguñá, and R. Ballester. 1970. First Report on an Asexual Form of the Planarian *Dugesia lugubris* s.l. *Rend. Accad. Nazl. Lincei*, **48**:282–284.

——— and G. Benazzi Lentati. 1945. Su di un fattore letale in *Dugesia* (*Euplanaria*) *gonocephala* Dugès. *Boll. Soc. Ital. Biol. Sper.*, **20**:1–3.

Benazzi Lentati, G. 1964. La polisomia nelle planarie. *Atti Soc. Toscana Sci. Nat.*, **71**:44–51.

Best, J. B., A. B. Goodman, and A. Pigon. 1969. Fissioning in Planarians: Control by the Brain. *Science*, **164**:565–566.

Brøndsted, H. V. 1969. *Planarian Regeneration*. New York: Pergamon Press.

Child, C. M. 1911. The Axial Gradient in *Planaria dorotocephala* as a Limiting Factor in Regulation. *J. Exptl. Zool.*, **10**:265–320.

Curtis, W. C. 1902. The Life History, the Normal Fission, and the Reproductive Organs of *Planaria maculata*. *Proc. Boston Soc. Nat. Hist.*, **30**:515–559.

Dahm, A. G. 1958. *Taxonomy and Ecology of Five Species Groups in the Family Planariidae*. Malmö: Nya Litografen.

Dugès, A. 1828. Recherches sur l'organisation et les môeurs des Planariées. *Ann. Sci. Nat.*, **15**:160.

Fedecka-Bruner, B. 1968. Etudes sur la régénération des organes génitaux chez la planaire *Dugesia lugubris*. *Bull. Biol. France Belg.*, **102**:3–44.

Hyman, L. H. 1951. *The Invertebrates: II. Platyhelminthes and Rhynchocoela. The Acoelomate Bilateria*. New York: McGraw-Hill.

Jenkins, M. M. 1967. "Aspects of Planarian Biology and Behavior." In W. C. Corning and S. C. Ratner (eds.), *Chemistry of Learning*, pp. 116–143. New York: Plenum Press.

Kenk, R. 1937. Sexual and Asexual Reproduction in *Euplanaria tigrina* (Girard). *Biol. Bull.*, **73**:280–294.

———. 1941. Induction of Sexuality in the Asexual Form of *Dugesia tigrina* (Girard). *J. Exptl. Zool.*, **87**:55–69.

Lender, T. and F. Zghal. 1968. Influence du cerveau et de la neurosécrétion sur la scissiparité de la Planaire *Dugesia gonocephala. Compt. Rend. Acad. Sci. Paris,* **267:**2008–2009.

Lobetti Bodoni, L. 1918. *Planaria subtentaculata* Drap. *Boll. Museo Zool. Anat. Comp. Univ. Torino,* **32:**1–12.

Okugawa, K. I. 1957. An Experimental Study of Sexual Induction in the Asexual Form of Japanese Fresh-water Planarian *Dugesia gonocephala* (Dugès). *Bull. Kyoto Gakugei Univ.,* no. 11, 8–27.

—— and M. Kawakatsu. 1954a. Studies on the Fission of Japanese Fresh-water Planaria, *Dugesia gonocephala* (Dugès). I. *Bull. Kyoto Gakugei Univ.,* no. 4, 25–34.

—— and ——. 1954b. Ibid III. *Bull. Kyoto Gakugei Univ.,* no. 5, 42–52.

—— and ——. 1956. Ibid. IV. *Bull. Kyoto Gakugei Univ.,* no. 8, 23–42.

—— and ——. 1957. Ibid. VI. *Bull. Kyoto Gakugei Univ.,* no. 10, 18–37.

Vandel, A. 1920. Le développement de l'appareil copulateur des Planaires est sous la dépendance des glandes génitales. *Compt Rend. Acad. Sci. Paris,* **170:**249–251.

——. 1921. La régénération des glandes génitales chez les Planaires. *Compt. Rend. Acad. Sci. Paris,* **172:**1072–1074.

——. 1922. Modes de reproduction des Planaires Triclades Paludicoles. *Bull. Biol. France Belg.,* **55:**343–518.

Vowinckel, C. 1970. Stimulation of Germ Cell Proliferation in the Planarian *Dugesia tigrina* (Girard). *J. Embryol. Exptl. Morphol.,* **23:**407–418.

Chapter 21

Relationships between Reproductive Activity and Parental Age in a Sexual Race of *Dugesia dorotocephala*

Marie M. Jenkins
Department of Biology, Madison College,
Harrisonburg, Virginia

Whether or not planarians undergo a true senescence has not thus far been established. Child, using susceptibility to dilute cyanide as a criterion, found that larger, presumably older, fissioning planarians were more resistant to the chemical, and he equated the change with a decrease in rate of metabolic processes. He considered this decrease to be the primary manifestation of senescence and thought it possible that agamic reproduction with accompanying rejuvenescence might occur simply as a result of senescence. Child did not study the occurrence of natural death in planarians, nor did he investigate possible relationships between increase in age and rate of fission (1915).

According to Comfort, a terminal increase in liability to die, as shown by a life table of an adequate population sample, is essential for demonstrating that senescence does occur (1954, 1964). He recognizes, however, that life tables are unavailable for most organisms, and that secondary criteria are, of necessity, widely used in practice and, if judiciously chosen, acceptable. One useful criterion mentioned by Comfort is the decline in reproductive capacity with increasing age, provided the complete life cycle of the organism is known.

Few studies have presented evidence that an increase in age in planarians can be correlated with any phase of sexual activity. Vandel (1922) reported that old sexual planarians, living under conditions which were not unfavorable for other, younger animals, developed lesions before death. Both Voigt (1928) and Abeloos (1930) noted a decrease in cocoon production with age, and the latter reported that both morphological and physiological changes occurred as sexually reproducing planarians grew older. In his opinion, however, true evidence of senility included only irreversible changes which could appear in animals of very different sizes after growth had been completed. More recently, Balázs and Burg (1962a, b) have shown that both cocoon production and fertility decreased in *Dugesia lugubris* as individuals grew older, and that death in older planarians was preceded by an involution of the entire organism, but Reynoldson *et al.* found no evidence over a period of 2 years that aging affected cocoon production in this species (1965).

Child's extensive studies on senescence in *Dugesia dorotocephala* were restricted entirely to asexual individuals, for the rare sexual animals that appeared in his cultures failed to reproduce. He believed that sexual reproduction in this species did not occur in nature (1915), but during recent years several sexual races have become known (Jenkins, 1960; Jenkins and Brown, 1963). It now appears that the condition of exclusive asexuality may be correlated with the locality from which the animals are obtained (Jenkins, 1970) for sexual strains have been found only in cold, constant-temperature springs, in certain areas in southern United States. The animals become sexual when 4 to 6 months old, whatever the season, and produce numerous viable cocoons during all months of the year, both in the laboratory and in the natural habitat. Fission may or may not occur prior to the onset of sexuality, but once maturity has been reached, sexual *D. dorotocephala* mate and produce cocoons continuously. They show no evidence of alternating sexual and asexual phases (Jenkins and Brown, 1963, 1964).

The secondary criterion advanced by Comfort as acceptable for a study of senescence, although not useful for studying propagation by fission, can be as well applied to sexually reproducing races of planarians as to other sexual animals. This criterion, the change in reproductive capacity with increasing age, is the basis for a series of investigations on reproductive activity in *D. dorotocephala* which have been under way for the past several years.

In the present paper, evidence is presented regarding the effect of parental age, during the lifetime of the individuals, on the number of cocoons deposited, the proportion that hatched, and the number of offspring produced. Consideration is also given to other factors correlated with aging in planarians, including the presence or absence of fission; the place of fission in the life cycle of the sexually reproducing race; and the life-span of sexual planarians of this species.

MATERIALS AND METHODS

Animals used in this study were descendants of a sexual race of *Dugesia dorotocephala* collected originally from Buckhorn Springs, Murray County, Oklahoma (Jenkins and Brown, 1963). Cocoons from stock cultures were isolated and allowed to hatch in individual bowls, and offspring that emerged were fed, counted, and examined on a set schedule, three times every 2 weeks, for the remainder of their lives, with exceptions noted below. Planarians thus obtained constituted first-generation adult groups. Animals from some groups were inbred; others were crossbred.

In inbred groups (I) siblings emerging from any one cocoon were not, at any time, placed in a container with animals that had hatched from another cocoon. If fission products of immature planarians appeared, they were removed at the examination period, and heads and tails were separated. Animals which became sexual without having fissioned previously (S) were thus kept separated from fission-sexual (FS) animals, that is, anterior pieces of planarians which had fissioned before the onset of sexuality. Tail pieces which later became sexual are not included in this report.

After experimental animals reached maturity, cocoons they deposited were removed at each feeding period, isolated, and allowed to hatch individually until a minimum of 200 F_2 offspring in each group was obtained. Except in two F_1 inbred groups, cocoons deposited thereafter were kept in groups of one to five, and all offspring that emerged were counted and returned to stock. In these two inbred groups a detailed record was kept of the development of all offspring produced throughout the lives of the parents.

All F_2 offspring were fed, counted, and examined on the set schedule used for the F_1 animals, and were in every way accorded the same treatment as that given their parents. When F_2 planarians reached maturity, they were placed in bowls labeled to show their reproductive history, as follows:

I-S/S: F_2 generation inbred; individuals of both generations became sexual without prior fission.

I-FS/S: F_2 inbred individuals became sexual without prior fission, but members of the F_1 generation fissioned before becoming sexual.

I-S/FS: F_2 individuals fissioned before becoming mature, but F_1 animals became sexual without fissioning.

I-FS/FS: Individuals in both F_1 and F_2 generations fissioned before becoming sexually mature.

In crossbred groups (X), siblings that emerged from one cocoon were kept separated from other groups of siblings until each individual either fissioned or developed a gonopore. Offspring from six cocoons were then combined and the resulting bowls of sexual specimens were labeled either X-S (crossbred, sexual without fission) or X-FS (crossbred, sexual after fission). An F_2 generation was obtained as for inbred specimens. Similar formulas were used to label the bowls except that an X replaced the initial I.

During the fourth and fifth years of life, when it became obvious that cocoons produced were sterile, most of the animals that remained were returned to stock. Two inbred groups were kept, however, until the death of all animals of which they were composed.

All experimental animals were kept in covered glass finger bowls of 250 m capacity, with not more than eight to ten planarians per bowl. Trays of bowls were kept either in a BOD incubator or in a laboratory room furnished with a heat pump–air conditioner, both set to maintain a temperature of $19\,°C$. Variation in the laboratory temperature proved to be $\pm3\,°C$, in contrast with $\pm0.5\,°C$ in the BOD, and after the first 18 months all animals were kept in BOD incubators.

Planarians were fed before each examination and were allowed access to the food for several hours to be sure that adequate nourishment was obtained by each individual. After food remains were removed, each group of animals was rinsed several times and placed in a clean bowl containing freshly aerated culture solution. Once a month, in addition to the examination, the gliding length of all animals was recorded.

The culture solution used was prepared according to the formula given by McConnell (1967) except that 10 ml of natural springwater was added per gallon in order to provide trace elements. The resulting solution was allowed to set overnight and aerated a minimum of 10 min before being used.

Data obtained in this study were subjected to statistical analysis by means of Student's "T" test (Ostle, 1954).

RESULTS AND DISCUSSION

Ten groups of animals were studied, five inbred and five crossbred. Owing to the length of time covered by the study, data are presented on the basis of four 13-week intervals per year rather than monthly periods. The beginning date for each group is that of emergence from the cocoon, and interval dates in various groups do not coincide. Data for each of the groups (Table 1) are a compilation of results obtained from two or more replicates or subgroups, with the exception of I-FS/S, which is composed of only two individuals. The parent group of these two, I-FS, which contained fifteen adults, produced cocoons steadily, after a rather late start, but only two sexual individuals survived long enough to produce viable offspring.

The great variation in the total number of adults in each of the ten groups was due to high juvenile mortality in some groups and to an exceptionally high number of sexual individuals in others. In the latter case, mature animals above the number indicated in Table 1 were returned to stock cultures and are not considered in this report.

Onset of Sexuality

Nearly two-thirds of the adult planarians in this study reached maturity during the second or third 13-week interval. A few became sexual before the end of the first interval, but only two of these produced cocoons. During the fourth and fifth intervals another 27 percent became mature, and the remaining 6 percent began sexual reproduction at various times during the last seven intervals, two as late as the final 3 months of the 3-year period.

Only five nonfissioning (*S*) planarians became sexual after the close of the first year. These five were in the same group, *X-S*, and all reached maturity shortly after the second year began. All other animals which matured during the second or third year of life had undergone fission before becoming sexual. The delay can be ascribed in part to a certain amount of time being required for reconstitution after fission, but this explanation does not account for the wide fluctuation noted. Evidence indicates that reproductive characteristics in *D. dorotocephala* and other planarians are essentially dependent on genetic factors (Benazzi, 1966), and it appears probable that the marked differences in age at the onset of maturity in *S* and *FS* groups are a manifestation of inherited variations.

Size/Age Relationships

Sexual maturity is not a function of size in these animals. At the time the first cocoons are deposited, most *D. dorotocephala* have a gliding length of 20 to 25 mm, although a length of only 15 to 16 mm is not unusual. Growth continues after maturity until a gliding length of 30 to 35 mm or longer is attained. Members of asexual races of the species often reach this length but, except for the occasional ones which discontinue fission and develop gonads, are usually more slender. The average weight of a sexual *D. dorotocephala* ranges between 30 and 50 mg.

The largest animal in the present experiment measured 37 mm gliding length during the sixth interval of its life; the average length of all worms at this time was 33.88 mm. Maximum size was maintained for most of the second year, but during the third and following years of life a gradual decrease occurred in both gliding length and volume. Average length at death, judged by measurements of the preceding month, was 25 mm. Although starvation commonly causes shrinkage in planarians, this decrease in size with increasing

Table 1 Overview of Events Occurring during the First Three Years of Life in Ten Groups of Sexually Mature *Dugesia dorotocephala*

Each year is divided into four intervals of 13 weeks each, but interval dates for various groups do not coincide because each beginning date was that of emergence of planarians from the cocoon. Variation in the number of adults in a group during one interval is shown by numerals separated by dashes. Interval averages were obtained from individual counts made every 4 to 5 days during the lifetime of the animals. Refer to text for explanation of group symbols.

Group		Intervals, first year				Intervals, second year	
	1		2	3	4	5	6
1. (*I-S*)	(22)		5–8(7)*	†15–20–18 (D+D*)	†18–20–15 (5DC)	15–14 (D)	14–12 (2D*)
2. (*X-S*)	(25)	(4)*	4–5(2)*	†7–20–19 (D)	19–18 (D)	†18–23	23–18 (D+4DH)
3. (*I-FS*)	(15)			(4–3)* (D)	†3–8–7 (D)	†7–9–7 (D-JF)	7
4. (*X-FS*)	(68)		8–19–18	†18–38	†38–43–41 (2D)	†41–55–54 (D)	†54–59–50 (3D+6DH)
5. (*I-S/S*)	(19)		1–19	19	19	19	19–18 (D)
6. (*X-S/S*)	(41)	2	2–18	†18–25–24 (D)	†24–40–27 (3D)	37–35 (2D)	35–29 (6D)
7. (*I-FS/S*)	(2)		1–2	2	2	2–1 (D)	1
8. (*X-FS/S*)	(37)	(6)*	6–27–25 (D+D*)	†25–35–34 (D)	34–32 (2D)	32	32–30 (2D)
9. (*I-S/FS*)	(36)		(2)*	†2–7–5 (D+D*)	†5–12	†12–30–29 (D)	29–26 (3D)
10. (*X-FS/FS*)	(35)	(2)*	2–15	†15–26	†26–31–30 (D)	†30–32	†32–34–31 (3D)
Average no. adults/interval	(300)	2	82.19	194.59	206.15	249.51	237.76

() Initial numbers in parentheses indicate highest total number of adults in a group.
()* Planarians which became sexually mature during the interval but produced no cocoons; not included in interval averages.
BS Subgroup returned to stock cultures at termination of experiment.
D Died; death due to unknown causes.
D* Died; death due to planarians' crawling out of water and drying; remnants found at next feeding period.
DC Died; death one of many that occurred in 1 week in one BOD incubator; cause unknown.

Intervals, second year		Intervals, third year				D	J
7	**8**	**9**	**10**	**11**	**12**		
12	12	12	12	12	12–7 (5DC)	12D +3D*	
18–17 (D)	17–15 (2DC)	15–13 (D+BS)	13–12 (JD)	12–4 (D+3DC+4BS)	4	15D	JD
7	7	7	†7–8	†8–7–10 (D)	10	4D	JF
†50–53–52 (D*)	52–49 (2D+JS)	49–31 (2D+16BS)	31–26 (3D+D*+JS)	†26–27–24 (D+JS+JF)	†24–25–23 (D+JS)	22D +2D*	4JS JF
18–15 (3D)	15–5 (10DC)	5–3 (2D)	3	3–2 (D)	2	17D	
29–28 (D)	28–25 (3D)	25–23 (D+D*)	23	23–19 (3D+JF)	19–18 (D)	21D +D*	JF
1	1–0 (D)					2D	
30–27 (3D)	27–25 (D+JF)	25–24 (D)	24–23 (D)	23–16 (7D)	16–15 (JF)	19D +D*	2JF
26–25 (D*)	†25–26–25 (D*)	25–12 (D+11DC+JD)	12–9 (D+2JD)	†9–11–7 (2D+2JD)	†7–8–4 (2D+2BS)	22D +3D*	
31–26 (2D+3DC)	26–24 (D+JD)	24–23 (D*)	23–20 (3D)	20–15 (5D)	15–14 (D)	19D +D*	JD
						164	16
214.68	198.45	164.62	139.37	119.34	102.85		Total

DH Died; death one of several that occurred among planarians kept in laboratory when cooling system failed during an August heat wave.

† Planarians, recently sexual, added to group after sixth month.

J "Juvenile"; planarian lost gonopore and diminished in size.

JS "Juvenile" which became sexual again within 1 to 3 months

JF "Juvenile" which divided one or more times after becoming juvenile; all products died.

JD "Juvenile" which died within 1 to 4 months without further change.

age was not due to inadequate food. Beef liver, which has been shown to be nutritionally adequate for planarians, and superior to many other foods usually provided (Wulzen, 1923, 1927), was consistently used as the food of choice, and sufficient exposure allowed at each feeding period. The interval used, three times every 2 weeks, was chosen in preference to twice-a-week feeding because digestion of a meal in these animals requires approximately 5 days (Willier, Hyman, and Rifenburgh, 1925), and if food is presented more often, some of the planarians may fail to ingest any.

Results of this experiment in regard to size/age relationships agree with the findings of Balázs and Burg (1962a, c) who report that growth in *D. lugubris* is intensive during the first 2 to 3 months of life. The rate then gradually decreases and, as the worms grow old, an actual reduction in size occurs. Abeloos also noted a gradual and prolonged decrease in weight of sexual planarians during the course of continued cocoon production (1930). In both *D. lugubris* and *D. dorotocephala* sexual maturity occurs before the worms reach their maximum volume, but in the former species the average age at maturity is slightly earlier than in the latter, being but 3.6 months.

Although in *D. dorotocephala* greatest size was attained early in the second year of life, many of the animals lived well into the fifth and sixth years, with no fission occurring in any sexual worm. This fact alone makes it obviously impossible to correlate size with age, since the largest size may be reached when the animal has lived less than one-fourth its life-span. Even in juveniles, individual size is sufficiently variable that no dependence can be placed on an age/size correlation (Jenkins, 1967). Abeloos also reported for *D. gonocephala* a great variation in the size of young hatched from the same cocoon (1930). The range of gliding lengths in *D. dorotocephala* at emergence is 1 to 10 mm and neither extreme is uncommon.

Comfort (1960) stated that few species can be aged by inspection, and a careful study of size/age relationships in sexual planarians will show that his conclusion is applicable to these animals. Both Comfort, and Balázs and Burg (1962c) are of the opinion that discrepancies found in the literature are often due to differences between young and "old" animals being described which are, in fact, differences between infants and young adults. This confusion in regard to the relationship of age to size seems to be particularly prevalent in planarian studies.

Loss of Sexuality

It has been assumed that, if sexual races of *D. dorotocephala* were found, a pattern of seasonal alternation between sexual and asexual phases, similar to that known for *D. tigrina*, would obtain (Hyman, 1951; de Beauchamp, 1962;

Lange, 1968), but no evidence thus far supports this view. Immature specimens of the sexual race may undergo fission one or more times without respect to season, particularly if conditions are unfavorable, but the process does not continue indefinitely in the sexual race and does not, therefore, serve as an effective means of reproduction.

Occasionally, sexually mature *D. dorotocephala* become juvenile in appearance. The gonopore becomes indistinct, then disappears, and the entire animal may shrink, sometimes to less than one-half its former size. These changes take place while other animals of the same age and in the same culture dish, under optimal conditions of food and temperature, continue normal cocoon production.

In the present study, sixteen of the 300 animals became, externally, nonsexual (Table 1). In four of the sixteen, all in the *X-FS* group, the condition was temporary. The worms became sexual again within 1 to 3 months, and soon could not be distinguished from other mature adults. Seven of the remaining twelve died within 1 to 4 months without further change. The other five divided spontaneously, four of them one time and one several times, but all "fission" products soon died.

In fifteen of the sixteen cases the changes occurred during the last five intervals of the 3-year period, when the animals were 22 to 36 months old. This points to the possibility that an apparent return to an immature state may be a manifestation of senescence. Since all animals so affected either became sexual again within a short period, or died, with or without having divided, such loss of reproductive ability and decrease in size was certainly not correlated with either rejuvenescence or agamic propagation.

Benazzi (1967) considers that sexuality, in general, is an irreversible state, but he has found that ex-fissiparous individuals may become nonsexual and undergo a spontaneous division which is not identical with asexual reproduction. Data from the present study indicate that this phenomenon may occur either in animals which have become sexual after fission (*FS*) or in those which reached maturity without prior fission (*S*), provided the race is one in which fission may occur.

In the first study of sexual activity in *D. dorotocephala* (Jenkins and Brown, 1963) it was noted that fission products began to appear in the cultures when the animals were nearing 2 years of age. There was no evidence of rejuvenation, and it was assumed that the occurrence of fission before death in a number of individuals was a part of the aging process. It now appears that the nonrejuvenatory "fission" observed in some of the older animals at that time was a type of spontaneous division not used by the animal for asexual propagation, such as has been found in the present investigation. A study of planarian sexuality during the third to the sixth year of life might provide data which would assist materially in determining whether or not the occasional reversion to an apparently immature state, with or without accompanying spontaneous division, is correlated with the aging process in this species.

Cocoon Production

In most groups, cocoon production began during the second interval, when the planarians were 4 to 6 months of age (Table 2), but two inbred groups, *I-FS* and *I-S/FS*, did not deposit egg capsules until the third and fourth intervals, and animals in one crossbred group, *X-S/S*, began cocoon production during the third month of life. The number of cocoons produced per adult worm increased rapidly during the first year (Fig. 1), but the rise thereafter was slight. The overall peak of cocoon production occurred near the end of the third year, over $1\frac{1}{2}$ years after the maximum production of offspring. Although not shown in Table 2, cocoon deposition decreased after the third year and, in most groups, ended during the fourth year.

Haranghy and Balázs (1964) noted a similar initial rise in cocoon production in *Dugesia lugubris* which they attributed to asynchronous sexual maturation. Maximum production was reached when the animals were 10 months old; then a gradual decrease followed, but the authors consider this decline not to be statistically significant. Highest production in *D. lugubris* was 2.27 cocoons per worm per month (Balázs and Burg, 1962a); by the time the animals had reached the age of 19 months, the average had decreased to 0.94. Haranghy and Balázs (1964) reported that planarians 3 and 4 years old continued to deposit cocoons, the average being 1.1 to 1.4 cocoons per individual. Voigt (1928) found a lower production in *Polycelis nigra*. These animals were 10 months old before the first cocoon was deposited, and they produced an average of 5.3 cocoons during the breeding period of the first year, or 1.9 cocoons per worm per month. The second year, during a 2-month breeding period, the monthly average was 0.95 cocoons per adult. No cocoons were deposited the third year. Other reports give figures for cocoons produced by planarians collected from natural habitats. The age of the animals is unknown and valid comparisons cannot be drawn, but yearly production seems to be quite low in most species studied.

Cocoon production in *D. dorotocephala* was higher than in any other species for which data are available. Maximum monthly output, an average of 2.76 cocoons per worm, was reached during interval 11, when the animals were 31 to 33 months old. During interval 4, at 10 months of age, *D. dorotocephala* produced an average of 2.1 cocoons per adult per month, and during interval 7, when the planarians were 19 months old, the average was 1.87. The latter figure is nearly twice the average for *D. lugubris* at the same age.

Variation in cocoon production among individual groups was quite marked. In the two first-generation inbred groups, *I-S* and *I-FS*, the maximum occurred at the end of the second year, and in one of the second-generation inbred, at the end of the first year (Table 2). The latter group, *I-FS/S*, had the fewest adult animals but produced at the peak period, per adult worm, close to 20 percent more cocoons than, and nearly twice as many offspring as, any other group. Yet, its reproductive history was by far the shortest in the study, for the animals died soon after their brief explosion of reproductive activity. A number of theories (Comfort, 1964) have attempted to relate senescence with death due to exhaustion induced by reproductive processes. Salmonids, for example, die after spawning, owing to endocrine changes which accompany migration, and in a number of planarian species (Okugawa, 1957; Vandel, 1922; de Beauchamp, 1931), high mortality occurs immediately after a breeding period is completed. It may be in this case that the burst of reproductive activity culminated in physiological exhaustion, and that the factor or factors that caused the high rate of reproduction also caused early aging and death, but there is no evidence from any other group that would support this idea.

If a comparison is made of the highest points reached at peaks of cocoon production in the various groups (Table 2), it is found that the two second-generation inbred groups with mixed ancestry, *I-FS/S* and *I-S/FS*, are at the head and foot of the list, with 14.50 and 3.83 cocoons respectively per adult worm, during intervals 4 and 12. Except for the anomalous *I-FS/S*, the highest producers are all crossbred. The only crossbred group excluded, *X-FS*, had such a consistent high level of production that no one interval is outstanding.

If the groups are arranged, instead, according to the lowest cocoon production during a single interval, and the initial low production of the first year is excluded, all five inbred groups constitute the lowest producers and all crossbred the highest. Furthermore, the two contrasting second-generation inbred groups, *I-FS/S* and *I-S/FS*, are the two lowest of all. The latter is consistently low throughout, but the former is so inconsistent in its output that it is lowest in one interval but significantly higher than any other group during another interval.

In comparison with *D. lugubris*, monthly cocoon production in *D. dorotocephala* at the peak period is higher in nine of the ten groups. The only group with a lower peak production than the 2.27 of *D. lugubris* is *I-S/FS* with 1.28 cocoons per adult per month during the twelfth interval.

Fertility in Inbred and Crossbred Groups

Of the 11,313 cocoons deposited during the first 3 years, 77.66 percent were fertile and gave rise to 45,185 living young. Most of the offspring, slightly over 87 percent, emerged before the third year began. The average number of embryos per fertile cocoon for the entire 3-year period was 5.14. During the

Table 2   Effect of Parental Age on Production of Cocoons and Living Young in Ten Groups of a Sexual Race of Dugesia dorotocephala.
Results are given in terms of average production per adult worm per 13-week interval. Total adults = 300; total cocoons deposited = 11,313; total young hatched = 45,185. Refer to text for explanation of symbols.

Adult groups	Intervals, first year				Intervals, second year	
	1	2	3	4	5	6
1. (*I-S*)						
Cocoons		0.17	1.92	4.01	3.52	3.93
Offspring		0.52	10.18	14.89	18.60	22.58
2. (*X-S*)						
Cocoons		10.93	5.36	6.90	4.29	6.84
Offspring		20.19	20.39	39.22*	25.00	34.90
3. (*I-FS*)						
Cocoons				1.76	3.72	4.43
Offspring				—	1.24	13.43
4. (*X-FS*)						
Cocoons		1.84	5.44	6.87	3.71	6.23
Offspring		3.43	16.48	38.56	28.86	38.93*
5. (*I-S/S*)						
Cocoons		0.24	1.56	5.86	3.63	3.25
Offspring		—	6.69	58.54*	22.16	13.90
6. (*X-S/S*)						
Cocoons	1.50	3.41	5.82	4.47	6.37	6.73
Offspring	4.00	9.71	25.33*	23.88	21.26	20.37
7. (*I-FS/S*)						
Cocoons		7.33	11.50	14.50*	7.70	1.00
Offspring		—	109.50	201.00*	57.70	—
8. (*X-FS/S*)						
Cocoons		3.90	6.30	10.27	8.37	6.70
Offspring		6.96	26.67	106.10*	57.37	46.02
9. (*I-S/FS*)						
Cocoons			1.79	0.41	1.80	2.48
Offspring			—	0.63	9.90	19.59*
10. (*X-FS/FS*)						
Cocoons		3.38	5.87	7.87	7.21	7.73
Offspring		7.30	31.81	46.68	40.60	44.73
Average of all per adult worm:						
Cocoons	1.50	4.14	5.32	6.30	4.39	5.62
Offspring	4.00	7.82	24.16	45.78*	28.02	31.12

*Peaks of production.

Intervals, second year		Intervals, third year				
7	8	9	10	11	12	Average
5.25	6.83*	3.83	3.75	2.50	2.20	37.8
56.08	62.92*	51.92	28.25	12.08	7.04	275.6
6.58	6.97	5.74	7.44	8.74	9.75*	60.7
25.78	36.24	12.35	9.89	1.75	2.75	231.2
3.57	7.43*	3.00	4.76	3.27	1.60	32.3
22.86	34.43*	18.86	7.98	14.81	1.50	108.5
4.96	4.75	5.44	6.44	6.98*	6.30	53.5
30.77	27.38	18.87	24.13	16.11	10.11	258.7
2.47	2.81	1.50	1.00	1.80	8.70*	34.7
3.77	1.98	3.40	—	—	2.61	187.8
5.31	7.37	7.54	9.30*	7.92	3.17	71.3
13.50	13.69	9.25	6.70	3.86	2.84	167.9
5.00	—	—	—	—	—	54.1
—	—	—	—	—	—	483.5
8.86	10.40	10.40	8.85	11.58*	11.22	94.5
15.74	31.95	18.22	4.47	7.72	3.57	375.9
2.85	2.04	.91	1.08	1.76	3.83*	18.6
11.91	2.31	.39	1.38	0.27	4.66	69.6
9.02	10.71	9.47	8.95	11.90*	10.43	98.6
28.32	47.18*	17.33	4.99	4.66	5.77	339.3
5.70	6.56	6.23	6.93	7.59*	6.31	
22.57	27.32	16.03	11.14	8.33	5.32	

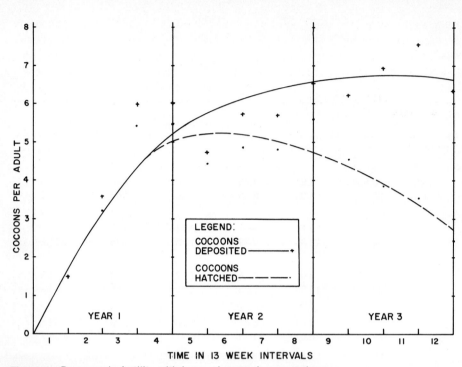

Figure 1 Decrease in fertility with increasing age in a sexual race
of *Dugesia dorotocephala* as shown by a comparison of cocoons
deposited and cocoons hatched over a 3-year period. Adult planarians numbered 300. Approximately one-half the animals did not
fission during their lifetime. The others underwent fission before
becoming mature, but the sexual condition was continuous thereafter.

first year the great majority of cocoons produced were fertile, and in several
subgroups, both inbred and crossbred and both *S* and *FS*, 100 percent fertility was found. Highest fertility occurred during the fourth interval (Table 2).
During this 13-week period over 91 percent of the cocoons were fertile and
each adult worm produced an average of 45.78 offspring. Thereafter, fertility
decreased markedly (Fig. 1).

During the eighth interval, at the close of the second year, the number of
fertile cocoons per worm decreased from 5.78 to 5.65, and the number of
offspring to 27.32. A year later, during interval 12, the number of fertile
cocoons decreased to 2.38 and the number of offspring to 5.32 per worm.
Fertility was significantly lower than cocoon deposition during the second year,
but during the third year the difference was highly significant, beyond the .005
confidence level.

Fertility in *D. lugubris* appears to be lower than in *D. dorotocephala*.

Balázs and Burg (1962b) found the average number of embryos per cocoon in *D. lugubris* to be 3.47, with a range from 2.68 to 5.15. The average number of descendants per worm per month in that study was as follows: 4 months, 0.414; 10 months, 4.046; and 20 months, 0.712. Maximum production of offspring was reached at approximately the same age in both *D. lugubris* and *D. dorotocephala*, but the decrease in fertility which followed the peak was much more pronounced in the European species. Balázs and Burg stated they expected cocoons from parents over 2 years old to be perfectly barren; in the American species viable young were produced in some animals as late as the beginning of the fifth year.

An overall picture of offspring production in first- and second-generation inbred and crossbred groups is presented in Fig. 2. Both inbred groups are significantly lower than the average of all groups, and the second-generation

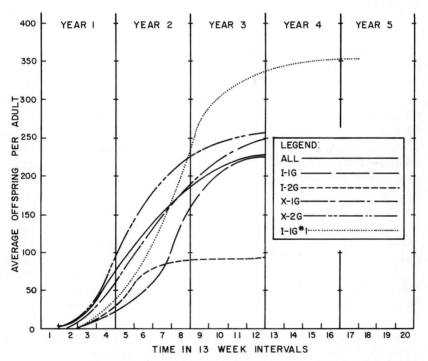

Figure 2 Comparison of cumulative totals of offspring produced by first- and second-generation inbred and crossbred groups of *Dugesia dorotocephala*. *I* = inbred; *X* = crossbred; *1G* and *2G* = first and second generation respectively; *All* = average of all groups in experiment; *I—1G #1* = one inbred subgroup with exceptionally high production for which a 5-year record was kept. Total number of adult planarians = 300. Alternation of asexual with sexual phases did not occur.

crossbred is significantly higher, all beyond the .005 confidence level. Particularly striking is the low level of production shown by the second-generation inbred compared with all other groups.

A comparison of the lowest producers of offspring in all groups for the full 3 years shows that the five inbred groups produced less than any crossbred group, and the three second-generation inbred groups were the three lowest of all. This points to the likelihood that sterility, partial or complete, is essentially dependent on genetic factors, as suggested by Benazzi (1966). Low cocoon production or lack of it, death of embryos, and early death of offspring could all be expressions of lethal factors, and in second-generation inbred strains the probability of any inherited trait appearing would be significantly increased.

One subgroup, *I-1G #1*, a member of the *I-S* group, departed radically from the other inbred groups (Fig. 2). This subgroup contained a total of eight sexual adults, all hatched from one cocoon. Two of the animals died near the close of interval 12 (2DC; see Table 1), and another within the month, shortly after interval 13 began. Four of the remaining five died at various times during the two following years; the final one lived to be 6 years 7 months old. None of the eight fissioned at any time. The last viable cocoon in the group was deposited during interval 17, when the animals were 4 years 1 month old.

The *I-1G #1* planarians produced 380 cocoons, an average of 44.5 per adult; 287 of these were fertile. Total offspring numbered 2,785, but only 80 individuals were produced after the end of the third year. Ninety-two juveniles became sexual without fission; 132 fissioned and 84 of these later became sexual. Twenty-five of the remaining 2,561 were accidentally discarded; 446 developed lesions and were isolated (15 of these later recovered and 7 of the 15 became sexual); and 2,090 died as juveniles. The average number of offspring per adult for the 3-year period was 459, a figure exceeded by no other group except the short-lived, high-yielding *I-FS/S*, which produced 483 young per adult. Next highest were two crossbred second-generation groups, *X-FS/FS* and *X-FS/S*, with 339 and 376 young per adult respectively.

The history of this subgroup substantiates the idea that reproductive characteristics of planarians are under genetic control since, in inbred lines, whatever factors are present are likely to be manifested to a greater degree. Data from this group indicate that parental longevity, fertility, and juvenile mortality may be heritable, but that the two latter characteristics are attributable to distinct factors, since high juvenile mortality and low parental fertility do not necessarily coexist.

Effect of Fission on Fertility

A comparison of total offspring produced by sexual and fission-sexual lines, in crossbred groups only, is shown in Fig. 3. First-generation groups departed

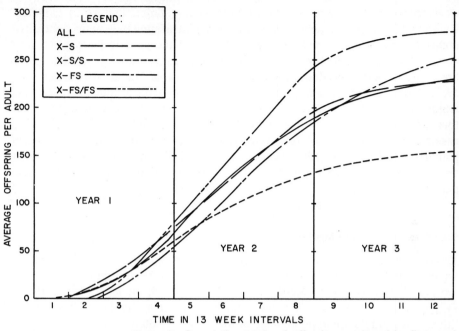

Figure 3 Cumulative totals of offspring produced by first- and second-generation crossbred groups of *Dugesia dorotocephala*, compared with each other and with the average of all groups in the experiment. *X-S* and *X-S/S* = planarians which became sexual without prior fission; first and second generations respectively. *X-FS* and *X-FS/FS* = first- and second-generation planarians which fissioned before becoming sexual. No alternation of fissioning and sexuality occurred.

little from the average of all groups, whether or not the individuals fissioned before becoming sexual, but second-generation groups differed significantly from the average at a level of confidence well beyond .005. Offspring were produced in far greater numbers by animals which, for two successive generations, had undergone fission before becoming sexual, and by contrast, animals which did not fission for two successive generations produced far fewer offspring per adult worm. It appears that fission, while not an effective means of reproduction in the sexual race, may still serve an important function by imparting vigor to those animals which undergo the process before becoming sexual. Vigor is an indefinite term, not suited to exact definition, but its presence or absence in a planarian race is obvious. Vigorous strains grow and reproduce; others do not.

Studies of the *D. gonocephala* group (Benazzi *et al.*, 1970) show that ability to fission is a reproductive mechanism, genetically controlled, and that ex-fissiparous animals can transmit fission-determining factors in gametes. Evidence from the present study indicates that a similar genetic control of ability to fission exists in *D. dorotocephala*, but the genetic factors involved apparently are not dominant, at least in this species. Groups of mixed ancestry are found in both inbred and crossbred strains (Table 1) and, as shown by the *S/FS* and *FS/S* groups, fission-sexual animals can be produced by parents which have not fissioned, and the reverse is also true. How these genetic factors act has not yet been demonstrated, but recent studies indicate that neurosecretory hormones are involved in the development of sexual organs in planarians (Ghirardelli, 1965; Grasso, 1965) and the onset and continuance of sexual activity (Lender, 1964), and it is possible that neurosecretions also play a role in the genetic control of fission and fertility.

Sterility and Juvenile Mortality

Deposition of sterile cocoons by planarians, particularly those species in which both sexual and asexual races occur, has been reported by many workers (Hyman, 1925; Kenk, 1935; Dahm, 1958; Benazzi, 1966, 1967). Often cocoon production appears to be normal but no embryos develop. In other cases only a small percentage of the cocoons hatch, or only a few young emerge. In some instances, as in the *I-1G #1* subgroup, the majority of the egg capsules hatch and many juveniles are produced, but mortality is high.

All these conditions have been found in the present study. In the two inbred first-generation groups, *I-S* and *I-FS*, which constitute the *I-1G* line in Fig. 2, 71 percent of the juveniles died without either fissioning or becoming sexual. In the corresponding crossbred groups, *X-S* and *X-FS*, mortality of juveniles was 66 percent. All hatched offspring died in 17 percent of the cocoons deposited by all inbred groups, but a similar 100 percent mortality did not occur in any crossbred group. By contrast, in one crossbred subgroup of *X-FS/S*, 19 young emerged from one cocoon and all 19 became sexual without prior fission.

Benazzi has suggested that much of the sterility in various planarian populations may be due to embryonal lethality (1967). This explanation is in accord with many of the findings of the present study. Effects of lethal genes would be expected to show up to a much greater degree in inbred lines than in crossbred, and to involve a greater number of individuals in the second inbred generation than in the first. Genes which acted before cocoons opened would account for much of the sterility, and the high incidence of juvenile mortality could be due to similar genes which exerted their effect after the young emerged from the cocoon. A different explanation would have to be sought, however, for the type of sterility found in some planarian populations in which sexual animals appear to be unable to produce cocoons.

Mortality and Senescence in Sexual Planarians

During the 3-year period shown in Table 1, 55 percent of the adult worms died, 64 percent of these deaths being due to unknown causes. A further 6 percent of deaths occurred when individuals crawled up on the sides or covers of the culture dishes, became stranded, and died. Remaining deaths are attributed to two catastrophic occurrences: 6 percent died when the heat pump–air conditioner failed to function properly during an August heat wave (DH), and 24 percent died within 1 week in one BOD incubator (DC). No cause of death in the latter instance could be found. In both cases, several subgroups at various interval ages were affected (for example, see intervals 4 and 12 in *I-S*). It is possible that some of the later deaths in affected groups were due to a lowered resistance resulting from either of these two unfortunate mishaps.

During the first 18 months of life approximately 18 percent of the sexual adults died, in contrast with only 5 percent reported for the same period of time by Balázs and Burg (1962c) for *D. lugubris*. Total mortality after maturity during the 3-year period was much higher in the inbred groups than in the crossbred, being 67 percent and 49 percent respectively. The higher number of deaths in the inbred group suggests that lethal genes which exert their effect during adult life, as well as those which act on the embryo or the juvenile, may be present in this sexual race.

Natural deaths varied from three to twenty-one per 13-week interval, with no significant indication of increasing likelihood of the animals' dying as time passed. This may be due to the comparatively small sample, but in view of the continued production of young in all groups (Table 2), it appears more probable that the worms were merely "middle-aged" during much of their second or third year of life, and that the rapid rise in mortality which is characteristic of senescence had not yet set in.

Although most of the 97 planarians that remained alive at the end of the 3-year period were still producing viable cocoons, the number of nonfertile cocoons had increased significantly (Fig. 1). The decline and eventual loss of fertility, which Medawar considers to be an ingredient of the aging process (1958), is characteristic of a number of invertebrate groups. Callahan found that as adult houseflies (*Musca domestica*) aged, a reduction in the number of eggs laid occurred, batches of viable eggs were interspersed with nonviable, and fewer offspring survived to adulthood (1962). In the mealworm *Tenebrio molitor*, the percentage of eggs which hatched decreased with an increase in parental age at four different temperatures (Ludwig and Fiore, 1960); and hatchability of eggs and survival rate of larva of the southern stinkbug *Nezara viridula* decreased with increasing parental age (Kiritani and Kimura, 1967). The rotifer *Philodina*, when well-fed and maintained at a temperature of 25°C, produces eggs for 7 to 8 days, then enters a postreproductive period which may

last a little over 3 weeks (Fanestil and Barrows, 1965). Postreproductive sterility is also characteristic of the marine gastropod *Gibbula* (= *Trochus*) *um-
bilicalis*, which produces young during its second, third, and fourth years
of life, but often becomes sterile during the fifth year (Pelseneer, 1933). During
this last year the animals do not grow, although the shell may increase in
thickness.

The above invertebrates have certain characteristics in common, in addition to a decline in fertility with age and a period of postreproductive sterility. All reach a determinate size, all experience a decline in rate and final cessation of growth, and all, barring accidents, live a more or less predetermined
period of time which is characteristic of the species.

Evidence from the present study shows that planarians which remain
continuously sexual exhibit a decline in fertility with age; undergo a period
of postreproductive sterility; reach a maximum size fairly early in life; and
decrease slightly in size after the maximum is attained although conditions
remain optimal. Since regular periods of fission and accompanying rejuvenation do not occur, it is logical to assume that the life-span in the sexual race is
also determinate.

Life-span

The normal span of life is not known for individuals of any sexual race of
planarians. Seasonal intervention of fission in many turbellarian species in-
troduces a problem to which there is, as yet, no solution. Haranghy and Balázs
(1964) advocate the use of two definitions for life-span in planarians. They
suggest that the period of time from emergence from the cocoon until death
be used for sexually reproducing species, and, in agamic species, the time from
fission to fission, that is, from the formation of the individual to its cessation.
This suggestion has much to recommend it, but it does not solve the problem
of chronological age in such animals as the catenulid, *Stenostomum* (Sonneborn, 1930), in which head portions undergo senescence but tails propagate
indefinitely. Nor does it provide a solution for computing age in triclads which
are seasonally sexual, such as certain races of *D. tigrina* (Curtis, 1902), or in
planarians such as sexual *D. dorotocephala*, which may or may not fission
before becoming sexual, but which characteristically remain continuously
mature once that condition has been established.

In several instances in the present study, in both inbred and crossbred
groups, sexual and fission-sexual specimens issued from the same cocoon.
It seemed illogical to calculate the age of an *FS* adult from the date of fission,
and at the same time to record siblings from the same cocoon as being 3 to
4 months older. All ages given here, therefore, are computed from the date
the individual emerged from the cocoon.

Based on the evidence provided by the present study, it appears that the normal span of life in sexually reproducing *D. dorotocephala* is 4 to 5 years, although some individuals may live, without undergoing fission, for a longer period.

Life History

The life history of the sexual race of *D. dorotocephala* can be divided into stages and summarized as follows:

1 *Immature Period.* Average duration, 4 to 6 months from date of emergence from cocoon, at a constant temperature of 19°C. Range in *S* animals, 3 to 13 months, and up to 3 years in *FS* individuals. Average gliding length at emergence, 5 to 6 mm, with a range of 1 to 10 mm. Fission may or may not occur during this period. Mortality is high.

2 *Period of Sexual Maturity.* Average duration, 2 to 3 years, although fertile cocoons may be produced as late as the beginning of the fifth year of life. Average gliding length at maturity, 20 to 25 mm, but extremes of 15 to 16 mm and 25 to 30 mm are not unusual.

Maximum size is normally attained early in the second year of life, at which time animals may average 30 to 35 mm in length, or longer. Copulation and cocoon production are continuous during this period, but fertility decreases slightly during the second year and markedly during the third year of life.

A small percentage of individuals may, without regard to season, become nonsexual. Such individuals may become sexual again, may die without further change, or may undergo spontaneous division, then die. No periods of asexual reproduction occur.

3 *Postreproductive Period.* Average duration, 1 to 2 years. Cessation of cocoon production usually occurs during the fourth year of life, although sexual planarians may live another 1 or 2 years. Other senescent changes which take place during the postreproductive period are not described in this paper.

Conclusion

The sexual race of *D. dorotocephala* undergoes a true senescence, as shown by the following secondary criteria: (1) Continuous sexual activity during maturity is followed by a postreproductive period which is characterized by the decline and eventual cessation of cocoon production; (2) the animals reach a determinate size which, under optimal conditions, is followed by a gradual volume decrease as aging continues; and (3) seasonal periods of fission and consequent rejuvenation do not occur.

Further verification of the occurrence of senescence in sexually reproducing planarians could be provided by the following investigations:

    1   A detailed study of physiological, morphological, and histological changes that may occur in individual animals during and after the third year of life.

    2   A life table of an adequate sample of sexual animals, covering a period of six or more years, as needed, with data on all deaths. Segregation of sexual and fission-sexual animals, and permanent separation of fragments produced by nonrejuvenatory spontaneous division would be necessary.

ACKNOWLEDGMENTS

I wish to thank Sarah Faulconer for assisting in compiling data; Homer Austin for helping with statistical analyses; and Betty Leeth for her diligence in caring for planarians. This research was supported in part by a grant from Sigma Xi and NIH Grant # HD 02217.

BIBLIOGRAPHY

Abeloos, M. 1930. Recherches expérimentales sur la croissance et la régénération chez les planaires. *Bull. Biol. France Belg.*, **64**:1–140.

Balázs, A., and M. Burg. 1962a. Quantitative Data to the Changes of Propagation according to Age. I. Cocoon Production of *Dugesia lugubris*. *Acta Biol. Acad. Sci. Hung.*, **12**(4):287–295.

—— and ——. 1962b. Quantitative Data to the Changes of Propagation according to Age. II. Fertility and Number of Embryos in *Dugesia lugubris*. *Acta. Biol. Acad. Sci. Hung.*, **12**(4):297–304.

—— and ——. 1962c. Span of Life and Senescence of *Dugesia lugubris*. *Gerontologia*, **6**:227–236.

Beauchamp, P. de. 1931. Races et modes de reproduction chez la Planaire *Fonticola vitta* (Dugès). *Compt. Rend. Soc. Biol.*, **107**:1001–1003.

——. 1962. "Classe des Turbellariés." In P. P. Grassé (ed.), *Traité de Zoologie*, vol. IV, pt. 2, pp. 35–212. Paris: Masson et Cie.

Benazzi, M. 1966. Cariologia della planaria americana *Dugesia dorotocephala*. *Accad. Nazl. Lincei*, (VIII)**40**(6):999–1005.

——. 1967. Considerazioni sui rapporti tra moltiplicazione agamica e sessualità. *Accad. Nazl. Lincei*, (VIII)**42**(6):742–746.

——, J. Baguñá, and R. Ballester. 1970. First Report on an Asexual Form of the Planarian *Dugesia lugubris* s. l. *Accad. Nazl. Lincei*, (VIII)**43**(2):42–44.

Callahan, R. F. 1962. Effects of Parental Age on the Life Cycle of the House Fly, *Musca domestica* Linnaeus (Diptera: Muscidae). *J. N.Y. Entomol. Soc.*, **70**(3):150–158.

Child, C. M. 1915. *Senescence and Rejuvenescence*. Chicago: University of Chicago Press.

Comfort, A. 1954. Biological Aspects of Senescence. *Biol. Rev.*, **29**(3):284–329.

——. 1960. "The Study of Mortality in Populations." In B. L. Strehler (ed.), *The*

Biology of Aging, A Symposium, pp. 154–161. American Institute of Biological Sciences, Washington, D.C.

———. 1964. *Ageing, The Biology of Senescence*. New York: Holt, Rinehart and Winston.

Curtis, W. C. 1902. The Life History, the Normal Fission and the Reproductive Organs of *Planaria maculata*. *Proc. Boston Soc. Nat. Hist.*, **30**(7):515–559.

Dahm, A. G. 1958. *Taxonomy and Ecology of Five Species Groups in the Family Planariidae (Turbellaria: Tricladida, Paludicola)*. Malmö: Nya Litografen.

Fanestil, D. D., and C. H. Barrows. 1965. Aging in the Rotifer. *J. Gerontol.*, **20**(4): 462–469.

Ghirardelli, E. 1965. "Differentiation of the Germ Cells and Regeneration of the Gonads in Planarians." In V. Kiortsis and H. A. L. Trampusch (eds.), *Regeneration in Animals and Related Problems*, pp. 177–184. Amsterdam: North-Holland Publishing Company.

Grasso, M. 1965. Dimostrazione di cellule neurosecretrici in *Dugesia tigrina*. *Accad. Nazl. Lincei*, (VIII)**38**(5):712–714.

Haranghy, L., and A. Balázs. 1964. Ageing and Rejuvenation in Planarians. *Exptl. Gerontol.*, **1**:77–91.

Hyman, L. H. 1925. The Reproductive System and Other Characters of *Planaria dorotocephala* Woodworth. *Trans. Am. Microscop. Soc.*, **44**(2):51–89.

———. 1951. *The Invertebrates*. II. *Platyhelminthes and Rhynchocoela*. New York: McGraw-Hill.

Jenkins, M. M. 1960. Respiration Rates in Planarians. I. The Use of the Warburg Respirometer in Determining Oxygen Consumption. *Proc. Oklahoma Acad. Sci.*, **40**:35–40.

———. 1967. "Aspects of Planarian Biology and Behavior." In W. C. Corning and S. C. Ratner (eds.), *Chemistry of Learning*, pp. 116–143. New York: Plenum Press.

———. 1970. Sexuality in *Dugesia dorotocephala*. *J. Biol. Psychol.*, **12**(1):71–80.

——— and H. P. Brown. 1963. Cocoon Production in *Dugesia dorotocephala* (Woodworth) 1897. *Trans. Am. Microscop. Soc.*, **82**(2):167–177.

——— and ———. 1964. Copulatory Activity and Behavior in the Planarian *Dugesia dorotocephala* (Woodworth) 1897. *Trans. Am. Microscop. Soc.*, **83**(1):32–40.

Kenk, R. 1935. Studies on Virginian Triclads. *J. Elisha Mitchell Sci. Soc.*, **51**(1): 79–125.

Kiritani, K., and K. Kimura. 1967. Effects of Parental Age on the Life Cycle of the Southern Green Stink Bug, *Nezara viridula* L. (Heteroptera: Pentatomidae). *Appl. Entomol. Zool.*, **2**(2):69–78.

Lange, C. S. 1968. A Possible Explanation in Cellular Terms of the Physiological Ageing of the Planarian. *Exptl. Gerontol.*, **3**:219–230.

Lender, T. 1964. Mise en évidence et rôle de la neurosécrétion chez les Planaires d'eau douce (Turbellariés. Triclades). *Ann. Endocrinol.*, **25**(5):61–65.

Ludwig, D., and C. Fiore. 1960. Further Studies on the Relationship between Parental Age and the Life Cycle of the Mealworm, *Tenebrio molitor*. *Ann. Entomol. Soc. Am.*, **53**(5):595–600.

McConnell, J. V. 1967. *A Manual of Psychological Experimentation on Planarians.* Ann Arbor, Mich.: Planarian Press.

Medawar, P. B. 1958. *The Uniqueness of the Individual.* New York: Basic Books.

Okugawa, K. I. 1957. An Experimental Study of Sexual Induction in the Asexual Form of Japanese Fresh-water Planarian, *Dugesia gonocephala* (Dugès). *Bull. Kyoto Gakugei Univ.*, ser. B, no. 11, 8–27.

Ostle, B. 1954. *Statistics in Research*, chap. 5. Ames: Iowa State College Press.

Pelseneer, P. 1933. La durée de la vie et l'âge de la maturité sexuelle chez certaines Mollusques. *Ann. Soc. Roy. Zool. Belg.*, **64**:93–104.

Reynoldson, T. B., J. O. Young, and M. C. Taylor. 1965. The Effect of Temperature on the Life Cycle of Four Species of Lake-dwelling Triclads. *J. Animal Ecol.*, **34**:23–43.

Sonneborn, T. M. 1930. Genetic Studies on *Stenostomum incaudatum* (nov. spec.). I. The Nature and Origin of Differences among Individuals Formed during Vegetative Reproduction. *J. Exptl. Zool.*, **57**(1):57–108.

Vandel, A. 1922. Recherches expérimentales sur les modes de reproduction des Planaires triclades paludicoles. *Bull. Biol. France Belg.*, **56**:343–518.

Voigt, W. 1928. Verschwinden des Pigmentes bei *Planaria polychroa* und *Polycelis nigra* unter dem Einfluss ungünstiger Existenzbedingungen. *Zool. Jahrb.*, **45**: 293–316.

Willier, B. H., L. H. Hyman, and S. A. Rifenburgh. 1925. A Histochemical Study of Intracellular Digestion in Triclad Flatworms. *J. Morphol.*, **40**(2):299–340.

Wulzen, R. 1923. A Study in the Nutrition of an Invertebrate, *Planaria maculata. Univ. Calif. Publ. Physiol.*, **5**(15):175–187.

———. 1927. The Nutrition of Planarian Worms. *Science*, **65**(1683):331–332.

Chapter 22

Epilogue

Nathan W. Riser
Marine Science Institute, Northeastern University,
Nahant, Massachusetts

In 1959 when Dr. Hyman reviewed advances in turbellariology, she noted that the greatest activity in work on the phylum Platyhelminthes was in descriptions of new species. This work has continued and has been expanded through recent interest in the bionomics of interstitial fauna. Major investigations have included those of Dörjes (1968) on acoels, Rieger (1971a, b) on macrostomids, and Schilke (1970a, b) on kalyptorhynchs. A new generation of investigators has entered the field collaborating with, and continuing the work of, their teachers. It can be anticipated that the next 10 years will see not only an expansion of knowledge, but also a comprehensive integra-

517

tion of the accumulated information. The discovery of a freshwater acoel *Oligochoerus limnophilus* Ax and Dörjes, 1966 is significant at this time.

Riedl (1969) elevated the Gnathostomulida to either class or phylum status, depending upon the hierarchy system to be applied by the individual worker. In addition to descriptions of new species and genera from North America, he described hypodermic transmission of sperm packets and their migration into the bursa in *Gnathostomula jenneri* Riedl, 1969. Penetration of the mature egg by sperm migrating through the bursal mouthpiece is postulated. The egg is discharged through the dorsal body wall, and in this species adheres "slightly" to the substrate. Cleavage was equal, holoblastic, and of the typical spiral pattern. The annelid cross occurred prior to gastrulation, which was described as by epiboly without the occurrence of a blastocoel or blastopore. Following gastrulation, two large cells remained which were considered to be $4d^1$ and $4d^2$. Development in this species was direct. Müller and Ax (1971) described the postembryonal development of *Gnathostomula paradoxa* Ax, 1956. This species is protandric with the stylet forming first and then the remainder of the reproductive organs as previously reported for *G. jenneri* by Riedl (1969). *G. paradoxa* demonstrates the sterile phase postulated by Riedl following his study of *G. jenneri* and clearly shows the initial resorption of the male genitalia followed by the female reproductive organs, and a new cycle of organogenesis occurring in the same basic pattern of the original. Additional North American species were described by Riedl (1970a, 1970b) and Sterrer (1970). The cells of the epithelium of *Gnathostomula paradoxa* and *Gnathostomaria lutheri* Ax, 1956 each bear a single flagellum, and in the latter species each cell contains glandular packets at its distal end as described by Ax (1964). He reported that specific staining indicated no relationship of these secretion granules to rhabdites. He also questioned the statements by Mamkaev (1961) about the occurrence of a flagellated epithelium in a species of *Catenula*. Sterrer (1968) noted the similarity of the sperm of gnathostomulids to those of gastrotrichs as had Ax (1965), although the latter author also stressed the similarity to nematode sperm. Sterrer (1968) described the mature sperm of several species of the genus *Pterognathia*. These sperm have a long spirally coiled head, a relatively structureless middle piece, and a thin tail. He described sperm motion in *P. swedmarki* Sterrer, 1966 and *P. sorex* Sterrer, 1968 as a rotation about the long axis resulting from undulation of the tail. These sperm differ markedly from the teardrop-shaped sperm of *Gnathostomula paradoxa* and the small bleblike sperm of *Labidognathia longicollis* Riedl, 1970, which bear a group of bristles on one end. At present there appears to be a consensus that conuli are spermatophores.

As indicated in Chap. 4, interstitial organisms are highly specialized. We cannot anticipate connecting links under these circumstances, and while these organisms may produce enigmas and phylogenetic confusion and bring on controversy as to what is plesiomorph and what is apomorph, the fact remains that missing links will increase, and history has all but obliterated connecting links between major units of organisms which lack fossilizable hard parts. The systematic position of the gnathostomulids still remains unclear.

A significant modification of the male copulatory organs has been described by Reisinger (1968) in the prorhynchids *Xenoprorhynchus steinboecki* Reisinger, 1968, an aquatic species, and *X. brasiliensis* Reisinger, 1968, a terrestrial species. The testes open into a common genital atrium and the copulatory organs (stylet, penis bulb, and associated structures) lose their connection with the other male reproductive organs and become a venom-producing, stylet-bearing, prey-capturing device. The excretory system in the genus is unusual in the relationship of the paranephrocytes to the canals. The flame bulbs are not capped by the nucleus, the latter organelle being located at the beginning of the capillary. The evolution of cyrtocytes and rhabdites displayed by EM studies was elaborated by Reisinger (1969).

Dörjes (1968) described a number of unique new species of acoels among which were *Archocelis macrorhabditis* with rhabdite cells up to 100 μm long and *Simplicomorpha gigantorhabditis* in which the rhabdite cells extend over halfway through the body.

Apelt (1969b) demonstrated the spiral-duet cleavage for thirteen species of acoels and commented on the probable reasons for the controversial statements of Bogomolov (1960) and Steinböck (1966) on the fusion of blastomeres. In the original description of *Oligochoerus limnophilus*, Ax and Dörjes included the embryology and development, and here too, the spiral-duet plan occurred. The majority of the species studied by Apelt hatched within 5 days, but *Archocelis macrorhabditis* with eggs 75 to 80-μm diameter and *Pseudohaplogonaria vacua* Dörjes, 1968 with eggs of 180 to 190-μm diameter required over 9 days. Internal brooding of embryos by *Diopisthoporus brachypharyngeus* Dörjes, 1968, a strictly female species, was described also.

Spiral-quartet cleavage has been described as occurring in the macrostomid *Macrostomum romanicum* Mack-Fira 1968 (= *M. salinum nomen nudum*) by Ax and Borkott (1970), the prorhynchid *Xenoprorhynchus steinboecki* by Reisinger (1970), and the proseriate *Monocelis fusca* Oersted, 1843 by Giesa and Ax (1965). A stereoblastula was described for *X. steinboecki*, and gastrulation was by epiboly. The embryology of *M. fusca* included an unequal coeloblastula (also recorded for *Coelogynopora biarmata* Steinböck, 1924 in the same publication) with small animal pole cells and larger yolk-filled vegetal pole cells. This development takes place in the interior of a yolk-cell stroma which fills the egg capsule. There is a flattening of the blastula obliterating the blastocoel and producing a two-cell layered plate. Cells at the margin of the plate develop processes which extend into the surrounding yolk stroma. These cells are the four primitive mouth cells. In the ventral plate area two primitive gut cells develop, also with pseudopods. The primitive mouth cells expand, forcing the primitive gut cells inward, and thus gastrulation occurs by epiboly. There is no blastopore or vegetal pole opening *per se*. The primitive mouth cells pass yolk cells intact into a primitive gut cavity. The primitive mouth cells do not form a definitive structure comparable to the embryonic pharynx of triclads. The yolk becomes totally engulfed, and organogenesis begins at the animal pole with early development of the head followed by the pharynx and then the tail. The statocyst appears early before differential growth has carried the tail to proximity with the head forming the typical U-shaped encapsulated embryo.

Parke and Manton (1967) isolated and characterized the alga *Platymonas convolutae* P. and M., 1967, symbiotic in *Convoluta roscoffensis*. EM studies showed the absence of the theca, flagellar pit, eyespot, and free flagella in algae within the host tissue. The flagellar end of the symbiont is directed inward, and the Golgi apparatus is reduced. Oschman and Gray (1965) reported the demonstration of projections from the algal symbionts extending into the epidermis of *C. roscoffensis*. They noted the occurrence of vacuoles in epidermal cells and postulated phagocytosis of algal extensions by epidermal cells. Apelt (1969a) isolated the diatom *Licmophora hyalina* and *L. communis* from *Convoluta convoluta*. This acoel is symbiont-free at the time of liberation from the egg capsule. In cultural experiments, infection was attained in 2 to 4 days. The ingested diatoms leave their tests and establish themselves as naked bionts between the cells of the peripheral parenchyma.

Bashiruddin and Karling (1970) described *Triloborhynchus astropectinis* from the intestinal ceca of *Astropecten irregularis*. Cecal tissue was observed drawn into the pharynx, and the authors postulated that the host tissue as well as cecal contents might be utilized as food by the worm. Both an unarmed prohaptor and opisthaptor occur and are utilized in locomotion. Karling (1970) reinvestigated *Pterastericola fedotovi* Beklemishev, 1916, a pterastericolid lacking prohaptor and opisthaptor, and was unable to clarify the function of the lumenless vagina ("pseudovagina"). On morphological grounds he noted that the pterastericolids could be phylogenetically placed between the dallyelloid turbellarians and the monogenetic trematodes.

Two other phylogenetic curiosities should be mentioned. Karling (1965) redescribed the macrostomid *Haplopharynx rostratus* Meixner, 1938, a species with a supraterminal anus and a proboscis armed with rhammites produced by postpharyngeal glands. He noted the intermediate position between this species and the nemertineans. It should be noted, however, that the proboscis of *H. rostratus* is terminal and that the tracts from the postpharyngeal glands pass lateral to the pharynx and ventral to the brain. Ax and Ax (1969) drew attention to the chorda intestinalis of *Nematoplana nigrocapitula*

Ax, 1966, as indicating a model for the evolution of the chorda dorsalis of chordates. In this species, the turgescent cells of the dorsal wall of the gut extend forward as a cecum over the brain and to the anterior end of the body as is characteristic of the genus. The figure of *Nematoplana coelogynoporoides* Meixner, 1938, reproduced by Ax (1969) from Strietzel (1967), shows a marked narrowing of the gut as it passes over the brain and then a bulbous expansion filling the anterior end. The authors note the analogy of this extension to the stomochord of hemichordates. [It should be noted that this cecum is functioning in a way analogous to the coelomic pouch in the hemichordate proboscis, and the vacuolated anterior parenchyma of *Gnathostomula paradoxa* (see Ax, 1965).] The cephalic cecum consisting of a core of cells occurs in other proseriates and has been discussed by Ax (1957) and well illustrated by Lanfranchi (1969) in the type figure of *Otoplana truncaspina*. The distribution of the chordoid tissue along the entire dorsal wall of the gut of *N. nigrocapitula* is proposed as a model for the intermediate stage necessary to evolve a separate chorda dorsalis (notochord).

The study of distribution and, eventually, speciation among marine turbellarians is in its infantile stages. W. Tomkiewicz of our laboratory has discovered the well-known European marine triclad *Uteriporus vulgaris* Bergendal, 1890 at Saint Andrews, New Brunswick. Riser (1970) reported the occurrence of the polyclad *Taenioplana teredini* Hyman, 1944 from Boynton Beach and Sanibel Island, Florida. The specimens upon which the type description of this polyclad was based were poorly fixed and measured 45 mm in length and 3 mm in width. Poulter (unpublished M.S.thesis) recorded a length of 90 mm and breadth of 4 mm in living specimens from Hawaii. The living Florida specimens attained a length of 110 mm and a breadth of over 6 mm. I have in culture a small specimen of the species recently obtained from wood bored by *Lyrodus massa* at Galeta Island, Panama. Thus, to date all records are from waters north of the equator: Hawaii, Florida, Israel, and Panama, but Dr. R. D. Turner of Harvard University has observed (pers. comm.) a similar polyclad in teredinid burrows in New Guinea. Dr. John Culliney of Harvard University has maintained the species in our laboratory using the shipworm *Lyrodus pedicellatus* as a food source. The specimens from Florida are avidly eaten by the polychaete *Nereis arenaceodentata*, which also inhabits shipworm burrows. The dispersal of these worms might have been originally by drifting wood, but spread by wooden ships is a possibility. Ax and Ax (1967) established subspecies for the otoplanid *Itaspiella armata* (Ax, 1951) on the basis of comparison of European with North American Pacific coast specimens. They further recorded the North Atlantic *Notocaryoplana arctica* Steinböck, 1935 [= *Notocaryoplanella glandulosa (Ax, 1951)*] from the same region. The latter species, though not previously recorded from the western Atlantic is common at Nahant, Massachusetts, as is *Bothriomolus balticus* Meixner, 1938 and *Archiloa unipunctata* (Fabricius, 1826). Malacologists have attacked the problem of demes and clines, but this remains to be systematically approached with other groups. The otoplanid *B. balticus* at Nahant shows a consistent pattern of zero to four anterior bristles compared to the six to seven recorded by Ax (1956) and three to twelve recorded by Karling and Kinnander (1953) from northern Europe. These limitations would appear to indicate isolation of genetic strains and the possibility of speciation in progress. One can visualize the spread of these forms from highly variable populations in the North Sea and the establishment of isolated demes. Whether these are treated as demes, subspecies, or species will depend upon the philosophy of the investigator.

BIBLIOGRAPHY

Apelt, G. 1969a. Die Symbiose zwischen dem acoelen Turbellar *Convoluta convoluta* und Diatomeen der Gattung *Licmophora*. *Marine Biol.*, **3**:165–187.

——. 1969b. Fortpflanzungsbiologie, Entwicklungszyklen und vergleichende Frühentwicklung acoeler Turbellarien. *Marine Biol.*, **4**:267–325.

Ax, P. 1956. Monographie der Otoplanidae (Turbellaria). Morphologie und Systematik. *Akad. Wiss. Lit. Mainz Abhandl. Math. Nat. Kl.*, 1955, **13**:499–796.

——. 1957. Ein chordoides Stützorgan des Entodermis bei Turbellarien. *Z. Morphol. Ökol. Tiere*, **46**:389–396.

——. 1964. Das Hautgeisselepithel der Gnathostomulida. *Verhandl. Deut. Zool. Ges. München*, 1963, pp. 452–461.

——. 1965. Zur Morphologie und Systematik der Gnathostomulida. Untersuchungen an *Gnathostomula paradoxa* Ax. *Z. Zool. Syst. Evolutionsforsch.*, **3**:259–276.

——. 1969. Populationsdynamik, Lebenszyklen und Fortpflanzungsbiologie der Mikrofauna des Meeressandes. *Verhandl. Deut. Zool. Ges. Innsbruck*, 1968, pp. 66–113.

—— and Renate Ax. 1967. Turbellaria Proseriata von der Pazifikküste der U.S.A. (Washington) 1. Otoplanidae. *Z. Morphol. Tiere*, **61**:215–254.

—— and ——. 1969. Eine Chorda intestinalis bei Turbellarien (*Nematoplana nigrocapitulata* Ax) als Modell für die Evolution der Chorda dorsalis. *Akad. Wiss. Lit. Mainz Abhandl. Math. Nat. Kl.*, 1969, **5**:135–147.

—— and H. Borkott. 1970. Organisation und Fortpflanzung von *Macrostomum salinum* (Turbellaria-Macrostomida). *Inst. Wiss. Film Göttingen*. Film C 947, pp. 3–11.

—— and J. Dörjes. 1966. *Oligochoerus limnophilus* nov. spec., ein kaspisches Faunenelement als erster Süsswasservertreter der Turbellaria Acoela in Flüssens Mitteleuropas. *Intern. Rev. Ges. Hydrobiol.*, **51**:15–44.

Bashiruddin, M., and T. G. Karling. 1970. A New Entocommensal Turbellarian (Fam. Pterastericolidae) from the Sea Star *Astropecten irregularis*. *Z. Morphol. Tiere*, **67**:16–28.

Bogomolov, S. I. 1960. The Development of *Convoluta* in Relationship to the Morphology of the Turbellaria. *Tr. Obshch. Est. Imp. Kazan. Univ.*, **63**:155–208. (In Russian.)

Dörjes, J. 1968. Die Acoela (Turbellaria) der Deutschen Nordseeküste und ein neues System der Ordnung. *Z. Zool. Syst. Evolutionsforsch.*, **6**:56–452.

Giesa, S., and P. Ax. 1965. Die Gastrulation der Proseriata als ein ursprünglicher Entwicklungsmodus der Turbellaria Neoophora. *Verhandl. Deut. Zool. Ges. Kiel*, 1964, pp. 109–122.

Hyman, L. H. 1959. *The Invertebrates*. V. *Smaller Coelomate Groups*. New York: McGraw-Hill. pp. 731–735.

Karling, T. G. 1965. *Haplopharynx rostratus* Meixner (Turbellaria) mit den Nemertinen verglichen. *Z. Zool. Syst. Evolutionsforsch.*, **3**:1–18.

———. 1970. On *Pterastericola fedotovi* (Turbellaria), Commensal in Sea Stars. *Z. Morphol. Tiere*, **67**:29–39.

——— and H. Kinnander. 1953. Några virvelmaskar från Östersjön. *Särtryck ur Svensk Faunistisk Revy*, no. 3, 73–79.

Lanfranchi, A. 1969. Nuovi Otoplanidi (Turbellaria Proseriata) delle coste della Liguria e della Toscana. *Boll. Zool.*, **36**:167–188.

Mamkaev, Y. V. 1961. A New Representative of the Gnathostomulids: *Gnathostomula murmanica* sp. n. *Dokl. Akad. Nauk SSSR, Biol. Sci. Sect. English Transl.* (1962), **141**:1115–1117.

Müller, U., and P. Ax. 1971. Gnathostomulida von der Nordseeinsel Sylt mit Beobachtungen zur Lebenweise und Entwicklung von *Gnathostomula paradoxa* Ax. *Akad. Wiss. Lit. Mainz Abhandl. Math. Nat. Kl.*, 1971:311–349.

Oschman, J. L., and P. Gray. 1965. A Study of the Fine Structure of *Convoluta roscoffensis* and Its Endosymbiotic Algae. *Trans. Am. Microscop. Soc.*, **84**:368–375.

Parke, M., and I. Manton. 1967. The Specific Identity of the Algal Symbiont in *Convoluta roscoffensis*. *J. Marine Biol. Assoc. U.K.*, **47**:445–464.

Reisinger, E. 1968. *Xenoprorhynchus* ein Modellfall für progressiven Funktionswechsel. *Z. Zool. Syst. Evolutionsforsch.*, **6**:1–55.

———. 1969. Ultrastrukturforschung und Evolution. *Ber. Phys.-Med. Ges. Würzburg*, n.f. **77**:5–47.

———. 1970. Zur Problematik der Evolution der Coelomaten. *Z. Zool. Syst. Evolutionsforsch.*, **8**:81–109.

Riedl, R. J. 1969. Gnathostomulida from America. This Is the First Record of the New Phylum from North America. *Science*, **163**:445–452.

———. 1970a. On *Labidognathia longicollis*, nov. gen., nov. spec., from the West Atlantic Coast (Gnathostomulida). *Intern. Rev. Ges. Hydrobiol.*, **55**:227–244.

———. 1970b. *Semaeognathia*, a New Genus of Gnathostomulida from the North American Coast. *Intern. Rev. Ges. Hydrobiol.*, **55**:359–370.

Rieger, R. 1971a. Die Turbellarienfamilie Dolichomacrostomidae nov. fam. (Macrostomida) 1 Teil, Vorbemerkungen und Karlingiinae nov. subfam. 1. *Zool. Jahrb. Syst.*, **98**:236–314.

———. 1971b. Die Turbellarienfamilie Dolichomacrostomidae Rieger. II. Teil. Dolichomacrostominae 1. *Zool Jahrb. Syst.*, **98**:569–703.

Riser, N. W. 1970. Biological Studies on *Taenioplana teredini* Hyman 1944. *Am. Zool.*, **10**:553 (Abst.)

Schilke, K. 1970a. Kalyptorhynchia (Turbellaria) aus dem Eulitoral der deutschen Nordseeküste. *Helgoländer Wiss. Meeresuntersuch.*, **21**:143–265.

———. 1970b. Zur Morphologie und Phylogenie der Schizorhynchia (Turbellaria, Kalyptorhynchia). *Z. Morphol. Tiere*, **67**:118–171.

Steinböck, O. 1966. Die Hofsteniiden (Turbellaria acoela). Grundsätzliches zur Evolution der Turbellarien. *Z. Zool. Syst. Evolutionsforsch.*, **4**:58–195.

Sterrer, W. 1968. Beiträge zur Kenntnis der Gnathostomulida. I. Anatomie und Morphologie des Genus *Pterognathia* Sterrer. *Arkiv Zool.*, (2)**22**:1–125.

———. 1970. On Some Species of *Austrognatharia, Pterognathia* and *Haplognathia* nov. gen. from the North Carolina Coast (Gnathostomulida). *Intern. Rev. Ges. Hydrobiol.*, **55**:371–385.

Strietzel, W. 1967. "Lebenszyklus und Populationsdynamik von *Nematoplana coelogynoperoides* Meixner am Sandstrand von Sylt." Staatsexamensarbeit Göttingen.

Index

Page numbers in **boldface** indicate figures.

Acmostomum, 49
Adenorhynchus, 48
Agigea Lake, 280
Allostomum, 253, 273, 281, 284, 285
Anaperus, 159, 161
Annulorhynchus, 166, 167
Anthopharynx, 239

Archaphanostoma, 250, 273, 284
Archetype, Turbellarian: of Ax, 5–6
    of Karling, 2–5
Archilina, 254, 273, 281, 284
Archiloa, 522
Archilopsis, 238
Archimonotresis, 238